Lecture Notes in Physics

New Series m: Monographs

W0051090

Springer-Verlag Berlin Heidelberg GmbH

The Editorial Policy for Monographs

The series Lecture Notes in Physics reports new developments in physical research and teaching - quickly, informally, and at a high level. The type of material considered for publication in the New Series m includes monographs presenting original research or new angles in a classical field. The timeliness of a manuscript is more important than its form, which may be preliminary or tentative. Manuscripts should be reasonably self-contained. They will often present not only results of the author(s) but also related work by other people and will provide sufficient motivation, examples, and applications.

The manuscripts or a detailed description thereof should be submitted either to one of the series editors or to the managing editor. The proposal is then carefully refereed. A final decision concerning publication can often only be made on the basis of the complete manuscript, but otherwise the editors will try to make a preliminary decision as definite as they can on the basis of the available information.

Manuscripts should be no less than 100 and preferably no more than 400 pages in length. Final manuscripts should preferably be in English, or possibly in French or German. They should include a table of contents and an informative introduction accessible also to readers not particularly familiar with the topic treated. Authors are free to use the material in other publications. However, if extensive use is made elsewhere, the publisher should be informed. Authors receive jointly 50 complimentary copies of their book. They are entitled to purchase further copies of their book at a reduced rate. As a rule no reprints of individual contributions can be supplied. No royalty is paid on Lecture Notes in Physics volumes. Commitment to publish is made by letter of interest rather than by signing a formal contract. Springer-Verlag secures the copyright for each volume.

The Production Process

The books are hardbound, and quality paper appropriate to the needs of the author(s) is used. Publication time is about ten weeks. More than twenty years of experience guarantee authors the best possible service. To reach the goal of rapid publication at a low price the technique of photographic reproduction from a camera-ready manuscript was chosen. This process shifts the main responsibility for the technical quality considerably from the publisher to the author. We therefore urge all authors to observe very carefully our guidelines for the preparation of camera-ready manuscripts, which we will supply on request. This applies especially to the quality of figures and halftones submitted for publication. Figures should be submitted as originals or glossy prints, as very often Xerox copies are not suitable for reproduction. For the same reason, any writing within figures should not be smaller than 2.5 mm. It might be useful to look at some of the volumes already published or, especially if some atypical text is planned, to write to the Physics Editorial Department of Springer-Verlag direct. This avoids mistakes and time-consuming correspondence during the production period.

As a special service, we offer free of charge LaTeX and TeX macro packages to format the text according to Springer-Verlag's quality requirements. We strongly recommend authors to make use of this offer, as the result will be a book of considerably improved technical quality.

Manuscripts not meeting the technical standard of the series will have to be returned for improvement.

For further information please contact Springer-Verlag, Physics Editorial Department II, Tiergartenstrasse 17, D-69121 Heidelberg, Germany.

Stephen Childress   Andrew D. Gilbert

Stretch, Twist, Fold:
The Fast Dynamo

 Springer

Authors

Stephen Childress
Courant Institute of Mathematical Sciences
New York University, 251 Mercer Street
New York, NY 10012, USA

Andrew D. Gilbert
Department of Mathematics, Laver Building
University of Exeter, North Park Rd.
Exeter, EX4 4QE, United Kingdom

Cataloging-in-Publication Data applied for

Die Deutsche Bibliothek - CIP-Einheitsaufnahme

Childress, Stephen:
Stretch, twist, fold: the fast dynamo / Stephen Childress ;
Andrew D. Gilbert.
  (Lecture notes in physics : N.s. M, Monographs ; Vol. 37)
    ISBN 978-3-662-14014-7   ISBN 978-3-540-44778-8 (eBook)
    DOI 10.1007/978-3-540-44778-8
NE: Gilbert, Andrew D.:; Lecture notes in physics / M

ISBN 978-3-662-14014-7

© Springer-Verlag Berlin Heidelberg 1995
Originally published by Springer-Verlag Berlin Heidelberg New York in 1995
Softcover reprint of the hardcover 1st edition 1995

Typesetting: Camera-ready by authors using T$_E$X
SPIN: 10481143   55/3142-5 43210 - Printed on acid-free paper

Preface

The study of the magnetic fields of the Earth and Sun, as well as those of other planets, stars, and galaxies, has a long history and a rich and varied literature, including in recent years a number of review articles and books dedicated to the dynamo theories of these fields. Against this background of work, some explanation of the scope and purpose of the present monograph, and of the presentation and organization of the material, is therefore needed. Dynamo theory offers an explanation of natural magnetism as a phenomenon of magnetohydrodynamics (MHD), the dynamics governing the evolution and interaction of motions of an electrically conducting fluid and electromagnetic fields. A natural starting point for a dynamo theory assumes the fluid motion to be a given vector field, without regard for the origin of the forces which drive it. The resulting kinematic dynamo theory is, in the non-relativistic case, a linear advection–diffusion problem for the magnetic field.

This kinematic theory, while far simpler than its magnetohydrodynamic counterpart, remains a formidable analytical problem since the interesting solutions lack the easiest symmetries. Much of the research has focused on the simplest acceptable flows and especially on cases where the smoothing effect of diffusion can be exploited. A close analog is the advection and diffusion of a scalar field by laminar flows, the diffusion being measured by an appropriate Péclet number. This work has succeeded in establishing dynamo action as an attractive candidate for astrophysical magnetism.

In recent years, however, it has been realized that the very singular limit of small diffusion, or large magnetic Reynolds number, is the relevant one for most of the problems of interest. That is to say, it is appropriate to regard the fluid as an almost-perfect conductor. These developments coincide with an interest in advection–diffusion by chaotic fluid flows, coming from the classic problem of computing turbulent transport. Interestingly but perhaps not surprisingly, dynamo action and chaotic motion turn out to be closely linked in the limit of small diffusion, a result of the stretching experienced by fluid elements in such flows. The perfectly conducting limit in chaotic motion causes the magnetic field to inherit the complexity of the Lagrangian chaos, and this leads to a special class of problems linking dynamo theory with the growing body of work on chaotic dynamics. The present volume arose from recent attempts to develop a theory of this class of dynamo problems.

This is not to say that the point of view taken here is new to dynamo theory. The class of 'fast' dynamos was identified at least twenty-five years ago and the importance of the perfectly conducting limit has long been realized and emphasized in the work of Alfvén, Elsasser, Parker, Braginsky, and many others. What is new is the attempt to analyze this limit at what might be termed the 'microscopic' level, using methods and results from dynamical systems theory. The idea is to approach the dynamo problem via the perfect conductor and the use of Lagrangian maps of the magnetic field. The full array of analytical tools for maps is thereby brought into the analysis.

This infusion of ideas has led to rapid progress at the basic level of understanding how fast dynamos might work, and has suggested a number of promising approaches for constructing a rigorous theory, but it has also revealed the intrinsic difficulty of analysing small-scale magnetic structures in the perfect conductor. To date, the research has consisted primarily of constructing various plausible fast dynamos, studying them using approximate techniques when possible, and testing them by numerical simulation. In this way the outlines of a theory have begun to emerge, albeit one wanting of more than a handful of theorems.

The present monograph grew from a suggestion by Uriel Frisch that it would be useful to have an account of the present status of this theory, however tentative, in order to collect, on the one hand, the constructive efforts of the modellers, and, on the other, the theoretical concepts which seem to be most relevant to the analysis of fast dynamos, along with a summary of results to date. As we looked over this material, it became clear that it would be difficult and premature to try to impose too much order on results which consist largely of a set of illustrative examples. We have accordingly emphasized the computational evidence and used this as a basis for very simple approximate models, for which some analysis is possible. This led to a natural division of our writing into two parts. The first three chapters, making up Part I, introduce the problem and present examples of fast dynamo action in flows and maps. The remaining nine chapters constitute Part II, and deal with various analytical approaches and model systems. The reader will thus encounter many of the most useful concepts several times, first as they arise in numerical experimentation, and again as they are brought into the formulation and analysis of models. The introductory overview of Chap. 1 should be paired with the discussion of Chap. 4, for example. In the latter we attempt to summarize the computational evidence in the form of a number of conjectures, and also introduce terminology and definitions which appear useful.

The topics treated in Part II include integrable flows (Chap. 5), which link the present study to classical dynamo theory, dynamical systems methods applied to the estimation of growth rate (Chaps. 6 and 7), and examples of the calculation of eigenvalues and eigenfunctions in the perfectly conducting limit (Chap. 9). These ideas are complemented by specific applications involving nearly integrable flows (Chap. 8), strongly chaotic flows (Chap. 10), and random flows (Chap. 11). In Chap. 12 we consider some aspects of the dynamical or MHD dynamo problem under the conditions of fast dynamo action.

That final chapter links the present studies to the problems of physical interest, such as stellar and galactic dynamos, which will surely be the focus of future numerical calculations. In the dynamical context, and for ordinary conducting fluids, fast dynamo theory becomes indistinguishable from the theory of fully developed MHD turbulence. The special features of fast dynamo theory, with its emphasis on growth rates and the magnetic structure of eigenfunctions, are replaced by an equilibrium theory of dynamically self-consistent magnetic structure. There is nonetheless interest in the implications which the kinematic theory might have for these saturation fields. We have addressed these issues here in what we term 'post-kinematic theory'. It is intriguing that the small-scale structures of MHD turbulence are a kind of nonlinear counterpart to the linear fine structure of the magnetic field of a fast dynamo, but the methods needed are quite different from those used in kinematic theory.

We would like to take this opportunity to thank the many colleagues who have helped us in the writing and editing of this work. Keith Moffatt offered early encouragement and, by suggesting the title of the monograph, gave some immediate coherence to the project. We would especially like to acknowledge the long collaborations with Paul Roberts and Andrew Soward, which have spanned the last three decades of research in dynamo theory. Our work with Bruce Bayly and Isaac Klapper laid the basis for some of the ideas discussed here, and they have helped substantially in our review of this work. Some of the early research was initiated at the 1987 Woods Hole Summer Program in Geophysical Fluid Dynamics, and we thank Director Willem Malkus for his enthusiasm and encouragement. We thank Eric Aurell, Konrad Bajer, Axel Brandenberg, Fausto Cattaneo, Pierre Collet, Dave Galloway, Rainer Hollerbach, Ralf Kaiser, Phil Kronberg, Liz Mansfield, Phil Morrison, Gene Parker, Annick Pouquet, Mike Proctor, Kurt Riedel, Ed Spiegel, Hank Strauss, Pierre-Louis Sulem, Samuel Vainshtein, and Misha Vishik for their assistance, through discussion and correspondence before or during the writing, or through their comments on a preliminary draft. The preparation of the color plates is supported financially by the University of Exeter. Ed Friedman and Estarose Wolfson of the Courant Institute were of immense help in the creation of these color images. We are grateful to Wolf Beiglböck and his team at Springer-Verlag, who were helpful and dealt efficiently with our problems and queries. We shall maintain a list of errata for this book and related information which may be obtained through the world wide web address http://www.maths.ex.ac.uk/~adg/.

ADG would like to thank Gonville and Caius College for a Research Fellowship during the inception of the project and the Nuffield Foundation for additional support in its final phase. We are especially grateful for the research support which has been provided to us by the National Science Foundation.

New York   S. Childress
Exeter   A.D. Gilbert
June 1995

Contents

1. The Fast Dynamo Problem

This book is about the structure of magnetic fields in electrically conducting fluids. Most ordinary fluids, such as air or water at familiar temperatures and pressures, are not good conductors of electricity, and so offer little direct experience of the kind of phenomena we shall be examining. In more extreme environments — the liquid core of the Earth or the atmospheres of stars, for example — electrically conducting fluids or plasmas are common. And there magnetic fields of surprising complexity are observed. Although we shall be concerned with models that are very simple and idealized, the motivation for the work described in the following pages lies in these examples of magnetic fields in astrophysics.

In the theory to be described below, the magnetic field reflects closely the motions of the fluid, just as the swirls of cream in a cup of coffee reveal the pattern of eddies stirred in by a spoon. Symbolically we have

magnetic field   +   movement   →   change in magnetic field.

In fact, it could be said that our subject is really the geometry of fluid flow as revealed by an embedded magnetic field. There is thus a close link to the Lagrangian properties of flow fields and, since a magnetic field may be regarded as consisting of magnetic field lines, to the behavior of material lines. The concept of a *fast dynamo* will be related to the growth of average line length in a flow and to the stretching and folding properties of the flow. The search for fast dynamos will then lead naturally to chaotic flows, and from the chaos will emerge the complex structure of magnetic fields typical in fast dynamos.

We begin this introductory chapter with a brief outline of the reasons for considering dynamo action in astrophysical bodies and summarize the equations satisfied by the magnetic field. We then (in §§1.3–5) give a heuristic discussion of slow and fast dynamo action and describe the simplest models of the latter process. Several key concepts dealing with the stretching of lines in flows are introduced in §1.6.

1.1 Magnetic Fields in Astrophysics

In its broadest sense, the study of the interactions between a magnetic field and a flow of an electrically conducting fluid, including the key issue of the maintenance of field against the losses of dissipation, comprises the magnetohydrodynamic (MHD) *dynamo problem*. We shall often refer to these as yet unspecified processes of sustaining or exciting a magnetic field as *dynamo action*. Dynamo theory was originally conceived as a model for the solar field (Larmor 1919), but the main stimulus to the subject came from a paradox connected with the Earth's magnetic field. Over 50 years ago it was estimated that the time of natural decay of the magnetic field (roughly the parameter L^2/η discussed in §1.2.2) is about 150,000 years (see the discussion in Roberts 1967). That is, if the Earth were made of a solid material with an electrical conductivity equal to that of the fluid core, any magnetic field would decay in a time of this order. By 'any field' we mean one defined initially throughout the conducting body, and extending continuously into the surrounding non-conducting space as the usual dipole field together with higher harmonics. The reason for this decay lies in the high temperature of the core, which does not support permanent magnetization but allows the free flow of electrical currents. The currents associated with the field generate heat because of electrical resistance and magnetic energy is thus dissipated.[1]

The paradox comes from the history of the geomagnetic field, which is provided by paleomagnetic records. This data, taken from cores of ocean sediment and other sources, yield a continuous record of magnetic activity for over 10^8 years. Since there has been no decay of the field there must therefore be a sustaining source of magnetic energy, and the most plausible theory invokes dynamo action in the fluid core.

Because of the paradox, dynamo theory has replaced William Gilbert's celebrated model of the Earth: a gigantic spherical lodestone with permanent North and South poles aligned with the axis of rotation. Gilbert, who was court physician to Queen Elizabeth I, gathered and organized the compass measurements of the seafarers of his day. From these observations, Gilbert formulated his theory and published it in his famous treatise *de Magnete* in 1600. Gilbert's theory correctly explains the basic dipole structure of the geomagnetic field.

However MHD dynamo theory offers the only generally accepted explanation for the persistence of the field, and for other observable properties such as the temporal evolution of the dipole component and the field's multipole structure. Variations of the field occur on many time-scales, from aperiodic reversals of polarity on the scale of 10^6 years, to pulsations on the scale of

---

[1] Looking ahead to §1.2.1, with $\mathbf{u} = 0$ in equation (1.2.1*c*), the current, whose existence follows from (1.2.1*a*), is associated with a non-zero electric field (provided σ is finite). From (1.2.1*b*) this causes a change of $\mathbf{B}$, which on general thermodynamic grounds must decrease magnetic energy.

a few hours, and some of these changes, on scales exceeding 10 years, are believed to be linked directly to motions of the core fluid (Hide & Roberts 1961).[2]

Observations of the magnetic fields of the planets in our solar system, the Sun, and numerous stars support the view of an active, ever-changing magnetic climate driven by some kind of dynamo action (Moffatt 1978, Parker 1979, Zeldovich, Ruzmaikin & Sokoloff 1983). Of the planets, Jupiter stands out as similar in many respects to the Earth, and the dipole field has a surface strength about ten times larger. Estimates of decay again suggest dynamo action within a fluid core, which for Jupiter probably occupies most of the planet. Jupiter is also similar to the Earth in the inclination of the axis of its dipole relative to the axis of rotation. Mercury, Mars, and Venus are also endowed with dipole fields, but their strengths are considerably weaker than the geomagnetic field, by a factor of about 10^{-3}. All are nevertheless good candidates for planetary dynamo action. Uranus is an unusual example of planetary magnetism in that the axis of its dipole is inclined at approximately $60°$ to its rotation axis (Hide 1988).

As Larmor (1919) originally suggested, the Sun offers another likely instance of dynamo action, and a dramatic example of astrophysical magnetic activity at comparatively close range. The magnetic field can be measured from Zeeman splitting of spectral lines. The well-known 22-year solar cycle (comprising two roughly 11-year cycles of opposite mean polarity), as well as sunspots, prominences, flares, and a host of other events, offer a rich mixture of magnetohydrodynamic phenomena (Priest 1984). In contrast to the Earth, the existence of a dynamo in the Sun is not implied by an estimate of dissipative decay. Rather, we must infer some continuous process of creation and destruction of field from the cycle itself, given the change of parity occurring each 11 years. The exact mechanism of dynamo action is not understood, but sunspot activity is an important symptom of the process (Babcock 1959). The visible light from the Sun is emitted from the *photosphere*, with a thickness of only 10^{-3} of the solar radius. This layer lies on top of the *convection zone*, having a thickness of perhaps one-third the solar radius. The turbulent motions within the convection zone offer one plausible source of dynamo action. There is also the possibility of significant activity at the base of the convection zone, where there exists a thin stably-stratified layer (see Spiegel 1994). In this scenario, the layer acts as a storage region for field gradually built up by

---

[2] The reader interested in dynamo theories of the Earth's magnetic field will find extensive discussion in the books by Roberts (1967), Moffatt (1978), Parker (1979) and Krause & Rädler (1980), and may draw on numerous review articles, including Busse (1978), Childress (1984), Cowling (1957a), Ghil & Childress (1987) Hide & Roberts (1961), Roberts (1971, 1987) and Weiss (1971). Other reviews with an emphasis on fast dynamo theory are Roberts & Soward (1992), Bayly (1994) and Soward (1994b). For a readable overview see also Moffatt (1989), reprinted in Childress et al. (1990b).

dynamo action, until instabilities form tubes of flux which rise through the convection zone to create bipolar pairs of sunspots.

Regardless of how the magnetic cycle operates, a remarkable property of the solar field emerges from studies of its small-scale structure. Although the average photospheric field is very modest, roughly comparable to the terrestrial field, the lines of force are not at all evenly distributed. Rather, they are bunched into small tubes of intense field approximately 10^3 times the mean field (Schüssler 1990). This *fibrillation* or *intermittency* of the field seems to occur in all regions of magnetic activity and may extend to vertical tubes penetrating well into the convection zone (Parker 1979). The reason for this fibrillated structure is not known but one explanation invokes a kind of thermal instability associated with a superadiabatic environment (Parker 1979, Kerswell & Childress 1992). We shall argue here that the fine-scale structure of a fibrillated magnetic field is typical of fast dynamo action. The feature of a weak average magnetic field dominated by intense, small-scale magnetic structures will be basic to many of the models we consider below.

Certain of the distant stars, called *magnetic stars*, have a magnetic field sufficiently strong to be sensed from Earth using the Zeeman effect. The observations integrate across the disc of the star, and so tend to average the field. If our Sun is typical, then this average field will be a small multiple of the strongest field. It is thus likely that many stars have a magnetic field, even though few have one strong enough to be detectable through the average value.

A similar difficulty applies to the measurement of galactic and intergalactic fields. Estimates give average field strengths of the order of 10^{-5}–10^{-6} of the Earth's field. While weak, this field is nevertheless dynamically active through interactions with both the intergalactic gas and cosmic rays. Of the more exotic distant objects with magnetic fields, we note only that pulsars have been identified with rotating neutron stars supporting an inclined dipole field (Gold 1968). We remark that our discussion of electromagnetism will be entirely non-relativistic and therefore inappropriate for certain massive objects such as black holes.

This brief summary of the magnetic cosmos suggests that magnetic activity, like the matter with which it is associated, is very unevenly distributed. Underlying this structure is a weak background field, perhaps a relic of the birth of the Universe. But this background is punctuated by the violent activity which may be caused by dynamo action: movements of matter that intensify magnetic fields by many orders of magnitude, and so appear almost as singularities on the background field. The mechanisms of dynamo action, which we will investigate quantitatively, offer the promise of understanding magnetic structure on the scales of many astronomical objects: planets, stars, galaxies, and intergalactic clouds. Whether or not the dynamo model is correct for any given natural occurrence of magnetism will rest on more detailed physical models, and it is hoped that the general ideas to be developed here will prove to be useful in the construction and elaboration of these models.

1.2 Kinematics of the Magnetic Field

Although we shall consider the MHD dynamo problem in Chap. 12, the focus of this monograph is the interaction of magnetic and electric fields with a *given* fluid motion. In this section we shall encounter the simplest formulation of this *kinematic* problem, examine the principal dimensionless parameter of the theory, and derive some immediate consequences. We conclude with a critique of the kinematic approach to a dynamical subject.

1.2.1 Pre-Maxwell Equations for a Moving Medium

We shall suppose that an electrically conducting fluid fills all of three-dimensional space, and that the fluid moves with a velocity $\mathbf{u}(\mathbf{x}, t)$ defined throughout space and for all $t \geq 0$. We take $\mathbf{u}$ to be endowed with characteristic scales U, L and T of velocity, length and time, respectively. Generally in astrophysics an electrically conducting body such as a star is surrounded by space essentially devoid of matter. The boundary conditions for magnetic activity then include conditions for matching the internal magnetic field of the body with an external vacuum magnetic field. However for the fast dynamo problem it is convenient to suppose that space consists entirely of conducting matter and to consider one of two cases: (a) $\mathbf{u}$ vanishes except within a bounded domain $\mathcal{D}$ of diameter L, or (b) $\mathbf{u}$ is periodic in space with periodicity cube $\mathcal{D}$ of side L: $\mathbf{u}(\mathbf{x} + L\mathbf{e}, t) = \mathbf{u}(\mathbf{x}, t)$ for any integer vector $\mathbf{e}$. For discussion of these boundary conditions see §1.2.5; here we simply note that (a) is the more physically reasonable, while (b) is often convenient mathematically and computationally. Both of these idealizations miss some of the interesting aspects of bounded conductors. However the ideas to be developed regarding fast dynamo action can be illustrated adequately with flows of these two types.

For the time-dependence of $\mathbf{u}$ there will again be several possibilities. In a *steady* flow, $\mathbf{u}$ is independent of time. We also consider *time-periodic* flow, with a given period T, $\mathbf{u}(\mathbf{x}, t + T) = \mathbf{u}(\mathbf{x}, t)$. Finally we shall consider, in Chap. 11, flows which change in time in a random way.

Another property of interest is the *compressibility* of the fluid. In virtually all of our examples the fluid will be incompressible, with $\nabla \cdot \mathbf{u} = 0$. While this is an unreasonable assumption for astrophysical plasmas, the effects of compressibility in the *kinematic* theory are significant only in the diffusive term; see (1.2.9) below. Since diffusion will be small in our models, compressibility is a secondary consideration in fast dynamo theory. However in the present chapter we shall consider the effects of compressibility on the governing equations, in order to exhibit the terms which will ultimately be dropped.

A good approximation for the modelling of planetary, stellar, and galactic magnetic fields is that U, $L/T \ll c$, where c is the speed of light. In this case a reduced form of Maxwell's equations is valid, in which currents associated with charge separation are neglected, as are relativistic corrections for the motion of the fluid. This *pre-Maxwell* system can be written in terms of the magnetic field $\mathbf{B}(\mathbf{x}, t)$, electric field $\mathbf{E}(\mathbf{x}, t)$ and current $\mathbf{J}(\mathbf{x}, t)$ as follows:

Ampère's law:	$\nabla \times \mathbf{B} = \mu \mathbf{J},$	(1.2.1a)
Faraday's law:	$\nabla \times \mathbf{E} = -\partial \mathbf{B}/\partial t,$	(1.2.1b)
Ohm's law:	$\mathbf{J} = \sigma(\mathbf{E} + \mathbf{u} \times \mathbf{B}),$	(1.2.1c)
	$\nabla \cdot \mathbf{B} = 0, \qquad \nabla \cdot \mathbf{E} = q/\varepsilon.$	(1.2.1d, e)

Here μ is the magnetic permeability in vacuum, σ is the electrical conductivity, ε is the dielectric constant, and q is the charge density. For simplicity we take μ, σ, and ε to be constant. The charge density q is a scalar function of $(\mathbf{x}, t)$ which can be derived from the divergence of Ohm's law. The term $\sigma(\mathbf{u} \times \mathbf{B})$ is called the *induced current*. In order to formulate useful problems we must add to (1.2.1) boundary conditions appropriate to $\mathcal{D}$; see §1.2.5.[3]

If $(1.2.1c)$ is substituted into $(1.2.1a)$, the curl taken, and $(1.2.1b)$ used, we obtain an equation for $\mathbf{B}$ called the *induction equation*:

$$\frac{\partial \mathbf{B}}{\partial t} - \nabla \times (\mathbf{u} \times \mathbf{B}) - \eta \nabla^2 \mathbf{B} = 0. \tag{1.2.2}$$

Here and elsewhere it is understood that $\mathbf{B}$ has zero divergence $(1.2.1d)$. The parameter η, having dimensions L^2/T, is the *magnetic diffusivity*,

$$\eta \equiv \frac{1}{\mu\sigma} = \text{magnetic diffusivity.} \tag{1.2.3}$$

In the kinematic theory the induction equation is a linear evolution equation for the magnetic field, balancing the effects of diffusion (the final term in (1.2.2)) and distortion by the flow. The distortion or *advection* of the magnetic field can be studied with the aid of the vector identity

$$\nabla \times (\mathbf{u} \times \mathbf{B}) = \mathbf{B} \cdot \nabla \mathbf{u} - \mathbf{u} \cdot \nabla \mathbf{B} + (\nabla \cdot \mathbf{B})\mathbf{u} - (\nabla \cdot \mathbf{u})\mathbf{B}. \tag{1.2.4}$$

Using $(1.2.1d)$, and assuming for the moment that $\nabla \cdot \mathbf{u} = 0$, $(1.2.4)$ may be used in $(1.2.2)$ to put the induction equation in the form

$$\frac{d\mathbf{B}}{dt} - \eta \nabla^2 \mathbf{B} = \mathbf{B} \cdot \nabla \mathbf{u}, \qquad (\nabla \cdot \mathbf{u} = 0), \tag{1.2.5}$$

---

[3] Maxwell's equations contain the additional term $\mu\varepsilon\, \partial \mathbf{E}/\partial t$ on the right-hand side of $(1.2.1a)$. This is the *displacement current*. Since $1/\mu\varepsilon = c^2$, the square of the speed of light, we have a relativistic correction under the conditions given earlier, $L/T \ll c$.

where we introduce the *material derivative*, or time derivative taken relative to a point moving with the fluid:

$$\frac{d}{dt} \equiv \frac{\partial}{\partial t} + \mathbf{u} \cdot \nabla. \tag{1.2.6}$$

Thus the advection of $\mathbf{B}$ has two components: (1) parallel transport of the field by the flow, and (2) distortion of the field by the velocity derivative tensor $\mathbf{D}$ defined by

$$D_{ij} = \frac{\partial u_i}{\partial x_j} \qquad \text{or} \qquad \mathbf{D} = (\nabla \mathbf{u})^{\mathrm{T}}. \tag{1.2.7}$$

It is the second process which is of primary interest, for it is the gradients of the flow velocity which enable the stretching, folding, and twisting of field essential to the dynamo mechanism.

As noted earlier the convenient assumption of *incompressibility*, $\nabla \cdot \mathbf{u} = 0$, is made frequently and will be adopted in most of our discussion. However it is important to see what effects fluid compressibility brings to the kinematic dynamo problem. To treat a compressible material, the fluid density $\rho(\mathbf{x}, t)$ is introduced. Assuming no sources of matter, the equation of conservation of mass is

$$\frac{\partial \rho}{\partial t} + \nabla \cdot (\rho \mathbf{u}) \equiv \frac{d\rho}{dt} + \rho \nabla \cdot \mathbf{u} = 0. \tag{1.2.8}$$

If this is used to eliminate $\nabla \cdot \mathbf{u}$ from (1.2.4), the induction equation takes the form

$$\frac{d(\mathbf{B}/\rho)}{dt} - \frac{\eta}{\rho} \nabla^2 \mathbf{B} = (\mathbf{B}/\rho) \cdot \nabla \mathbf{u}. \tag{1.2.9}$$

In the kinematic theory both $\mathbf{u}$ and ρ are assumed known; so a compressible material involves an induction equation for $\mathbf{B}/\rho$ with a modified diffusion operator $\rho^{-1}\nabla^2(\rho \, \cdot \,)$.

1.2.2 The Perfect Conductor

A *dimensionless* form of the induction equation (1.2.2) is obtained by choosing L, U, and L/U as units of length, velocity, and time; in other words, quantities are scaled with *advective* length- and time-scales.[4] The equation that results is again (1.2.2) but with η^{-1} replaced by the dimensionless parameter

$$R \equiv \frac{UL}{\eta} = \text{magnetic Reynolds number.} \tag{1.2.10}$$

---

[4] In some sources quantities are scaled using diffusive length- and time-scales, taking L, η/L and L^2/η as measures of length, velocity and time.

The name comes from its similarity to the Reynolds number of fluid dynamics, where the kinematic viscosity replaces η. A small magnetic Reynolds number emphasizes diffusion of flux, advection having a small effect. At large values of R, diffusion can be significant only where $\nabla^2 \mathbf{B}$ is large, i.e., where $\mathbf{B}$ possesses small scales or has discontinuities in its value or its gradient. If the field varies only on the large scale, L, the distortions of the field come primarily from advection. We show in table 1.1 some characteristic values of R for various examples of astrophysical dynamo action discussed in §1.1 (Roberts 1967, Parker 1979, Zeldovich $et\ al.$ 1983).

Table 1.1 Magnetic Reynolds numbers in astrophysics.

Object	L (m)	U (m/s)	η (m^2/s)[a]	τ_{T} (s)[b]	R
Core of Earth	10^6	10^{-3}	1	10^{12}	10^3
Core of Jupiter	10^7	10^{-1}	1	10^{11}	10^6
Solar conv. zone	10^8	10^3	10^3	10^8	10^8
Solar corona	10^9	10^3	1	10^9	10^{12}
Interstellar medium[c]	10^{17}	10^3	10^3	10^{17}	10^{17}
Galaxy[c]	10^{18}	10^4	10^3	10^{17}	10^{19}

[a] Molecular value.
[b] Using η_{T} such that $R_{\mathrm{T}} = 10^3$.
[c] Average value. Values of R are dramatically reduced when R is based on ambipolar diffusion, a dynamical effect; in this case Zeldovich $et\ al.$ (1983) cite $R \sim 10^6$ in galaxies.

The magnetic Reynolds number R based on $molecular$ values of η is relevant since turbulence, which usually has the effect of reducing the effective turbulent Reynolds number to a thousand or so, is of interest to us here as a domain of dynamo action. The Kolmogorov scaling of velocity and length in the turbulent inertial range yields an effective value R_e of R for an eddy of size L_e equal to $(L_e/L)^{4/3}$ times the value in table 1.1, which means that R_e is large over a substantial range of L_e. On the other hand turbulence can destroy field as well amplify it, so for estimating the decay of field in the $absence$ of dynamo action it is important to replace the molecular value of η by its effective turbulent value η_{T}. As we noted above, typically the turbulent resistivity reduces a Reynolds number to effective values of 10^3 or less. We introduce the turbulent decay time $\tau_{\mathrm{T}} = L^2/\eta_{\mathrm{T}}$, and estimate η_{T} by setting $R_{\mathrm{T}} \equiv UL/\eta_{\mathrm{T}} = 10^3$. The resulting times for decay are included in table 1.1. Since the age of the solar system is about 10^{15} s and that of the galaxy about 10^{17} s, we see that the estimates suggest dynamo action except possibly for interstellar and galactic fields. Certain detailed estimates for a galactic disc are supportive of a dynamo (Zeldovich $et\ al.$ 1983), but recent data has challenged this viewpoint; see Chap. 12.

Regardless of the ultimate scope of dynamo theory, we are led to the limit $R \to \infty$ as the relevant one for astrophysical studies of dynamo action (Vainshtein & Zeldovich 1972). We shall call this the *perfectly conducting limit*, because R is proportional to the electrical conductivity σ of the medium, a perfect conductor being one of infinite σ. This conventional terminology is somewhat misleading, however, since from table 1.1 we note that over the vast range of scales in astrophysics the conductivity changes far less that the scale of length. It is the great size of astrophysical objects which provides the large value of R.

Given the importance of the perfectly conducting limit, it is natural to introduce the idealization of a perfect conductor. (Here the alternative limit of infinite size becomes awkward!) For the perfect conductor the diffusion of $\mathbf{B}$ is identically zero and induction is governed by the reduced equation

$$\frac{\partial \mathbf{B}}{\partial t} - \nabla \times (\mathbf{u} \times \mathbf{B}) = 0. \tag{1.2.11a}$$

This equation takes the form

$$\frac{d(\mathbf{B}/\rho)}{dt} = (\mathbf{B}/\rho) \cdot \nabla \mathbf{u} \tag{1.2.11b}$$

in a compressible fluid, and

$$\frac{d\mathbf{B}}{dt} = \mathbf{B} \cdot \nabla \mathbf{u}, \qquad (\nabla \cdot \mathbf{u} = 0) \tag{1.2.11c}$$

in an incompressible one. From the pre-Maxwell system (1.2.1) we see that a well-defined current exists in a perfect conductor, since the curl of $\mathbf{B}$ has a value, but the corresponding electromotive force, or e.m.f., is infinitesimal. The electric field and its induced counterpart then cancel out identically,

$$\mathbf{E} + \mathbf{u} \times \mathbf{B} = 0. \tag{1.2.12}$$

The reduced equation (1.2.11) has an important implication for the evolution of the magnetic field. The lines of force are *tied* or *frozen* into the conducting fluid. This terminology is due to Alfvén (see Cowling 1957a). More precisely, we say that a line is *material* if every point on the line moves with the fluid. It follows that the lines of force in a perfect conductor are material lines. Two nearby points on a material line can separate, in which case the line is said to be *stretched* by the flow. This can occur even in incompressible flow. For example, if $\mathbf{u} \equiv (u, v) = (x, -y)$ in two dimensions, any line segment initially on the x-axis is stretched exponentially under the flow. This simple example is actually of great interest, since it represents flow in the neighborhood of a critical point (or stagnation point), where the velocity vanishes and the fluid moves along hyperbolae $xy = $ const. Such *hyperbolic* (or *X-type*) points occur in many steady flows, and so stretching emerges as a common feature of fluid motion. Other examples of stretching of lines of force in two-dimensional flow are easily found. If the flow $(u, v) = (0, \sin 2\pi x)$ acts on a line of force initially coinciding with the x-axis, the line is deformed into a sine-wave and hence stretched. If the flow is $\mathbf{u} = (u_r, u_\theta) = (0, r \exp(-r^2))$ in polar coordinates (r, θ), the same initial line is stretched by being wound up into a spiral.

To establish the property that lines of force are material lines in a perfect conductor, suppose that at time t two close points $\mathbf{x}$ and $\mathbf{x} + \delta\mathbf{l}$ are marked, and that subsequently these points move with the fluid. At time $t + dt$ the points have moved to $\mathbf{x} + \mathbf{u}(\mathbf{x}, t)dt$ and $\mathbf{x} + \delta\mathbf{l} + \mathbf{u}(\mathbf{x} + \delta\mathbf{l}, t)dt$, and so the vector $\delta\mathbf{l}$ is stretched to $\delta\mathbf{l} + \delta\mathbf{l} \cdot \nabla\mathbf{u}\, dt + O(\delta\mathbf{l}^2)$. We thus see that $\delta\mathbf{l}$ satisfies

$$\frac{d\delta\mathbf{l}}{dt} = \delta\mathbf{l} \cdot \nabla\mathbf{u}(\mathbf{x}, t) + O(|\delta\mathbf{l}|^2). \tag{1.2.13}$$

Now take the limit $\delta\mathbf{l} \to 0$ and replace $\delta\mathbf{l}$ by an infinitesimal line element $d\mathbf{l}$; comparing $(1.2.11b)$ and $(1.2.13)$ we see that $d\mathbf{l}$ and $\mathbf{B}/\rho$ satisfy the same equation. A magnetic line of force, given say by $\mathbf{y}(s)$ for $-\infty < s < +\infty$, satisfies

$$\frac{d\mathbf{y}}{ds} = \frac{\mathbf{B}}{\rho}(\mathbf{y}(s), t). \tag{1.2.14}$$

It follows that $d\mathbf{y}$ and $\mathbf{B}$ are parallel. Identifying $d\mathbf{y}$ with $d\mathbf{l}$ establishes the material property of lines of force. The term *magnetic line* is also used.[5]

The most elegant approach to the perfectly conducting limit uses *Lagrangian coordinates*, often called Lagrangian variables. In the Lagrangian representation of a fluid, particles are given a label $\mathbf{a}$, usually identified with the *initial* position of the particle. The Lagrangian coordinates are then defined by the function $\mathbf{x}(\mathbf{a}, t)$, giving the position at time t of a fluid particle initially at $\mathbf{a}$. Lagrangian coordinates are therefore determined as solutions to the system of ordinary differential equations

$$\frac{d\mathbf{x}}{dt} = \mathbf{u}(\mathbf{x}, t), \qquad \mathbf{x}(\mathbf{a}, 0) = \mathbf{a}. \tag{1.2.15}$$

We denote the *Lagrangian map* $\mathbf{a} \to \mathbf{x}$ by M, or M^t to exhibit the time. Note that for an incompressible fluid, $\nabla \cdot \mathbf{u} = 0$, the map M^t is volume-preserving.

Define the Jacobian matrix $\mathbf{J}^t(\mathbf{a}) \equiv \mathbf{J}(\mathbf{a}, t)$ of the map M by

$$J_{ij}(\mathbf{a}, t) = \frac{\partial x_i(\mathbf{a}, t)}{\partial a_j} \quad \text{or} \quad \mathbf{J} = \partial\mathbf{x}/\partial\mathbf{a}. \tag{1.2.16}$$

At time t, $d\mathbf{l} = \mathbf{J}(\mathbf{a}, t)\, d\mathbf{a}$ is tangent to the material line which initially had tangent $d\mathbf{a}$ at $\mathbf{a}$. This is the same $d\mathbf{l}$ as emerged from the limit of $(1.2.13)$ as a

---

[5] This result is identical to *Helmholtz's theorem* concerning the material property of a vortex line in inviscid fluid flow.

solution of the induction equation.[6] It follows that $\mathbf{B}(\mathbf{x}, t)$ can be represented in the form

$$\frac{\mathbf{B}(\mathbf{x}(\mathbf{a}, t), t)}{\rho(\mathbf{x}(\mathbf{a}, t), t)} = \mathbf{J}(\mathbf{a}, t) \frac{\mathbf{B}(\mathbf{a}, 0)}{\rho(\mathbf{a}, 0)}. \tag{1.2.17}$$

This is sometimes called *Cauchy's solution* of the induction equation. As $\rho(\mathbf{a}, 0)/\rho(\mathbf{x}, t) = \det \mathbf{J}(\mathbf{a}, t)$, $\mathbf{B}(\mathbf{x}, t)$ can be expressed entirely in terms of the initial field and the Jacobian. It is therefore a representation in terms of Lagrangian coordinates, which transfers the difficulty of solving the induction equation to the difficulty of solving the Lagrangian system (1.2.15). We can think of (1.2.17) as defining an operator $\mathbf{T}$ on an initial magnetic field,

$$(\mathbf{TB})(\mathbf{x}, t) = \frac{\rho(\mathbf{x}, t)}{\rho((M^t)^{-1}\mathbf{x}, 0)} \mathbf{J}((M^t)^{-1}\mathbf{x}, t)\mathbf{B}((M^t)^{-1}\mathbf{x}, 0). \tag{1.2.18}$$

We shall call $\mathbf{T}$ the *induction operator* for a perfect conductor. Note that the ratio of densities in (1.2.18) is unity for an incompressible flow.

As an example, we introduce a flow which will frequently be useful to us. The *Beltrami wave* $\mathbf{u}^{(x)}$ is defined by

$$\mathbf{u}^{(x)} = (0, \sin qx, \pm \cos qx). \tag{1.2.19}$$

This wave possesses the *Beltrami property* (Milne-Thomson (1955), p. 76), that the curl of $\mathbf{u}$ is parallel to $\mathbf{u}$ everywhere in space, since $\nabla \times \mathbf{u}^{(x)} = \pm q\mathbf{u}^{(x)}$. This is a strong form of a Beltrami field, where the function relating $\nabla \times \mathbf{u}$ to $\mathbf{u}$ is in fact a constant. We define the *parity*, or *relative helicity*, of the wave as sign$(\pm q)$. The concept of helicity is discussed in §1.2.3. Beltrami waves will be used often in the constructions to follow.

The Lagrangian coordinates for this wave are

$$\mathbf{x}(\mathbf{a}, t) = (a_1, a_2 + t \sin qa_1, a_3 \pm t \cos qa_1). \tag{1.2.20}$$

Thus

$$\mathbf{J}(\mathbf{a}, t) = \begin{pmatrix} 1 & 0 & 0 \\ qt \cos qa_1 & 1 & 0 \\ \mp qt \sin qa_1 & 0 & 1 \end{pmatrix}. \tag{1.2.21}$$

The first column of the matrix tells us the image of the initial field $(1, 0, 0)$ at time t. Every field line is mapped onto a helical line with a pitch angle having tangent qt. The field in the (y, z)-plane increases with t as the line is stretched, leaving unaffected the original field component and flux in the x-direction.

---

[6] To show directly that $\mathbf{J}$ satisfies the induction equation (1.2.11c) as a tensor equation, take the gradient with respect to $\mathbf{a}$ of the system (1.2.15), and use the fact that the derivative on the left is a material derivative. Then apply the chain rule on the right.

Because of the simplicity of the Lagrangian coordinates in this example, the Eulerian form of the magnetic field is easily seen to be $\mathbf{B} = (1, qt \cos qx, \mp qt \sin qx)$, and it can be checked that $(1.2.11\,a)$ is satisfied. The *induced e.m.f.* is $\mathbf{u} \times \mathbf{B} = (\mp qt, \pm \cos qx, -\sin qx)$. There is thus a linearly growing mean current parallel to the initial field, which can be viewed as creating the helical turns. The relation between mean field and mean current is important in certain smoothing approximations used in dynamo theory, which lead to the identification of the *alpha effect* (see Moffatt 1978, Krause & Rädler 1980, Ghil & Childress 1987, and Chap. 5).

1.2.3 Invariants

When describing the evolution of physical properties of a continuous medium it is useful to work with finite neighborhoods rather than the idealizations of point and line. By a *flux tube* we shall mean a small bundle of magnetic lines, that is, a tubular domain made up entirely of lines of force. Suppose now that we consider a small finite section of a flux tube bounded by two cross sections S_1 and S_2, and a cylindrical surface S_3 (Fig. 1.1 a). Since $\nabla \cdot \mathbf{B} = 0$ everywhere, it follows from the divergence theorem applied to the interior of this region, and the fact that the field at points of S_3 is tangent to S_3, that the integrals of the normal component of the magnetic field over the surfaces S_1 and S_2 are the same. This invariant of the flux tube is called the *magnetic flux* or simply the flux or *strength* of the tube.

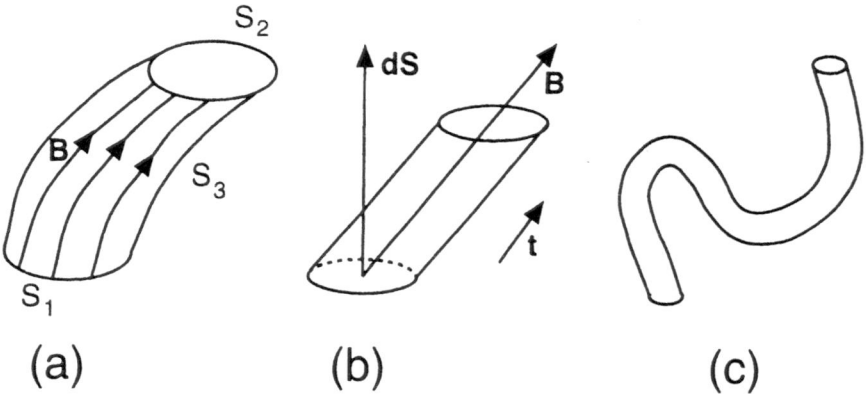

(a)   (b)   (c)

Fig. 1.1 Stretching in a perfect conductor increases the magnetic energy of flux tubes: (*a*) initial configuration of a flux tube, (*b*) an infinitesimal sub-tube, and (*c*) the stretched tube.

It follows from the material property of the constituent field lines that the points of a flux tube move with the fluid in a perfect conductor. More importantly, the flux of a tube is also invariant under the flow. To see this, consider an infinitesimal flux tube consisting of a bundle of field lines passing through the area element $d\mathbf{S}$ for a distance dL in the direction of the unit vector $\mathbf{t}$, which is proportional to $\mathbf{B}$ (Fig. 1.1b). The sub-tube, of volume $dV = dL\,d\mathbf{S}\cdot\mathbf{t}$, moves with the fluid and, by conservation of mass, $\rho\,dV$ is invariant under the flow. But the material property of magnetic lines allows us to identify $\mathbf{t}\,dL$ with a *constant* multiple of $\mathbf{B}/\rho$ at each Lagrangian point, and so $\mathbf{B}\cdot d\mathbf{S}$ is also invariant. Integrating over any cross-section S of the original tube, we then obtain invariance of flux,

$$\frac{d}{dt}\int_S \mathbf{B}\cdot d\mathbf{S} = 0. \tag{1.2.22}$$

The conservation of flux is a consequence of the material property of lines *and* mass conservation. By itself, the material property of lines does not imply conservation of flux. Since (1.2.11a) is not equivalent to the limit of (1.2.13), the lines of force and the intensity of the field defined along those lines are distinct fields.

This result has a crucial consequence for dynamo theory, because it tells us how a magnetic field can be amplified by the movement of a perfect conductor. It is necessary for the area of the cross-section of a flux tube to decrease, so that the intensity of the magnetic field will increase to maintain a constant flux. In a compressible material this decrease in area can be achieved by simple compression, but a far more interesting possibility is the stretching of the tube (Fig. 1.1c), which can occur in the simplest incompressible flows (e.g., the flows with a stagnation point mentioned above). In fluid flows with *chaotic* streamlines, the stretching of fluid elements is particularly intense, and the term *stretching flow* is often applied when the stretching is over a region; see §1.6. Thus a link is established between stretching flows and dynamo action, a connection first emphasized by Arnold and his collaborators (Arnold *et al.* 1981).

In addition to the flux carried by a tube, there are other invariants of the flow of a perfect conductor associated with the *topology* of the magnetic field (Moffatt 1969). These invariants characterize the knottedness of the magnetic lines. Since material lines cannot break or pass through one another under the action of a smooth flow, any number determined uniquely by the initial knottedness of the field is a topological invariant. An example of such an invariant is the *magnetic helicity* $H_{\mathrm{M}}(\mathcal{D})$, defined by

$$H_{\mathrm{M}}(\mathcal{D}) = \int_{\mathcal{D}} \mathbf{A}\cdot\mathbf{B}\,dV \tag{1.2.23}$$

(Woltjer 1958, Moffatt 1969). Here we have introduced a *magnetic potential* $\mathbf{A}(\mathbf{x},t)$ of $\mathbf{B}$, which is a vector field such that $\nabla\times\mathbf{A} = \mathbf{B}$. If $\mathbf{B}$ satisfies (1.2.11a) it follows that $\mathbf{A}$ satisfies

$$\frac{\partial \mathbf{A}}{\partial t} - \mathbf{u} \times \mathbf{B} + \nabla\phi = 0, \tag{1.2.24}$$

for some scalar field $\phi(\mathbf{x}, t)$. For the case where $\mathbf{B}$ is periodic in all coordinates it is straightforward to show that the time derivative of $H_M(\mathcal{D})$ vanishes if $(1.2.11a)$ and $(1.2.24)$ are satisfied. For a finite domain $\mathcal{D}$ with $\mathbf{u}$ non-zero only in the interior it is possible to define ϕ to be zero on the boundary of $\mathcal{D}$, in which case the proof again follows easily (Berger & Field 1984).

These topological invariants are generally expressed as integrals over the domain, that is as *functionals* of $\mathbf{B}$. These functionals each capture some part of the topology, but not all of it. For example, the helicity can be shown to represent a summed signed linking number (Moffatt 1969), but there are arrangements of discrete flux tubes that have zero helicity but cannot be separated. An example is the Borromean rings; their knottedness cannot be characterized by the helicity, which is zero, but rather according to a different, more complicated topological invariant (Berger 1990).

We have thus shown that stretching of flux tubes provides a way of increasing $\mathbf{B}$ in a perfect conductor, subject to topological constraints. A convenient measure of field intensity is the *magnetic energy* $E_M(\mathcal{D})$, defined by

$$E_M(\mathcal{D}) = \frac{1}{2\mu} \int_{\mathcal{D}} B^2 \, dV. \tag{1.2.25}$$

The increase of E_M from a given movement of fluid may be accounted for by the work done by the forces applied to the fluid to effect the movement. These forces are balanced by electromagnetic or *Lorentz* forces. The Lorentz forces lie outside the concerns of the kinematic theory but are an essential part of the full MHD dynamo problem (see Chap. 12).

If the magnetic Reynolds number is large but finite, it is reasonable to expect the equations for field in a perfectly conducting fluid to provide approximations to the diffusive solutions. The nature of this approximation is a central issue of fast dynamo theory (see §1.5): we note here only that in the diffusive problem magnetic lines are no longer material lines, and topological invariants may be broken, because of diffusive cancellation of field. For nearby flux tubes of differing orientation can partially cancel, just as cold and hot fluids mix diffusively, and tubes can be severed and reconnected in various ways. This has far-reaching implications for magnetic structure. Of particular interest is the time required to cut and reconnect tubes of flux, relative to the natural time scale of the fluid motion, in the limit of large R (see Biskamp 1993). If reconnection occurs on the advective time scale, then topological constraints are effectively broken in the perfectly conducting limit. We will not deal explicitly with reconnection, although it will always be a part of the diffusive processes which we consider.

We conclude this section with mention of helicity as applied to the velocity field $\mathbf{u}$. The *vorticity field* $\boldsymbol{\omega}$ for a given flow $\mathbf{u}$ is defined to be $\boldsymbol{\omega} = \nabla \times \mathbf{u}$. The *helicity density* of the vorticity is then $\mathbf{u} \cdot \boldsymbol{\omega}$. The *vorticity helicity* H_V of the flow is defined analogously to H_M by

$$H_V = \int \mathbf{u} \cdot \boldsymbol{\omega} \, dV. \qquad (1.2.26)$$

The relative helicity or parity, introduced in §1.2.2, is just the helicity density divided by $|\mathbf{u}||\boldsymbol{\omega}|$. The vorticity helicity is, under certain conditions, a dynamical invariant of an Euler flow (Moffatt 1969), although this property plays no role in kinematic dynamo theory. Flows with non-zero helicity are, however, prime candidates for dynamo action, through creation of the alpha effect mentioned at the end of §1.2.2. For the Beltrami wave (1.2.19), the helicity density is uniform and equal to $\pm q$. This is one reason for the use of such waves as basic components of fast dynamos.

1.2.4 Limitations of Kinematic Theory

Before we embark on the analysis of various applications, a note of caution is appropriate regarding the scope of the kinematic theory. In kinematic theory velocity and magnetic fields are decoupled, and so it can provide a physical theory of dynamo action only during an initial phase, when the magnetic field is so weak that there is no dynamical coupling to the matter. That is to say, the existence of a (kinematic) dynamo ensures that a non-magnetic state is unstable. The instability is excited by introducing a weak *seed* magnetic field, far too feeble to affect the flow. The (usually exponential) growth of the seed field will then ensue, until the magnetic stresses build up to a point where the flow is affected and a full MHD theory is required to determine the subsequent evolution. All the kinematic theory can decide is whether or not the system is capable of dynamo action when no field is present! If a driven hydrodynamic system is capable of multiple equilibrium states when the field vanishes, then all of these states need to be tested for dynamo action.

Nevertheless this is a powerful result, similar to a local analysis of stability near critical points in the phase space of a dynamical system. Indeed, we can introduce such a phase space for the MHD system as a whole, and consider the possibility of establishing the existence of a dynamo-like *attractor*.[7] By this we mean that there should exist a solution of the MHD system, with a non-decaying magnetic field, which attracts general solutions in the phase space. Given a fixed rate of energy input to a system, thermodynamic considerations limit the size of the magnetic field that can be developed in any physical, that is to say, dissipative, system. (For examples of these arguments applied to the Earth's core see Backus 1975 and Hewitt, McKenzie & Weiss 1975.) Thus an attractor that is confined to a bounded region of the phase space in the absence of a field should stay bounded when extended to the MHD system. In principle then, the existence of magnetic fields can be explained by the instability of the non-magnetic state and so can be decided by kinematic considerations.

---

[7] We shall take up the idea of an attractor again in Chap. 12 in connection with dynamic equilibration.

There are, however, important questions which cannot be addressed within the kinematic framework. Foremost among them is the strength of the magnetic field for fully-developed or dynamically equilibrated solutions on the attractor. Also of interest is the nature of the magnetic field, whether cyclic, steady or chaotic in time, and how it is distributed over different length-scales. The solar magnetic field, for example, has in the past had epochs of many decades with little or no magnetic activity (Eddy 1976), after which a regular solar cycle resumed. The Earth's dipole field is known from the paleomagnetic data to have reversed in polarity hundreds of times over the last 10^8 years (see, e.g., Moffatt 1978). These events cannot be explained except in the context of MHD theory and the theory of nonlinear systems.

The choice of velocity field in kinematic induction is obviously important for applications to real problems, but the advance of theory can be triggered by considering special flows chosen to aid analysis or computation. Famous examples come from pioneering kinematic studies of the Earth's dynamo. Motions believed reasonable in a convective core were adopted by Bullard and Gellman, but the early proofs of dynamo action, by Herzenberg, Backus and others used somewhat artificial choices. Subsequent studies, by Braginsky and others, then combined the analytically interesting flows with the physics of fluids in rotating cores to bridge partially the gap between kinematic and MHD theory. For a more detailed account of these developments see Roberts (1971) and Moffatt (1978). It is with a view to the astrophysical applications of dynamo theory that we discuss, in Chap. 12, some preliminary investigations which include dynamics of the velocity field.

1.2.5 Boundary Conditions

We insert here a a few details concerning boundary conditions appropriate to kinematic and MHD dynamo theories. The derivation of these conditions is discussed in detail by Roberts (1967). They are of special concern to dynamo theory since the idea is to describe self-excitation of magnetic field by local processes involving interaction with a fluid flow. The boundary conditions must therefore be such as to isolate the system from spurious sources of magnetic energy, most importantly from infinity. One of our possible choices of $\mathcal{D}$ in §1.2.1, as periodicity cube for $\mathbf{u}$, leads immediately to the same periodicity conditions on the magnetic field: $\mathbf{B}(\mathbf{x} + L\mathbf{e}, t) = \mathbf{B}(\mathbf{x}, t)$. This implies that electromagnetic energy transferred into $\mathcal{D}$ across one face must be balanced by the same transfer out of $\mathcal{D}$ on another face, thus allowing wave-like dynamo action. Although in a sense any one cube is not really isolated from other cubes, collectively the periodic structure ensures that the energy in one cube is not growing at the expense of energy in another. If steady dynamo action at finite R is studied, it is necessary in a periodic geometry to impose the additional condition that the magnetic field have *zero mean*, since otherwise a uniform field qualifies for dynamo action.

It should be mentioned that alternative periodicity conditions on the electromagnetic field are possible and have been used in slow dynamo theory (see Childress 1970, Roberts 1970, Moffatt 1978). These allow a magnetic field $\mathbf{B}(\mathbf{x}, t) = e^{i\mathbf{n}\cdot\mathbf{x}}\mathbf{b}(\mathbf{x}, t)$ where $\mathbf{b}$ is periodic in the above sense and $\mathbf{n}$ is an independent wavenumber vector, usually taken to have components of magnitude smaller than $2\pi/L$. This is identical to a Bloch wave representation; see Brillouin (1946). A related device is used in Chaps. 2 and 3 where models are introduced involving a velocity field independent of z and a magnetic field of the form $\mathbf{B} = e^{ikz}\mathbf{b}(x, y, t)$. Here $\mathbf{B}$ satisfies the natural periodicity conditions only if k is a multiple of $2\pi/L$. Nevertheless we shall there take k as arbitrary, thereby effectively separating out the z coordinate and treating $\mathcal{D}$ as a two-dimensional periodicity square. Three-dimensional Bloch wave representations are another matter, however. Since we shall always set $\mathbf{n} = 0$ in our three-dimensional periodic models an explanation for this apparently restrictive assumption is needed.

In slow dynamo theory $\mathbf{n}$ is often taken to be small compared to $2\pi/L$, in which case the periodicity cubes can be thought of as units of a 'small-scale' flow, collectively producing magnetic structure on a much longer length scale. This turns out to be a very useful viewpoint for constructing dynamos, provided that the magnetic Reynolds number R based upon the size L of the cube is of order unity or smaller. In the present examples, R is large, and this changes completely the nature of the field, from one having structure only on scales comparable to the cube size, to one having structure over a range of scales, including very small scales. In effect, then, the cube size becomes the natural length for what we may call the 'large-scale' field of a fast dynamo. To state this somewhat differently, the role of the periodicity cube has changed from an element of the flow to the entire domain of the flow. In such a situation the Bloch wave structure, while certainly providing a larger class of magnetic fields, is no longer compelling. This is explicit in the examples just mentioned having two-dimensional periodicity, where the choice of k turns out to be immaterial provided it is of order unity; see, e.g., §3.2.

A cube of periodicity is, however, artificial as a model of an isolated astrophysical system such as a planetary core or a stellar convection zone. For example, the transition region where the Earth's dipole field interacts with the solar wind and connects up with the interplanetary magnetic field does not provide a simple boundary condition and of course has no relation to a periodicity condition. A much better model for the Earth is that of a homogeneous conducting sphere $\mathcal{D}$ (that part of the Earth consisting of the fluid core and the solid sub-core) surrounded by a vacuum having zero conductivity. Here the isolation of the dynamo is ensured by requiring that the vacuum field decay at infinity like a dipole. In this model, a discontinuity of electrical conductivity as well as in velocity is allowed at the boundary $\partial\mathcal{D}$ of the conducting region. The conditions which link the magnetic field within $\partial\mathcal{D}$ to the vacuum field are then that $\mathbf{B}$ and the tangential component $\mathbf{E}_t$ of the electric field $\mathbf{E}$ be continuous on $\partial\mathcal{D}$. The continuity of the normal component B_n is a property of any solenoidal field. The continuity of the tangential component $\mathbf{B}_t$ reflects the absence of current sheets on $\partial\mathcal{D}$. Finally, the continuity of $\mathbf{E}_t$ expresses the absence of non-zero flux in the boundary surface. The normal component of the electric field is in general not continuous, reflecting the presence of a surface charge distribution. Indeed, since current is a solenoidal field which vanishes in the vacuum, its normal component is continuous and vanishes on $\partial\mathcal{D}$. Since $\mathbf{J}$ has no normal component there, $E_n = -\mathbf{n} \cdot \mathbf{u} \times \mathbf{B}$ on the inner boundary. The value obtained on the outer boundary from the vacuum fields need not be the same.

For the case of an isolated, bounded conductor, the perfectly conducting limit leads to non-uniformities at $\partial\mathcal{D}$. Suppose, for example, that an initial field is confined within $\mathcal{D}$ and the limit $\sigma \to \infty$ is taken. Now let the field evolve under a flow, which we may assume to be a fast dynamo (§1.3). From the Cauchy solution (1.2.17) it follows that the vacuum field will remain zero; so the dynamo action cannot be observed from the exterior domain. However for any finite σ, field will instantly diffuse to the boundary and a vacuum field will be created which will grow under dynamo action. If the initial field extends into the vacuum region and the limit is taken first, the tied lines penetrating the boundary can still shift under the tangential flow there and so affect the exterior dipole moment, but the moment will in fact remain bounded and therefore cannot reflect the dynamo action occurring within (see §4.2.3).

1.2.6 The Planar Problem

The simplest geometry for the study of kinematic induction is the planar flow $\mathbf{u} = (u_x(x,y,z,t), u_y(x,y,z,t), 0)$ interacting with a planar magnetic field $\mathbf{B} = (B_x(x,y,z,t), B_y(x,y,z,t), 0)$. We summarize now the form taken by the induction equation, since it will be used several times below. For these planar fields[8] the magnetic potential of (1.2.24) reduces to a scalar,

$$\mathbf{B} = (A_y, -A_x, 0), \tag{1.2.27}$$

and (1.2.2) may be integrated (or (1.2.24) simplified) to give

$$\frac{dA}{dt} \equiv \frac{\partial A}{\partial t} + \mathbf{u} \cdot \nabla A = \eta \nabla^2 A. \tag{1.2.28}$$

In fact the integration leaves an arbitrary function of time on the right of (1.2.28), but this may be absorbed into A.

Now (1.2.28) is an *advection–diffusion* equation for A. In the perfectly conducting limit we have

$$\frac{\partial A}{\partial t} + \mathbf{u} \cdot \nabla A = 0 \tag{1.2.29}$$

---

[8] A distinction between planar and two-dimensional problems will be drawn in §2.1.

and the magnetic potential is a material invariant. In this limit the magnetic lines $A = $ const. are distorted by the flow, but the flux contained between any two lines is invariant. Under the action of the flow such lines may approach each other, and gradients of A increase, together with the magnetic field. Diffusion smooths out highly sheared field when the flow folds the field lines and brings oppositely oriented field close together. Not surprisingly planar flows fail as dynamos. We shall prove this, allowing a general magnetic field with a z-component, in §4.2.2.

1.3 Slow and Fast Dynamos

Given a motion $\mathbf{u}$ defined on $\mathcal{D}$, and a corresponding density ρ, the induction equation (1.2.9) (or (1.2.5) for incompressible flow) is a linear partial differential equation for $\mathbf{B}$. We have defined the kinematic dynamo problem as the study of the growth or decay of the magnetic field for a prescribed flow. The magnetic Reynolds number R is a given non-negative parameter, and indeed is the only dimensionless parameter in the kinematic problem at the outset. We define

$$\epsilon = 1/R, \tag{1.3.1}$$

so that we deal with the limit $\epsilon \to 0$ as an alternative to formulations involving η or R. We shall refer to this as the 'diffusionless limit', 'dissipationless limit' or 'perfectly conducting limit'; with a slight abuse of language we shall often refer to ϵ or $1/R$ as the 'diffusivity'.

It is natural to measure growth of field in terms of an exponent, or rate of exponential growth, which we denote by $\gamma(\epsilon)$. Given an initial magnetic field $\mathbf{B}_0$, we measure the field strength by the magnetic energy E_M given by (1.2.25), and define

$$\gamma(\epsilon) = \sup_{\mathbf{B}_0} \limsup_{t \to \infty} \frac{1}{2t} \log E_M(t). \tag{1.3.2}$$

When ϵ is finite and the magnetic field diffuses, $\gamma(\epsilon)$ measures the mean exponential growth of energy of a *smooth* field evolving under $\mathbf{u}$. In order that this growth rate exist as an average property of $\mathbf{u}(\mathbf{x}, t)$, we shall want $\mathbf{u}$ to be in some sense stationary in time. The simplest example is a steady flow, $\partial \mathbf{u}/\partial t = 0$ or $\mathbf{u} = \mathbf{u}(\mathbf{x})$. Here time is separable in the induction problem through the substitution

$$\mathbf{B}(\mathbf{x}, t) = e^{p(\epsilon)t} \, \mathbf{b}(\mathbf{x}). \tag{1.3.3}$$

Solving the resulting eigenvalue problem gives a set of complex growth rates $p_i(\epsilon)$ and a corresponding set of eigenfunctions. From these $\gamma(\epsilon)$ is determined by

$$\gamma(\epsilon) = \sup_i \operatorname{Re} p_i(\epsilon). \tag{1.3.4}$$

If $\mathbf{u}$ is periodic in time with period T, we seek solutions of the induction equation satisfying the Floquet condition

$$\mathbf{B}(\mathbf{x}, t + T) = e^{p(\epsilon)T} \mathbf{B}(\mathbf{x}, t), \tag{1.3.5}$$

in which case $\gamma(\epsilon)$ is again given by (1.3.4). More generally, a suitable $\mathbf{u}$ could be restricted, as a function of time, to some chaotic attractor. The examples studied below will be mostly steady or time-periodic flows, but in Chap. 11 we also study some random flows.

Supposing we have found $\gamma(\epsilon)$, we shall say that the flow $\mathbf{u}$ is a *kinematic dynamo* (or simply a *dynamo*) if $\gamma(\epsilon) > 0$ for some $\epsilon > 0$. That is, to say that $\mathbf{u}$ operates as a dynamo at the magnetic Reynolds number ϵ^{-1} means that for some initial magnetic field the magnetic energy grows exponentially.

We can establish at once that all growth rates in these definitions have an absolute upper bound in terms of parameters of the given flow. We consider the case in which the domain $\mathcal{D}$ of §1.2.1 is a cube of periodicity in $\mathbf{x}$, and multiply the induction equation (1.2.2) by $\mathbf{B}$ to obtain a local equation for magnetic energy. Integrating this over $\mathcal{D}$ and using (1.2.4), we obtain

$$\frac{d}{dt} \int_{\mathcal{D}} B^2 \, dV = \int_{\mathcal{D}} \Big[2 B_i B_j e_{ij} - \nabla \cdot (\mathbf{u} B^2) \\ - B^2 \nabla \cdot \mathbf{u} - 2\eta \mathbf{B} \cdot \nabla \times \nabla \times \mathbf{B} \Big] \, dV, \tag{1.3.6}$$

where

$$e_{ij} = \frac{1}{2} \left(\frac{\partial u_i}{\partial x_j} + \frac{\partial u_j}{\partial x_i} \right) \tag{1.3.7}$$

is the symmetric rate-of-strain matrix for the flow. Using the divergence theorem and the spatial periodicity of the fields, we can bring (1.3.6) into the form

$$\frac{dE_{\mathrm{M}}}{dt} = \mu^{-1} \int_{\mathcal{D}} \Big[B_i B_j \left(e_{ij} - \frac{1}{2} \delta_{ij} \nabla \cdot \mathbf{u} \right) - \eta (\nabla \times \mathbf{B})^2 \Big] \, dV. \tag{1.3.8}$$

Suppose that

$$\Lambda = \sup_{\mathbf{x} \in \mathcal{D}} \sup_{t \geq 0} \max_{i=1,2,3} \lambda_i(\mathbf{x}, t), \tag{1.3.9}$$

where $\{\lambda_i(\mathbf{x}, t)\}$ are the eigenvalues of $2e_{ij} - \delta_{ij} \nabla \cdot \mathbf{u}$ at $(\mathbf{x}, t)$. It then follows from (1.3.8) that

$$\gamma(\epsilon) \leq \Lambda. \tag{1.3.10}$$

This estimates growth in terms of rate of strain. The energy cannot grow any faster than the maximal local straining rate which the flow develops.[9]

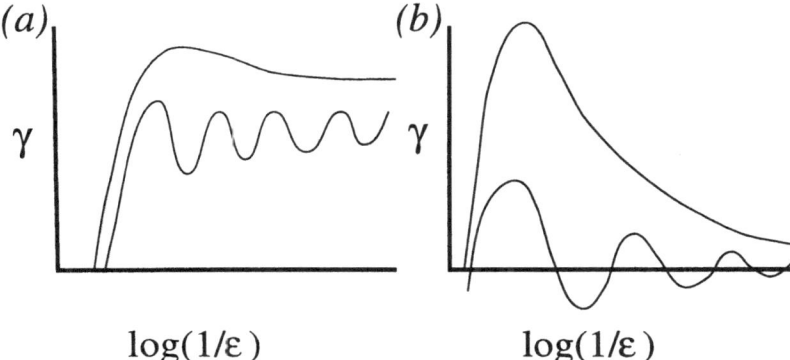

Fig. 1.2 Schematic picture of possible growth rates for fast and slow dynamos: $\gamma(\epsilon)$ against $\log R = \log(1/\epsilon)$ for (a) fast dynamos, and (b) slow dynamos.

If the flow and fluid are such that the magnetic Reynolds number is extremely large, as in the astronomical systems discussed earlier, then we can distinguish two types of dynamo depending on the behavior of the growth rate $\gamma(\epsilon)$ in the limit $R \to \infty$ or $\epsilon \to 0$. If $\gamma(\epsilon)$ remains *positive and bounded away from zero* in this limit we call the dynamo *fast*. Since we have taken dimensionless variables based on the turn-over time of $\mathbf{u}$, this means that the magnetic field developed by a fast dynamo is amplified at a rate comparable with the convective time-scale of the flow. More formally the property

$$\gamma_0 \equiv \liminf_{\epsilon \to 0} \gamma(\epsilon) > 0 \qquad (1.3.11)$$

defines a *fast dynamo* and γ_0 is called the *fast dynamo growth rate*; see Fig. 1.2(a). A dynamo is called *slow* if $\gamma_0 \leq 0$; in this case the growth rate is not bounded above zero for sufficiently small ϵ (see Fig. 1.2b). A typical slow dynamo is a smooth helical flow (Ponomarenko 1973) for which $\gamma(\epsilon) = O(\epsilon^{1/3})$ (§5.4.1) and so growth of field occurs on a time-scale somewhere between that of convection and the much longer $O(1/\epsilon)$ diffusive time-scale.[10]

---

[9] The same estimate applies when the domain $\mathcal{D}$ is a bounded region outside of which lies unbounded conductor at rest. We then impose continuity of $\mathbf{B}$ on the boundary and require that $\mathbf{B} \times (\nabla \times \mathbf{B})$ decay as $o(|\mathbf{x}|^{-2})$ as $|\mathbf{x}| \to \infty$ to make a distant surface integral vanish; see §1.2.5. We also assume the magnetic energy in all space is finite. The growth rate then refers to this total energy.

[10] In fluid dynamics and MHD 'fast' is often used to refer to time scales smaller than the natural diffusion time for the system, but not necessarily as small as the

In physical terms, a fast dynamo has the property that dynamo action occurs at a rate which becomes independent of ϵ in the perfectly conducting limit. The term 'fast' is used here as one interpretation of the physical meaning of a large magnetic Reynolds number R, namely that the characteristic velocity U is large. This terminology has persisted in spite of the fact, noted earlier, that in astrophysical problems R is generally large because of large physical size L.

The distinction between slow and fast dynamos was made by Vainshtein & Zeldovich (1972), and is largely motivated by the solar dynamo, which appears to have the fast dynamo property. For the solar convection zone $R \sim 10^8$ (table 1.1), and yet the field evolves on a convective time-scale of months for sunspots and years for the solar cycle. This oversimplifies the situation, since the solar dynamo is certainly dynamic rather than kinematic; nevertheless it motivates the study of fast amplification mechanisms at large R.

The infimum limit in (1.3.11) allows for the possibility that a limit does not exist, because of oscillatory behavior, for example. The infimum makes this a conservative definition, in that it ensures that for sufficiently small ϵ, $\gamma(\epsilon)$ is bounded away from zero by a fixed number. We point out that a supremum rather than an infimum limit is sometimes used in the definition of a fast dynamo. The supremum limit seems acceptable physically, since it ensures that growth rates exceeding a fixed positive number are realized for arbitrarily small ϵ, but it allows also for the existence of arbitrarily small values of ϵ where the growth rate is close to zero or even negative. We adopt the stronger (1.3.11), with the remark that in the examples that have been studied up to now the difference has not been important, since the straightforward limit $\lim_{\epsilon \to 0} \gamma(\epsilon)$ usually exists.

The concept of fast dynamo action raises a number of interesting questions. First, do fast dynamos exist? This involves subsidiary questions. Can fast dynamo action be proved for various classes of flows and maps? What special properties of flows and maps are needed for fast dynamo action? Are 'most' flows fast dynamos? Second, what magnetic structure is typical of the fields developed by dynamo action? How are the geometrical properties of flow and field linked? Finally and perhaps centrally, is it possible to formulate a theory for the perfect conductor which will decide for or against fast dynamo action for a given flow or map? We stress at once that few of these questions have been answered satisfactorily, and then only for very restrictive models.

---

advective time-scale. In the case of the induction equation the diffusion time is L^2/η. In the present problem such usage would imply that some dynamos defined here as 'slow' would qualify as 'fast'. In a practical sense the distinction between a fast dynamo and a slow dynamo with, say, $\gamma \sim 1/\log \epsilon$, is not essential. But it is worth indicating the origins of fast dynamo action in stretching flows by adopting the strict time-scale of the fluid, namely L/U. We disregard for the moment additional time scales distinct from this, as might be introduced by a time-periodic component with a frequency much different from U/L. An example of such a secondary frequency is given in Chap. 8.

We shall be led in our studies of these matters to a number of areas of mathematics with an extensive literature and a growing list of applications to MHD, for example ergodic theory of flows and maps, spectral theory of linear operators, dynamical systems and Hamiltonian chaos, and geometric measure theory. We shall do no more than touch on these subjects, to the extent needed to indicate their possible significance and suggest interesting directions for future research.

The example which we shall introduce in the next section, as well as the various models discussed in Chaps. 2 and 3, serves to indicate some of the properties of fast dynamos, and Chaps. 5 and 7 illustrate some of the differences between slow and fast varieties of dynamos. The two key ingredients for a fast dynamo are *significant line stretching* and *constructive folding*. By good line stretching we mean that a finite length of material line in the flow stretches exponentially in time. This implies positive *topological entropy* (for smooth flows) and is a signature of flows possessing chaotic trajectories — values of $\mathbf{a}$ for which the Lagrangian trajectory $\mathbf{x}(\mathbf{a}, t)$ defined by (1.2.15) shows recurrent, but apparently random or chaotic, behavior. Even simple flows $\mathbf{u}(\mathbf{x}, t)$ can have this property of *Lagrangian chaos* (Arnold 1965, Hénon 1966). Fast dynamos will usually exhibit Lagrangian chaos over extensive regions of space.

In view of our picture of the stretching of flux tubes (Fig. 1.1), we can see that good stretching of lines is needed to increase magnetic energy on a convective time-scale. However equally important is the way in which the flow folds the field. If it folds field so as to bring together flux in opposing directions at small scales, such field will be vulnerable to diffusive destruction and the corresponding dissipation of magnetic energy. If on the other hand the folding is constructive in the sense that it tends to bring stretched tubes into alignment, there will be less dissipation. This will be made clearer when the stretch–twist–fold model is introduced in §1.4.

In the limit of large R, the complexity inherent in a chaotic flow will carry over to the magnetic field: over finite regions of space the flow will create fine structure in the field over a range of scales, since the smoothing effect of diffusion will only be seen on extremely small length scales. This complexity in the nature of the magnetic structure sets the tone of fast dynamo theory. The primary mathematical problems are connected with the manner in which the idealization of the perfect conductor, $\epsilon = 0$, can be used to identify, study, and prove the requisite properties for fast dynamo action in the limit $\epsilon \to 0$. Since magnetic lines are frozen into a perfect conductor, the theory can also be thought of as concerned with the average properties of material lines in a flow with complex Lagrangian structure.

The issue of convergence to a perfectly conducting limit can be compared with the relationship of classical boundary-layer theory of Navier–Stokes flows to the class of Euler limits. (We discuss magnetic boundary layer theory in §5.5.) While both theories deal with *singular perturbations* associated with small diffusion, the singular sets tend to be very different. Classical boundary layers are generally confined to curves or surfaces, while in chaotic flow the diffusive structures essentially fill a region, $\epsilon\nabla^2\mathbf{B}$ being nowhere negligible! Convergence in fast dynamo theory cannot therefore be in the strong sense of pointwise approximation over open sets. One of the main themes of our discussion will be the appropriate weak sense of convergence to the perfectly conducting dynamo. We amplify on this point in §1.5, and return to it frequently in subsequent chapters.

1.4 Stretch–Twist–Fold: the STF Picture

The distinction between slow and fast dynamos was first drawn by Vainshtein & Zeldovich (1972) in a paper on astrophysical magnetic fields. In this paper they describe the 'rope' or stretch–twist–fold (STF) fast dynamo, which is the archetype of the elementary models of the process. We show it in Fig. 1.3.

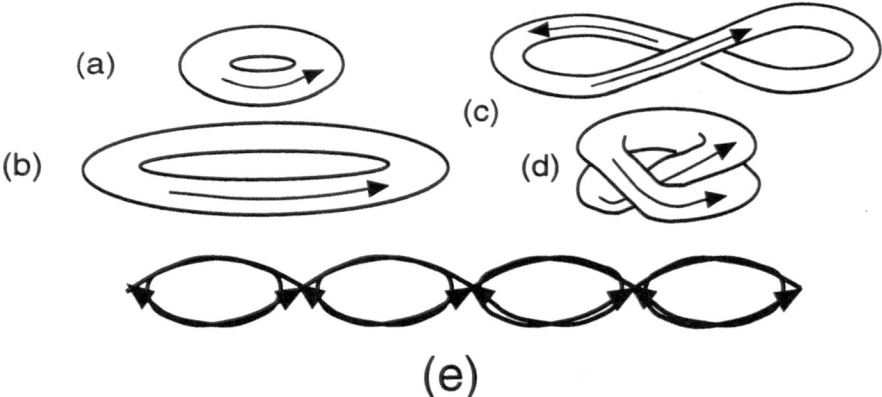

Fig. 1.3 The stretch–twist–fold (STF) fast dynamo. A torus (a) carrying magnetic flux F is stretched to twice its length (b), then twisted to give (c), and finally folded to give (d). The process preserves orientation, thus yielding a torus of roughly the original shape, but now carrying flux $2F$. (e) The process applied to an array of identical tori. We show the result of the stretch–twist as in (c) applied to a one-dimensional array of tori, resulting in flux doubling of every torus.

The flux tube shown here is regarded as being distorted by the flow of an incompressible perfect conductor. Ignoring for the moment the obvious mismatch of field which occurs at the crossover point following the twist and fold, the process essentially returns the magnetic field to the original configuration but with twice the flux and thus roughly twice the field strength

and four times the energy. Moreover, no tearing or cutting of flux tubes has
occurred. The sequence of moves is somewhat misleading in that the fold
operation, which is perhaps the most difficult to realize with simple flows, is
actually unnecessary provided that one considers a collection of tori which can
share tubes of flux. We indicate in Fig. 1.3(e) how the stretched and twisted
tori can overlap in an array of identical moves.

We may repeat the STF sequence again and again on the same domain.
The length of the original tube doubles each time, and so material lines stretch
exponentially: the flow generating the STF sequence must be chaotic. The
other ingredient in a fast dynamo, constructive folding, is also present as
the field is always brought approximately back into alignment after the STF
cycle. Because of this, weak diffusion will tend to smooth the bundle of field,
but not to destroy much flux or energy. Each time the sequence is repeated,
we double the flux thanks to the constructive reinforcement of the stretched
field, but with presumably minor penalties at the region of mismatch. If this
assessment is correct, then the STF procedure determines a fast dynamo with
a growth rate

$$\gamma_0 \simeq \frac{1}{T} \log 2 \qquad\qquad (1.4.1)$$

if the cycle has period T. The change of energy over one cycle is equal to the
work done against magnetic forces, minus the small dissipative losses.

The STF picture represents the 'optimal' reinforcement of stretched field
available in a three-dimensional domain. The key feature is the lack of any
appeal to diffusion, except as would aid to smooth out the internal structure.
The mechanism is fundamentally a *flux doubling* operation. In earlier work
on magnetic enhancement Alfvén (1950) introduced a number of similar ideas
which involve in part the diffusive separation of flux tubes. In Fig. 1.4(a) we
show one such model, based upon division of a stretched tube into two tubes,
which are then combined to double the field. This process would fail in the
limit of infinite R. the time required for this separation being at least some
positive power of R (Vainshtein & Zeldovich 1972). Thus Fig. 1.4(a) depicts
a slow dynamo.

A second model considered by Alfvén (1950) is shown in Fig. 1.4(b). As
Roberts & Soward (1992) have noted, the use of a twist makes this essentially
equivalent to the STF mechanism, since the separation of the loop from the
main flux tube is now an inessential part of the process. Indeed in Fig. 1.4(c)
we show a variant of 1.4(b) where the main tube is a torus. In this form it is
identical to a map later analyzed in more detail by Finn & Ott (1988a, b).
By varying the size of the smaller loop, a flux-doubling process is established
with a highly intermittent distribution of field over the cross section. Although
Alfvén did not make a formal distinction between slow and fast dynamos, his
crucial use of a twist in the second model is compelling. We propose (following
Ghil & Childress 1987) to call this the Alfvén–Vainshtein–Zeldovich or AVZ
model, leading to the STF picture and its variants.

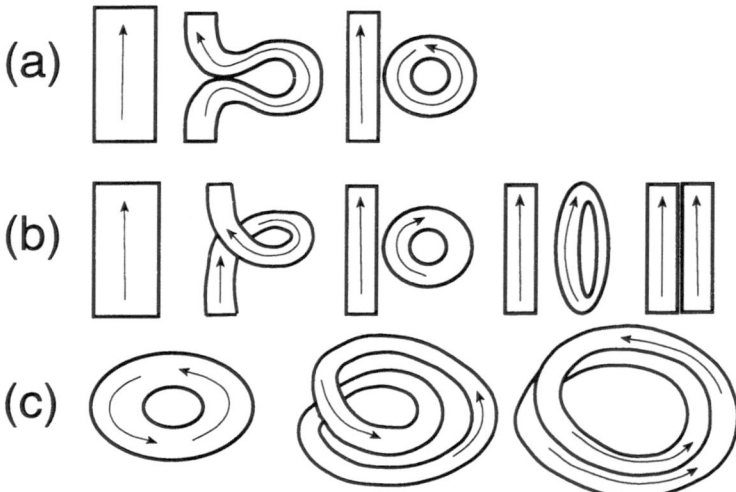

Fig. 1.4 (a) and (b) are Alfvén's (1950) models of dynamo action at large magnetic Reynolds number. (a) Slow splitting of a flux tube. (b) A twist is introduced. In (c), we show a variant of (b) where a twist is introduced without splitting.

The simplicity of the STF picture masks a number of interesting details, as well as non-trivial problems, which emerge when one attempts to provide a precise formulation of the steps involved. First, observe that the field is directed not only along the main axis of the torus. Significant transverse fields are generated by the twisting of the tube; these are of the order of the axial field times the ratio of minor to major radius (Moffatt & Proctor 1985). However the transverse fields tend to be suppressed because they are compressed during the 'stretch' operation of STF. Second, the crossover regions multiply and are eventually distributed throughout the domain which, together with the internal twisting, makes even weak diffusion significant everywhere as a smoothing effect. Third, the STF model is defined here in terms of a finite region embedded in three-dimensional space. Exactly how the required map is effected by a moving conductor in typical fluid flows is not obvious. Since the STF movements are realizable 'pretzel making' moves, there exists a flow and a corresponding map which would underlie the model: indeed one has been written down by Moffatt & Proctor (1985). A natural question then is, if and how does this local behavior arise in actual fluid motions, and how does the domain localized in Fig. 1.3 relate to other nearby regions of the flow?

A related question, raised by Arnold (1994), concerns the field-free voids that are inevitably trapped within the toroidal structure as the STF moves are repeated over and over. We can never bring the two tubes of flux into contact using a smooth flow (in a finite time) during the fold operation, and so we must always entrain fluid into the torus and, correspondingly, lose fluid containing field from the torus into the rest of the volume.[11] It is helpful to take the flow which performs the basic STF moves to be restricted to a bounded domain, $\mathcal{D}$. Let $\mathcal{D}$ contain a small torus $\mathcal{D}_0$ filled by the initial magnetic field. Now it is plausible that we can define efficient STF movements in the neighborhood of $\mathcal{D}_0$ but, as we have noted, after one sequence of moves the field must be distributed over a slightly larger torus $\mathcal{D}_1$, containing both field and voids. For the second sequence of moves, $\mathcal{D}_1$ will map into a still larger torus $\mathcal{D}_2$, and so on, until the field is spread throughout $\mathcal{D}$. Therefore it does not seem to be possible to 'hold' the torus in some local part of a more extensive flow, a region where the motion is especially close to the STF picture. Inevitably the field will be swept into less efficient movements, which could significantly alter the growth of flux and the toroidal shape of its support. This raises a major objection to any simple realization of the STF picture, and suggests that the almost totally constructive folding it represents will never occur in practice.

It can be argued that the smaller scales of the void fraction might be 'squeezed out' by the continual stretching (and casual observation of a taffy pulling machine would suggest that this is possible). However Lorentz forces would be needed to drive out the fluid, and so the process could only occur in a full MHD model, where the nature of the driving forces must be taken into account. The outcome is not at all obvious.

Many of the specific flows and maps studied below offer some answers to these questions. The examples will suggest that numerous fast dynamo mechanisms exist, and many share features of the AVZ idea and the STF picture. The STF process remains an idealized picture of how a fast dynamo could operate, with *perfectly constructive folding*, whereas real flows fall somewhat, perhaps substantially, short of this. The title of this monograph was chosen, at the suggestion of Prof. H.K. Moffatt, to reflect the importance of the physical picture embodied in this simple but basic model.

1.5 The Nature of the Perfectly Conducting Limit

From §1.3 we know that fast dynamo theory concerns the limit of small ϵ (or large $R = 1/\epsilon$) of a diffusive induction operator, $\mathbf{T}_\epsilon$, $\epsilon > 0$, with the property that $\mathbf{T}_\epsilon \mathbf{B}_0(\mathbf{x}) = \mathbf{B}(\mathbf{x}, t)$.[12] The induction operator is determined by

---

[11]If the flow is allowed to contract volumes indefinitely, then it is possible to ensure that the STF cycle smoothly maps the torus into itself. This gives a model known as the 'solenoid' (e.g., Falconer 1990), which has a strange attractor. Iterating the map leads to flux concentrated on this attractor. However in such a map volumes are contracted indefinitely, leading to unbounded fluid densities; such a map cannot model dynamo action in a physical fluid, even if it is compressible.

[12]We use the term operator here to indicate that $\mathbf{T}_\epsilon$ acts on initial field, for some particular value of t. We are in fact dealing with a semi-group of operators defined

$$\frac{\partial \mathbf{B}}{\partial t} - \nabla \times (\mathbf{u} \times \mathbf{B}) - \epsilon \nabla^2 \mathbf{B} = 0, \qquad \mathbf{B}(\mathbf{x}, 0) = \mathbf{B}_0(\mathbf{x}). \qquad (1.5.1)$$

For a stretching flow, characterized by scales U and L, there will be other scales of magnetic structure arising from the Lagrangian chaos. Familiar singular perturbation techniques suggest that the smallest undiffused magnetic scales are of size $O(\sqrt{\epsilon})$ (see Moffatt & Proctor 1985 and Chap. 5). As we have already noted, at these scales diffusion is not formally negligible over finite regions in the flow. Thus, in the perfectly conducting limit $\epsilon \to 0$, arbitrarily small scales of magnetic structure will emerge.

We emphasize that this has fundamental implications for fast dynamo action. For any positive ϵ, we expect (1.5.1) to have smooth eigensolutions (for steady or time-periodic flows), but these will exhibit structure down to the $\sqrt{\epsilon}$ diffusive cut-off. On any given eigensolution $\mathbf{B}_p$ with complex growth rate $p(\epsilon)$, the action of $\mathbf{T}_\epsilon$ at time t is thus well-defined and has the property

$$\mathbf{T}_\epsilon(t + T)\mathbf{B}_p = e^{p(\epsilon)T}\mathbf{T}_\epsilon(t)\mathbf{B}_p, \qquad (1.5.2)$$

where T is the period of a time-periodic flow and is arbitrary for a steady flow. Then all scales present in the eigenfunction grow at the same rate, and *any* measure of field strength (not just the energy) can be used to define $\gamma(\epsilon)$ and γ_0.

On the other hand, for induction in a perfect conductor we have introduced the operator $\mathbf{T}$ determined by the Cauchy solution (1.2.18). In a stretching flow, we expect this operator to produce, over time, ever smaller magnetic scales. Thus, starting from any smooth initial condition, the action of $\mathbf{T}$ is to introduce ever finer magnetic structure; so different scales grow at different rates. We therefore see that $\mathbf{T}$ cannot possess smooth eigenfunctions! This point was first emphasized in the context of steady flows by Moffatt & Proctor (1985); see §4.3.

Putting $\epsilon \equiv 0$ thus changes completely the nature of the induction problem. In the perfect conductor magnetic energy grows at the rate at which work is done by the motion. The emergence of small scales means that measures of growth based upon norms involving derivatives of the field will be larger than the growth rate based on energy. In other words there is no longer an obvious single measure of growth that reflects what actually happens when $0 < \epsilon \ll 1$. We shall (somewhat arbitrarily) define $\gamma(0)$ to be half the growth rate of the energy in a perfect conductor. Because of diffusive cancellation we will see that $\gamma(0) \geq \gamma_0$ provided the flow is smooth; see §6.6. This means that in this case growth of energy must occur in the perfect conductor if the motion is a fast dynamo. Strict inequality occurs in practice because of field cancellation, that is, energy can be put into the perfect conductor which cannot be realized in the growth of an eigenfunction for $\epsilon > 0$, however small ϵ may be.

---

on the interval $0 < t < \infty$ (Kato 1966). The term *monodromy operator* is also used (Yudovich 1989).

A crucial problem, then, is to identify the appropriate measure of growth in a perfect conductor that will agree with the limit γ_0. The various proposals which have been advanced center on the computation of *average magnetic flux* (as in the discussion of STF) or a related linear functional of **B**, mathematically a 'weak' measure of growth rate. We will generally use the symbol Γ, often with arguments, to denote one of these weak measures of magnetic intensity. A good definition will have the property that $\Gamma = \gamma_0$ for the flow of interest. An understanding of how Γ is to be chosen amounts to a theory of fast dynamo action based on the equations for a perfect conductor. The examples discussed in Chaps. 2 and 3 will indicate how these ideas work out in practice and will suggest what kinds of flow fields allow a perfectly conducting approach to fast dynamo action. Conjectures based upon these examples are discussed in Chap. 4.

A schematic picture of a fast dynamo, summarizing these properties, is given in Fig. 1.5. There is a process at the scale L of the flow which amplifies flux. It does so imperfectly and creates some small-scale structure and cancellations, which cascade passively to smaller and smaller scales until they are annihilated diffusively. Reducing ϵ increases the length of the cascade. Setting $\epsilon = 0$ means there is nothing to terminate the cascade, and the field never settles down. However the amplification of flux still continues, and can be detected by a flux average at scale L.

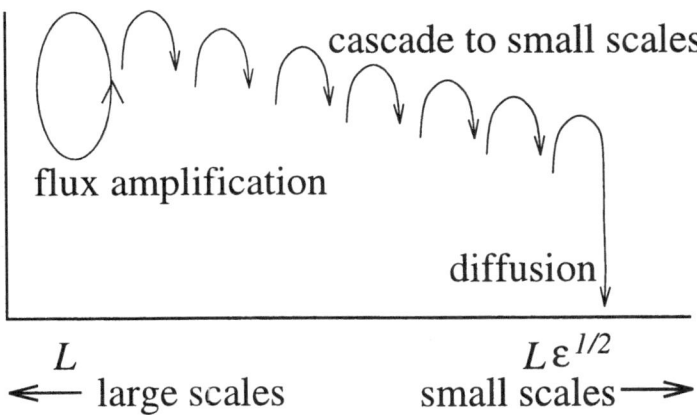

Fig. 1.5 A schematic picture of a fast dynamo.

1.6 Measures of Stretching

In §1.3 we have identified stretching as a necessary ingredient for a fast dynamo. In this section we discuss briefly different measures of stretching in fluid flows, in preparation for the examples of Chaps. 2 and 3; more discussion is given in §6.1. Using Lagrangian coordinates, start with a vector $\mathbf{v}$ at a point $\mathbf{a}$ in the flow, and allow the vector to be carried passively, frozen in incompressible fluid flow. After a time t the resulting vector is $\mathbf{J}(\mathbf{a}, t)\mathbf{v}$, from (1.2.17), and has length $|\mathbf{J}(\mathbf{a}, t)\mathbf{v}|$. We define the *Liapunov exponent at* $\mathbf{a}$ by

$$\Lambda_{\mathrm{Liap}}(\mathbf{a}) = \max_{\mathbf{v}} \limsup_{t \to \infty} \frac{1}{t} \log |\mathbf{J}(\mathbf{a}, t)\mathbf{v}|. \tag{1.6.1}$$

This is the asymptotic rate of stretching of vectors at $\mathbf{a}$, maximised over all initial vectors. We also define the *maximum Liapunov exponent* by

$$\Lambda_{\mathrm{Liap}}^{\max} = \sup_{\mathbf{a}} \Lambda_{\mathrm{Liap}}(\mathbf{a}), \tag{1.6.2}$$

which measures the maximum stretching anywhere in the flow; this is an upper bound on fast dynamo growth rates, though generally not a very tight one (see Chap. 6).

In a chaotic flow trajectories have 'sensitive dependence on initial conditions' and since $\Lambda_{\mathrm{Liap}}(\mathbf{a})$ measures time-asymptotic stretching, it depends sensitively on the starting point $\mathbf{a}$. Nevertheless for 'almost all' initial conditions in a given chaotic region, the value of $\Lambda_{\mathrm{Liap}}(\mathbf{a})$ is the same, and this is called the *Liapunov exponent* Λ_{Liap} of the flow or map (Oseledets 1968). This is usually measured numerically by following a vector along a randomly chosen trajectory and observing its exponential rate of growth.

Another measure of stretching throughout a chaotic region of the flow, more useful in fast dynamo studies, is the rate of stretching of *material lines*, rather than infinitesimal vectors. Let C be a smooth curve, of finite length at $t = 0$, carried frozen in the flow. At time t the curve is carried to a convoluted curve $M^t C$, of length $|M^t C|$. The maximum rate of stretching of material lines is then given by the *line-stretching exponent*,

$$h_{\mathrm{line}} \equiv \sup_{C} \lim_{t \to \infty} \frac{1}{t} \log |M^t C|. \tag{1.6.3}$$

The line-stretching exponent is related to a quantity called the *topological entropy* h_{top} for a smooth flow, discussed further in §6.1; here we mention that $h_{\mathrm{line}} = h_{\mathrm{top}}$ for two-dimensional flows, while in three dimensions, $h_{\mathrm{line}} \leq h_{\mathrm{top}}$. If we think of the material lines as magnetic field lines, h_{line} (and so h_{top}) is a plausible upper bound for fast dynamo growth rates, as conjectured by Finn & Ott (1988a, b). This will be seen in numerous examples, and an upper bound of h_{top} has been proved by Vishik (1992) and Klapper & Young (1995); see §6.1.

2. Fast Dynamo Action in Flows

Much of the evidence for fast dynamo action in physically reasonable fluid flows comes from careful numerical simulations. These can never give a completely definite answer as to whether a fast dynamo exists; the numerical value of ϵ that can be achieved is limited by the computer's memory since solutions of the induction equation have a smallest scale of order $\sqrt{\epsilon}$ (see §1.5 and Chap. 5), and to verify the process all sensible scales need to be resolved. Therefore in the absence of an analytical theory a simulation can only suggest the asymptotic behavior of the growth rate as $\epsilon \to 0$. On the other hand there is as yet no mathematical theory powerful enough to prove that fast dynamo action can occur in realistic flows, and much of the progress in the theory of fast dynamos has come from interplay between numerical simulation, modelling, approximation and tentative analysis.

It therefore seems appropriate to precede any detailed discussion of models and analysis with an account of some of these numerical experiments. In this chapter we concentrate on flows, as being the most reasonable examples physically. In Chap. 3 we consider, with the same motivation and purpose, examples using maps.

2.1 Computing a Fast Dynamo

The reader, especially if he or she is an astrophysicist, will be wondering what we meant above by 'reasonable'! In an ideal world one might simulate kinematic dynamo action in Rayleigh–Bénard convection in a rotating spherical shell, modelling the solar convection zone. Such a simulation would be expensive in human and computer time, but it would not be of direct relevance to the Sun without inclusion of Lorentz forces and all the complications of full nonlinear MHD, not to mention additional physics pertinent to the convection zone. Thus to focus on the basic question of the existence of fast dynamos, most simulations use flows which do not arise from thermal convection, and often use simplified geometry. We therefore take reasonable flows to mean flows that are smooth, having no discontinuities or singularities in space or time, and which exist in ordinary three-dimensional space. We shall usually adopt a velocity field with a single well-defined length scale, even though the fluid turbulence appropriate to astrophysical problems involves a wide range of scales; see Chap. 12.

We often use flows that are periodic in space, and we include these in the class of reasonable flows; the magnetic field is usually taken to have the same periodicity. This is because very efficient computer codes may be written for this geometry. However there are some caveats when periodic space is used (see §1.2.5, §4.2.3) since for $\mathbf{u} = 0$ any initial field that is uniform in space remains constant. This is not true in a more realistic geometry; for example in a conducting sphere with an insulating vacuum outside any field decays when $\mathbf{u} = 0$. Nevertheless in the search for growing fields and fast dynamos, little is lost by using periodic geometry, and the fast amplification mechanisms that we shall discuss could equally well occur in a sphere.

One way to proceed would be to write down a flow, such as that of Moffatt & Proctor (1985), which performs the STF sequence of operations, and to implement it numerically. However this is still quite difficult to compute, as the STF process is three-dimensional and unsteady. Instead we adopt a minimalist and ahistorical approach: we ask which general classes of flows might include fast dynamos, choose members of these classes and test them numerically.

To obtain a dynamo at all, fast or slow, in ordinary or periodic three-dimensional space, none of the components (u_x, u_y, u_z) of $\mathbf{u}$ can be identically zero (§4.2.2). From our discussion of the STF model in §1.4, exponential stretching and constructive folding of field are identified as key ingredients. To obtain good exponential stretching of material lines we need a chaotic flow, and this means that $\mathbf{u}$ must depend non-trivially on at least *three* of x, y, z and t. This gives us the following classes of possible chaotic fast dynamos, in order of increasing complexity:

1) two-dimensional unsteady flows, $\mathbf{u}(x, y, t)$,
2) three-dimensional steady flows, $\mathbf{u}(x, y, z)$,
3) three-dimensional unsteady flows, $\mathbf{u}(x, y, z, t)$.

We use the term *two-dimensional* in a precise sense throughout this monograph. It will always mean *depending on two coordinates only*. A two-dimensional flow can have as many as three non-trivial components in three-dimensional space, of the form $\mathbf{u} = (u_x(x, y, t), u_y(x, y, t), u_z(x, y, t))$. We refer to flows of the form $\mathbf{u} = (u_x(x, y, z, t), u_y(x, y, z, t), 0)$ as *planar* flows. Given a two-dimensional flow $\mathbf{u}(x, y, t)$ it is often convenient to split it up into its *horizontal part* $\mathbf{u}_H \equiv (u_x, u_y)$, which is a planar flow, and its *vertical part* u_z; we sometimes write this as $\mathbf{u} = (\mathbf{u}_H, u_z)$.

Of these three classes of flow, the first class, of unsteady two-dimensional flows, is by far the easiest to deal with, and in §§2.2–5 we explore a space-periodic example of Otani (1988, 1993) in detail; similar examples are studied by Galloway & Proctor (1992) and, in a sphere, by Hollerbach, Galloway & Proctor (1995). In §2.6 we look at examples of three-dimensional steady flows. There has been little exploration of the third class of flows, from the point of view of dynamos, or even as dynamical systems (but see Feingold, Kadanoff & Piro 1988). Other possibilities include the random two- or three-dimensional flows studied in Chap. 11.

2.2 Numerical Evidence for a Fast Dynamo in Otani's MW+ Flow

We explore fast dynamo action in an unsteady two-dimensional flow,

$$\mathbf{u}(x, y, t) = 2\cos^2 t\,(0, \sin x, \cos x) + 2\sin^2 t\,(\sin y, 0, -\cos y), \qquad (2.2.1)$$

first studied by Otani (1988, 1993).[13] We call this the MW+ flow for brevity, meaning 'modulated waves, positive parity'. The flow is made up of two Beltrami waves

$$\mathbf{u}^{(x)} = (0, \sin x, \cos x), \quad \mathbf{u}^{(y)} = (\sin y, 0, -\cos y), \qquad (2.2.2)$$

modulated by the oscillatory terms $2\cos^2 t$ and $2\sin^2 t$, which switch the waves off and on in a smooth way.[14] Otani's modulation is one of many that might be considered by assigning general amplitudes $\alpha(t)$ and $\beta(t)$ to the waves, and so parameterising an oriented closed curve in the (α, β)-plane.

The MW+ flow is two-dimensional but has all three components of velocity. The flow is periodic in x and y, period 2π, and in t, period π. We consider only magnetic fields having the same spatial periodicity. Because the flow is independent of z, we follow the usual practice of separating out the z-dependence of a magnetic field, writing

$$\mathbf{B}(x, y, z, t) = \exp(ikz)\mathbf{b}(x, y, t), \qquad (k \in \mathbb{R}), \qquad (2.2.3)$$

where k is a wavenumber, which must be non-zero for dynamo action (§4.2.2). From the induction equation (1.5.1), the components b_x and b_y satisfy:

$$\partial_t b_x + (u_x\partial_x + u_y\partial_y + iku_z)b_x = (b_x\partial_x + b_y\partial_y)u_x + \epsilon\nabla^2 b_x, \qquad (2.2.4a)$$
$$\partial_t b_y + (u_x\partial_x + u_y\partial_y + iku_z)b_y = (b_x\partial_x + b_y\partial_y)u_y + \epsilon\nabla^2 b_y, \qquad (2.2.4b)$$

with $\nabla^2 = \partial_x^2 + \partial_y^2 - k^2$. From $\nabla \cdot \mathbf{B} = 0$,

$$\partial_x b_x + \partial_y b_y + ikb_z = 0. \qquad (2.2.5)$$

---

[13]This is equivalent to Otani's flow after a translation.
[14]The model is based on earlier fast dynamo studies of Bayly & Childress (1987, 1988, 1989), who *pulsed* the waves and diffusion (see §3.1.2). Bayly and Childress obtained good numerical evidence for fast dynamo action in their pulsed flow, and Otani extended this by switching on waves smoothly and allowing diffusion to act continuously.

With $k \neq 0$ this equation can be used to calculate b_z from b_x and b_y. The equations (2.2.4a, b) do not involve b_z, and so the problem may be reduced to one for two complex fields $(b_x, b_y)(x, y, t)$. The only complication is the wavenumber k; for each value of k one obtains a growth rate $\gamma(\epsilon, k)$, and one should maximize this over k to obtain the maximum growth rate $\gamma(\epsilon)$ in (1.3.2). We shall find this less of a difficulty than it might appear.

The equations for (b_x, b_y) can be solved numerically for finite ϵ to find the asymptotic growth rate. Typically b_x and b_y are written as Fourier series in (x, y) and (2.2.4a, b) are stepped in time; diffusion acts continuously. The simulation must resolve scales of order $\sqrt{\epsilon}$ and so the storage space needed increases as ϵ^{-1}. For a three-dimensional flow one needs $O(\epsilon^{-3/2})$ storage space and so the saving in considering two-dimensional flows is significant.

Note that the u_z component of the flow enters into equations (2.2.4a, b) for b_x and b_y in a very limited way through the terms $iku_z b_x$ and $iku_z b_y$. Were we to remove these terms from the equations, we would obtain equations for the evolution of field in the horizontal part of (2.2.1),

$$\mathbf{u}_H(x, y, t) = (u_x, u_y, 0) = (2\sin^2 t \sin y, 2\cos^2 t \sin x, 0). \tag{2.2.6}$$

This planar flow cannot be a dynamo (§4.2.2); therefore the terms $iku_z b_x$ and $iku_z b_y$ are crucial in generating field. They do not affect the magnitude of the field directly, but they affect its complex phase, because of the action of u_z in moving fluid elements in the z-direction. These *phase shifts* are needed to avoid cancellation and achieve fast dynamo action, as we shall now establish numerically.

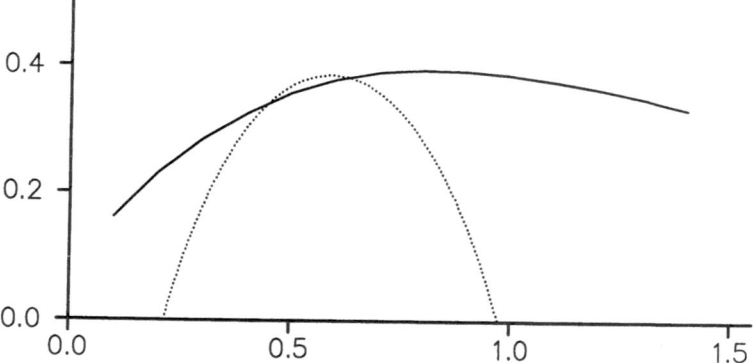

Fig. 2.1 Growth rate $\gamma(\epsilon, k)$ (solid) plotted against k for MW+ with $\epsilon = 10^{-3}$. The dotted curve gives the approximation (2.3.11). (After Otani 1993.)

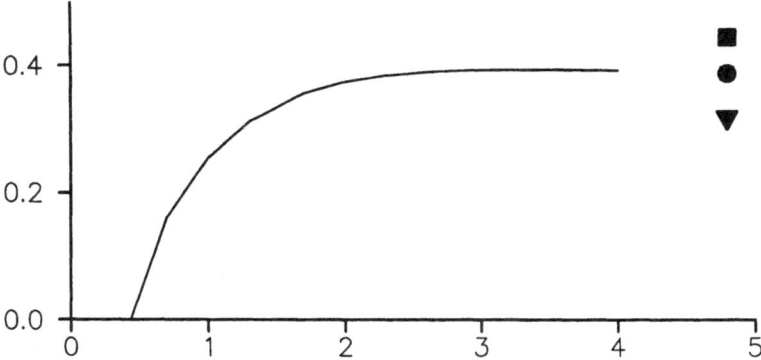

Fig. 2.2 Growth rate $\gamma(\epsilon, k)$ plotted against $\log_{10} 1/\epsilon$ for MW+ with $k = 0.8$. Also shown are Γ_C (circle), h_{line} (square) and Λ_{Liap} (triangle).

With a code to solve (2.2.4a, b) for the flow MW+, we fix k, ϵ and some initial condition, say,

$$\mathbf{B}_0 = (i, 1, 0)\exp(ikz). \tag{2.2.7}$$

We run the code until the fastest growing magnetic field eigenmode emerges and the growth rate settles down to $\gamma(\epsilon, k)$. The solid line in Fig. 2.1 shows $\gamma(\epsilon, k)$ for $\epsilon = 10^{-3}$ and values of k from 0.1 to 1.4. We observe positive growth rates, with values in the range 0.2 to 0.4. There is a gentle dependence on the vertical wavenumber k, with a maximum around $k = 0.8$. To see how these growth rates vary in the fast dynamo limit $\epsilon \to 0$, we fix $k = 0.8$ and in Fig. 2.2 plot $\gamma(\epsilon, k)$ against $\log_{10} 1/\epsilon$. As $\epsilon \to 0$ the growth rate $\gamma(\epsilon, k)$ appears to asymptote around the positive value 0.39. From the computational evidence, the dynamo appears to be fast (Otani 1988, 1993). Note that a further optimisation over k, which is implied by the supremum over all initial fields in equation (1.3.2) for $\gamma(\epsilon)$, can only *increase* the growth rate.

2.3 Eigenfunctions and the Stretch–Fold–Shear Mechanism

Although the dynamo growth rate for MW+ shows clear asymptotic behavior as $\epsilon \to 0$ with $k = 0.8$ in Fig. 2.2, the corresponding eigenfunctions become wild in this limit. Fig. 2.3 shows eigenfunctions for decreasing values of ϵ. We plot $\mathrm{Re}\, b_x(x, y)$ at time $t = 40\pi$ on a color scale with yellow/red for

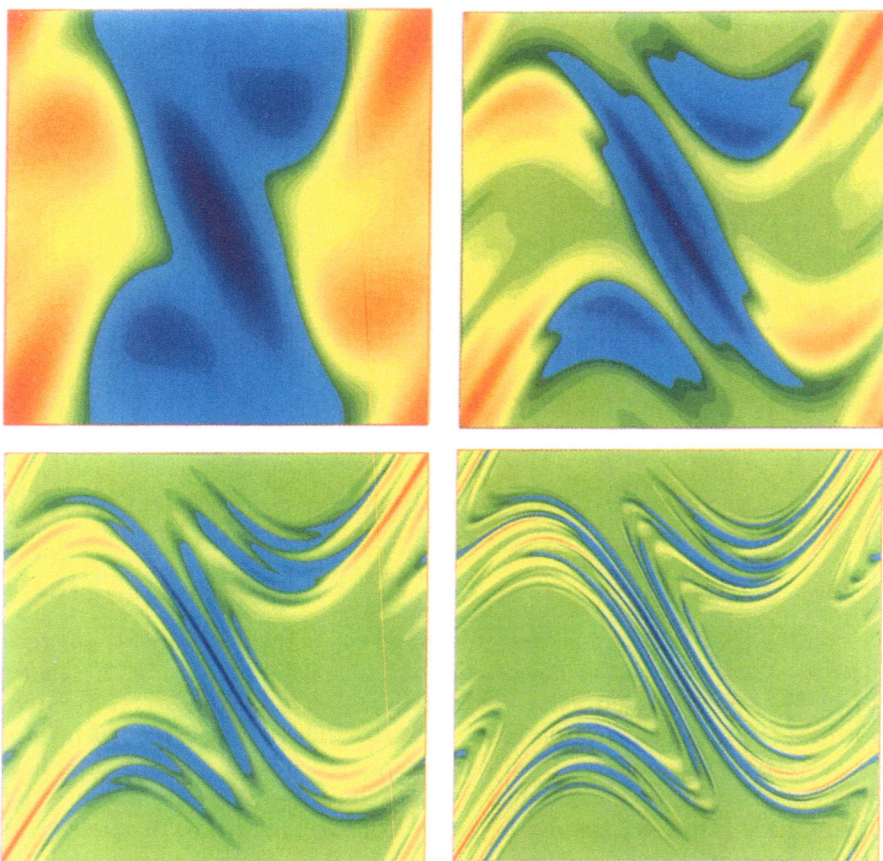

Fig. 2.3 Eigenfunctions for $k = 0.8$ and $\epsilon = $ (a) 10^{-1}, (b) 10^{-2}, (c) 10^{-3}, and (d) 10^{-4} (shown by row, from top to bottom). The real part of $b_x(x, y)$ is plotted for $0 \leq x, y \leq 2\pi$. (After Otani 1993.)

positive $\operatorname{Re} b_x$ and torquoise/blue for negative $\operatorname{Re} b_x$; values near to zero are shown by green.

What can be said about this emerging magnetic structure as $\epsilon \to 0$? First, the field is concentrated in certain regions of space and is not spread uniformly over the box. These are the regions where the flow is most chaotic. Figure 2.4 shows a *Poincaré section* for the flow.[15] A Lagrangian particle trajectory $\mathbf{x}(\mathbf{a}, t)$ is followed in the flow MW+ and its (x, y)-position is marked whenever t is a multiple of π, the period of the flow; the motion of the particle

---

[15]This is a Poincaré section relative to the time variable; for a flow with period one, it is often called a *time-one map*.

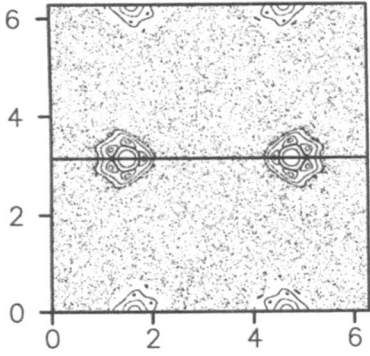

Fig. 2.4 Poincaré section of the flow MW+, with $0 \le x, y \le 2\pi$. The horizontal line, $y = \pi$, shown is the line C used for averaging field in §2.4.

in z is not represented. This is done for several different starting points **a**. In chaotic regions the points appear to fill areas densely. Elsewhere the points appear to lie on invariant curves, the *KAM tori* (see, e.g., Lichtenberg & Lieberman 1983), but in between are further thin bands of chaos. Comparing Figs. 2.3(d) and 2.4 we see that the magnetic field is strong in the main chaotic region, and it weakens near the islands; this is direct pictorial confirmation that exponential stretching and chaos are important for fast dynamos (Arnold *et al.* 1981).

The second ingredient identified from the STF picture is constructive folding. In Fig. 2.3(d) there are fine bands of strong magnetic field, indicating the presence of stretching in the chaotic flow. Most of the bands in the center of the picture have the same sign of b_x and point roughly in the $(-1, 1)$ direction. Mingled with these, however, are bands of field pointing in the opposite direction. This shows that the constructive folding is imperfect, unlike the idealised STF picture where field is always brought into perfect alignment. This feature, of partial but not perfect folding, appears to be typical of chaotic fluid flows. Note that while the small scales of the eigenfunctions become ever more complicated as $\epsilon \to 0$, the large-scale structure of the field remains similar. As ϵ is reduced, the scale at which diffusion is active is lowered, but the same large-scale folding mechanism still amplifies field at the same rate.

Figure 2.5 shows the folding during one period of the flow: bands of field are stretched, folded and compressed. The field is dominated at this stage by the fastest growing eigenfunction, which takes the Floquet form (1.3.5):

$$\mathbf{b}(x, y, t) = \exp(p(\epsilon)t)\,\mathbf{c}(x, y, t), \qquad (\gamma(\epsilon) = \mathrm{Re}\,p(\epsilon)), \qquad (2.3.1)$$

where **c** is periodic in t with period π. The form of the eigenfunction repeats every π-period, amplified by a complex factor (which happens to be real in this case). This may be seen from Figs. 2.5 and 2.3(d).

Fig. 2.5 Field evolution with $k = 0.8$ and $\epsilon = 10^{-4}$ for $t - 40\pi = (a)$ $\pi/5$, (b) $2\pi/5$, (c) $3\pi/5$, and (d) $4\pi/5$ (shown by row, from top to bottom). A complete period of the flow is given by $(2.3d) \to (a) \to (b) \to (c) \to (d) \to (2.3d)$.

Note that if only the horizontal part of the flow, $\mathbf{u}_H$ in (2.2.6), were acting, the folding in the plane would form alternating bands of field of opposite orientation, which would be destroyed effectively by diffusion (Fig. 2.6). Instead, the vertical component u_z plays a crucial role in changing the phase of magnetic field and so prevents complete cancellation. Diffusion operates continuously during the simulation and at each stage the finest bands of field are diffusing, nearby bands are merging if they have the same sign, and cancelling if not.

To understand the dynamo mechanism at work here, the key is to look at magnetic field evolution with $\epsilon = 0$ using the Lagrangian setting of §1.2.2. Consider the action of a single Beltrami wave,

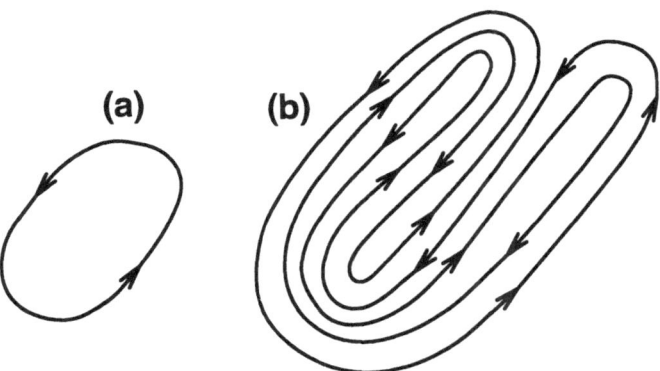

Fig. 2.6 Folding in the plane creates alternating field: (a) initial field, (b) folded field. (After Finn & Ott 1988b.)

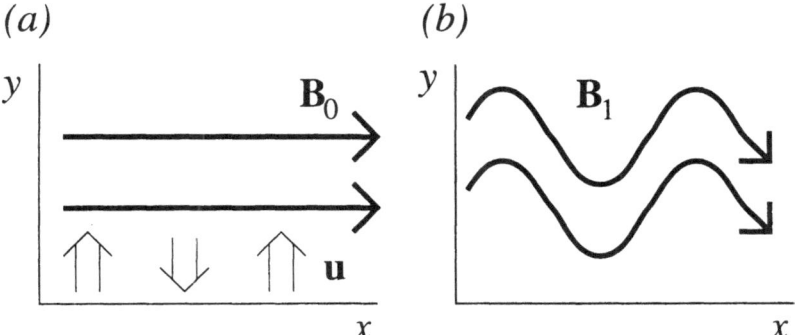

Fig. 2.7 Schematic picture of field evolution under M_1: (a) initial field $\mathbf{B}_0$ (2.3.5) and (b) field $\mathbf{B}_1$ (2.3.6). (After Bayly & Childress 1988.)

$$\mathbf{u}_1 = 2\cos^2 t\,(0, \sin x, \cos x), \tag{2.3.2}$$

of the flow (2.2.1) acting for a time period of π. During this time a Lagrangian particle at $\mathbf{x}$ is moved to $M\mathbf{x}$ with

$$M\mathbf{x} = (x, y + \pi \sin x, z + \pi \cos x) = M_2 M_1 \mathbf{x}, \tag{2.3.3}$$

where we have split M into two simpler maps,

$$M_1\mathbf{x} = (x, y + \pi \sin x, z), \qquad M_2\mathbf{x} = (x, y, z + \pi \cos x). \tag{2.3.4}$$

The first of these two maps, M_1, arises from the horizontal part of $\mathbf{u}_1$ and has a *stretch–fold* action, pulling out loops of field in the y-direction. Its action on a constant magnetic field,

$$\mathbf{B}_0(\mathbf{x}) = (1, 0, 0), \qquad (2.3.5)$$

is to generate a field

$$\mathbf{B}_1(\mathbf{x}) = (1, \pi \cos x, 0); \qquad (2.3.6)$$

see Figs. 2.7(a, b). This stretching has increased the energy of the field, but has not brought field into alignment since the y-component fluctuates. For example, if we average the field over x and y, we obtain the original field, with no increase: $<\mathbf{B}_1>_{x,y} = (1, 0, 0)$.

To obtain folding that is constructive we need M_2, from the vertical part of $\mathbf{u}_1$, which moves fluid elements up and down in z. This is only effective if the field has some variation in z, and so instead of (2.3.5) start with the field

$$\mathbf{B}_0 = (1, 0, 0) \cos kz \qquad (2.3.7)$$

(Fig. 2.8a). Under the action of M_1 we obtain

$$\mathbf{B}_1 = (1, \pi \cos x, 0) \cos kz, \qquad (2.3.8)$$

shown in Fig. 2.8(b). The action of M_2 on $\mathbf{B}_1$ is to *shear* field in the z-direction, moving it up and down as shown by the open arrows in Fig. 2.8(b). The resulting field is

$$\mathbf{B}_2 = (1, \pi \cos x, -\pi \sin x) \cos(kz - k\pi \cos x). \qquad (2.3.9)$$

Field in the $z = 0$ plane with a component in the $+y$-direction is moved up, and field in the $z = \pi/k$ plane with a component in the $+y$-direction is moved down. There is therefore a tendency for $+y$-field to accumulate on the intermediate plane $z = \pi/2k$ and similarly for $-y$-field to accumulate on the plane $z = 3\pi/2k$, as shown in Fig. 2.8(c).

The map M has achieved amplification of field by the stretch–fold operation M_1, and constructive folding — accumulation of like-signed field — by the shear operation M_2. This combination is called the *stretch–fold–shear* (SFS) fast dynamo mechanism (Bayly & Childress 1988) and appears to be operating in the MW+ flow. The mechanism goes back to studies by Roberts (1972) and was identified by Soward (1987) in the modified Roberts cell fast dynamo (see §5.5.2).

Note that the shear operation also generates z-directed field, which is not depicted in Fig. 2.8(c) and because of (2.2.5) can be ignored. Since particles move up and down at most a distance π under M_2, we would expect, very roughly, that $\pi/2k \approx \pi$ would give optimal reinforcement, i.e., $k \approx 1/2$. This is not far from the optimal value of $k \simeq 0.8$ seen numerically (Fig. 2.1). For an estimate of the growth rate, average $\mathbf{B}_2$ over x and y (Otani 1993):

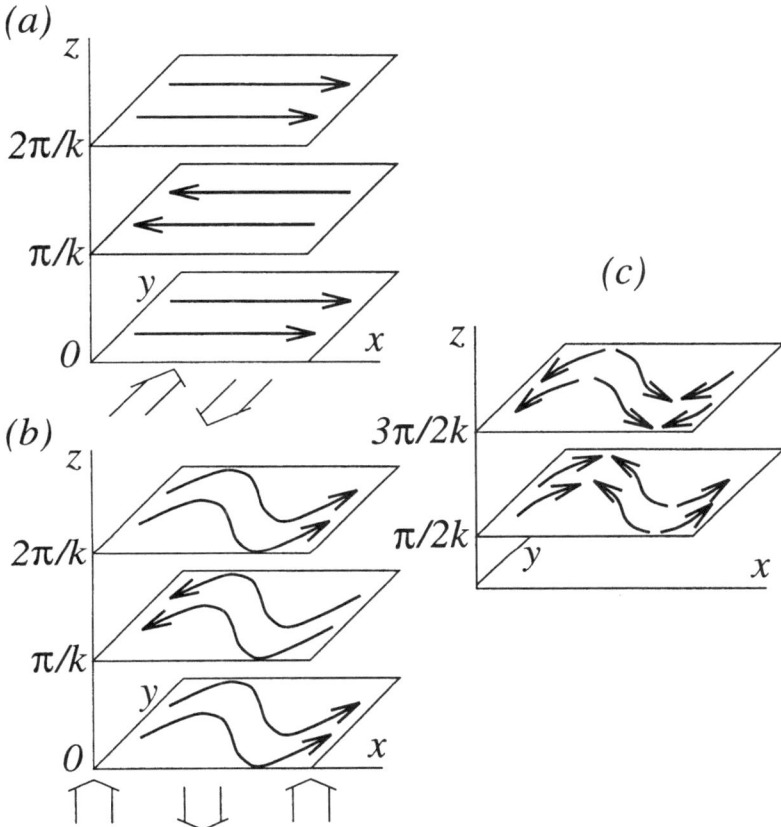

Fig. 2.8 Schematic picture of field evolution under M_1 and M_2: (a) initial field $\mathbf{B}_0$ (2.3.7), (b) field $\mathbf{B}_1$ (2.3.8), and (c) field $\mathbf{B}_2$ (2.3.9), omitting the z-component. (After Otani 1993.)

$$<\mathbf{B}_2>_{x,y} = (J_0(k\pi)\cos kz, \pi J_1(k\pi)\sin kz, 0); \qquad (2.3.10)$$

here J_0 and J_1 are Bessel functions. The average x-component is reduced, as $J_0(k\pi) < 1$, but an average y-component is generated by the SFS mechanism, varying as $\sin kz$. Since $\pi J_1(k\pi)$ can be greater than one, the average y-field generated can be bigger than the initial x-field. Thus the picture we have of amplification in the original MW+ flow is that average x-field is converted by the first wave into stronger average y-field; the second wave then generates yet stronger average x-field from the y-field, and so on. A rough estimate of the growth rate may be found from (2.3.10) by taking $\pi J_1(k\pi)$ as the amplification factor for the field over an average time per wave of $\pi/2$, yielding a growth rate of the average field:

$$\Gamma \approx (2/\pi) \log \pi J_1(k\pi) \tag{2.3.11}$$

(Otani 1993). This is plotted as a dotted line in Fig. 2.1; considering the crudeness of the calculation, the agreement is acceptable and indicates that the dynamo is indeed operating by an SFS mechanism. For pulsed Beltrami waves, this kind of approximation appears to be asymptotic in a limit in which the waves act for very long times; see Chap. 10.

2.4 Magnetic Field Evolution with $\epsilon = 0$

In the last section we discussed the evolution of the magnetic field in the MW+ flow with small non-zero diffusion $0 < \epsilon \ll 1$. We gained some understanding by ignoring diffusion and modelling the evolution of field by the stretch–fold–shear mechanism. In this section we set $\epsilon = 0$ and study field evolution without diffusion. Our aim is to see to what extent one can relate evolution with $\epsilon = 0$ and evolution with $0 < \epsilon \ll 1$.

Figure 2.9 shows evolution of field in the MW+ flow for $\epsilon = 0$ from the initial condition (2.2.7) with $k = 0.8$; the color coding of Fig. 2.3 is used. As time increases the chaotic stretching and folding generates finer and finer bands of field which pile up in the center of the picture. Although the field becomes ever more complicated, it does begin to resemble the diffusive eigenfunctions (Fig. 2.3) in terms of general structure, suggesting some relation between them. There is evidence of imperfect constructive folding taking place; the central bands of field largely point in the $(1, -1)$ direction (as in Fig. 2.3), but between are fields pointing in the opposite direction. Because of exponential stretching in the flow, very fine bands of extremely strong field are generated, which would be smeared out by diffusion for $\epsilon > 0$. Their strength is disguised in Fig. 2.9, as we have not followed Fig. 2.3 and plotted $\mathrm{Re}\, b_x$, but instead shown

$$F(\mathrm{Re}\, b_x) = \max(0, \log(10|\,\mathrm{Re}\, b_x|)) \,\mathrm{sign}(\mathrm{Re}\, b_x), \tag{2.4.1}$$

which preserves the sign of the field while reducing peak values.

There is visual evidence for constructive folding in Fig. 2.9, which needs to be quantified. There are two approaches: the first, introduced by Finn & Ott (1988a, b), is to consider the growth rate of magnetic flux with and without magnetic diffusion. Choose some fixed, well-behaved, orientable surface S (with boundary) and measure the *magnetic flux* through S,

$$\Phi_S(t) = \int_S \mathbf{B}(\mathbf{x}, t) \cdot d\mathbf{S}, \tag{2.4.2}$$

for some initial condition $\mathbf{B}_0(\mathbf{x})$, as a function of time (or number of iterations for a map instead of a flow). Its asymptotic growth rate is defined by:

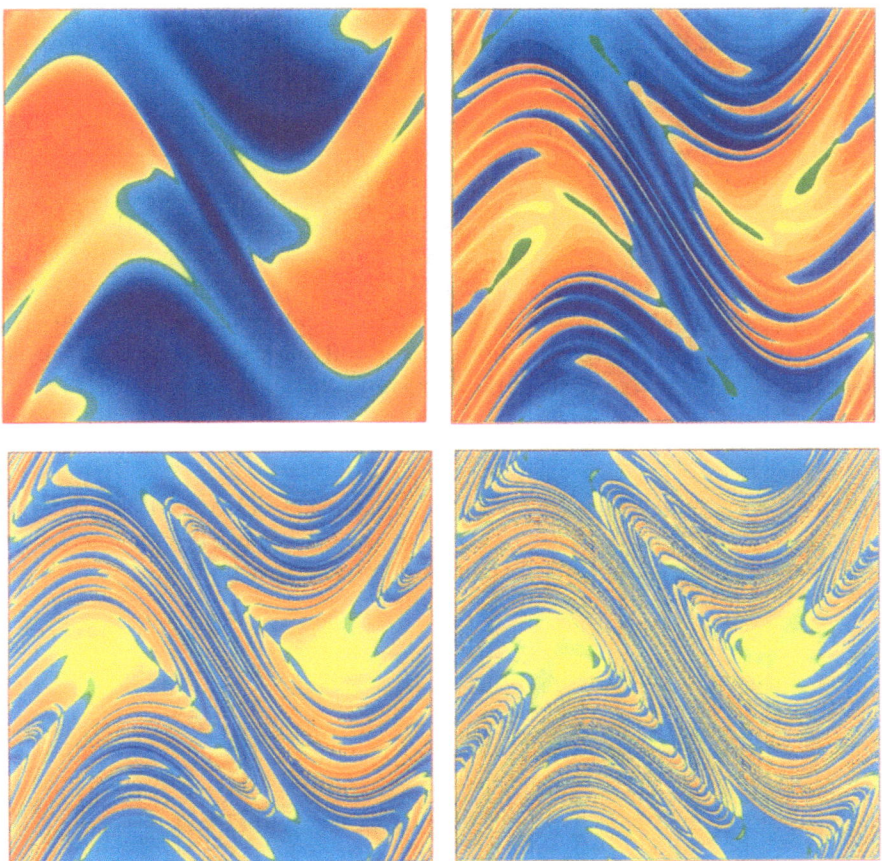

Fig. 2.9 Magnetic field evolution with $\epsilon = 0$ in the MW+ flow for $0 < x, y < 2\pi$. The quantity $F(\operatorname{Re} b_x)$ (2.4.1) is plotted at $t = (a)$ π, (b) 2π, (c) 3π, and (d) 4π (shown by row, from top to bottom).

$$\Gamma(S, \epsilon) = \limsup_{t \to \infty} \frac{1}{t} \, \log |\Phi_S(t)| \qquad\qquad (2.4.3)$$

(the lim sup allows for possible oscillations in Φ_S). For $\epsilon > 0$ this growth rate will simply be $\gamma(\epsilon)$,

$$\Gamma(S, \epsilon) = \gamma(\epsilon), \qquad (\epsilon > 0) \qquad\qquad (2.4.4)$$

for *general* surfaces S and initial fields $\mathbf{B}_0$, since after long periods of time the fastest growing eigenfunction will dominate.

If $\epsilon = 0$, the growth rate $\Gamma_S \equiv \Gamma(S, 0)$ may still exist, despite the increasing complexity of the magnetic field and lack of smooth eigenfunctions. Finn &

Ott (1988 *a, b*) conjecture that Γ_S, the flux growth rate with *zero* diffusion, is *equal* to the fast dynamo exponent γ_0, involving the limit $\epsilon \to 0$ of *vanishing* diffusion, that is,

$$\Gamma_S = \gamma_0 \quad (\equiv \lim_{\epsilon \to 0} \gamma(\epsilon) \equiv \lim_{\epsilon \to 0} \Gamma(S, \epsilon)), \tag{2.4.5}$$

for general surfaces and initial fields; we call this the *flux conjecture*.[16] Clearly Γ_S is a good measure of how the flow stretches and folds fields in the absence of any diffusive smoothing and the conjecture formalises our discussion of the STF dynamo in §1.4 and the estimate (1.4.1) of γ_0. The significance of the conjecture, if correct, is that it allows one to calculate the desired quantity, the fast dynamo growth rate γ_0, by setting $\epsilon = 0$ and calculating Γ_S. A more precise and conservative form of the flux conjecture is

$$\Gamma_{\text{sup}} \equiv \sup_{\mathbf{B}_0} \sup_S \Gamma_S = \gamma_0. \tag{2.4.6}$$

The second, parallel approach, of Bayly & Childress (1989), to quantifying constructive folding in a flow is to consider the growth of *smooth functionals* of the evolving magnetic field. This is discussed further in §4.1.

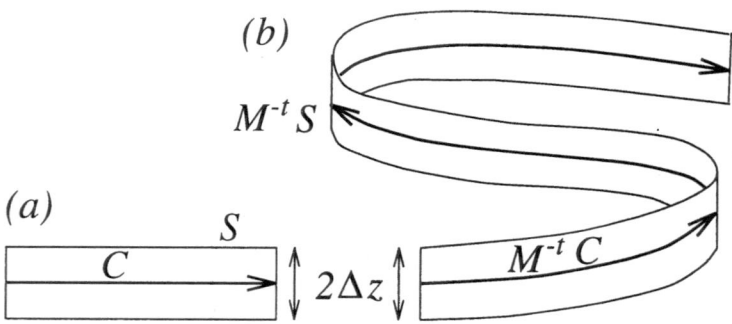

Fig. 2.10 Schematic picture showing (*a*) initial curve C and surface S, and (*b*) $M^{-t}C$ and $M^{-t}S$.

Evidence for the validity of the conjecture may be obtained by testing it numerically for the MW+ flow. Define the line

---

[16]It is assumed here that the limit of $\gamma(\epsilon)$ as $\epsilon \to 0$ exists, as appears to be the case in examples, and so the 'lim inf' in (1.3.11) is not needed. If the limit did not exist, this would imply a controlling influence of vanishing diffusion, and it would seem unlikely that useful information could be gained by setting $\epsilon = 0$.

$$C = \{0 \leq x \leq 2\pi, y = \pi, z = 0\}, \qquad (2.4.7)$$

shown in Fig. 2.4. (This can be considered a closed curve, given the periodicity of the flow.) We measure the *flux through the line C*:

$$\Phi_C(t) = \int_C \mathbf{B}(\mathbf{x}, t) \cdot \mathbf{n} \, ds; \qquad (2.4.8)$$

here $\mathbf{n} = (0, 1, 0)$ is the normal to the line in the (x, y)-plane. To relate $\Phi_C(t)$ to a true flux, extend the line C between $-\Delta z$ and Δz in the z-direction to make a surface S (Fig. 2.10a); the flux through this surface is $\Phi_C(t)(2/k)\sin k\Delta z$ from (2.2.3).

To find $\Phi_C(t)$ at $t = n\pi$ — some multiple of the period of the flow — carry C and S back in the flow from time t to time 0 to obtain a curve $M^{-t}C$ and surface $M^{-t}S$ (Fig. 2.10b). The flux of $\mathbf{B}(\mathbf{x}, t)$ through S is the same as that of $\mathbf{B}_0(\mathbf{x})$ through $M^{-t}S$ by conservation of flux (1.2.22). Since $M^{-t}S$ is a ribbon of width $2\Delta z$ in the z-direction it may be seen that

$$\Phi_C(t) = \int_{M^{-t}C} \mathbf{B}_0(\mathbf{x}) \cdot \mathbf{n} \, ds. \qquad (2.4.9)$$

Here $\mathbf{n}$ is the normal to the curve in the (x, y)-plane and $ds = (dx^2 + dy^2)^{1/2}$ is its arc-length projected onto an (x, y)-plane. So to calculate $\Phi_C(t)$ all we do is follow C back in time, leading to a very convoluted curve, and integrate the initial condition along this curve.

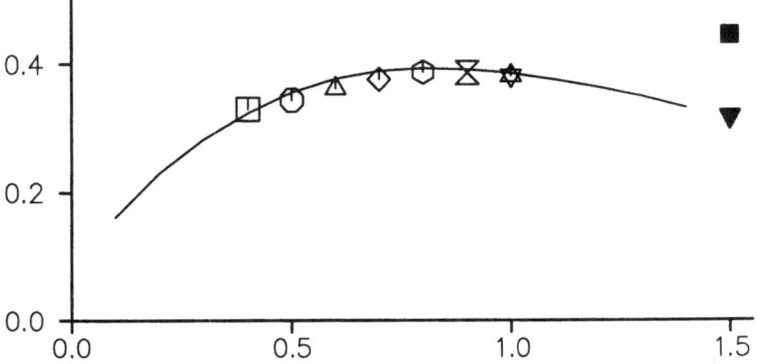

Fig. 2.11 Growth rate $\gamma(\epsilon, k)$ (solid) for MW+ with $\epsilon = 10^{-3}$ (Otani 1993) and flux growth rates Γ_C (markers), plotted against k. Also shown are h_{line} (solid square) and Λ_{Liap} (solid triangle).

When $\Phi_C(t)$ is calculated numerically for $k = 0.8$ there is a clear exponential increase in time with $\Gamma_C \simeq 0.39$. This growth rate is shown by a solid

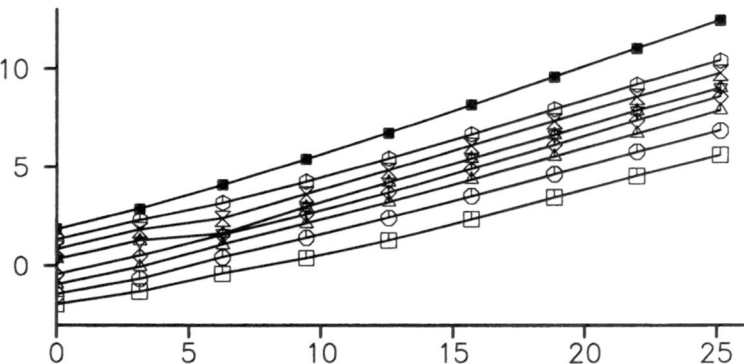

Fig. 2.12 Growth of flux: $\log |\Phi_C(t)|$ (open markers) plotted against time t for MW+. The markers correspond to those in Fig. 2.11; additive constants are used to separate the curves. The length of $M^{-t}C$, used to calculate h_{line}, is shown by solid squares.

circle on Fig. 2.2 and agrees well with the asymptote of $\gamma(\epsilon)$ as $\epsilon \to 0$. Thus there is not only visual evidence of constructive folding, but the rate of growth of flux for $\epsilon = 0$ appears to agree with the fast dynamo growth rate γ_0, giving numerical support for the flux conjecture (2.4.5). Further evidence is shown in Fig. 2.11, which shows $\gamma(\epsilon, k)$ plotted against k for $\epsilon = 10^{-3}$ (taken from Fig. 2.1), together with $\epsilon = 0$ flux growth rates plotted with various markers. Again there is good agreement between the small-ϵ and the $\epsilon = 0$ growth rates. Figure 2.12 shows the growth in $\Phi_C(t)$ for each value of k, plotted with corresponding markers; the flux growth rates in Fig. 2.11 were calculated as $\Gamma_C = \log |\Phi_C(8\pi)/\Phi_C(3\pi)|/5\pi$.

Assuming the flux conjecture holds and we can obtain the fast dynamo growth rate γ_0 from the flux growth rate Γ_{sup}, can we place upper or lower bounds on γ_0 from knowledge of the dynamical properties of the chaotic flow? It is difficult to obtain lower bounds, since planar chaotic flows fail to be fast dynamos, but an upper bound can be found; from (2.4.9) we have

$$|\Phi_C(t)| \le |M^{-t}C| \max_{\mathbf{x}} |\mathbf{B}_0(\mathbf{x})|, \tag{2.4.10}$$

where $|M^{-t}C|$ is the length of the curve $M^{-t}C$ when projected onto the plane $z = 0$. Thus

$$\Gamma_{\text{sup}} \le h_{\text{line}} \equiv \sup_C \lim_{t \to \infty} \frac{1}{t} \log |M^{-t}C|; \tag{2.4.11}$$

h_{line} is the line-stretching exponent introduced in §1.6.[17] For these two-dimensional flows it is the same as the *topological entropy* h_{top} of the flow **u**, discussed in §6.1 and conjectured to be an upper bound on γ_0 by Finn & Ott (1988*a*, *b*). Proofs of this conjecture have been obtained by Vishik (1992) and Klapper & Young (1995) (see §6.1); these are decidedly non-trivial as they involve careful consideration of the limit $\epsilon \to 0$.

The length of $M^{-t}C$ is easily measured numerically at the same time as the flux $\Phi_C(t)$, and its growth is shown in Fig. 2.12. Its growth rate h_{line} is shown in Figs. 2.2, 11 by a solid square and gives an upper bound on growth rates. The discrepancy between the actual growth rate of flux and h_{line} is a measure of the cancellation in calculating the integral (2.4.9) and so of the folding properties of the flow (Finn & Ott 1988*a*, *b*). Quantities closely related to h_{line}, but which build in the cancellations, have been constructed and it is conjectured that these give the fast dynamo growth rate γ_0 (§§6.2, 7.2).

Figures 2.2, 11 also show the Liapunov exponent Λ_{Liap} of the flow (solid triangle), defined in §§1.6, 6.1, which gives the rate of stretching of *typical* vectors in the flow. Surprisingly, perhaps, this is lower than both γ_0 and h_{line}. The intuitive reason is that the growth of field (and of the length of finite curves) is generally dominated by vectors which stretch *unusually* quickly in the flow (Bayly 1986, Finn & Ott 1988*a*, *b*). The Liapunov exponent is therefore not an upper bound on fast dynamo growth rates. This will be seen in models based on maps (§3.3) and was first observed numerically in the smooth flows of Galloway & Proctor (1992).[18]

2.5 Discussion of the Flux Conjecture

The idea of the flux conjecture, that one can set $\epsilon = 0$ and seek a fast dynamo by focusing on fluxes or other spatial averages, is of such theoretical importance that we take the space to discuss it further. The emphasis on flux arises since the application of weak diffusion for a finite time rearranges flux and cancels plus and minus flux, but has a small effect on the total flux in a region (see Finn & Ott 1988*b*). On the other hand energy, being a positive quantity, cannot cancel and so tends to be dissipated. Consider a very simple example: set

$$\mathbf{B}_0(\mathbf{x}) = \left(1 + C\cos(x/\sqrt{\epsilon})\right)\mathbf{e}_z, \tag{2.5.1}$$

---

[17]Strictly, we have defined this as the line-stretching exponent of $\mathbf{u}_{\text{H}}$ in (2.2.6); which flow is used is irrelevant as exponential stretching is not possible in the z-direction for such two-dimensional flows. Note also that h_{line} is defined as the line-stretching exponent of the time-reversed flow in (2.4.11); this is also unimportant, as the flows in this chapter possess time-reversing symmetries.

[18]In this paper the Liapunov exponents given are too large by a factor of 2π, as noted by Y. Ponty, A. Pouquet and P.-L. Sulem.

which is the sum of a mean field and a fluctuating field on an $O(\sqrt{\epsilon})$ length-scale typical for fast dynamos. Let $\Phi_S(t)$ be the flux through the surface $S = \{0 \leq x, y \leq 1\}$ and let $B_{\text{rms}}(t)$ be the root-mean-square (r.m.s.) field on S. We constrain C by

$$1 \ll C \ll \epsilon^{-1/2}. \tag{2.5.2}$$

The first inequality, $1 \ll C$, arises since the fluctuating fields are generally stronger than the mean in a fast dynamo; the second is discussed shortly. Initially

$$\Phi_S(0) = 1 + O(C\sqrt{\epsilon}), \qquad B_{\text{rms}}(0) = C/\sqrt{2} + O(1) + O(C^2\sqrt{\epsilon}). \tag{2.5.3}$$

Thus Φ_S is dominated by the mean field, and B_{rms} by the fluctuating field. For a fast dynamo typically $B_{\text{rms}}^2/\Phi_S^2 \sim \epsilon^{-\varrho_{\text{E}}}$ with $0 < \varrho_{\text{E}} < 1$ (see §12.2) from which we obtain the second inequality for C in (2.5.2) above.

Now let diffusion ϵ act for a unit time; the field again takes the form (2.5.1) but with C replaced by C/e. The flux remains the same to leading order, but the r.m.s. field has decreased by an order-one factor. This is correct even though the mean field is quite weak relative to the fluctuations. The flux does not remain exactly constant, since some flux of order $C\sqrt{\epsilon} \ll 1$ leaks out of the surface S. These results can be summarized by saying that flux is *robust* to the effects of diffusion, unlike B_{rms} or other field averages.

The flux conjecture can be proved for a number of idealised models; see the discussion in §4.1. However these models are rather special since they are all based on *hyperbolic* maps, for which the chaos is relatively simple; for example such maps do not admit the mixture of chaos and integrable islands seen in Fig. 2.4 for MW+. Hyperbolic maps are not easily embedded in smooth flows in ordinary space, and no mathematical proof of the flux conjecture is available for any flows of the complexity of MW+.

The difficulty in making the flux conjecture rigorous is that the simple argument above is for a finite time only; over long times, diffusion has the potential to rearrange the whole field, no matter how small ϵ may be. As an example, in a planar flow with $0 < \epsilon \ll 1$ the magnetic energy grows for a time, because of stretching, but at the same time scales are reduced; at some time t_{diff} diffusion intervenes to destroy small-scale alternating field, and cause a catastrophic decay of magnetic energy (see §4.2.2). The time t_{diff} increases as $\epsilon \to 0$ and so this effect is not captured by a finite-time argument. Fast dynamos for which diffusion is important in organising field structure in the limit $\epsilon \to 0$ are termed 'diffusive' in §4.1.1; for these $\Gamma_{\text{sup}} < \gamma_0$.

No examples of diffusive fast dynamos are known in *smooth* flows, however, and it appears that when flux growth, $\Gamma_{\text{sup}} > 0$, or fast dynamo action, $\gamma_0 > 0$, is obtained in a smooth flow, the flux conjecture $\Gamma_{\text{sup}} = \gamma_0$ holds; see the discussion and conjectures in §4.1.1. Thus for smooth fast dynamos, the picture in Fig. 1.5 applies: a chaotic stretching and folding process amplifies a large-scale field and determines the structure of the field, while the action of diffusion is entirely passive, smoothing structure and destroying fluctuations at small scales. The destructive potential of diffusion is not realised, because the field grows too quickly and is able to maintain a large-scale field, albeit weak, through constructive folding.

There are many studies of flux growth for $\epsilon = 0$ in patched and pulsed flows (Finn & Ott 1988a, b, Bayly & Childress 1988, Gilbert & Childress 1990, Finn $et\ al.$ 1991, Gilbert 1992, Klapper 1992b, c), smooth flows (Finn $et\ al.$ 1991, Gilbert 1991a, 1992), and convective cells (Schmidt, Chernikov & Rogalsky 1993). These show clear evidence of flux growth and indicate fast dynamo action if the flux conjecture holds. However there are almost no studies checking the flux conjecture for general maps or flows by comparing these flux growth rates with the limiting $\epsilon \to 0$ dynamo growth rate γ_0. A notable exception is Klapper's (1992b, c) study of a patched, pulsed flow; see also Lau & Finn (1993a). We therefore present further numerical results which support the flux conjecture for smooth two-dimensional flows; three-dimensional steady flows are addressed in §2.6.

Figure 2.13 shows growth rates $\gamma(\epsilon, k)$, with $\epsilon = 10^{-3}$, and $\epsilon = 0$ flux growth rates for Otani's (1993) parity-reversing flow MW$-$,

$$\mathbf{u} = 2\cos^2 t\,(0, \sin x, \cos x) + 2\sin^2 t\,(\sin y, 0, \cos y). \tag{2.5.4}$$

The line (2.4.7) and the initial condition (2.2.7) are used. Again good agreement between growth rates is obtained where the growth in flux can be measured reliably, for $0.4 \leq k \leq 0.7$. Figure 2.13 indicates h_{line}, which is a comfortable upper bound, and the Liapunov exponent.

For some values of k an accurate flux growth rate is not obtained: we do not see clear exponential behavior before running out of numerical resolution. The following difficulties can arise.

$a)$ There may be a transient, dependent on the initial conditions, before clear exponential growth emerges.

$b)$ Chaotic stretching causes $M^{-t}C$ to grow exponentially in length and so the computer time required to calculate the flux at time t increases exponentially with t.

$c)$ There will generally be cancellation in calculating the integral (2.4.8).

Measuring flux growth can thus be something of a hit-and-miss affair.

Finally we follow flux in two two-dimensional flows of Galloway & Proctor (1992). Figure 2.14(a) shows results for $\gamma(\epsilon, k)$ and Γ_C for the 'circularly polarized' or CP flow:

$$\begin{aligned}
\mathbf{u}(\mathbf{x}, t) &= \sqrt{3/2}\,(0, \sin(x + \cos t), \cos(x + \cos t)) \\
&\quad + \sqrt{3/2}\,(\cos(y + \sin t), 0, \sin(y + \sin t)),
\end{aligned} \tag{2.5.5}$$

using $k = 0.57$, the initial condition (2.2.7) and the line

$$C = \{x = 0.8\pi, 0 \leq y \leq 2\pi, z = 0\}. \tag{2.5.6}$$

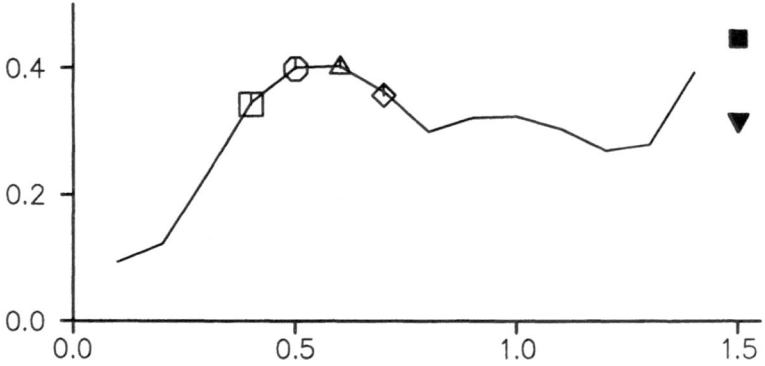

Fig. 2.13 Growth rate $\gamma(\epsilon, k)$ (solid) for MW$-$ with $\epsilon = 10^{-3}$ (Otani 1993) and flux growth rates Γ_C (markers) plotted against k. Also shown are h_{line} (solid square) and Λ_{Liap} (solid triangle).

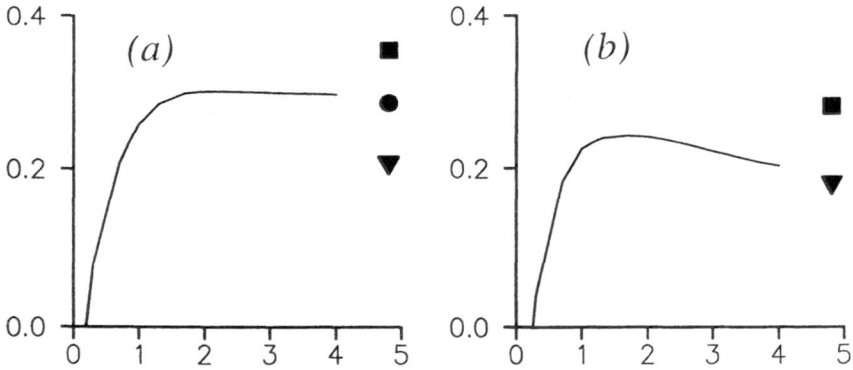

Fig. 2.14 Growth rate $\gamma(\epsilon, k)$ for (a) the CP flow, with $k = 0.57$, and (b) the LP flow, with $k = 0.62$, plotted against $\log_{10} 1/\epsilon$ (Galloway & Proctor 1992). Also shown are h_{line} (square), Λ_{Liap} (triangle) and, in (a) only, Γ_C (circle).

The flux growth rate agrees with the $\epsilon \to 0$ asymptote and is larger than the Liapunov exponent (Galloway & Proctor 1992), but smaller than h_{line}.

Measuring flux for Galloway and Proctor's 'linearly polarized' or LP flow,

$$\mathbf{u}(\mathbf{x}, t) = \sqrt{3/2}\,(0, \sin(x + \cos t), \cos(x + \cos t))$$
$$+ \sqrt{3/2}\,(\cos(y + \cos t), 0, \sin(y + \cos t)), \qquad (2.5.7)$$

is less successful. Here $k = 0.62$; the initial condition (2.2.7) and line

$$C = \{x = 0, 0 \le y \le 2\pi, z = 0\} \qquad (2.5.8)$$

are used. The flux $\Phi_C(t)$ is convex for $0 \le t \le 8\pi$, with decreasing slope, and we are unable to determine a growth rate. However h_{line} can be measured and is shown by a square in Fig. 2.14(b): it appears to be an upper bound for any asymptotic fast dynamo growth rate. Perhaps the difficulty in measuring the flux growth rate is connected with the fact that the asymptotic approach to the fast dynamo growth rate γ_0 as $\epsilon \to 0$ is quite slow in this case.

2.6 Steady Three-dimensional Flows

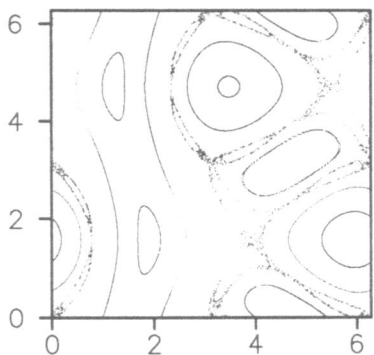

Fig. 2.15 Poincaré section of the ABC flow with $A = B = C = 1$, for $x = 0$ and $0 \le y, z \le 2\pi$. (After Dombre *et al.* 1986.)

The most studied example in the second class introduced in §2.1, that of steady three-dimensional flows, is the ABC flow:

$$\mathbf{u} = A(0, \sin x, \cos x) + B(\cos y, 0, \sin y) + C(\sin z, \cos z, 0). \qquad (2.6.1)$$

This periodic flow is the sum of three steady Beltrami waves, parameterised by A, B and C. These flows have been studied in the context of alpha-effect dynamos (Childress 1970) and the topology of Euler flows (Arnold 1965, Hénon 1966); they were first written down by Beltrami (1889). The flow has the Beltrami property $\nabla \times \mathbf{u} \propto \mathbf{u}$ and is periodic in space. Steady Beltrami flows can also be contained in spheres (with a $\mathbf{u} \cdot \mathbf{n} = 0$ boundary condition); see Zheligovsky (1993) for a detailed study of their topology and for numerical results at large R suggesting fast dynamo action.

If one or more of the parameters is zero, the flow is integrable and is not a fast dynamo. However the flow with $A = B = 1$, $C = 0$ is extraordinarily close to being fast (§5.5.2). If all three parameters are non-zero the flow is non-integrable (Arnold 1965, Hénon 1966, Dombre et al. 1986) and contains a mixture of chaotic regions and regular islands. Figure 2.15 shows a Poincaré section for the case $A = B = C = 1$; there are regular regions, or vortices, which snake their way through periodic space, and thin bands of chaos.

Dynamo action in ABC flows has been studied by Arnold & Korkina (1983), Galloway & Frisch (1984, 1986) and Lau & Finn (1993a). In these studies the field is taken to have the same 2π-periodicity as the flow, although allowing field to have a larger scale can lead to an enhancement of growth rate, as observed by Galanti, Pouquet & Sulem (1993). Results for the highly symmetrical case $A = B = C = 1$, abbreviated as the '111 flow', and field of the same periodicity, are shown in Fig. 2.16, which gives the dynamo growth rate γ as a function of magnetic Reynolds number $R = 1/\epsilon$. There are two windows of dynamo action. The first window, found by Arnold and Korkina, is in the range $8.9 \lesssim R \lesssim 17.5$. Galloway and Frisch discovered a second window in which dynamo action begins at $R \simeq 27$ and continues at least as far as $R = 550$; Lau and Finn extended this to $R = 1000$. The two windows of dynamo action have fastest-growing eigenfunctions of different symmetries. The field (2.6.4) below has the same symmetries as eigenfunctions in the first window (Arnold & Korkina 1983), while in the second window the eigenfunctions are triply degenerate (Galloway & Frisch 1986).

Around $R = 200$ to 250 Galloway and Frisch found it difficult to measure the growth rate, this being reflected by the error bars in Fig. 2.16. For $R \lesssim 200$ the complex dynamo eigenvalue $p(R)$ (1.3.3) has an imaginary part, corresponding to periodic behavior superposed on the exponential growth. Around $R \simeq 200$ the imaginary part is small, the corresponding period large, and the real part γ of the growth rate is hard to measure. For $R \gtrsim 300$, the imaginary part appears to be zero. The cause is a mode-crossing around $R = 300$–350 from an oscillatory growing mode to a steady growing mode (Lau & Finn 1993a).

The largest values of R achieved are considerably smaller than those for the MW+ flow in §2.2, since the 111 flow is fully three-dimensional. At the largest values of R, in the range $350 \leq R \leq 1000$, the growth rate appears to have saturated; this range, in which R varies by a factor of 3, is of reasonable size and so these results give support for the 111 flow as a fast dynamo with $\gamma_0 \simeq 0.077$.[19] It is worthwhile to check this, or alternatively to test the flux conjecture, by looking at the evolution of flux with $\epsilon = 0$. This paints rather a different picture, but before we do this we need to discuss the Lagrangian structure of the flow a little more; see Childress & Soward (1985) and Dombre et al. (1986).

---

[19]The apparent tailing-off of the curve for $R \simeq 1000$ is probably accounted for by numerical error. The numerical codes exhibit differences even at $R = 100$, for which Lau & Finn (1993a) obtain $\gamma = 0.055$, slightly lower than the value $\gamma = 0.057$ of Galloway & Frisch (1984).

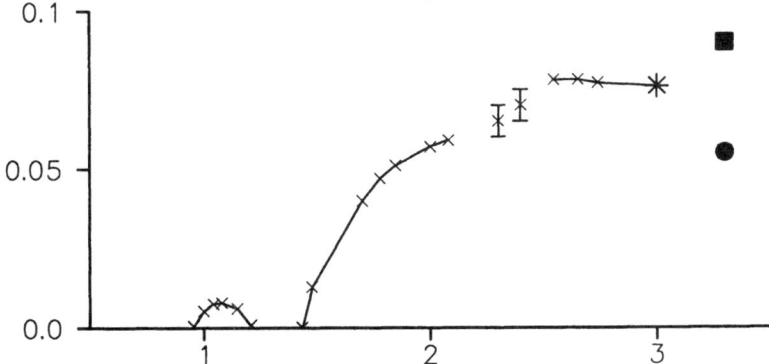

Fig. 2.16 Dynamo growth rate $\gamma(R)$ plotted against $\log_{10} R$ for the 111 flow. The data points for $R \leq 550$ (times signs) are from Galloway & Frisch (1986) and Arnold & Korkina (1983), and the $R = 1000$ point (asterisk) is from Lau & Finn (1993a). The flux growth rate Γ_S for the initial condition (2.6.4) is shown by a circle and the rate of line-stretching, h_{line}, by a square.

The 111 flow is highly symmetric, possessing 24 spatial symmetries[20] (see Arnold & Korkina 1983, Arnold 1984, Dombre *et al.* 1986, Gilbert 1992), and so quite special from the point of view of dynamical systems. The flow has unstable stagnation points at

$$4\mathbf{r}/\pi = (-3, 1, -1),\ (1, -1, -3),\ (-1, -3, 1),\ (3, 3, 3),$$
$$(1, -3, 3),\ (-3, 3, 1),\ (3, 1, -3),\ (-1, -1, -1), \tag{2.6.2}$$

in one periodicity box $-\pi < x, y, z < \pi$. The first (second) set of four have two-dimensional unstable (stable) manifolds and one-dimensional stable (unstable) manifolds.

These stagnation points are connected by *heteroclinic orbits*, that is orbits that are asymptotic to one stagnation point as $t \to -\infty$ and to another as $t \to \infty$. Some of these heteroclinic orbits are straight line segments. An example is the line segment joining $\mathbf{r}_1 = -\pi/4(1, 1, 1)$ to $\mathbf{r}_2 = \pi/4(3, 3, 3)$ (Fig. 2.17); this segment lies in both the one-dimensional unstable manifold of $\mathbf{r}_1$ and the one-dimensional stable manifold of $\mathbf{r}_2$. This linking of two one-dimensional manifolds in three-dimensional space is non-generic — almost any small perturbation would destroy the link. The eigenvalues of the rate-of-strain matrix at $\mathbf{r}_1$ are $\sqrt{2}$ and $-1/\sqrt{2}$ twice, corresponding to a two-dimensional stable manifold. The eigenvalues of the rate-of-strain matrix at

---

[20]This becomes 48 if one is allowed to reverse time; however these additional symmetries are not relevant for the kinematic dynamo problem, which involves the irreversible process of diffusion (or flux averaging).

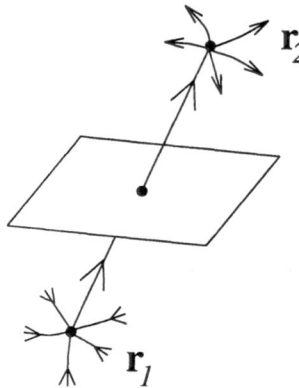

Fig. 2.17 Geometrical structure of 111 flow. The stagnation points $\mathbf{r}_1$ and $\mathbf{r}_2$ are joined by a heteroclinic orbit which is a straight line. The plane placed in between is given by $\tilde{z} = 0$ in (2.6.3).

$\mathbf{r}_2$ are $-\sqrt{2}$ and $1/\sqrt{2}$ twice, and this point has a two-dimensional unstable manifold.

The thin bands of chaos in Fig. 2.15 are associated with the stagnation points and their connections. Simulations of dynamo action at high R also reveal that the fastest-growing magnetic eigenfunction possesses ropes (or cigars) of magnetic field localised near the connections. Near a hyperbolic point such as $\mathbf{r}_1$ field tends to be squashed into a rope-like structure because of exponential contraction (Galloway & Frisch 1986); see §5.6 and Dobrokhotov *et al.* (1993). The rope centered at $\mathbf{r}_1$ extends out towards $\mathbf{r}_2$ (and also in the opposite direction) where the local expansion in the flow spreads the field out into more diffuse structures; similar ropes are seen in other steady flows (Zheligovsky 1993).

The natural place to look at the behavior of magnetic flux is across one of the ropes, say on the plane $x + y + z = 3\pi/4$. This plane is midway between $\mathbf{r}_1$ and $\mathbf{r}_2$, as shown schematically in Fig. 2.17. Define coordinates $(\tilde{x}, \tilde{y})$ on this plane $\tilde{z} = 0$ by setting:

$$\mathbf{x} = \mathbf{c} + \tilde{x}\mathbf{e}_{\tilde{x}} + \tilde{y}\mathbf{e}_{\tilde{y}} + \tilde{z}\mathbf{e}_{\tilde{z}}$$
$$\equiv \frac{\pi}{4}(1,1,1) + \frac{\tilde{x}}{\sqrt{2}}(-1,1,0) + \frac{\tilde{y}}{\sqrt{6}}(-1,-1,2) + \frac{\tilde{z}}{\sqrt{3}}(1,1,1). \tag{2.6.3}$$

Figure 2.18 shows a Poincaré section for this plane with $-0.2 < \tilde{x}, \tilde{y} < 0.2$. The section shows a three-fold symmetry, corresponding to the symmetry $x \to y \to z \to x$ of the 111 flow. The straight heteroclinic orbit in Fig. 2.17 intersects the axis of symmetry of Fig. 2.18.

The initial condition that shows clearest behavior in the 111 flow is

$$\mathbf{B}_0(\mathbf{x}) = (\sin z - \cos y, \sin x - \cos z, \sin y - \cos x), \tag{2.6.4}$$

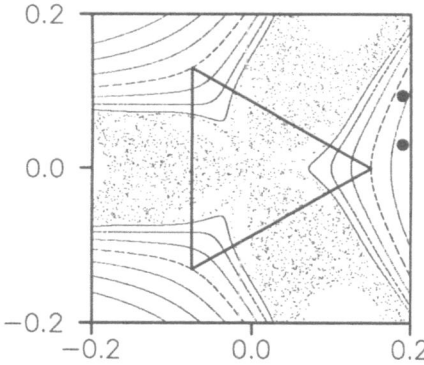

Fig. 2.18 Poincaré section for the 111 flow, using the plane $\tilde{z} = 0$ with $-0.2 < \tilde{x}, \tilde{y} < 0.2$. The symmetrically placed equilateral triangle (with one vertex at $\tilde{x} = 0.15$, $\tilde{y} = 0$) is the triangle C used for flux measurements. The two circles indicate the spacing of grid points in Lau & Finn's (1993a) simulations.

and Fig. 2.19(a, b) shows the evolution for $\epsilon = 0$ of the component $B_{\tilde{z}} = \mathbf{e}_{\tilde{z}} \cdot \mathbf{B}$ perpendicular to the section in Fig. 2.18 (actually $F(B_{\tilde{z}})$ is plotted). There are bands of field piling up in the chaotic region, and evidence of constructive folding.

For the initial condition (2.6.4) with $\epsilon = 0$ we measure the flux $\Phi_S(t)$ through the surface S bounded by the triangle C shown in Fig. 2.18. The flux is calculated up to $t = 90$ using flux conservation:

$$\Phi_S(t) = \int_S \mathbf{B}(\mathbf{x}, t) \cdot d\mathbf{S} = \int_{M^{-t}S} \mathbf{B}_0(\mathbf{x}) \cdot d\mathbf{S} = \int_{M^{-t}C} \mathbf{A}_0(\mathbf{x}) \cdot d\mathbf{r}, \quad (2.6.5)$$

where $\mathbf{A}_0(\mathbf{x})$ is a vector potential for $\mathbf{B}_0(\mathbf{x})$. To find $M^{-t}C$ the code places points around C, follows orbits back for a time t, and tries to ensure that the spacing of points on $M^{-t}C$ is less than some given tolerance, 0.1, say. At some point this runs into difficulties. The rate of strain at a stagnation point is as much as $\sqrt{2}$ and so for $t = 90$ infinitesimal vectors can be stretched by as much as $\exp(90\sqrt{2}) \sim 10^{55}$. However points cannot be placed as close together as 10^{-55} on C without being swamped by rounding error. To avoid this problem, no points are ever placed closer than 10^{-10} apart on C; whenever points diverge by more than 10^{-6} in the flow, new points are inserted in between. The method seems robust and gives results which agree in detail with Gilbert (1991a, 1992), where a different method is used.

The flux $\Phi_S(t)$ (solid) is shown in Fig. 2.20 and shows clear exponential growth with oscillations. The sharp downward spikes correspond to zero-crossings of $\Phi_S(t)$. Also shown is the length of the contour $M^{-t}C$ (dashed). The flux growth rate $\Gamma_S \simeq 0.055$ is shown in Fig. 2.16 by a solid circle together with growth rates for finite R. The comparison is disappointing; the flux growth rate is smaller than the apparent $R \to \infty$ asymptote. The situation is in fact even worse than this! The growing fields in the second window

Fig. 2.19 Magnetic field evolution in steady flows with $\epsilon = 0$. On the top row, field $F(B_{\bar{z}})$ in the ABC 111 flow (2.6.1) is shown at (a) $t = 30$ and (b) $t = 60$ (after Gilbert 1991a). The region corresponds to that in Fig. 2.18 and the initial condition is (2.6.4). Field $F(B_z)$ in the Kolmogorov flow (2.6.6) is shown (bottom row) for (c) $t = 10$ and (d) $t = 20$. The initial condition is (2.6.7) and the area shown is $z = \pi/2$, $0 < x, y < \pi$.

of Fig. 2.16 belong to a different symmetry class from our initial condition (2.6.4); Galloway & O'Brian (1993) find that beginning with this initial condition, the magnetic field actually *decays*, with $\gamma(R) \simeq -0.03$ in the moderate range $100 \leq R \leq 400$, whereas we have seen flux growth for $R = \infty$.

Two possible reasons for this discrepancy are:

$a)$ The magnetic Reynolds number in the diffusive simulations is not yet high enough for the asymptotic behavior of $\gamma(R)$ to emerge.

$b)$ The flux conjecture does not hold for this flow.

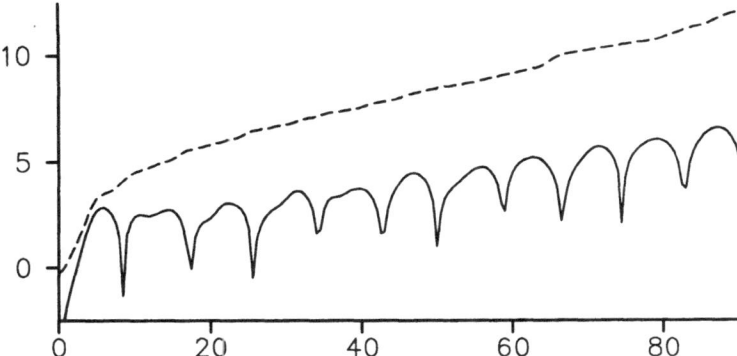

Fig. 2.20 Growth of flux and line-length for the 111 flow. The solid line shows $\log |\Phi_S(t)|$ plotted against time t. The dashed line gives the evolution of $\log |M^{-t}C|$.

If (b) is correct, diffusion will continue to affect the growth rate as $R \to \infty$. Certainly diffusion appears to be significant at the largest R attainable numerically (Galloway & O'Brian 1993), the magnetic ropes being dominated by diffusion (see also Zheligovsky 1993). Note that since the 111 flow has non-generic heteroclinic connections which might perhaps lead to some efficient but diffusive amplification mechanism, it is possible that the flux conjecture may fail for this flow, but might hold for classes of 'generic' flows.

In support of (a), note that the bands of chaos shown in Figs. 2.15, 18 are very thin and so R must become quite large before the magnetic field 'senses' that this flow is chaotic at all, i.e., before the $O(R^{-1/2})$ fast dynamo length-scale is much smaller than the width of the chaotic regions. Lau & Finn (1993a) use a resolution of up to 100^3 points for $R = 1000$, which would be spaced by $2\pi/100$; this spacing is indicated on Fig. 2.18, and is not that much smaller than the width of the chaotic regions. Also note that the pictures of magnetic field in Galloway & Frisch (1986) and Lau & Finn (1993a) do not show much of the fine-scale structure one would expect if the field could sense the chaos in the flow.

Further support for (a) is given by measuring the rate of line stretching, h_{line}, from Fig. 2.20; the result, $h_{\text{line}} \simeq 0.09$, is shown by a square on Fig. 2.16. For steady three-dimensional flows, h_{line} is equal to the topological entropy h_{top}, which is an upper bound for γ_0 (see §6.1). This is larger than all growth rates in Fig. 2.16, and so does not offer support for option (a) above. However results of Galanti et al. (1993) show that magnetic field with spatial periodicity twice that of the flow has enhanced growth rates. For example for $R = 250$ the growth rate is approximately 0.17; since this is rather larger than h_{top} this indicates that certainly $R = 250$ is not yet in the asymptotic regime. This result lends weight to (a).

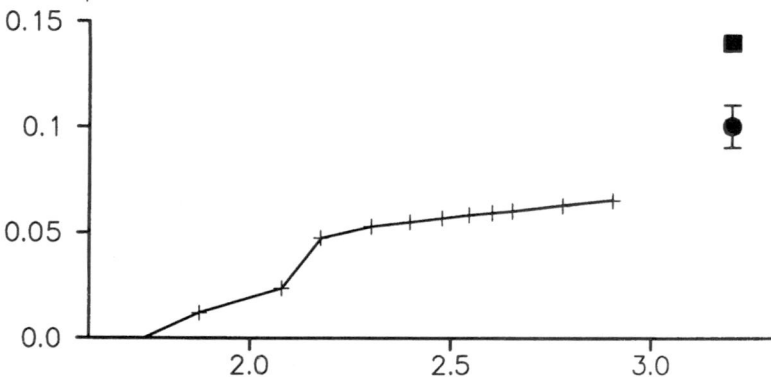

Fig. 2.21 Dynamo growth rate $\gamma(R)$ plotted against $\log_{10} R$, for the Kolmogorov flow (2.6.6) (after Galloway & Proctor (1992), numerical results courtesy D.J. Galloway). Also shown is a range of flux growth rates Γ_S (circle) and h_{line} (square).

Therefore with the values of R attainable numerically at present, it is unclear whether or not the flux conjecture holds for the 111 ABC flow.[21] If chaos does play a crucial role in fast dynamos, it makes sense to look at a flow which has rather larger chaotic regions than the 111 flow, as recognised by Galloway & Proctor (1992), who study a 'Kolmogorov flow'

$$\mathbf{u} = (\sin z, \sin x, \sin y). \tag{2.6.6}$$

This is the 111 flow without the cosine terms; it has zero fluid helicity and a Poincaré section shows chaos everywhere. It is again a highly symmetric flow, having 24 symmetries and 8 possible symmetry classes or representations for the magnetic field (Gilbert 1991b). Galloway and Proctor followed dynamo action numerically up to $R = 800$ and the results are shown in Fig. 2.21. The growth rate appears to show asymptotic convergence as R increases, and the results are highly suggestive of fast dynamo action.

The flow has stagnation points at $(x, y, z) = (m\pi, n\pi, p\pi)$, for integers m, n and p. If $m + n + p$ is even (odd) the point has a one-dimensional unstable (stable manifold) and a two-dimensional spiralling stable (unstable) manifold.

---

[21]Galloway & Frisch (1986) and Galloway & O'Brian (1993) have also studied the '522' ABC flow, which has no stagnation points. Numerical simulation indicates a vigorous fast dynamo with magnetic field taking the form of sheets; the flux conjecture has not yet been tested for this flow.

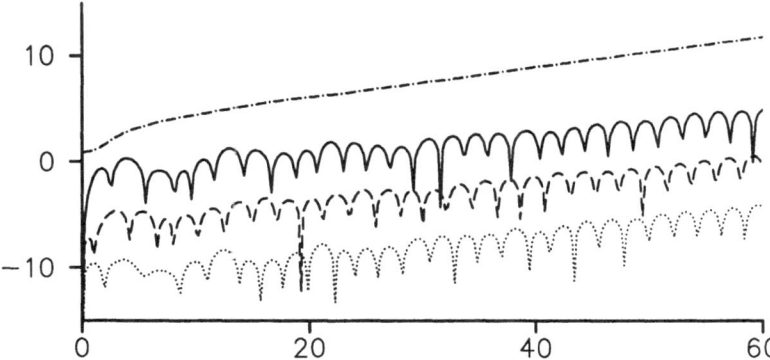

Fig. 2.22 Flux growth for the Kolmogorov flow (2.6.6); $\log |\varPhi_S(t)|$ is plotted against time t for the initial conditions, (2.6.7) (solid), (2.6.9a) (dashed) and (2.6.9b) (dotted). The growth in $\log |M^{-t}C|$ is also shown (dot–dash). The curves are separated by the addition of constants.

As for the 111 ABC flow, there are non-generic straight-line heteroclinic connections; for example the straight unstable manifold of $(0, 0, 0)$ is connected to the stable manifold of (π, π, π). Galloway & O'Brian (1993) show the fastest growing field for $R = 450$, containing strong double ropes aligned with the heteroclinic connections.

Figures 2.19(c, d) show magnetic field evolution with $\epsilon = 0$ from the initial condition:

$$\mathbf{B}_0 = (2 \sin z, - \sin x, - \sin y); \tag{2.6.7}$$

the field takes on a spiral form. To measure flux growth we use the surface

$$S = \{3\pi/8 \le x \le 5\pi/8, -\pi/16 \le y \le \pi/16, z = \pi/2\}, \tag{2.6.8}$$

which intersects the bottom of Figs. 2.19(c, d). Figure 2.22 shows flux growth for three initial conditions, (2.6.7) (solid),

$$\mathbf{B}_0 = (2 \cos y, - \cos z, - \cos x), \quad \mathbf{B}_0 = (\cos y, \cos z, \cos x) \qquad (2.6.9a, b).$$

(dashed and dotted); these initial conditions belong to different symmetry classes. There is clear growth with oscillations.[22] The growth in $|M^{-t}C|$ is

---

[22]The growth rates obtained by Galloway and his colleagues are *non-oscillatory*. However results for model steady flows, discussed in §7.3, indicate that for $\epsilon = 0$ there can be many fast-growing modes with (almost) the same real growth rate but different frequencies. Thus it appears that one cannot apply the flux conjecture to the *complex* growth rate but only to its real part.

also shown (dot–dash). Figure 2.21 shows the results of Galloway & Proctor (1992), together with a range of flux growth rates (circle) and h_{line} (square), obtained from Fig. 2 22. While the flux growth rates perhaps appear somewhat high (depending on how far to the right they are plotted!), they are not obviously incompatible with the curve for $\gamma(R)$, given that it is still rising at $R = 800$. This gives some support for the flux conjecture in three-dimensional steady flows, although plainly further studies of these flows are needed.

3. Fast Dynamos in Maps

As we have seen in Chap. 2, numerical experiments using two-dimensional unsteady flows strongly suggest the existence of fast dynamos, and point toward a connection between computation with $\epsilon \equiv 0$ and the limit of computations with $\epsilon > 0$. In particular the flux conjecture appears justified in all cases of fast dynamo action that have been examined. For three-dimensional steady flows there is some support for fast dynamo action, but there are still perplexing gaps in our understanding of magnetic structure and its measurement. The flux conjecture seems to work for the Kolmogorov flow but has not been verified for the ABC 111 flow of §2.6.

The results certainly support the view that in the limit $\epsilon \to 0$ the magnetic fields produced by fast dynamo action are largest in the regions of chaos and acquire structure on all scales of length. This complexity reflects the Lagrangian structure of the flow as revealed by Poincaré maps. In one case, namely the MW+ flow studied by Otani (1988, 1993), we have identified a mechanism for fast dynamo action, the SFS mechanism, which is fairly simple and has roots in well-understood examples of slow dynamo action.

In the present chapter we turn to an analogous study of *maps*, both with and without diffusion. The use of maps in dynamo theory is relatively new, and has been reviewed by Bayly (1994). We will be dealing with maps that are sufficiently simple to allow some explicit analysis — this is their main advantage over flows. The formulation of the fast dynamo problem in the context of maps allows us to put some of the results obtained for flows on a more substantial mathematical footing. However, there will be some unphysical features, e.g., loss of continuity, which will take us a step backward in realistic modelling.

To construct useful models based on maps, we must deal with two issues. One is the selection of maps. The other is the manner in which diffusion is incorporated. Not all of the maps we consider arise from flows, in the sense that they are equivalent to a Lagrangian map of a certain fluid flow. Rather, they can be chosen to *model* a flow. This is the case with our first example, which attempts to model the stretch–fold–shear mechanism, discussed in §2.3. We formulate the SFS mechanism in terms of Beltrami waves in §3.1, and approximate it by a simpler map, the SFS map, in §3.2. This example leads us to still simpler maps based upon the baker's map. In §3.3, we discuss various examples. Finally, in §3.4 we introduce models based upon the famous *cat map* of Arnold & Avez (1967), which offer significant advantages for analysis.

3.1 Pulsed Beltrami Waves

3.1.1 Modelling Stretch–Fold–Shear

Beltrami waves were introduced in §1.2.2 to give an example of a Lagrangian map, in §2.2 as a basic element for the construction of useful flows, and in §2.3 to analyze the SFS mechanism. The Lagrangian maps $M^{(x)}$, $M^{(y)}$ will be defined for the waves $\mathbf{u}^{(x)}$, $\mathbf{u}^{(y)}$ given by (2.2.2). If the waves act for time α, we have

$$M^{(x)} : (x, y, z) \to (x, y + \alpha \sin x, z + \alpha \cos x), \qquad (3.1.1a)$$

$$M^{(y)} : (x, y, z) \to (x + \alpha \sin y, y, z - \alpha \cos y). \qquad (3.1.1b)$$

These will be taken as maps on a $2\pi \times 2\pi \times 2\pi$ cube of periodicity. The induction operator corresponding to $M^{(x)}$ is defined by (1.2.18) and is given by

$$\mathbf{T}^{(x)}\mathbf{B}(M^{(x)}\mathbf{x}) = \mathbf{J}^{(x)}(\mathbf{x}) \cdot \mathbf{B}(\mathbf{x}), \qquad (3.1.2a)$$

$$\mathbf{J}^{(x)} = \begin{pmatrix} 1 & 0 & 0 \\ \alpha \cos x & 1 & 0 \\ -\alpha \sin x & 0 & 1 \end{pmatrix}. \qquad (3.1.2b)$$

We may take the magnetic field to have the same periodicity as the flow. The formulation is thus in terms of maps on a 3-torus, $\mathbb{T}^3$.

Although $\mathbf{T}^{(x)}$ and the analogous map $\mathbf{T}^{(y)}$ are quite simple, if applied alternately to a magnetic field, they produce nevertheless a complicated structure, one which is awkward to express explicitly because of the iterated trigonometric functions. (This problem is taken up in the limit of small α in Chap. 8 and large α in Chap. 10.) We therefore seek to mimic the action of the Beltrami waves with simpler transformations acting on the same cube of periodicity.

Now if $\mathbf{T}^{(x)}$ is applied to the initial field $(1, 0, 0)$, we obtain the distorted field $(1, \alpha \cos x, -\alpha \sin x)$. Field lines which are initially parallel to the x-axis are mapped into helices. Note that this helical field does not depend upon y or z. As discussed in §2.3, we can view its creation as consisting of two steps. First, the horizontal part of the flow, $\mathbf{u}_{\mathrm{H}}^{(x)} = (0, \sin x, 0)$, stretches and folds field lines in the (x, y)-plane. Because of the action of the shear of this horizontal flow the largest y-directed field is developed near its zeros $x = m\pi$. Now consider the action of the vertical flow $(0, 0, \cos x)$. We observe that the vertical motion is largest at the position of the largest horizontal y-field. It is this vertical distortion of the stretched horizontal field that produces the helical structure (cf. Figs. 2.7, 8).

If now $\mathbf{B} = e^{ikz}\mathbf{b}(x,y)$ and $\mathbf{b}$ is initially $(1,0,0)$, the dependence on z means that the vertical flow gives a phase shift which can modify average flux, as discussed in §2.3. What we emphasize here is that the vertical flow and horizontal shear have extrema which coincide. This shearing by the vertical flow can be regarded as a quite separate distortion of lines which have already been stretched and folded by a horizontal flow.

We can then try to mimic these distortions with maps which are simpler, more discrete, and less smooth! In addition we shall separate the horizontal and vertical motions in (3.1.1), and model them in different ways. We emphasize this point because the trigonometric functions in (3.1.1) do not distinguish horizontal from vertical motion in any significant way, and it is important to keep in mind the change in the way flux is stretched and folded in the simplified maps.

We shall first consider the modelling of the horizontal flow, $\mathbf{u}_{\mathrm{H}}^{(x)}$. The sine function on $0 < x < \pi$ stretches and folds field lines which are initially parallel to the x-axis. In the process variations of the field in y are converted into variation in x. We seek to model this process with a simple discontinuous map. A model of a full period of the sine function will be constructed in a square of side 2. However for describing a single 'fold' it is sufficient to restrict attention to the unit square.

The simplest stretch–fold operation is the *folded baker's map* (or folded baker's transformation), which has the form

$$(x,y) \rightarrow \begin{cases} (2x, y/2), & (0 \leq x < 1/2), \\ (2 - 2x, 1 - y/2), & (1/2 \leq x < 1), \end{cases} \quad \text{(folded)} \quad (3.1.3)$$

and is shown in Fig. 3.1. The action of the map on the unit square is as follows: the square is compressed in y and stretched in x to yield a $2 \times 1/2$ rectangle, which is then cut at the line $x = 1$, and the right half folded over and stacked on the left half, thus reassembling the original square. This is an area-preserving map of the unit cube onto itself, of the kind a baker (equipped with a knife!) might use to prepare pastry. It is the simplest two-dimensional representation of the idea of complete cancellation of flux following stretching.

We remark that the usual 'baker's map' is defined without the fold, that is, after the stretching and cutting the right-hand part is stacked on the left without first turning it over. This version of the map, which we shall call the *stacked* baker's map, is defined by

$$(x,y) \rightarrow \begin{cases} (2x, y/2), & (0 \leq x < 1/2), \\ (2x - 1, (1 + y)/2), & (1/2 \leq x < 1), \end{cases} \quad \text{(stacked)} \quad (3.1.4)$$

and is shown in Fig. 3.2. The stacked baker's map is the simplest model of completely constructive assembly of stretched field lines, and is a rudimentary

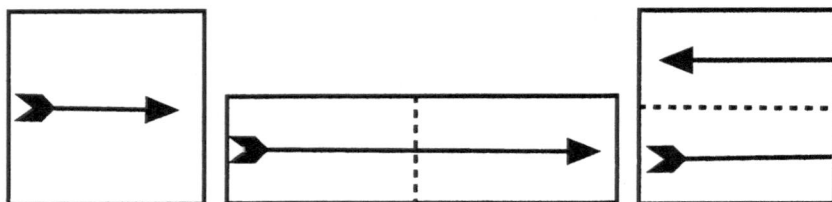

Fig. 3.1 The stacked baker's map.

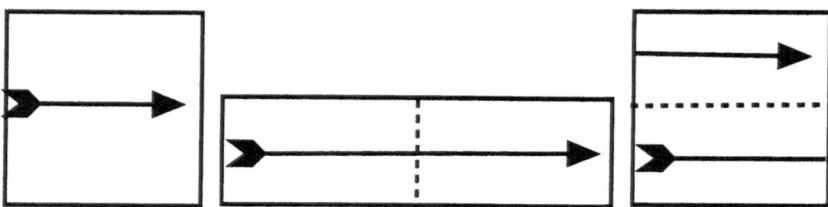

Fig. 3.2 The folded baker's map.

model of the field evolution in the three-dimensional STF dynamo of §1.4 (see Finn & Ott 1988a, b).[23]

If we take the initial magnetic field to be $\mathbf{B} = (1,0)$ and perform the stacked map successively, then after m applications the field developed is $(2^m, 0)$. That is, the map is a flux-doubling map, with

$$\mathbf{T}^m(1,0) = (2^m, 0). \tag{3.1.5}$$

This property was pointed out by Davis (1958) for a similar model in a sphere. Here, and elsewhere in our discussion of maps which are repeatedly applied, we shall take the 'time' between applications to be unity in our dimensionless models.

Since the field structure is invariant and constant, we seem to have in the stacked baker's map a simple but robust fast dynamo, with growth rate of total flux given by $\Gamma_S = \log 2$, S being a unit square in the (y, z)-plane. The problem with this example lies, of course, in the discontinuity of the map.

---

[23]The 'stacking' operation can be made somewhat more plausible as a model of a flow if a line of squares extends along the x-axis and the top half of the image map comes from, say, the right half of the neighbor to the left.

Any cutting operation must be effected with diffusion, and without a direct calculation of this operation it is impossible to verify fast dynamo action from a limit of small diffusion. Even with diffusion, planar flows cannot give dynamo action, and so a stacked baker's map has to be achieved by a three-dimensional process, such as STF, which raises other difficulties (as discussed in §1.4).

The folded baker's map is similar but there is a significant difference, since the fold reverses the orientation of the field. After n steps the field is $(\pm 2^n, 0)$, the $+$ sign applying to $0 \leq y < 2^{-n}$ and thereafter reversing with each layer of thickness 2^{-n}. The growth of energy is the same as for the stacked map, but the total flux through the square, computed on any line $x = \text{const.}$, is zero after the first step. On the other hand, the evolution of flux through an interval $p < y < q$ depends upon the numbers p, q. If $q - p = 2^{-n}m$ for integers $m, n > 0$, the flux is zero after $n + 1$ iterations. However for general p and q the flux will be $O(1)$ at each iteration; for example with $p = 0$ and $q = 1/3$, the flux remains $2/3$ after the first iteration, as may be verified by writing $\frac{1}{3} = \frac{1}{4} + \frac{1}{16} + \cdots$.

This example for the folded baker's map is analogous to the failure of the flux conjecture in planar flow; see §4.2.2. It raises the issue of exactly how we are to compute the 'cancellation' of field in a perfect conductor. A simple flux average, normally an integral over a surface of the normal component of the magnetic field as discussed in Chap. 2, is appealing on physical grounds. But the oscillation on a scale 2^{-n} of a field of magnitude 2^n can produce averages (much smaller than 2^n) which depend very much on how the average value is calculated, as we shall see below. These examples deal however with flows which are not exhibiting dynamo action. It will turn out that when fast dynamo action does occur the success of the flux conjecture noted for the unsteady flows of Chap. 2 is also observed in maps; see §3.2.

With small diffusion, the rapid oscillation of field would be smoothed out once $\sqrt{\epsilon} \sim 2^{-n}$, a fact we shall verify explicitly below. But since the flux is zero, the average which should emerge from smoothing must be zero. Thus the folded baker's map fails doubly as a fast dynamo, first from the discontinuities, and second from the cancellation of field. The second is the more forceful, since one can imagine a slightly different map where the cut is replaced by a smooth fold (see Fig. 3.3 below).[24]

We now compare the folded baker's map and the horizontal part of $\mathbf{T}^{(x)}$. We need to relate the variable amplitude of a Beltrami map, depending on the time α over which the flow acts, with the discrete, integer amplitude of the

---

[24]We shall not, in this monograph, make use of *wavelets* in the representation of the magnetic field obtained under iterated Lagrangian maps. However such a representation could be very useful in certain cases. The most natural connection is between the folded baker's map and the Haar wavelet; see Strang (1989). The image of the field $(1, 0)$ under the folded baker's map is, as a function of y, a Haar wavelet. A second mapping produces two smaller wavelets, and the map is thus simply represented in a Haar wavelet decomposition.

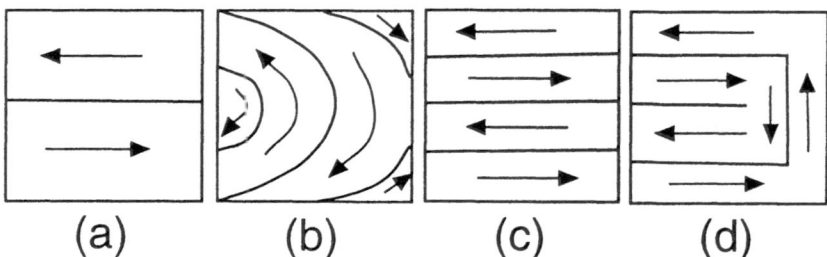

Fig. 3.3 The folded baker's map compared with a sinusoid flow: (a) the initial field, piecewise constant with the indicated orientation, (b) the image of (a) under the map $(x, y) \rightarrow (y - 0.83 \sin \pi x, -x)$, ($c$) the image of ($a$) under the folded baker's map (3.1.3), and (d) the folds actually created in a continuous version of the folded baker's map.

baker's map, which is 2 here. The curve $y = \alpha \sin x$, when interpreted as a field line which initially coincides with the x-axis, represents stretching by a factor $(2\pi)^{-1} \int_0^{2\pi} (1 + \alpha^2 \cos^2 x)^{1/2} \, dx$, which is equal to 2 when $|\alpha| \approx 2.6$. Now rescale the Beltrami wave to the unit square, $\mathbf{u}^{(x)} = (0, \sin \pi x, \cos \pi x)$, and consider the horizontal part only. The baker's map then models an amplitude of $\alpha / \pi \simeq 0.83$. The flow $\mathbf{u} = (0, -\sin \pi x)$ in two dimensions, acting for time 0.83 and followed by the rotation $(x, y) \rightarrow (y, -x)$, produces the pattern shown in Fig. 3.3(b). We start with the configuration 3.3(a), and 3.3(c) shows the corresponding effect of the folded baker's map. The simplest periodic extension (assuming the map is on the unit square) is then with period unity in both directions, rather than period 2. The space can thus be viewed as the torus (x, y) mod 1. The change from two to four magnetic layers is to be visualized as a result of stretching and folding by a sinusoidal y-motion followed by a rotation of the entire plane.

The points of discontinuity in field from the baker's map correspond to folds in field from the wave, as we suggest in a continuous approximation to the folded baker's map shown in Fig. 3.3(d). The hidden 'fold' is of course the main defect of the map, but one can argue as in the case of the 'crossover' region of the STF map that diffusion should act simply to connect up flux already more or less aligned, and to dissipate the small-scale y-directed field components. Also, the field in the fold will tend to be suppressed by compression during subsequent mappings.

We now turn to modelling the vertical flows in the Beltrami waves. The essential point is to obtain different z-displacements in the two halves of the baker's map, in order that the phases developed in the layers of opposite orientation are not the same. We consider a general shear given by the map

$$(x, y, z) \to (x, y, z + \alpha f(y)), \tag{3.1.6}$$

where $f(y)$ is some specified and normalized function, and α is a shear amplitude, distinct from the amplitude parameter of the Beltrami flow.[25] A simple continuous vertical shear, which is easy to work with analytically, is a constant, i.e., the displacement is a linear function, which we choose to be $f(y) = y - 1/2$. In this case one can argue that $\alpha = 1$ in (3.1.6) corresponds to a vertical component of a Beltrami wave, $(0, 0, \cos x)$, with an amplitude of about $\pi/2$. An amplitude of 2.6 then requires an α of about 1.66. Thus the folded baker's map, followed by the vertical shear with linear displacement and $\alpha = 1.66$. should correspond roughly to the action of the Beltrami map $M^{(x)}$ of displacement amplitude 2.6, followed by a rotation. We shall see below that calculations of fast dynamo action give somewhat higher growth rates for Beltrami waves than this comparison would suggest.

We call the combination of folded baker's map and vertical shear with linear $f(y)$ the *stretch–fold–shear* or SFS map.[26] The SFS map is thus defined by

$$(x, y, z) \to \begin{cases} (2x, y/2, z + \alpha f(y/2)), & (0 \le x < 1/2), \\ (2 - 2x, 1 - y/2, z + \alpha f(1 - y/2)), & (1/2 \le x < 1), \end{cases} \tag{3.1.7}$$

with $f(y) = y - 1/2$. We show in Fig. 3.4 the map for $\alpha = 1$ applied to a piecewise constant initial field.

3.1.2 Diffusion in Maps

We now consider the inclusion of diffusion in a model based upon a map such as the SFS map. The map represents deformation of a perfect conductor, and so the only way to introduce diffusion is to 'turn on' diffusion between the application of maps. We call this procedure *pulsed diffusion*, and refer to a model using it with a map or flow as a 'pulsed map' or 'pulsed flow'. This splitting of diffusion and motion in kinematic models has a long history in dynamo theory. It was a key element in the method used by Backus (1958) to prove the existence of kinematic dynamos in three dimensions. It is also reminiscent of numerical splitting methods.

Physically, the splitting makes sense because it allows the smoothing of small-scale structures developed during the mapping stage. Let us suppose that diffusion acts for a time T between actions of a map M. The diffusive time-scale associated with magnetic structures of size l is l^2/η; so the

---

[25]We use α for the amplitude in a flow or map where only one such parameter intrudes. When two parameters are needed, such as for two independent Beltrami waves, α and β will be used.

[26]We distinguish carefully between the SFS *mechanism*, which appears to operate in smooth flows, and the SFS *map*, which is designed to model the SFS mechanism, but is idealized, being based on the discontinuous folded baker's map.

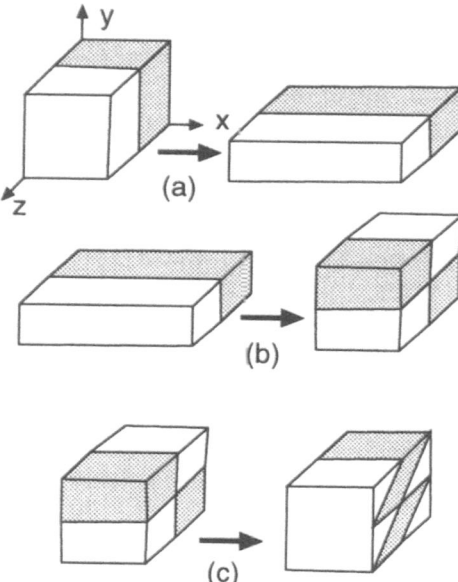

Fig. 3.4 The stretch–fold–shear (SFS) map (3.1.7) with $\alpha = 1$. The sequence $(a) \rightarrow$ (b) is the stretch–fold of the folded baker's map (3.1.3). The step (c) is the shear step. Here the sign of the x-directed field is positive in the black regions and negative in the white regions.

smallest structure which can survive an interval of diffusion will have size of order $\sqrt{\eta T}$. If we set $R = L^2/\eta T$ where L is the domain size, we have a large parameter as in the case of a diffusive flow. Now, however, the small scales are developed impulsively at times $T, 2T, \ldots$, when the map is applied. When spectral methods are used the diffusive phase is especially simple to implement and amounts to a suppression of high wave-number modes, essentially a spectral cut-off.

For the stacked baker's map (3.1.4), diffusion has no effect since the field is uniform between maps; the flux conjecture holds. For the folded map (3.1.3) the situation is more complicated. Since repeated application of the folded baker's map relentlessly decreases any vertical scales of variation of the magnetic field, we anticipate that the effect of small diffusion will be dramatic. Fortunately we can compute the effects exactly using Fourier series. Let the map be applied at times $t = 1, 2, 3, \ldots$, and suppose the initial field has the form $\mathbf{B}_0 = (b_0(y), 0)$ where $b_0(y) = -b_0(1 - y)$ and $b_0(0) = 0$; this symmetry is preserved by both map and diffusion. Let b be represented by a sine series with coefficients $\{a_k\}$. At positive integer times the y-scale is reduced by a factor of $1/2$ and the amplitude of the field increases by a factor of two. Thus at the completion of the nth interval of diffusion (at $t = n - 0$) the field is $(b_n(y), 0)$ where

$$b_n(y) = 2^{n-1} \sum_{k=1}^{\infty} a_k \sin (2^n k\pi) e^{-\epsilon k^2 \pi^2 4(4^n - 1)/3}; \qquad (3.1.8)$$

the exponential includes the cumulative effect of diffusion. The field energy grows exponentially if $\epsilon = 0$, simply by stretching. If $\epsilon > 0$ exponential growth still occurs until such time that $\epsilon 4^t \approx 1$ or $t = O(\log 1/\epsilon)$, after which the field energy drops rapidly to zero (as the negative exponential of an exponential of time). Thus $\gamma_0 = -\infty$, while $\gamma(0) = 2$. The flux growth rate depends on how the flux is measured, being $\Gamma_S = -\infty$ for the total flux, where S now denotes a line $x = \text{const.}$, but can be zero for other measures of flux. The flux conjecture fails here, as discussed in §4.1.4.

The folded baker's map thus fails as a dynamo and as a fast dynamo. Similar conclusions can be reached in smooth flows which fail to develop dynamo action, such as the planar flows discussed in §4.2.2. One can distinguish between models that destroy all flux, as is the case here, and those that preserve the original flux without causing it to grow. The planar flow preserves net flux, and if it is zero the field is destroyed diffusively. If the domain carries non-zero flux the flow will tend to concentrate it into thin boundary layers; see §5.5. The behavior of the folded baker's map is somewhat special in that the result will depend upon how in the model the field is extended as a periodic function of y. If the period is 2, finite flux can be supported, but if it is 1 then, as we have just seen, the flux is annihilated. The relevant solutions of the diffusive problem are discussed in §9.5.

3.2 The SFS Map

We now consider calculations of fast dynamo action, both with and without diffusion, in the SFS map. The advantage of this model is the essentially one-dimensional representation of the magnetic field. Since the field at any time has the form $\mathbf{B} = e^{2\pi i k z}(b(y), 0, 0)$, now taking all periods as unity, we have

$$b(y) = \sum_{n=-\infty}^{\infty} b_n e^{2\pi i n y}. \qquad (3.2.1)$$

The Lagrangian map induced on $b(y)$ by (3.1.7) is

$$T^{\text{SFS}} b(y) = 2 \operatorname{sign}(1/2 - y) e^{-2\pi i k \alpha(y - 1/2)} b(\tau(y)), \qquad (3.2.2)$$

where we introduce the *tent function* or tent map

$$\tau(y) = \min(2y, 2 - 2y). \qquad (3.2.3)$$

The diffusion is best described as the map on Fourier coefficients

$$H_\epsilon : \; b_n \to b_n \, \exp(-4\pi^2\epsilon(n^2 + k^2)), \qquad\qquad (3.2.4),$$

where as usual $\epsilon = 1/R$. For most of the calculations for the SFS models and its variants we shall put $k = 1$.

3.2.1 Numerical Results

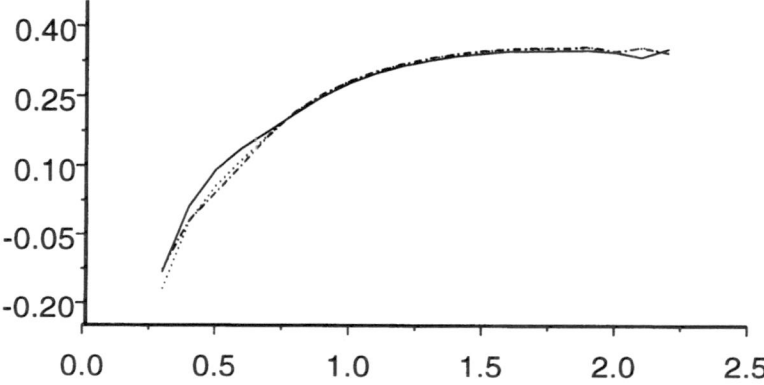

Fig. 3.5 Growth rate $\Gamma(S, \epsilon)$ of total flux, b_0, in the SFS map as a function of α for $k = 1$, with $\epsilon = 10^{-5}$ (solid) and $\epsilon = 10^{-6}$ (dash). The dot–dash curve is for the adjoint SFS map with $\epsilon = 0$; see §3.2.2.

Calculations of Bayly & Childress (1988) using the map $T_\epsilon^{\text{SFS}} \equiv H_\epsilon T^{\text{SFS}}$ were carried out over a range of α and for several values of ϵ. The total flux

$$\Phi = \int_0^1 b(y)\, dy = b_0 \qquad\qquad (3.2.5)$$

was used as a measure of dynamo action. This is analogous to the flux integral over the layers of the generalized baker's map of §3.3 as used by Finn & Ott (1988a, b), and the generalized delay flows of Finn et al. (1991) (see §7.3.1). The physical idea was discussed in §2.5: the ultimate effect of small diffusion is to smooth the tiny structures while leaving averages such as flux essentially unchanged. As we have seen from the examples in Chap. 2, in practice the value of the method can be limited by numerical accuracy, since maximum field grows generally at a faster rate than the flux. Thus the one-dimensional form of the present model is valuable for allowing extremely fast, accurate computations. In Fig. 3.5 we plot the growth rate $\Gamma(S, \epsilon)$ of total flux Φ, S being $0 \le y \le 1$, as a function of α for several choices of ϵ. The convergence

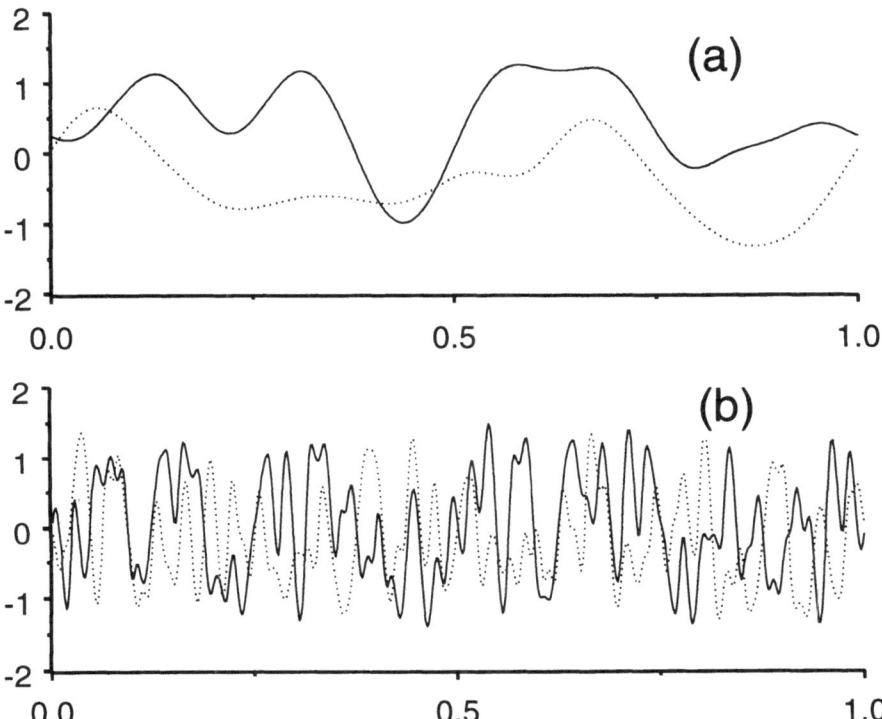

Fig. 3.6 Normalized eigenfunctions of $T_\epsilon^{\mathrm{SFS}}$, with $\alpha = 1$ and $k = 1$. The solid and dotted lines are the real and imaginary parts, respectively, of $b(y)$. In (a) $\epsilon = 10^{-3}$, and in (b) $\epsilon = 10^{-5}$.

to the perfectly conducting limit is indicated for a range of α above 0.5. There is also clear indication of a critical α for fast dynamo action. We show the corresponding eigenfunctions in Fig. 3.6. It appears that the eigenfunction becomes ever more complex in the perfectly conducting limit.

In the course of these calculations B.J. Bayly considered the operator $S_\epsilon^{\mathrm{SFS}}$ which is the formal adjoint of $T_\epsilon^{\mathrm{SFS}}$ and observed that its eigenfunctions are well-behaved in the perfectly conducting limit, a property that was subsequently proved (see §9.6.3). This allows far more accurate computations of growth rates than is possible using $T_\epsilon^{\mathrm{SFS}}$. The adjoint formulation, which we discuss in §3.2.2, also validates the flux conjecture of §2.4 for the SFS map. Using the adjoint formulation, Bayly & Rado (1993) and Rado (1993) have studied values of α out to 6. For $\epsilon = 0$ and $k = 1$, we show in Fig. 3.7 the growth rate as a function of α, and in Fig. 3.8 eigenfunctions of the adjoint operator $S_0^{\mathrm{SFS}} \equiv S^{\mathrm{SFS}}$.

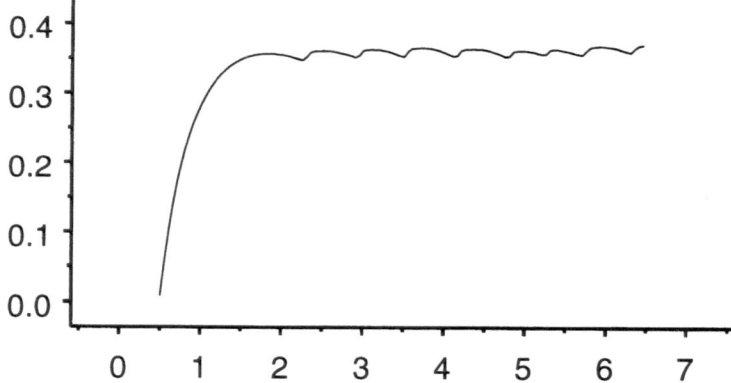

Fig. 3.7 Growth rate of flux in the SFS map, as a function of α. The results are for a perfect conductor, $\epsilon = 0$, and $k = 1$.

These calculations use the power method (Isaacson & Keller 1966). It is interesting that for α exceeding 2 there are points where apparently the branch of eigensolutions having maximum growth rate changes. This is another example of the *mode crossing* noted in §2.6. For the adjoint SFS operator, the power method breaks down near such points, and more refined methods for extracting eigenvalues are needed. We mention the interesting fact that the onset of fast dynamo action in the SFS map with $\epsilon = 0$ occurs at $\alpha = 0.5$ (for $k = 1$), and the associated adjoint eigenfunction is known explicitly (see §3.2.2). A somewhat lower threshold was indicated from the iterations of $T_\epsilon^{\mathrm{SFS}}$; see Fig. 3.5. We describe a possible explanation of this discrepancy in §9.6.2.

Note that for general k growth rates depend only on αk for $\epsilon = 0$; thus for any non-zero α, magnetic growth can be achieved for high-k modes, at sufficiently small ϵ that the $\exp(-4\pi^2 \epsilon k^2)$ factor in the diffusion operator (3.2.4) may be ignored. Within this family of SFS maps, this supports the conjecture that *almost all* flows, here all α exclusive of zero, are fast dynamos at sufficiently small ϵ; see §4.1.2.

We can now compare the growth rates developed by the SFS map with analogous flows based upon Beltrami waves, in the perfectly conducting limit. Otani's MW+ flow (2.2.1) uses waves with amplitudes $\alpha(t) = 2\cos^2 t$ and $\beta(t) = 2\sin^2 t$. This is analogous to piecewise constant waves of amplitude 2 each on for a half-period $\pi/2$, which give a maximum displacement in the (x, y)-plane of π. Now SFS is analogous to pulsed Beltrami waves with a displacement of 2.6 and half-period of 1, and so to make a comparison we need to multiply the velocity in Otani's flow field by $2.6/\pi = 0.83$ and multiply the growth rate by $\pi/2$. A numerical simulation (with $k = 1$ and $\epsilon = 10^{-3}$) gives

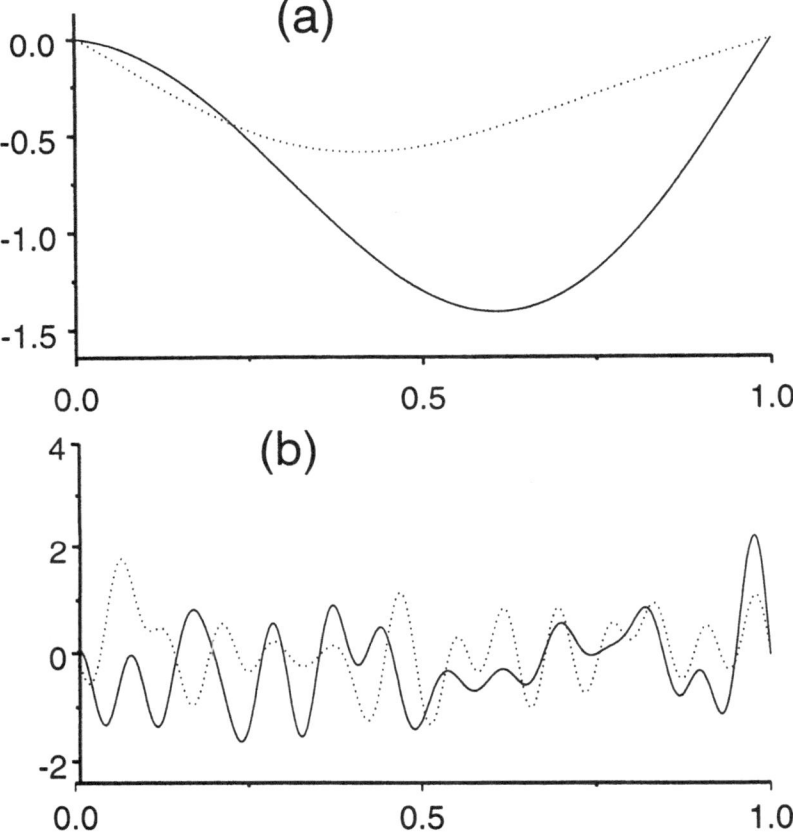

Fig. 3.8 Eigenfunctions for $\epsilon = 0$ of adjoint SFS map with $k = 1$. Shown are $\mathrm{Re}\,b$ (solid line) and $\mathrm{Im}\,b$ (dotted line), with (a) $\alpha = 1$, $\Gamma_S = 0.283$, and (b) $\alpha = 15$, $\Gamma_S = 0.365$.

a growth rate, corrected in this way, of about $\gamma_{\mathrm{OT}} = 0.46$. The dependence upon amplitude is quite sensitive. When the factor is 0.5 rather than 0.83, the corrected Otani growth rate is only 0.0838. (In fact a calculation using the appropriate Beltrami waves gives a growth rate of 0.10, in agreement with this.) For a factor 0.7 the value is 0.352. These values should be compared with the maximum SFS growth rates for $\epsilon = 0$, $k = 1$, α exceeding 1, of about 0.35. The Otani flow thus would appear to be a quite efficient example of the SFS mechanism, the enhanced growth beyond that predicted by the SFS map being a consequence of the differences of the modelling of shear as opposed to stretch–fold in the two cases.

3.2.2 The Adjoint SFS Map

We now consider the adjoint formulation of the SFS map. Let $b(y)$ and $c(y)$ be square-integrable complex-valued functions, and define an inner product,

$$(b, c) = \int_0^1 b(y)^* c(y)\, dy. \tag{3.2.6}$$

The adjoint map S^{SFS}, corresponding to the *direct* Lagrangian map T^{SFS} given by (3.2.2), is defined by the condition (see, e.g., Friedman 1956, Bollabás 1990)

$$(S^{SFS}b, c) = (b, T^{SFS}c) \tag{3.2.7}$$

for all square-integrable $b(y)$, $c(y)$. From this it is straightforward to obtain

$$S^{SFS}b(y) = e^{\pi i \alpha(y-1)}\, b(y/2) - e^{-\pi i \alpha(y-1)}\, b(1 - y/2), \tag{3.2.8}$$

where we set $k = 1$ here and from now on. As we shall see in Chap. 9, this is an example of a general adjoint formulation of dynamo theory, which involves an inverse Lagrangian map applied to the *vector potential* of the magnetic field, rather than to the field itself. For the SFS map, we can check that (3.2.8) is precisely the x-average of the inverse of the SFS map applied, not to field parallel to the x-axis, but rather to area elements with normal parallel to the x-axis (see Fig. 3.9). Indeed the inverse map first reverses the shear, $(x, y, z) \rightarrow (x, y, z - \alpha(y - 1/2))$, which introduces the phase factor $e^{2\pi i \alpha(y - 1/2)}$. Next, we apply the inverse of the folded baker's map (3.1.3), given by

$$(x, y) \rightarrow \begin{cases} (x/2, 2y), & (0 \le y < 1/2), \\ (1 - x/2, 2(1 - y)), & (1/2 \le y < 1). \end{cases} \tag{3.2.9}$$

This would introduce factors $\pm 1/2$ when the Jacobian of (3.2.9) is applied to a field parallel to the x-axis. Applied to area elements, however, the factors are ± 2. Since the field is evaluated using the inverse of the map (3.2.9) (cf. (1.2.18)), we carry the field with the forward map (3.1.3) to obtain $2b(y/2)e^{\pi i \alpha(y-1)}$ for $0 \le x < 1/2$ and $-2b(1-y/2)e^{-\pi i \alpha(y-1)}$ for $1/2 \le x < 1$. Averaging over x then yields (3.2.8).

We shall make use of S^{SFS} in Chap. 9 to establish the flux conjecture for the SFS map. We note here a few properties which bear on the growth rates shown in Fig. 3.7. B.J. Bayly (personal communication) found the exact eigenfunction of S^{SFS} marking the onset of fast dynamo action at $\alpha = 0.5$. It can be checked from (3.2.8) that if $\alpha = 0.5$, $b = \sin \pi y$ is a function satisfying $S^{SFS}b = -ib$. Careful calculations near this point (Rado 1993) indicate that the growth rate of flux is given locally by

$$\Gamma_S = 0.625(\alpha - 1/2) + O((\alpha - 1/2)^2). \tag{3.2.10}$$

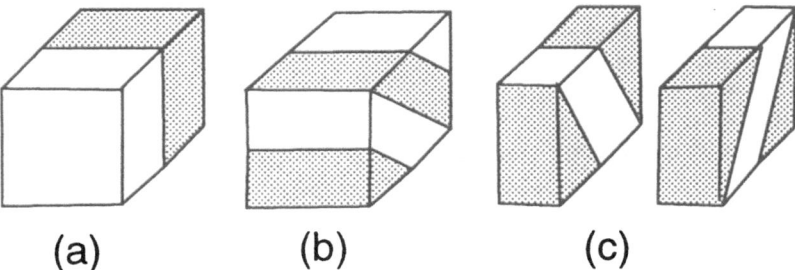

(a)   **(b)**   **(c)**

Fig. 3.9 The adjoint SFS map with $\alpha = 1$ and $k = 1$. The initial field (a) is ± 1, x-directed field which is positive (negative) indicated by white (black). The inverse shear is applied to obtain (b). The inverse baker's map produces two half-cubes (c). To obtain the final field the fields of the two half-cubes are averaged and multiplied by 4. The last step causes field to be doubled, rather than halved by the compression $(b) \rightarrow (c)$. This is equivalent to the map acting on oriented area elements rather than oriented line elements.

For larger α the growth rate tends to saturate at values between 0.345 and 0.365. One interpretation of this (Rado 1993) follows from the identity

$$\int_0^1 |S^{\mathrm{SFS}} b(y)|^2 \, dy = 2 \int_0^1 |b(y)|^2 \, dy$$
$$- 2 \operatorname{Re} \left[\int_0^1 e^{2\pi i \alpha (y-1)} b(y/2) b^*(1 - y/2) \, dy \right]. \tag{3.2.11}$$

The first term on the right will sum rapid oscillations, but the second term is, for large α, a product of three rapidly varying terms (cf. Fig. 3.8b). These will be three essentially uncorrelated signals with considerable cancellation and so the integral will be approximately zero. If b were an eigenfunction for the eigenvalue λ, then (3.2.11) would suggest that $\log |\lambda| \approx \frac{1}{2} \log 2 \approx 0.347$. Another way to think of this saturation is to view the large shear as a random shuffling of orientation. After the field strength is doubled, layers $y = \mathrm{const.}$ are randomly shuffled by the shear, with the result that with approximately equal probability an adjacent layer will have either the same or oppositely directed field. Flux then grows like the average displacement of a random walk, producing a growth rate approximately equal to $\log \sqrt{2}$. A similar argument is given by Klapper (1992c) in the context of a random SFS1 map.[27]

---

[27] It is interesting that a completely separate argument, given by Bayly & Childress (1988), leads to a similar estimate for flux growth in models based on a baker's map. We represent the field by a spectrum function $F(k)$ over arbitrary real wave numbers k, and replace the operator T^{SFS} by Q, where $QF(k) = \sqrt{2} F(k/2)$. With

3.3 Generalized Baker's Maps

3.3.1 Variable Stretching

Despite the relatively simplicity of the SFS map, the fast dynamo problem for this map remains difficult since the effect of many iterations of T^{SFS} has no simple analytical representation. Given the discontinuities already present in the baker's maps, it is natural to seek even simpler models which do allow explicit analysis The simplest examples, built from the baker's maps (3.1.3, 4), allow fast dynamo action, and illustrate the effects of variable stretching of field. These generalized baker's maps were introduced by Finn & Ott (1988*a*, *b*, 1990), who describe a number of examples and variants. The maps we consider here may be termed *cut and paste* models, for they freely allow arbitrary combinations of the stretching, cutting, pasting, stacking, and folding inherent in the baker's maps discussed above. They can be considered as being derived from the STF picture (§1.4) and its variants (see Finn & Ott 1988*a*, *b*), as well as from some steady flows (§7.3). In the context of STF, they tell us no more about the *existence* of fast dynamos than does the idealized STF picture, nor do they address its shortcomings; however they are useful in understanding magnetic field structure in fast dynamos and its relation to the underlying chaotic dynamics.

We start as before with the unit square and an initial magnetic field of unity aligned with the x-axis. Let $(x_0 = 0, x_1, x_2, \ldots, x_N = 1)$ be a partition of the lower edge, which divides the square into N vertical strips (see Fig. 3.10). Each strip is then stretched longitudinally until it has unit length, while preserving the area by compression transversely. The new strips are then stacked to reform the unit square. The magnetic field in the now horizontal strips can be reversed during the stacking according to some fixed rule, determined by the orientation factors $\epsilon_i = \pm 1$. If the magnetic field is initially unity everywhere, after one iteration the field strength (by conservation of flux in a perfect conductor) has values ϵ_i/p_i where $p_i = x_i - x_{i-1}$. Iteration of the map builds up, as before, a multilayered structure with field variations on arbitrarily small scales.

As a typical Lagrangian particle is tracked under iteration of this map, it will visit the N layers with probabilities p_i, this being a consequence of the *ergodic* property of baker's maps; see §3.3.2 below. Indeed a particle picked at random will lie in slice m_1 with probability p_{m_1}. Depending upon its location within the m_1th slice, it will next visit slice m_2 with probability p_{m_2}, etc. The temporal average rate of stretching of an x-directed (infinitesimal) vector attached to a typical Lagrangian point can thus be computed as

$$\Lambda_{\text{Liap}} = -\sum_{i=1}^{N} p_i \log p_i. \tag{3.3.1}$$

---

the L^2 norm on $(-\infty, +\infty)$, the operator Q doubles the norm but the flux, $F(0)$, grows by a factor $\sqrt{2}$.

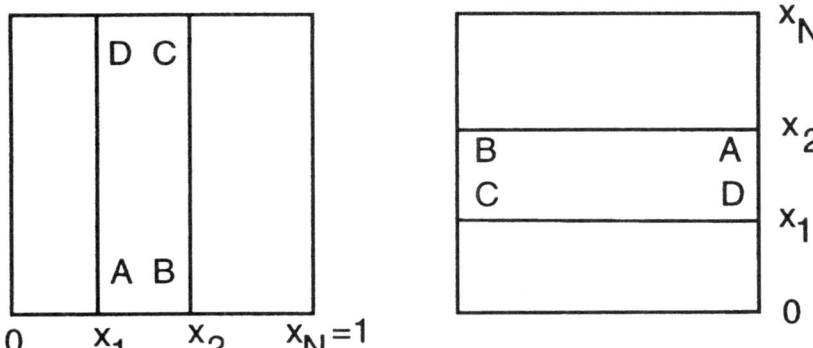

Fig. 3.10 A generalized baker's map of the unit square. N vertical strips, determined by a fixed partition of the lower edge, are stretched to width 1 while preserving area. They are then stacked, with possible reversals (folds), according to a fixed rule. The strip ABCD is reversed.

This Lagrangian average stretching rate is the *Liapunov exponent* of the map, defined in §§1.6, 6.1 (given that the time between applications of the map is again taken as unity).

Not all vectors stretch at this rate, however. Using symbolic dynamics (see §3.3.2) it may be seen that there is a fixed point of the map in each vertical strip of Fig. 3.10. An x-directed vector attached to the fixed point $\mathbf{x}_i$ in strip i will stretch by a factor $1/p_i$ per iteration, yielding $\Lambda_{\text{Liap}}(\mathbf{x}_i) = \log(1/p_i)$. This can be larger or smaller than Λ_{Liap} in (3.3.1). For generalized baker's maps it may be seen that $\Lambda_{\text{Liap}}^{\max} = \max_i \log(1/p_i)$, the maximum being achieved at the fixed point in the strip with the greatest value of $1/p_i$. In contrast, the folded and stacked baker's maps of (3.1.3, 4) have uniform stretching $p_1 = p_2 = 1/2$ and $\Lambda_{\text{Liap}} = \Lambda_{\text{Liap}}^{\max} = \Lambda_{\text{Liap}}(\mathbf{x}) = \log 2$.

Because of the small-scale layering of field with rapid variation in the y-direction, Finn & Ott (1988a, b) adopt the total x-directed magnetic flux through the square as the natural measure of the growth of the field, in other words the total flux through any vertical line $0 \le y \le 1$. This key step introduces the flux conjecture, discussed in §2.5. The supposition is that small diffusion of flux must smooth out the small-scale variations, leaving the average flux (just as the flux is dissipated entirely in the folded map considered above). Since each of the horizontal strips carries the same flux (up to an orientation), we see that flux increases at each step by the factor $\exp(\Gamma_S)$ where

$$\Gamma_S = \log \left| \sum_{i=1}^{N} \epsilon_i \right|. \tag{3.3.2}$$

Note that $\Lambda_{\mathrm{Liap}} = \Gamma_S$ for $p_i = 1/N$ and all ϵ_i of the same sign. Keeping like orientations, Λ_{Liap} will be smaller than Γ_S for any other choice of p_i. Thus non-uniform stretching makes the flux grow faster than the length of a typical vector, a behavior first noted by Bayly (1986). The reason is that measuring a flux implies an average weighted by the magnetic field, and this average emphasizes vectors which stretch unusually quickly. On the other hand we may also have $\Lambda_{\mathrm{Liap}} > \Gamma_S$ if not all ϵ_i have the same sign; that is, the cancellation of flux coming from folding of the field can lower growth below the Liapunov exponent.[28]

We thus have the rule: non-uniform stretching enhances flux, but cancellation reduces it, relative to the Lagrangian stretching of vectors. Both effects can be expected in real flows, and it is thus possible to obtain flux growth rates which are not too different from the Liapunov exponent even when cancellation occurs. Numerical evidence for the flux conjecture is given in Finn & Ott (1990).

Another measure of stretching focuses on the action of the map on an extensive geometrical object, such as a line. We observe that a line $y = y_0$, $0 < x < 1$ is stretched to length N before being cut into N pieces of unit length. The growth rate of line length h_{line} (as opposed to length of a line element or vector) is a global measure of stretching, introduced in §1.6, which can here be identified with the *topological entropy* h_{top} of the map. We have

$$h_{\mathrm{line}} \equiv h_{\mathrm{top}} = \log N \qquad (3.3.3)$$

for this map. The exponent h_{line} is greater than Λ_{Liap}, except for maps with uniform stretching $p_1 = p_2 = \cdots = p_N$. It, like the flux growth rate, tends to be dominated by vectors which stretch atypically fast. Plainly h_{line} is an upper bound on the flux growth rate (3.3.2), with equality when there are no cancellations and all the ϵ_i are the same (Finn & Ott 1988a, b). The flux conjecture would imply that h_{line} bounds the fast dynamo exponent γ_0, as observed for the flows of Chap. 2.

3.3.2 Mixing and Ergodicity

The baker's maps exhibit properties of mixing and ergodicity which will be present in all of our models of fast dynamos, and which are far easier to describe than the chaotic structure of smooth stretching flows. A *mixing* map M has the property that volumes become spread 'uniformly' over the domain $\mathcal{D}$ under repeated application of M. In the case of baker's maps, we follow the image of some subdomain U of the unit square $\mathcal{D}$. The measure $dx\,dy$ of area

---

[28]We emphasize here that flux is computed over the entire unit interval of the map. If sub-intervals are used to compute the flux, the difficulties with the flux conjecture which arise for the folded baker's map (§3.1.1) may also apply here; although the present structures are richer the details have apparently not been worked out.

applies here; we will use $\mu(U)$, the customary symbol for the measure of U, to represent area. The map is *mixing* if

$$\mu(M^n U \cap V) \rightarrow \mu(U)\,\mu(V) \quad \text{as } n \rightarrow \infty \tag{3.3.4a}$$

for any two subdomains U and V. An equivalent property is that for any two square-integrable functions f, g we have (Arnold & Avez 1967)

$$\lim_{n \to \infty} \int_{\mathcal{D}} f(M^n(x,y))\, g(x,y)\, dx\, dy = \int_{\mathcal{D}} f(x,y)\, dx\, dy \int_{\mathcal{D}} g(x,y)\, dx\, dy. \tag{3.3.4b}$$

Equations (3.3.4a) and (3.3.4b) are the same when f, g are the characteristic functions for the sets U, V in (3.3.4a). Since square-integrable functions may be approximated by linear combinations of such characteristic functions, (3.3.4b) for general f, g is a consequence of (3.3.4a).

To establish the mixing property is not a simple matter even for baker's maps, and the proof involves elements of measure theory (see, e.g., Ruelle 1989a). For baker's maps, however, the mixing property is fairly clear intuitively. Any small rectangle will be stretched out in x and stacked in y, in such a way that pieces of the rectangle are carried in thin layers. In this way the rectangle becomes distributed throughout the domain, and similarly for arbitrary subsets of the unit square.

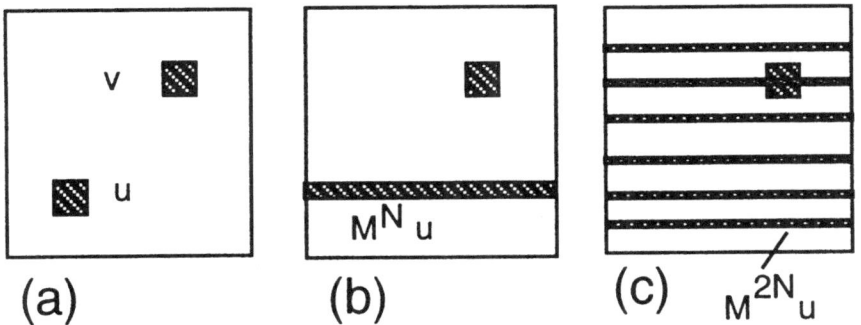

Fig. 3.11 Mixing in the stacked baker's map: (a) two squares of side 2^{-N} initially, (b) one square after N iterations, and (c) after $2N$ iterations.

We can at least make the mixing property plausible for the stacked and folded baker's maps by the following argument. Let us divide the unit square $\mathcal{D}$ into squares of side 2^{-N}. Consider two such squares, U and V (Fig. 3.11a). After N iterations the image $M^N U$ of U will be a thin rectangle, width 1 and height 2^{-2N} (Fig. 3.11b). After a further N iterations the image $M^{2N}U$ will comprise 2^N thin rectangles width 1 and height 2^{-3N}, exactly one of which will overlap V; this is true for either the stacked or folded baker's maps. Thus the area $\mu(M^{2N}U \cap V) = 2^{-4N} = \mu(U)\mu(V)$. Thus (3.3.4$a$) holds exactly at $n = 2N$ and also holds for $n \geq 2N$. Given that (3.3.4a) is correct for small squares, it is plausible that it can be extended to more general regions U and V by approximation.

Another way of viewing mixing is to use *symbolic dynamics* (see, e.g., Guckenheimer & Holmes 1983). We represent points of the square using binary expansions and writing $(x, y) = (0.a_1a_2a_3 \ldots, 0.b_1b_2b_3 \ldots)$, where each digit a_i or b_i is zero or one.[29] Put these two sequences back to back to form a doubly-infinite *symbol sequence*:

$$c = \ldots b_3b_2b_1.a_1a_2a_3 \ldots . \tag{3.3.5}$$

For the stacked baker's map the effect of M is to shift the decimal point one place to the right. For example 11.01 maps into 110.1, meaning that $(0.25, 0.75)$ maps into $(0.5, 0.375)$. Similarly M^{-1} shifts the decimal point to the left.

Points in a square U of side 2^{-N} mentioned above are specified by fixing the $2N$ binary digits $b_N \ldots b_2b_1.a_1a_2 \ldots a_N$ in (3.3.5), the remaining digits a_n, b_n for $n > N$ being arbitrary. A square V can be specified similarly. The action of M is to shift the list of specified digits to the left. Thus the area of $M^nU \cap V$ is the proportion of points that agree both with the specified digits of V and with those of M^nU. For $n \geq 2N$ iterations, the specified digits will not overlap, and the area will be precisely 2^{-4N} as before.

As Arnold & Avez (1967) stress, the representation of M as a shift of doubly-infinite sequences of zeros and ones establishes a formal isomorphism between the stacked baker's map and the shift of the decimal on the sets of outcomes of coin tosses with equal probabilities of $1/2$ (Bernoulli trials). If we define an *orbit* of M as the sequence $(x_n, y_n) = M^n(x, y)$ for $n = 0, \pm1, \pm2, \ldots$ for a point (x, y), then any sequence of coin tosses corresponds under this isomorphism to an orbit of M. In this sense the map is equivalent to a *shift map* on doubly infinite sequences of Bernoulli trials. We remark that a similar but somewhat more awkward correspondence with Bernoulli trials is possible for the folded baker's map given by (3.1.3) (see Klapper 1991). The fold is taken into account by doing a straightforward shift if $a_1 = 0$ in (3.3.5); if $a_1 = 1$, however, we do the shift, then interchange zeros and ones in the *new* $a_1a_2 \ldots$ and $b_2b_3 \ldots$ (but *not* b_1). Mixing very likely holds also in generalized baker's maps of the kind defined in §3.3, although we are not aware of an analysis.

---

[29]There is some ambiguity about this: for example, $0.011111\ldots$ and $0.100000\ldots$ represent the same value of $1/2$. However points with ambiguous binary expansions are exceptional, account only for zero area in the square, and so can be neglected.

This correspondence with random events brings us to the concept of *ergodicity* as it applies to baker's maps. For our purposes we may define an *ergodic map* M on points (x, y) in the unit square as follows. Let $f(x, y)$ be any continuous function defined on the unit square. Then M is ergodic (using the usual area as our measure) provided

$$\lim_{N \to \infty} \frac{1}{N} \sum_{n=0}^{N-1} f(M^n(x, y)) = \int_{\mathcal{D}} f(x, y)\, dx\, dy \qquad (3.3.6)$$

for almost all points (x, y) in the square. It is not difficult to show that any mixing map is also an ergodic map, but not conversely (see Arnold & Avez 1967). The proof rests on the fact that ergodicity is equivalent to the absence of sets invariant under M having area other than 0 or 1. Taking both the subsets U and V in $(3.3.4a)$ to be an invariant set A, the mixing property implies that $\mu(A)$ is 0 or 1.

We thus see that the baker's map not only represents something of a canonical 'stretch–fold' operation in our investigations into fast dynamo action but, because of its features of mixing and ergodicity, it illustrates how a deterministic process can be endowed with features of randomness. That these properties go hand in hand will shape much of our discussion.

3.4 The Slide or SFS1 Map

We now consider a model introduced in the context of generalized baker's maps by Finn *et al.* (1991), and independently by Klapper (1991, 1992c). This is a variant of SFS, with a simpler and analytically more tractable phase shift. The idea is to replace constant shear, or linear vertical velocity, by piecewise constant velocity, with infinite shear on the plane $y = 1/2$, using in (3.1.6),

$$f(y) = \frac{1}{2} \operatorname{sign}(y - 1/2). \qquad (3.4.1)$$

Because of the form of (3.4.1) we call this version of SFS the *slide map*, or the SFS1 map. Following the stretch and fold of the baker's map, we may imagine that the two halves of the cube (separated by the plane $y = 1/2$) are then slid in opposite directions parallel to the z-axis, producing piecewise constant phase factors in the mapping of the magnetic field. The latter is now given by

$$T^{\mathrm{SFS1}} b(y) = 2 \operatorname{sign}(1/2 - y)\, e^{-\pi i \alpha \operatorname{sign}(y - 1/2)}\, b(\tau(y)), \qquad (3.4.2)$$

where $\tau(y)$ is the tent map introduced in (3.2.3), and we have now set $k = 1$. The corresponding adjoint map is

$$S^{\mathrm{SFS1}} b(y) = e^{-\pi i \alpha}\, b(y/2) - e^{\pi i \alpha}\, b(1 - y/2). \qquad (3.4.3)$$

From (3.4.2) or (3.4.3) we observe that the slide map has the property that $\int_0^1 T^{\mathrm{SFSl}}b(y)\,dy = 2i\sin\pi\alpha\int_0^1 b(y)\,dy$. For $1/6 < \alpha \bmod 1 < 5/6$ the flux is increased in modulus and we have the prospect of fast dynamo action. This straightforward calculation of flux growth is the principal advantage of SFSl over the more realistic, but less easily analysed, SFS model.

While the slide map is discontinuous in both vertical and horizontal components, it has the advantage over the SFS map in being soluble explicitly (Klapper 1991, Childress 1992). The details will be given in Chap. 9. We note here only that the eigenvalues λ_k^* of S^{SFSl} form the discrete set

$$\lambda_k^* = 2^{-k}[e^{-\pi i\alpha} - (-1)^k e^{\pi i\alpha}]. \tag{3.4.4}$$

It follows that maximal growth is determined by the first two eigenvalues of T^{SFSl}, $\lambda_0 = 2i\sin\pi\alpha$ and $\lambda_1 = \cos\pi\alpha$, with

$$\Gamma_S = \log\max(2|\sin\pi\alpha|, |\cos\pi\alpha|). \tag{3.4.5}$$

We show this variation in Fig. 3.12. Here the mode crossing, mentioned in §2.6, is explicit.

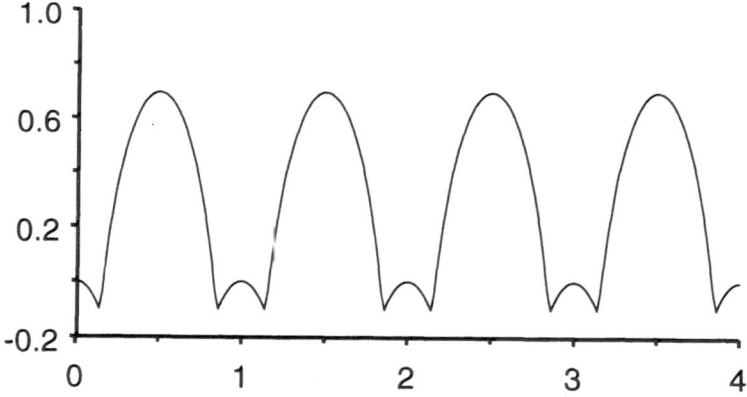

Fig. 3.12 Maximum growth rate Γ_S against α for the SFSl map, as defined by (3.4.5).

The eigenfunctions of S^{SFSl} corresponding to these leading eigenvalues are $b_0 = 1$ and

$$b_1 = \cos\pi\alpha - (2\cos\pi\alpha + 4i\sin\pi\alpha)(y - 1/2). \tag{3.4.6}$$

They are *not* orthogonal using the inner product (3.2.6). If total flux (through $0 \le y \le 1$) is used to evaluate flux growth for the slide map, we compute

$$\left(b_0, \left(T^{\text{SFS1}}\right)^n b\right) = \left(\left(S^{\text{SFS1}}\right)^n b_0, b\right) = \lambda_0^n (b_0, b). \tag{3.4.7}$$

This is fine when flux growth (and fast dynamo action) occurs, but does not give the maximal growth rate near the integer values of α where λ_1 dominates in modulus. Thus total flux is not the appropriate test of growth for all α; an inner product with b_1 is sometimes needed. Alternatively, a flux through a *general* surface $0 < p \leq y \leq q < 1$ would detect the maximal growth rate, since the characteristic function for this subinterval, when expanded in the adjoint eigenfunctions, will have non-zero projections onto both b_0 and b_1. These observations motivate testing for flux growth with the supremum of the inner product with an arbitrary smooth test function, and the supremum over surfaces as in (2.4.6). In §4.1 these ideas are discussed further and used to develop certain definitions of fast dynamo action.

On the other hand the flux conjecture holds in the SFS1 map in the following sense: if $|\lambda_0| > |\lambda_1|$ and we compute total flux in the diffusive problem (basing the value of γ_0 on this), we obtain $\gamma_0 = \Gamma_S = \log|\lambda_0|$. Indeed, if H_ϵ is again the diffusion operator for a period one extension, and $T_\epsilon^{\text{SFS1}} = H_\epsilon T^{\text{SFS1}}$, then we have

$$\left(b_0, \left(H_\epsilon T^{\text{SFS1}}\right)^n b\right) = \left(\left(S^{\text{SFS1}} H_\epsilon\right)^n b_0, b\right) = \lambda_\epsilon^n (b_0, b), \tag{3.4.8a}$$

where

$$\lambda_\epsilon = \exp(-4\pi^2 \epsilon k^2) \lambda_0. \tag{3.4.8b}$$

Here we have used the fact that $H_\epsilon b_0 = \exp(-4\pi^2 \epsilon k^2) b_0$ since b_0 is constant and therefore a solution of the heat equation under periodic extension with period one. The factor comes from the z-dependence of the field; see (3.2.4). We may then take the limit of small ϵ.

The situation is different when α is such that $|\lambda_1| > |\lambda_0|$, when some subtle effects of diffusion arise: growing eigenfunctions orthogonal to b_0 exist, which are not detected when the growth rate γ_0 is based on total flux, as above. We discuss this and other aspects of the diffusive slide operator in §9.6.

We have thus proved that for this particular measure of growth, namely total flux, when $\gamma_0 > 0$ we have $\Gamma_S = \gamma_0$ for the SFS1 map. For the perfect conductor we have the following. Let $c(y)$ be a smooth function with expansion $\sum_k P_k(c) b_k(y)$ in the eigenfunctions of S^{SFS1}; see §9.6. Then, if α is such that $|\lambda_0| > |\lambda_1|$,

$$\left(c, \left(T^{\text{SFS1}}\right)^n b\right) \approx \lambda_0^n P_0^*(c) \int_0^1 b \, dy, \tag{3.4.9}$$

when n is large. A similar formula, with the projection P_1 appearing, applies when $|\lambda_1| > |\lambda_0|$. Although these formulas resemble the mixing property (3.3.4b) we are not dealing here with a measure-preserving map on the fields $b(y)$.

One might think, because of the mixing property of the folded baker's map, that a flux integral of c would result since $(T^{\text{SFSl}})^n b$ would have the same small-scale structure everywhere. In fact the projection, P_0, whose calculation is non-trivial for a non-orthogonal basis of eigenfunctions, is needed.[30] For nearly integer α, the projection which dominates is P_1. There is so much oscillation in the field, evidently, that an $O(1)$ gradient in c can produce non-negligible deviations from a purely mixed state. To illustrate this last property we note that a computation at $\alpha = 0.9$ on 1024 points gives

$$\left|\text{Re}(1, (T^{\text{SFSl}})^{12}1)\right| = 0.0031, \quad \left|\text{Re}(y, (T^{\text{SFSl}})^{12}1)\right| = 0.084. \tag{3.4.10}$$

The first number is just $|2\sin \pi\alpha|^{12}$. For a purely mixed state, the second number should be $1/2$ times the first. Instead it is almost 30 times larger. The large oscillations of the field are systematically biased by the factor y, to produce the enhanced contribution to the integral. We consider these properties of the slide map further in §9.6.

In summary, the slide map provides a good example illustrating the importance of the form of the average on the resulting measure of mean field, while at the same time offering support for the flux conjecture.

3.5 Cat Maps

In the baker's maps and their variants a square domain is mapped onto itself and so represents a localized mixing process, like a highly idealized fluid 'eddy'. Another interesting class of maps is similarly localized, by adopting as the domain the torus $\mathbb{T}^2$. In this setting we obtain perhaps the simplest map capable of exhibiting fast dynamo action, and certainly the first to be exploited mathematically (Arnold *et al.* 1981). This is the well-known *cat map* of Arnold & Avez (1967), a map from the torus $\mathbb{T}^2$ to itself.

To construct a torus we take the plane $\mathbb{R}^2$ and identify all points modulo one:

$$(x, y) \sim (x + 1, y) \sim (x, y + 1). \tag{3.5.1}$$

The space we create with this identification is the two-torus $\mathbb{T}^2$; equivalently this can be thought of as the region $[0, 1] \times [0, 1]$ of $\mathbb{R}^2$ with the opposite edges identified: $(x, 0)$ with $(x, 1)$ and $(0, y)$ with $(1, y)$ (see Fig. 3.13).

Consider the class of linear area-preserving maps

$$M : \mathbf{x} \to \mathbf{Mx} \bmod 1 \quad \mathbf{x} \equiv (x, y), \tag{3.5.2}$$

---

[30]This corrects an error in equation (60) of Childress (1992), where a flux integral appears instead of a projection.

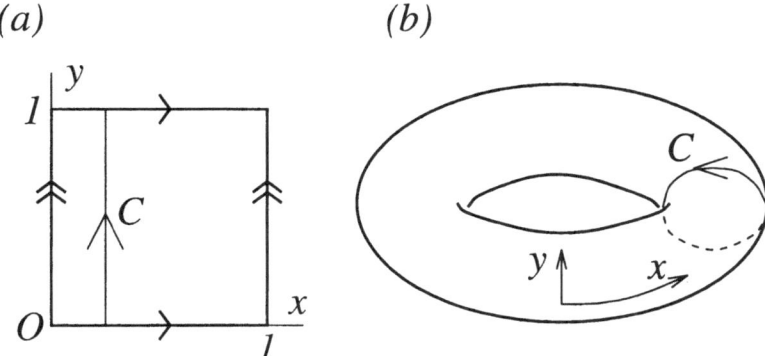

Fig. 3.13 The torus $\mathbb{T}^2$: (a) constructed as $[0,1] \times [0,1]$ with opposite edges identified; (b) after sewing up the edges we obtain the usual picture of a torus.

where $\mathbf{M}$ is a 2×2 matrix. $\mathbf{M}$ must have integer entries to preserve the identification (3.5.1) and so give a map on $\mathbb{T}^2$. We also need $\det \mathbf{M} = 1$, to preserve areas. For example

$$\mathbf{M} = \begin{pmatrix} 2 & 1 \\ 1 & 1 \end{pmatrix} \tag{3.5.3}$$

satisfies these conditions. The well-known picture showing the action of the map is shown in Fig. 3.14. The cat face is stretched out in one direction and compressed in the other, but its area is preserved.

The stretching and compression are related to the normalized eigenvectors $\mathbf{e}_+$ and $\mathbf{e}_-$ of $\mathbf{M}$ given by

$$\mathbf{e}_\pm = (1 \pm \sqrt{5}, 2) \, / \, (10 \pm 2\sqrt{5})^{1/2}; \tag{3.5.4}$$

the corresponding eigenvalues are λ and $1/\lambda$ with

$$\lambda = (3 + \sqrt{5})/2 > 1. \tag{3.5.5}$$

The product of the eigenvalues is unity since $\det \mathbf{M} = 1$. The cat is stretched in the $\mathbf{e}_+$-direction and compressed in the $\mathbf{e}_-$-direction.

Now suppose we have a smooth two-dimensional magnetic field $\mathbf{B} = (B_x(\mathbf{x}), B_y(\mathbf{x}))$ on $\mathbb{T}^2$. The Jacobian of the map M is the matrix $\mathbf{M}$ and so the Cauchy solution for $\epsilon = 0$ is:

$$\mathbf{TB}(\mathbf{x}) = \mathbf{MB}(M^{-1}\mathbf{x}). \tag{3.5.6}$$

This immediately gives us two eigenfunctions

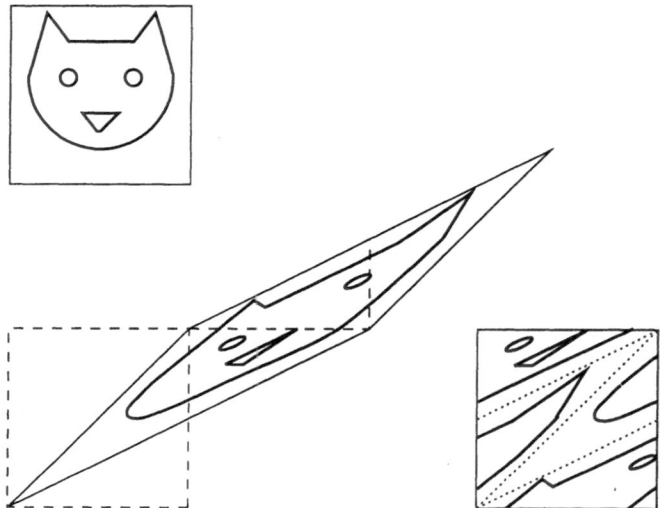

Fig. 3.14 The action of the cat map (3.5.2, 3) on points of $\mathbb{T}^2$. The action is shown on a cat's face.

$$\mathbf{B}_+(\mathbf{x}) = \mathbf{e}_+, \qquad \mathbf{B}_-(\mathbf{x}) = \mathbf{e}_-. \tag{3.5.7}$$

These are constant vector fields in the stretching and contracting directions, which grow and decay, respectively:

$$\mathbf{T}\mathbf{B}_+ = \lambda\mathbf{B}_+, \qquad \mathbf{T}\mathbf{B}_- = \lambda^{-1}\mathbf{B}_-.$$

We therefore have fast growth of the field $\mathbf{B}_+$ when $\epsilon = 0$ at a rate $\Gamma_S = \log \lambda$. For this map the line-stretching exponent and the Liapunov exponent are also $\log \lambda$.

To incorporate diffusion in a pulsed model, we apply the map M at times $t = 1, 2, \ldots$, and allow continuous diffusion ϵ to act in between.[31] Since the eigenfunctions are constant, diffusion has no effect and they remain unchanged during the diffusive step. Therefore we have a fast dynamo; the eigenfunction $\mathbf{B}_+$ has a positive growth rate $\gamma(\epsilon) = \log \lambda$ completely independent of diffusion.

This is a very simple and attractive fast dynamo — the growing eigenfunction has no small-scale structure and diffusion has no effect. However there is a difficulty in applying this model to explain astrophysical phenomena. The cat map cannot arise from a smooth flow on the torus. To see this consider the closed curve C on the torus given by $\mathbf{x} = (1/4, s)$ for $0 \leq s \leq 1$

---

[31]More formally we extend the field $\mathbf{B}_\pm = \mathbf{e}_\pm$ periodically to the whole plane and evolve the field using the diffusion equation in the plane with diffusivity ϵ.

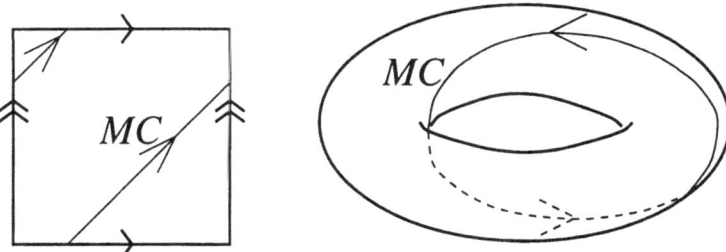

Fig. 3.15 The result of applying the cat map to the curve C in Fig. 3.13.

(Fig. 3.13). This curve goes once round the torus in the y-direction — the 'short way'. After applying the map M we obtain the curve MC given by $\mathbf{x} = (1/2 + s, 1/4 + s) \bmod 1$. This curve, shown in Fig. 3.15, loops once round the torus the 'short way' and once round the 'long way'. Now because C and MC are linked with the torus in different ways, a smooth flow plainly cannot distort one into the other. Thus the map M cannot arise from a smooth flow on $\mathbb{T}^2$.

It turns out that the cat map can be generated as the period-one map of a flow (Arnold *et al.* 1981); however this flow lives, not on $\mathbb{T}^2$, but on a manifold which cannot easily be embedded in ordinary or periodic three-dimensional space (see §6.4). Thus the cat map fast dynamo is rather special and omits many features we would expect in realistic fast dynamos, in particular partial cancellation of flux, complicated eigenfunctions, uneven stretching, and significant effects of diffusion. Nevertheless several other fast dynamo models are based on the cat map and it remains a mathematicians' paradigm of the fast dynamo, much as STF is the physicists' paradigm.

4. Dynamos and Non-dynamos

The purpose of this chapter is to link the discussion and examples of Chaps. 1–3 to the subsequent analysis and modelling. We reexamine a number of topics which were introduced in Chap. 1. Aided by the examples of fast dynamos given in Chaps. 2 and 3, we reconsider in §4.1 the formulation of the fast dynamo problem and the role of diffusion. In §4.2 we prove several anti-dynamo theorems, some of which have already been mentioned and used. These can be used to exclude classes of flows from consideration and to help determine possible fast dynamo mechanisms. Finally in §4.3 we discuss a result of Moffatt & Proctor (1985) concerning the non-existence of smooth eigenfunctions when $\epsilon = 0$.

4.1 Measures of Dynamo Action

4.1.1 Diffusive and Non-diffusive Fast Dynamos

We deal with the dimensionless induction equation,

$$\frac{\partial \mathbf{B}}{\partial t} - \nabla \times (\mathbf{u} \times \mathbf{B}) - \epsilon \nabla^2 \mathbf{B} = 0. \qquad (4.1.1)$$

The velocity $\mathbf{u}$ will be solenoidal. We first recapitulate the meaning of a dynamo in a magnetically diffusive material, as discussed in §1.3. For a given $\epsilon > 0$, the growth rate of magnetic field is

$$\gamma(\epsilon) = \sup_{\mathbf{B}_0} \limsup_{t \to \infty} \frac{1}{2t} \log E_{\mathrm{M}}(t), \qquad (4.1.2)$$

where $E_{\mathrm{M}}(t)$ is the magnetic energy at time t and $\mathbf{B}_0(\mathbf{x})$ the initial condition. It is understood that t denotes time in the case of a flow, but for a map t takes only integral values, corresponding to the iteration number. The parameter ϵ determines the diffusive smoothing of field. For evolution in a smooth flow as described by the induction equation (4.1.1), ϵ multiples the $\nabla^2 \mathbf{B}$ term. This is the customary formulation of a diffusive material and defines diffusion in the formulation of the fast dynamo problem. Our definition implicitly includes pulsed flows as instantaneous, infinitely fast movements separated by intervals of rest when diffusion is active. Random or turbulent flows (or maps) are included as an ensemble of such flows (or maps).

We remark that the smoothing of the field can be introduced in other ways. In computer codes using spectral methods, it is sometimes useful to introduce a fractional diffusion operator of the form $(\nabla^2)^q$, where q is a real number greater than unity, as a replacement for the Laplacian in (4.1.1). An example is *hyperdiffusion* with $q = 2$. The fourth-order operator lacks certain desirable features of the Laplacian, such as a maximum principle, but has the advantage, particularly for computations, of enhancing the decay of high wavenumbers. Another possibility is to use an integral operator S, which smooths over a ball of fixed radius ρ; in three dimensions:

$$S\mathbf{B}(\mathbf{x}, t) = \frac{3}{4\pi\rho^3} \int_{|\mathbf{y}-\mathbf{x}|\leq\rho} \mathbf{B}(\mathbf{y}, t)\, d\mathbf{y}. \qquad (4.1.3)$$

Here expansion of the integrand in a Taylor series gives the Laplacian as the first term of a differential operator of infinite order. Another possibility is to think of diffusion arising from random walks (§6.7).

In view of these various alternatives, it might be argued, bearing in mind the possibility of non-Fickian diffusion of astrophysical magnetic fields, that a theory based upon the perfectly conducting limit should be insensitive to the kind of smoothing. Indeed, such a view is the basis for the flux conjecture, and the search for criteria for fast dynamo action based on the diffusionless operator $\mathbf{T}$. On the other hand there are also situations where the form of diffusion could matter. In particular when diffusive boundary layers play a role in the dynamo (as in the Ponomarenko and modified Roberts dynamos of Chap. 5), more likely than not the nature of the diffusion will affect results such as the growth rate γ_0. Since the effect of smoothing, rather than the way smoothing is done, is the main concern here, we have adopted the classical Fickian diffusion model in (4.1.1) and will limit analysis to this model. This leaves open the possibility of a more general abstract characterization of the smoothing over small-scale magnetic structure.

Although the emphasis here will be on the use of the perfectly conducting limit in the analysis of fast dynamos, it should be understood that the physical effects of diffusion, however slight, are essential. Diffusion is needed to determine $\gamma(\epsilon)$, through which γ_0 is defined. If γ_0 can be recovered assuming a perfect conductor, it will be through some simulation of diffusion, in the sense that fine structure is filtered out, by flux averaging, for example. Such filtering introduces the arrow of time into the perfectly conducting theory, since averaging must be performed *after* the creation of small-scale structure and not before. In addition, topological changes associated with diffusion, principally the creation and destruction of magnetic helicity, will determine the equilibrated structure in any dynamical model; see Chap. 12.

We also remark that it is customary in near-ideal systems to invoke transport by turbulent diffusion, having the property that the effective diffusion coefficient remains finite in the limit of zero molecular diffusion. This would reflect the likely presence of turbulence. Such a procedure eliminates the fast dynamo limit in the average or mean fields. However this replaces the kinematic problem of dynamo action by one of averaging over small-scale dynamics, and the perfectly conducting limit must then apply to the turbulent field. Strauss (1986) has employed this idea to construct a fast dynamo based upon the tearing mode instability of MHD (see, e.g., Biskamp 1993).

The flow or map is a fast dynamo if the fast dynamo growth rate

$$\gamma_0 \equiv \liminf_{\epsilon \to 0} \gamma(\epsilon) > 0. \tag{4.1.4}$$

As pointed out in §1.3, this is a conservative definition; in practice the limit seems to exist. This definition involves the growth rate as a function of ϵ, and therefore an ordering where the limit $t \to \infty$ is followed by the limit $\epsilon \to 0$. These limits do not commute, and it is important also to consider the opposite order: $t \to \infty$ in a perfectly-conducting ($\epsilon = 0$) fluid. From the examples discussed in the preceding chapters, we anticipate that for $\epsilon = 0$ the magnetic field adopts complex small-scale structure at large time. In the limit $t \to \infty$, structure on all scales is realized and a weak measure of field amplitude is needed.

On the basis of the success of the flux conjecture and other weak measures of dynamo action in the examples of Chaps. 2 and 3, we introduce now some formal definitions. Let $\mathbf{f}(\mathbf{x})$ be a complex-valued C^∞ test field defined on $\mathcal{D}$; the choice of $\mathcal{D}$ has been discussed in §§1.2.1, 5. For any complex-valued solution $\mathbf{B}(\mathbf{x}, t)$ of the induction equation with $\epsilon \geq 0$, we define the linear functional on $\mathbf{B}(\mathbf{x}, t)$,

$$L(\mathbf{B}, \mathbf{f}) \equiv L_{\mathbf{f}}\mathbf{B} = \int_{\mathcal{D}} \mathbf{f}^* \cdot \mathbf{B} \, dx. \tag{4.1.5}$$

Then the *perfect growth rate relative to* $\mathbf{f}$ is defined by

$$\Gamma(\mathbf{f}, 0) \equiv \Gamma_{\mathbf{f}} = \sup_{\mathbf{B}_0} \limsup_{t \to \infty} \frac{1}{t} \log |L(\mathbf{B}, \mathbf{f})|, \qquad (\epsilon = 0); \tag{4.1.6}$$

the zero argument of Γ indicates that $\epsilon = 0$. We have formalized the notation introduced in Chap. 2: we use Γ for any measurement of growth of a flux, linear functional, or other linear spatial average.

Let us suppose that $\mathbf{u}$ is a fast dynamo according to (4.1.4). We say that $\mathbf{u}$ is a *non-diffusive* fast dynamo if

$$0 < \gamma_0 \leq \sup_{\mathbf{f}} \Gamma_{\mathbf{f}} \equiv \Gamma_{\mathrm{sup}}. \tag{4.1.7}$$

Otherwise we say that $\mathbf{u}$ is *diffusive*: $\Gamma_{\mathrm{sup}} < \gamma_0$. The idea behind a non-diffusive fast dynamo is that the growth of the eigenfunctions is associated in the limit of vanishing ϵ with a weak limit of the magnetic field. In a diffusive fast dynamo, the growth is dominated by a process which cannot be 'captured' by a single weak limit. Typically this occurs because the largest length-scale associated with the fastest-growing mode decreases when $\epsilon \to 0$, as in the Ponomarenko and modified Roberts dynamos discussed in Chap. 5. The decrease in scale makes the dynamo process diffusive for arbitrarily small ϵ, as the optimally growing eigenfunction shifts toward smaller scales to make use of the vanishing diffusion in rearranging and reconnecting field. Thus in the Ponomarenko dynamo $\Gamma(\mathbf{f}, 0) \leq 0$ for all $\mathbf{f}$. We illustrate the two types schematically in Fig. 4.1.

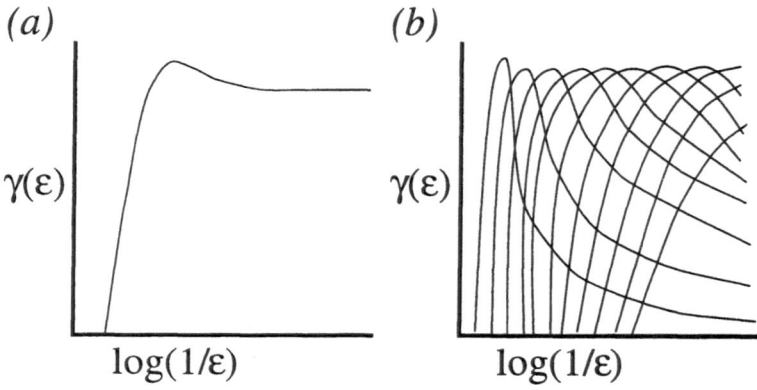

Fig. 4.1 Schematic picture of how a fast dynamo can be achieved in the $\epsilon \to 0$ limit. In plots of $\gamma(\epsilon)$ against $\log(1/\epsilon)$, (a) a single fast-growing mode acquires fine structure as $\epsilon \to 0$, leading to non-diffusive fast dynamo action, and (b) there is a series of mode-crossings, as for example in Ponomarenko's dynamo, leading to a diffusive fast dynamo. As ϵ decreases the length scale associated with the growing mode decreases.

Although we have allowed in (4.1.7) for the possibility that $\Gamma_{\text{sup}} > \gamma_0$, we observe that the statement $\Gamma_{\text{sup}} = \gamma_0$ amounts to a form of the flux conjecture (2.4.6) of Finn & Ott (1988a, b), recast using smooth linear functionals in place of fluxes. We state a variant of this as conjecture 4 below.

If $\mathbf{u}$ has the property that $\Gamma(\mathbf{f}, 0) > 0$ for some $\mathbf{f}$, then $\Gamma_{\text{sup}} > 0$ and we say that $\mathbf{u}$ is a *perfect dynamo*. A natural proposition worth considering is:

Conjecture 1 Every perfect dynamo is a fast dynamo.

Although unproven, this is a reasonable hypothesis. It is weaker than a flux conjecture since it does not require equality between Γ_{sup} and γ_0, but only that if there is flux growth for $\epsilon = 0$, the effect of diffusion will not be so drastic as to make the flow a slow dynamo or non-dynamo for small ϵ. It is clearly *not* true that every perfect dynamo is non-diffusive. The reason is that a flow could contain elements of both kinds of dynamo, diffusive and non-diffusive, with the diffusive component producing the larger growth rate. In this case Γ_{sup} need not be as large as γ_0.

All available examples of fast dynamos that are not perfect dynamos involve discontinuous flows or maps. This suggests the following conjecture.

Conjecture 2 Any smooth, fast dynamo is a perfect dynamo.

We formally define a *smooth* velocity field as one with continuous derivatives of all orders, i.e., belonging to the function space C^∞. It is also likely that any smooth velocity field **u** which is a fast dynamo is in fact a non-diffusive fast dynamo. This, taken with conjecture 1, would imply:

Conjecture 3 Any smooth, perfect dynamo is non-diffusive.

As noted above our generalized form of the flux conjecture is $\Gamma_{\text{sup}} = \gamma_0$. Specifically, we conjecture that equality holds for smooth flows:

Conjecture 4 For any smooth, perfect dynamo, $\Gamma_{\text{sup}} = \gamma_0$.

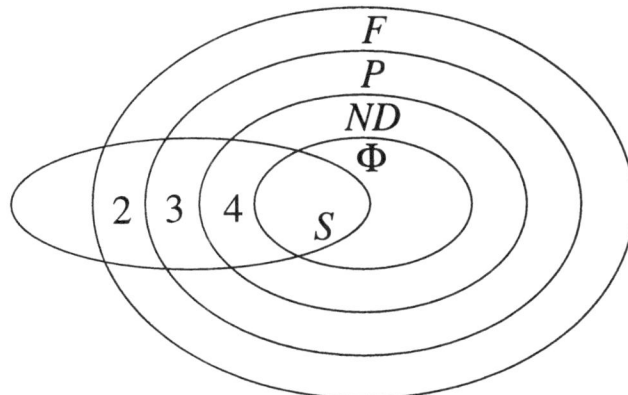

Fig. 4.2 Venn diagram summarizing conjectures; see discussion in the text.

Figure 4.2 shows a Venn diagram summarizing these conjectures. Here F is the set of fast dynamos, and P is the set of perfect dynamos, drawn inside F, assuming that conjecture 1 holds. The set ND is the set of non-diffusive fast dynamos, which lies inside P by definition, and Φ is the set of fast dynamos for which the flux conjecture holds, lying inside ND by definition. Overlapping all of these we have S, the set of smooth flows. If conjecture 2 holds the small region labelled 2 is empty; if 3 holds region 3 is empty, while if conjecture 4 holds both regions 3 and 4 are empty. The examples discussed in this monograph support these conjectures; the position is summarized in table 4.1. We have included a fifth conjecture, of Oseledets (1993), which is discussed in §6.2.

Table 4.1 Table of properties of models, based on the available evidence. 'Fast' means fast dynamo, 'Non-diff.' means non-diffusive fast dynamo and 'Perf.' means perfect dynamo; 'C1–5' refer to conjectures 1–4 of §4.1.1 and conjecture 5 of §6.2. A subscript 'c' in the entry indicates that the evidence is computational; '–' means not applicable. Otherwise the result has been established mathematically, either by rigorous analysis, or by formal asymptotic methods. Note that necessary conditions on a flow or map for Oseledets' conjecture, C5, to hold have not been formulated; for this reason we mark only models which are known to support this conjecture.

Model	Fast	Perf.	Non-diff.	C1	C2	C3	C4	C5
Stacked baker's map	y	y	y	$\checkmark$	–	–	–	$\checkmark$
SFSl map:[a] $1 < \|\lambda_0\|$	y	y	y_c	$\checkmark$	–	–	–	$\checkmark$
SFSl map:[a,b] $\|\lambda_0\| < \|\lambda_1\|$	y_c	n	n	–	–	–	–	
SFS map:[a] $\alpha k > 0.5$	y_c	y_c	y_c	$\checkmark_c$	–	–	–	$\checkmark_c$
SFS map:[a,b] $\alpha k < 0.5$	y_c	n_c	n_c	–	–	–	–	
Cat map	y	y	y	$\checkmark$	$\checkmark$	$\checkmark$	$\checkmark$	$\checkmark$
Cat flow	y	y	y	$\checkmark$	$\checkmark$	$\checkmark$	$\checkmark$	$\checkmark$
Diffeos. of $\mathbb{T}^{2c,d}$	y	y	y	$\checkmark$	$\checkmark$	$\checkmark$	$\checkmark$	$\checkmark$
Cat map with shear[e]	y	y	y	$\checkmark$	$\checkmark$	$\checkmark$	$\checkmark$	$\checkmark_c$
Pulsed Beltrami waves	y_c	y_c	y_c	$\checkmark_c$	$\checkmark_c$	$\checkmark_c$	$\checkmark_c$	
Ponomarenko flow	y	n	n	–	–	–	–	
Modified Roberts flow	y	n	n	–	–	–	–	
ABC 111 flow	?	y_c	?	?	?	?	?	
Kolmogorov flow	y_c	y_c	y_c	$\checkmark_c$	$\checkmark_c$	$\checkmark_c$	?	

[a] Note that for dynamo action in the SFS and SFSl maps, the shear α is fixed while the magnetic wavenumber k is free.
[b] For some subset of αk.
[c] When based on a hyperbolic matrix.
[d] C5 proven when the phase space is entirely chaotic.
[e] Established in the limit of strong stretching for suitable matrix and shear.

A theorem of Oseledets and Vishik (see §6.6) states that stretching somewhere in the flow, $\Lambda_{\text{Liap}}^{\max} > 0$, is a necessary feature of any smooth, steady fast dynamo. The stronger result of Vishik, Klapper and Young (§§6.1, 7) requires a positive topological entropy h_{top}. The indication is that smooth non-diffusive fast dynamos result from stretching over a region of finite volume, that is, a region of Lagrangian chaos which fills a substantial part of $\mathcal{D}$.

Diffusive fast dynamos should occupy a special niche in any theory of fast dynamo action.[32] The only examples of diffusive dynamos we have involve discontinuities in an essential way, either to stretch the field, as in the Ponomarenko and modified Roberts dynamos (§§5.4.1, 5.5.2), or else to rearrange it, as in SFS1 (§9.6.2). On the other hand, it is natural in a near-perfect conductor to allow flows of an *ideal fluid*, in particular an inviscid fluid, where discontinuities in the velocity field are allowed. Also in the Sun the fluid Reynolds number is much larger than the magnetic Reynolds number, and so one cannot rule out the existence of shear layers with width much smaller than the scale of the magnetic field, effectively a discontinuous flow field. The mixed flows mentioned above, where Lagrangian chaos coexists with diffusive dynamo action, represent the most general case, although the focus of most of our work is the chaotic component.

4.1.2 The Density of Fast Dynamos

An interesting question concerns the 'density' of fast dynamos among, say, flows with finite kinetic energy (L^2 velocity fields). A result of G.O. Roberts (1970) states that almost all spatially-periodic flows are (at least slow) dynamos of a certain class for almost all ϵ. In this result the admissible magnetic fields have the Bloch wave structure $\mathbf{B} = e^{i\mathbf{n}\cdot\mathbf{x}}\mathbf{b}(\mathbf{x}, t)$ where $\mathbf{b}$ is 2π-periodic on $\mathbb{T}^3$ and $\mathbf{n}$ has components in the interval $[0, 1]$. Roberts' result follows from the Fourier mode structure of the velocity field, and the fact that the condition for dynamo action may be reduced to a single scalar condition on the set of Fourier coefficients determining $\mathbf{u}$. This condition implies that non-dynamos occupy at most a hypersurface in the infinite-dimensional space of Fourier coefficients. Roberts' analysis simultaneously determines magnetic structure (in the context of first-order smoothing; see Ghil & Childress 1987) in terms of the velocity field.

Is it true that in three dimensions almost all fluid motions, in a similar way, are fast dynamos? This conjecture was advanced by Finn & Ott (1988a, b); see also Finn *et al.* (1989, 1991). Because of the close link between Lagrangian chaos and fast dynamo action, it is in part a statement about the extent of chaos in almost all fluid flows, and in part an assertion that most chaotic flows support fast dynamo action. Evidence for the conjecture may be found in, for example, the SFS map. With the notation of §3.2, set α to fix the form of the map, and take the physical magnetic field as a sum of modes with different values of k. For any non-zero α, there is a mode with, say, $k = 1/\alpha$, which is destabilized for ϵ sufficiently small. Thus all the SFS maps are fast dynamos except the special case $\alpha = 0$, when SFS reduces to the planar folded baker's map.

---

[32]The fast dynamos we classify as diffusive are termed 'intermediate' by Molchanov, Ruzmaikin & Sokoloff (1985). We prefer the former adjective as suggestive of the source of fast dynamo action in small-scale diffusive structures of the magnetic field.

A proof of any conjecture about the density of fast dynamos might involve, like Roberts' result, a complete characterization of magnetic structure in terms of the flow, to the extent needed to establish the positivity of γ_0. Alternatively, a proof might recognize that the perfect cancellation of flux seen in planar flows is very unusual in three dimensions, and that any slight perturbation is likely to allow partial constructive folding and amplification for ϵ sufficiently small. In other words the set of flows that have perfect cancellation may be in some sense sparse in the set of three-dimensional flows.

4.1.3 Perfect Eigenfunctions

Given the earlier discussion of flux growth rates, it is natural also to consider the possibility of defining a 'perfect' eigenfunction $\mathbf{d}(\mathbf{x})$ for $\epsilon = 0$. However a perfect eigenfunction is unlikely to be smooth or to possess finite energy, except in special cases where the stretching is entirely uniform, such as the cat map or stacked baker's map. Such an eigenfunction can typically exist only as a 'generalized function' and, in view of the use of smooth linear functionals in defining growth rates for $\epsilon = 0$, one might conjecture that $\mathbf{d}(\mathbf{x})$ is a distribution that can be integrated against a smooth test function. It would then satisfy the induction equation in a weak sense, that is, when that equation is integrated against a test function.[33]

Given this it is natural to conjecture that evolution with $\epsilon = 0$ from general initial conditions would approach $\mathbf{d}(\mathbf{x})$ in a weak sense, and that eigenfunctions with $0 < \epsilon \ll 1$ could be obtained from $\mathbf{d}(\mathbf{x})$ by smoothing, over scales of order $\sqrt{\epsilon}$ in the chaotic regions and over larger scales near to integrable islands. These two statements would imply that an eigenfunction with diffusion can be obtained by allowing the field to evolve with $\epsilon = 0$ for sufficiently long, and then smoothing (cf. discussion in Du & Ott 1993a, b).

Distributional eigenfunctions are a clear implication of the various models and numerical experiments. However we shall not attempt here to formulate a precise conjecture because there are few models for which the structure of the eigenfunction is clear, except where the eigenfunctions are trivial, such as in stacked baker's and cat maps. An exception is Oseledets' study of diffeomorphisms of $\mathbb{T}^2$ for which essentially all of the theorems contemplated here can be formulated and proved; see §10.1. Also some properties of such eigenfunctions may be established for the cat and pA maps with shear in §10.4. For pulsed Beltrami waves asymptotic properties of the perfect eigenfunctions are obtained by Soward (1993a, b, 1994a) and are discussed in §10.2.

Another difficulty arises from eigenfunction degeneracy. We have seen degeneracy in the SFSI map: for a given α, modes with different values of k have

---

[33]In terms of the discussion of §9.1.1 the corresponding eigenvalue is likely to be in the approximate spectrum of $\mathbf{T}$ in L^2, and the corresponding sequence of near-eigenfunctions in L^2 would tend to the distribution $\mathbf{d}$ in a weak sense (de la Llave 1993).

the same $\epsilon = 0$ growth rate — there is no single diffusionless eigenfunction. For SFS1 and SFS, such degeneracy can be removed by fixing the wavenumber k. However degeneracy (or near-degeneracy) for $\epsilon = 0$ appears to exist in steady three-dimensional flows and there is no obvious way to factor this out; see §7.3.

4.1.4 Fluxes versus Functionals

For simplicity the above discussion of eigenvalues and eigenfunctions is in terms of C^∞ test functions, but we should stress that we do not know the appropriate setting for a future general theory. Indeed for diffeomorphisms of $\mathbb{T}^2$ in §10.1, C^1 functions are sufficient, while in §10.4 we find ourselves using a space of functions analytic in a strip about the real axis. Also much of the discussion of the flux conjecture in Chaps. 2 and 3 is on the basis of fluxes, which do not fall within the scope of C^∞ linear functionals (4.1.5), although they can be considered a limit of such functionals. Fluxes fall naturally within a larger space of *currents* (see, e.g., Morgan 1988) which, with additional structure, has been used by M.M. Vishik (personal communication) to obtain the fast dynamo upper bound of h_{top}.

The question of whether fluxes or smooth functionals are used to measure growth for $\epsilon = 0$ becomes delicate when slow or non-dynamos, $\gamma_0 \leq 0$, are considered. Consider a planar flow $\mathbf{u}$ and planar field $\mathbf{B} = (A_y, -A_x)$, and measure flux Φ_C through a given fixed curve C in the plane connecting $\mathbf{x}_0$ to $\mathbf{x}_1$, say. For $\epsilon = 0$,

$$
\Phi_C(t) = \int_C \mathbf{B} \cdot \mathbf{n}\, ds = A(\mathbf{x}_1, t) - A(\mathbf{x}_0, t)
$$
$$
= A_0(M^{-t}\mathbf{x}_1) - A_0(M^{-t}\mathbf{x}_0),
$$

(4.1.8)

where $A_0(\mathbf{x})$ is the initial vector potential. Plainly for a general curve C and choice of A_0, Φ_C will not tend to zero for large times and so its growth rate, $\Gamma_C = 0$. However for $\epsilon > 0$ the flux $\Phi_C(t)$ must decay, with $\gamma(\epsilon) < 0$ and $\gamma_0 \leq 0$. In cases where there is good mixing in the flow, with $\gamma_0 < 0$ and fast decay of field in the $\epsilon \to 0$ limit, $\Gamma_C > \gamma_0$ and the flux conjecture in terms of fluxes (2.4.6) is, strictly, incorrect. Note however that this example does not contradict conjecture 4 since $\Gamma_{\text{sup}} > 0$ is required for a perfect dynamo.

This failure is essentially an edge-effect which we have already met for the folded baker's map in §3.1, and is associated with the particularly large level of fluctuations relative to the mean field when folding is not constructive. Despite this failure, one would not accidentally mistake a planar flow for a fast dynamo. For a smooth fast dynamo, the fluctuations are weaker, and we expect conjecture 4 to hold when fluxes (rather than smooth functionals) are used to define Γ_{sup}.

This difficulty probably disappears when a smooth functional is used to measure growth in planar flow, since we can write

$$L_f \mathbf{B} = \int (A_y f_1^* - A_x f_2^*)\, d\mathbf{x} = \int A(\partial_x f_2^* - \partial_y f_1^*)\, d\mathbf{x}, \qquad (4.1.9)$$

and if the flow is *mixing* in the support of $\mathbf{f}$ (see §3.3.2), then in the limit $t \to \infty$, $L_f(t)$ decays for $\epsilon = 0$, since

$$L_f \mathbf{B} \to \left(\int A\, d\mathbf{x} \right) \cdot \left(\int (\partial_x f_2^* - \partial_y f_1^*)\, d\mathbf{x} \right) = \left(\int A\, d\mathbf{x} \right) \cdot 0 = 0. \quad (4.1.10)$$

Nevertheless the equality of the decay rate Γ_{sup} and γ_0 has not been proved in this case, and any results here would be very interesting.

4.1.5 Dynamo Action in Other Norms

Because of the build-up of small-scale magnetic structure at small ϵ when magnetic field is evolved from an initial condition, the growth rate of different norms of $\mathbf{B}$, as well as the functionals $L(\mathbf{B}, \mathbf{f})$ defined by (4.1.5) will in general be different, until such time that diffusion becomes effective in smoothing over the small scales. This situation can be illustrated by comparing the growth of logarithms of the various measures on the same graph (Finn & Ott 1990). We sketch this in Fig. 4.3; see also Fig. 7.1.

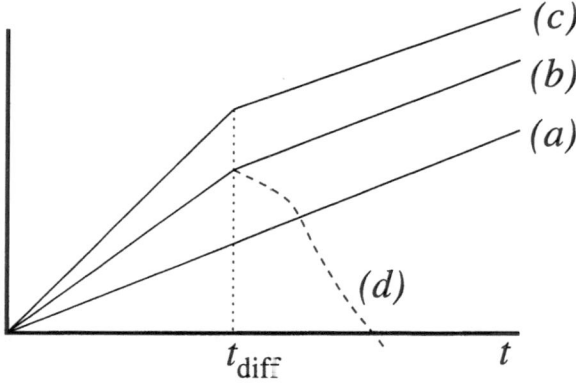

Fig. 4.3 Qualitative sketch of kinematic growth of magnetic field under various norms whose logarithms are plotted against time. The ordinates are (a) $L(\mathbf{B}, \mathbf{f})$, assuming the flux conjecture, (b) $\|\mathbf{B}\|$, the energy norm, and (c) $\|\mathbf{B}\|_1$, the first derivative or current norm (see §9.2.1). The time t_{diff} marks the onset of diffusion of the smallest scale magnetic structures. The dashed line (d) indicates the subsequent fall-off of energy in a non-dynamo.

Here we take $t_{\text{diff}}(\epsilon)$ to be a time when diffusive effects are realized. In order to verify dynamo action, growth must be measured for $t > t_{\text{diff}}(\epsilon)$. In practice there are limits on computation time and it is an interesting problem to determine if dynamo activity exists when there is limited information (Cattaneo, Hughes & Weiss 1991). This question is particularly important for dynamically self-consistent fast dynamos, since it bears on the observations of astrophysical magnetic fields; see Chap. 12.

4.2 Anti-dynamo Theorems

4.2.1 Formulation

Theorems concerned with the non-existence of dynamo action under various conditions have highlighted developments in the field beginning with the seminal work of Cowling (1934). Cowling argued on the basis of field line topology that dynamos having axisymmetric field and flow are impossible. The argument was strengthened by Braginsky (1964), Roberts (1967), and James, Roberts & Winch (1980). Ivers (1984) has contributed a comprehensive review and critique of the subject, as well as a number of improvements, extensions, and corrections to literature; see also Ivers & James (1984) and Kaiser, Schmitt & Busse (1994). Suffice it to say that, while the overall conclusions of the anti-dynamo literature are essentially correct, the supporting arguments have often been incomplete. The analysis is difficult, parallel to that of the global stability theorems of hydrodynamics, but applied to more complicated velocity fields, especially to toroidal and poloidal motions contained within a sphere. In §4.2.2 we give as an example a non-dynamo result for planar flows, and indicate in that case the mathematical difficulties that arise. In all of these results $\epsilon > 0$; an example of an anti-dynamo theorem for a perfect spherical conductor is discussed in §4.2.3.

We consider three-dimensional space with periodic boundary conditions, period unity in x, y and z, applied to both $\mathbf{u}$ and $\mathbf{B}$; this is equivalent to working on the three-torus, $\mathbb{T}^3$. As usual the flow is either steady or periodic in time. We take $\epsilon > 0$, and consider the initial value problem. We assume the magnetic field is dominated ultimately by the fastest growing or least quickly decaying eigenfunction, defined in the Floquet sense for time-periodic $\mathbf{u}$. There may in fact be a finitely degenerate set of eigenfunctions, particularly when the flow has symmetries. More important is the fact that the eigenfunctions do not generally form a mutually orthogonal set. This is a result of the non-normality of the induction operator; see Chap. 9. Although eigenfunctions form a complete set for steady $\mathbf{u}$, analogous results for time-periodic flows are lacking; see §9.2.2.

If we integrate the induction equation (4.1.1) over one periodic box $\mathcal{D}$, we obtain

$$<\mathbf{B}> = \int_{\mathcal{D}} \mathbf{B} \, d\mathbf{x} = \text{const.} \qquad (4.2.1)$$

The mean field neither decays nor grows, regardless of diffusion or the flow $\mathbf{u}$, and so the eigenvalue 0 is in the spectrum and triply degenerate (Arnold *et al.* 1982). Now (4.2.1) implies that a growing, or decaying, eigenfunction must have zero mean field, and therefore that we can set $<\mathbf{B}> = 0$ in our search for fast dynamos in spatially periodic flows. In any case the property that the mean field can be a non-zero constant is an unphysical feature of dynamos in periodic space and does not apply to more realistic geometry, such as a conducting sphere surrounded by vacuum.

Note that for small ϵ an evolving magnetic field will have small-scale structure, but it is important to realise that the integral in (4.2.1), over a periodicity box of $\mathbf{u}$ and $\mathbf{B}$, covers large-scale fields as well. There can be average field on scales which are 'large' but within the periodicity box. In effect the condition of zero average isolates the box from surrounding space. A non-zero average might represent the remnant 'seed' flux connecting two otherwise isolated systems, accompanied by large fields of zero average developed by dynamo action in each system. In more realistic geometry, for example a spherical conductor surrounded by non-conducting free space, the system is isolated by eliminating field at infinity.

Let us temporarily relax the condition $\nabla \cdot \mathbf{B} = 0$ and consider how $\nabla \cdot \mathbf{B}$ would evolve; indeed in a numerical simulation $\nabla \cdot \mathbf{B}$ might only be approximately zero. Take the divergence of the induction equation, in the form (4.1.1), to obtain

$$\partial_t \nabla \cdot \mathbf{B} = \epsilon \nabla^2 \nabla \cdot \mathbf{B}, \tag{4.2.2}$$

in which case $\nabla \cdot \mathbf{B}$ obeys a diffusion equation and decays away. If, however, we take the divergence of the induction equation in the form which replaces the second term in (4.1.1) by $\mathbf{u} \cdot \nabla \mathbf{B} - \mathbf{B} \cdot \nabla \mathbf{u}$ we obtain a slightly different equation

$$\partial_t \nabla \cdot \mathbf{B} + \mathbf{u} \cdot \nabla \nabla \cdot \mathbf{B} = \epsilon \nabla^2 \nabla \cdot \mathbf{B}, \tag{4.2.3}$$

which states that $\nabla \cdot \mathbf{B}$ obeys the advection–diffusion equation of a passive scalar, and again decays away for $\epsilon > 0$. Whichever form is used, the result is the same: growing eigenfunctions must have $\nabla \cdot \mathbf{B} = 0$. It also means this condition need not generally be imposed in computer codes, since $\nabla \cdot \mathbf{B}$ will tend to saturate at a low level when source terms (from inaccuracies in the discretization of the induction equation) balance the diffusion terms. Surprisingly the form (4.2.3) allows $\nabla \cdot \mathbf{B}$ to grow when $\mathbf{B}$ is a *weak solution* of the induction equation for $\epsilon = 0$ (Bayly 1992*b*), and even when $\epsilon > 0$ there can be significant transient growth in a norm that emphasizes derivatives of $\nabla \cdot \mathbf{B}$. (This follows from the fact that (4.2.3) is in general non-normal, whereas (4.2.2) is normal: see Chap. 9.) Rigorous results for Anosov maps are given by de la Llave (1993).

For the rest of this chapter we assume that we have zero mean field and $\nabla \cdot \mathbf{B} = 0$. It is convenient to use a vector potential $\mathbf{A}$ for the magnetic field $\mathbf{B} = \nabla \times \mathbf{A}$, introduced in §1.2.3. For a spatially periodic magnetic field having zero mean the potential is periodic in x, y and z. Uncurling the induction equation (4.1.1) yields:

$$\partial_t \mathbf{A} = \mathbf{u} \times (\nabla \times \mathbf{A}) + \epsilon \nabla^2 \mathbf{A} - \nabla \phi, \tag{4.2.4}$$

that is,

$$\partial_t \mathbf{A} + \mathbf{u} \cdot \nabla \mathbf{A} = u_i \nabla A_i + \epsilon \nabla^2 \mathbf{A} - \nabla \phi. \tag{4.2.5a}$$

The appearance of the scalar field ϕ corresponds to the gauge freedom to change $\mathbf{A} \rightarrow \mathbf{A} + \nabla \chi$ for any $\chi(\mathbf{x}, t)$. We can use this freedom to set $\phi = 0$, since this is equivalent to introducing a gauge and then solving $\chi_t - \epsilon \nabla^2 \chi = -\phi$. Note that once the gauge is fixed in this way, $\nabla \cdot \mathbf{A}$ cannot be specified independently.

We observe at this point that with the gauge $\phi = \mathbf{u} \cdot \mathbf{A}$, equation (4.2.5a) becomes

$$\partial_t \mathbf{A} + \mathbf{u} \cdot \nabla \mathbf{A} + A_i \nabla u_i = \epsilon \nabla^2 \mathbf{A}. \tag{4.2.5b}$$

For $\epsilon = 0$ this is the equation for an area element in the flow and has a solution analogous to the Cauchy solution (1.2.17) of the induction equation:

$$\mathbf{A}(\mathbf{x}(\mathbf{a}, t), t) = [\mathbf{J}(\mathbf{a}, t)^{-1}]^{\mathrm{T}} \mathbf{A}(\mathbf{a}, 0) \tag{4.2.6}$$

(reflecting the fact that $\mathbf{A}$ can be considered a one-form Lie-dragged in the flow; see, e.g., Schutz (1980)). It is then easily seen from (1.2.17) that in this gauge the magnetic helicity density $\mathbf{A} \cdot \mathbf{B}$ appearing in (1.2.23) is *materially conserved*:

$$(\mathbf{A} \cdot \mathbf{B})(\mathbf{x}, t) = (\mathbf{A} \cdot \mathbf{B})(\mathbf{a}, 0). \tag{4.2.7}$$

With this gauge $\mathbf{A}$ evolves like an area element, and so $\mathbf{A} \cdot \mathbf{B}$ is essentially a conserved flux. This result, which accounts for the invariance of magnetic helicity in a periodic box, will be used again.

4.2.2 Planar Flow

As an example of an anti-dynamo theorem, we now prove the impossibility of dynamo action in planar flow (Cowling 1957b, Zeldovich 1957, Zeldovich & Ruzmaikin 1980). We take a velocity field of the form $\mathbf{u} = (u_x, u_y, 0)(x, y, z, t)$ in periodic space, with $\mathbf{u}$ either steady or periodic in time. We will assume the magnetic field takes the form of a growing eigenfunction: this is sufficient for steady flows, but in the periodic case assumes a complete set of eigenfunctions (see §9.2). This assumption may be relaxed; see Ivers (1984) and Ivers & James (1984). For a non-dynamo theorem it is then sufficient to consider only eigenfunctions and eliminate all but decaying modes. Consider the component B_z which, from the induction equation in planar flow, obeys

$$\partial_t B_z + \mathbf{u} \cdot \nabla B_z = \epsilon \nabla^2 B_z; \tag{4.2.8}$$

this is a scalar advection–diffusion equation of a standard form. Multiplying (4.2.8) by B_z and integrating over $\mathcal{D}$, we obtain

$$\partial_t \int_{\mathcal{D}} B_z^2 / 2 \, d\mathbf{x} = -\epsilon \int_{\mathcal{D}} |\nabla B_z|^2 \, d\mathbf{x}. \tag{4.2.9}$$

But for this spatially periodic function with zero mean we have

$$\int_{\mathcal{D}} |\nabla B_z|^2 \, d\mathbf{x} \geq 4\pi^2 \int_{\mathcal{D}} B_z^2 \, d\mathbf{x}, \tag{4.2.10}$$

and so B_z decays to zero exponentially. Since we are here considering a growing eigenfunction, the only possible value of B_z is zero. Suppose now that we nevertheless have a non-decaying eigenfunction. The eigenfunction can then be written as $\mathbf{B} = (B_x, B_y, 0) = \nabla \times (0, 0, A)$, where $A(x, y, z, t)$ is periodic in space (as there is zero mean field). Now from (4.2.5) with $\phi = 0$ we see that A satisfies the same advection–diffusion equation (4.2.8) as does B_z. Repeating the argument just given, we find that A too will decay exponentially, which is a contradiction. Thus all eigenfunctions must decay — the flow cannot be a dynamo. A trivial corollary is that a two-dimensional planar flow $\mathbf{u} = (u_x, u_y, 0)(x, y, t)$ cannot amplify a similar two-dimensional planar magnetic field $\mathbf{B} = (B_x, B_y, 0)(x, y, t)$.

While flows with $u_z = 0$ cannot be dynamos, a two-dimensional flow of the form $\mathbf{u} = (u_x, u_y, u_z)(x, y, t)$ can amplify field, but the field $\mathbf{B}$ must have a non-trivial dependence on x, y and z. If $\mathbf{B} = \mathbf{B}(x, y, t)$, independent of z, amplification is impossible: in this case B_x and B_y obey

$$(\partial_t + (u_x, u_y) \cdot \nabla)(B_x, B_y) = (B_x, B_y) \cdot \nabla(u_x, u_y) + \epsilon \nabla^2 (B_x, B_y) \tag{4.2.11}$$

from (4.1.1). Since $\partial_x B_x + \partial_y B_y = 0$, we obtain the equations for the planar field $(B_x, B_y, 0)$ in the planar flow $(u_x, u_y, 0)$. Amplification of this field is impossible, as discussed above, and so any non-decaying eigenfunction must have $B_x = B_y \equiv 0$. Then by (4.1.1) the equation for B_z is

$$(\partial_t + \mathbf{u} \cdot \nabla)B_z = \epsilon \nabla^2 B_z \tag{4.2.12}$$

and so B_z must decay too. There are no growing solutions.

It is in fact true that these results also hold without the restriction to eigenfunctions, for a large class of planar flows with kinetic energy bounded in time, with growth measured by (1.3.2) from any solution of the initial-value problem (Ivers 1984, Ivers & James 1984). In this case, however, the decay of B_z or A in the L^2 norm, implied by (4.2.9) for example, does not directly imply *pointwise* decay. A more delicate argument is therefore needed, using the maximum principle for parabolic differential equations. This is not just a technicality, for it correctly points to the very extensive transient growth of energy which can occur (in the x- and y-components of the field), even in a non-dynamo; see Chap. 9.

For the models discussed in Chaps. 2 and 3 involving motions of the form $(u_x, u_y, u_z)(x, y, t)$, we have taken magnetic fields of the form $\mathbf{B} = \mathbf{b}(x, y, t) \exp(ikz)$ for some $k \neq 0$. This is probably the simplest class of models which break the constraints embodied in the anti-dynamo theorems.

Flows with relatively simple Lagrangian structure, in particular integrable flows, can be dynamos. Generally, however, the structure of the magnetic field is more complicated than that of the flow, and so a number of anti-dynamo theorems flow easily from constraints *on the magnetic field*, which are *ad hoc* assumptions about the form of the optimally growing eigenfunction. An example is Cowling's theorem as mentioned above, which requires that the magnetic field as well as the flow be axisymmetric. The slow dynamos obtained from axisymmetric flows always generate non-axisymmetric fields. A useful distinction is between anti-dynamo results that place conditions on the magnetic field, and those (such as the result for planar flow) that do not.

When we demand fast dynamo action, and stipulate that it be non-diffusive, the situation changes dramatically. Dynamo action is then closely associated with Lagrangian chaos, and it is thus the complexity of the Lagrangian structure of the flow that is striking. The theorem of Oseledets (1984) and Vishik (1989) mentioned earlier (see §6.6) states that if no initial infinitesimal vector is stretched exponentially, then the flow is not a fast dynamo. A related and immediate result is an anti-*perfect*-dynamo theorem. It is necessary for perfect dynamo action that there be a test function $\mathbf{f}$ such that, with $\epsilon = 0$,

$$|L(\mathbf{f}, \mathbf{B})| = \left| \int_{\mathcal{D}} \mathbf{f}^* \cdot \mathbf{B} \, d\mathbf{x} \right| \tag{4.2.13}$$

grows exponentially in time for some initial field. But by the Cauchy solution (1.2.17), we have $|L(\mathbf{f}, \mathbf{B})| \leq CM(t)$ where C is a positive constant and

$$M(t) = \sup_{\mathbf{a}} \max_{\{\mathbf{e} : |\mathbf{e}| = 1\}} |\mathbf{J}(\mathbf{a}, t)\mathbf{e}|. \tag{4.2.14}$$

Thus

$$\Gamma_{\mathbf{f}} \leq \lim_{t \to \infty} \sup_{\mathbf{a}} \max_{\{\mathbf{e} : |\mathbf{e}| = 1\}} |\mathbf{J}(\mathbf{a}, t)\mathbf{e}| \tag{4.2.15}$$

and the right-hand side must be positive for a perfect dynamo.

This is similar to but not identical to the condition in the Oseledets–Vishik theorem, the latter being that the maximum Liapunov exponent $\Lambda_{\mathrm{Liap}}^{\mathrm{max}}$ must be positive. Recall that $\Lambda_{\mathrm{Liap}}^{\mathrm{max}}$ is determined by taking a supremum over $\mathbf{a}$ *after* the limit in t, which is the opposite order to that in (4.2.15). It is not immediately obvious that this interchange of limits is possible, but this has in fact been established by Vishik (1988) using a compactness argument in a bounded domain (see also §6.6). Therefore $\Lambda_{\mathrm{Liap}}^{\max} > 0$ is a necessary condition for a perfect dynamo, as well as for a fast dynamo, in a bounded region.

4.2.3 Exterior Fields in the Perfectly Conducting Limit

The perfectly conducting limit has implications for the magnetic field generated *outside* a bounded fluid body; essentially as $\epsilon \to 0$ less and less field can escape. We consider here a result of Bondi & Gold (1950) concerning the magnetic field exterior to a body of perfectly conducting fluid $\mathcal{D}$, and its evolution under motions within $\mathcal{D}$. If the fluid is surrounded by vacuum, in the region $\hat{\mathcal{D}}$, and the magnetic field in the vacuum originates within the conductor, we may follow field lines back to the boundary $\partial \mathcal{D}$ and consider the normal component B_{n} of the field there. Note that for non-simply-connected bodies such as a torus, one can have closed field lines which are 'trapped' in the vacuum, for example threading the hole in a torus, and do not connect to the boundary. For field which is not trapped in this way, the following argument should apply irrespective of the topology of $\mathcal{D}$.

Assuming that B_{n} is a smooth function on $\partial \mathcal{D}$, the normal flux

$$\Phi_{\mathrm{n}} = \int_S B_{\mathrm{n}} \, dS \qquad\qquad (4.2.16)$$

though any subset S of $\partial \mathcal{D}$ frozen in the fluid flow will be an invariant of motions of $\mathcal{D}$. This follows from the frozen flux property of a perfect conductor, applied to the tangential flow on $\partial \mathcal{D}$. This holds in particular for subsets where B_{n} has one sign. By summing the absolute values of the fluxes of such subsets, it follows that the *pole strength*

$$F_{\mathrm{n}} = \int_{\partial \mathcal{D}} |B_{\mathrm{n}}| \, dS \qquad\qquad (4.2.17)$$

is an invariant of the motion. Thus field lines initially confined to $\mathcal{D}$ can never emerge into $\hat{\mathcal{D}}$.

Moffatt (1978) notes that the above observation of Bondi and Gold has a direct implication on the maximum dipole moment which can be developed by a perfectly conducting fluid sphere $\mathcal{D}$. Let the the radius of the sphere be r_0. In the vacuum the component of $\mathbf{B}$ that decays most slowly in space is the dipole:

$$\mathbf{B} = -(\mathbf{m} \cdot \nabla)(\mathbf{x}/r^3) + O(r^{-4}), \qquad (r \to \infty). \qquad (4.2.18)$$

Here $\mathbf{m} = \mathbf{m}(t)$ is the *dipole moment*. We may compute $\mathbf{m}$ as an integral over a sphere of radius r_0,

$$\mathbf{m} = \frac{3}{8\pi} \int_{r=r_0} \mathbf{x}(\mathbf{B} \cdot \mathbf{n}) \, dS, \qquad\qquad (4.2.19)$$

since the integral selects the relevant surface harmonic of the field. Here $\mathbf{n}$ is the outward normal of the boundary $\partial \mathcal{D}$. Using the divergence theorem, it follows from (4.2.19) that

$$\mathbf{m} = \frac{3}{8\pi} \int_{\mathcal{D}} \mathbf{B}\, d\mathbf{x}. \tag{4.2.20}$$

From (4.2.20) we may calculate the time derivative of the dipole moment. Using (1.2.1b) we have

$$\frac{8\pi}{3} \frac{d\mathbf{m}}{dt} = -\int_{\mathcal{D}} \nabla \times \mathbf{E}\, d\mathbf{x} = -\int_{\partial \mathcal{D}} \mathbf{n} \times \mathbf{E}\, dS. \tag{4.2.21}$$

Using $\mathbf{E} + \mathbf{u} \times \mathbf{B} = \epsilon \nabla \times \mathbf{B}$, and the vanishing of $\mathbf{u} \cdot \mathbf{n}$ on $\partial \mathcal{D}$, we have

$$\frac{8\pi}{3} \frac{d\mathbf{m}}{dt} = \int_{\partial \mathcal{D}} \mathbf{u}(\mathbf{B} \cdot \mathbf{n})\, dS - \epsilon \int_{\partial \mathcal{D}} \mathbf{n} \times (\nabla \times \mathbf{B})\, dS. \tag{4.2.22}$$

For a perfect conductor we drop the last term of (4.2.22), to obtain

$$\frac{8\pi}{3} \frac{d\mathbf{m}}{dt} = \int_{\partial \mathcal{D}} \mathbf{u}(\mathbf{B} \cdot \mathbf{n})\, dS. \tag{4.2.23}$$

Using now (4.2.23), we can divide the initial contribution to $\mathbf{m}$ into two parts, one coming from the flux out of the sphere, another coming from a balancing flux into the sphere. The motion will alter $\mathbf{m}$ but the maximum value of $|\mathbf{m}|$ is, by (4.2.19) and the frozen flux property, limited to the value $3Fr_0/4\pi$ obtained when the flux is concentrated in two delta functions at opposite poles.

Thus, as a consequence of Bondi & Gold (1950), in the 'frozen field' limit the dipole moment exterior to a body cannot exhibit dynamo action, in fact $|\mathbf{m}|$ cannot have any, even non-exponential, sustained increase with time. In fast dynamo theory we seek to use the perfectly conducting model to decide dynamo action, but we here have the curious situation that the dynamo is 'invisible' to an outside observer.[34] On the other hand a kinematic dynamo with diffusion will have an exponentially growing dipole moment as one component of the eigenfunction linked to the interior field by magnetic boundary layers.

The limit $\epsilon \to 0$ has been investigated by Hollerbach, Galloway & Proctor (1995) who follow fast dynamo action in a sphere numerically. When the field is normalized using the magnetic energy, they observe that its pole strength (4.2.17) decreases as $\epsilon \to 0$. (The dipole moment $\mathbf{m}$ is zero by symmetry for their fields.) In this limit fast dynamo action can occur, but the field becomes entirely trapped inside the sphere. We shall not consider this boundary effect

---

[34]This point was emphasized by H.K. Moffatt in a summarizing talk at the 1992 NATO Advanced Study Institute on Solar and Planetary Dynamos; see also Kraichnan (1979).

in this monograph, although it is of considerable importance in the detection of fast dynamo action in astrophysics. For the case of self-consistent MHD dynamos, discussed in Chap. 12, there is a related question concerning the size and structure of the vacuum field developed in the equilibrated state.

4.3 Non-existence of Smooth Eigenfunctions

Another kind of non-existence result emerges from an attempt to find *smooth* eigenfunctions in a perfect conductor. In the examples of Chaps. 2 and 3 we observed the emergence of fine structure in eigenfunctions of the induction equation (4.1.1) in the limit of small ϵ. This suggests that any attempt to find smooth eigenfunctions in realistic flows will fail. This interesting question was addressed by Moffatt & Proctor (1985) for the case of steady flow, and we follow their discussion. The question can be considered without prior knowledge of the spectrum of the induction operator (Chap. 9), since the result is one of non-existence of solutions of a particular form.

4.3.1 Vanishing of the Helicity Density

Let us suppose that there exists a smooth, i.e., infinitely differentiable, exponentially-growing eigenfunction,

$$\mathbf{B}(\mathbf{x}, t) = e^{pt}\,\mathbf{b}(\mathbf{x}), \qquad \mathrm{Re}\,p > 0, \tag{4.3.1}$$

for $\epsilon = 0$ in an incompressible steady flow. For simplicity we suppose that $\mathbf{u}$ is zero outside a bounded, simply connected region $\mathcal{D}$ of $\mathbb{R}^3$, say a sphere, and that $\mathbf{B} \cdot \mathbf{n} = \mathbf{u} \cdot \mathbf{n} = 0$ on $\partial\mathcal{D}$; Moffatt & Proctor (1985) also consider periodic boundary conditions.

First note that the magnetic helicity $H_{\mathrm{M}}(\mathcal{D})$ defined by (1.2.23) is invariant and so is zero for a growing eigenfunction. However a stronger result can be proved. For steady flow,

$$p\mathbf{b} = \nabla \times (\mathbf{u} \times \mathbf{b}), \qquad p\mathbf{a} = \mathbf{u} \times \mathbf{b} - \nabla\phi. \tag{4.3.2a, b}$$

Here $\mathbf{a}$ is a vector potential for $\mathbf{b} = \nabla \times \mathbf{a}$, which obeys the uncurled induction equation (4.3.2b). By a gauge transformation $\mathbf{a} + p^{-1}\nabla\phi \to \mathbf{a}$ (valid as $p \neq 0$), we obtain the simpler equation

$$p\mathbf{a} = \mathbf{u} \times \mathbf{b}, \tag{4.3.3}$$

which implies immediately that

$$\mathbf{a} \cdot \mathbf{b} \equiv \mathbf{a} \cdot \nabla \times \mathbf{a} = 0, \tag{4.3.4}$$

corresponding to zero helicity density $h_{\mathrm{M}} \equiv \mathbf{a} \cdot \mathbf{b} = 0$ in this gauge.

Another approach to this property, which applies also to time-periodic flow, is to invoke the material invariance of $\mathbf{A} \cdot \mathbf{B}$ given by (4.2.7) with the appropriate gauge of $\mathbf{A}$. Since the initial values of $|\mathbf{A} \cdot \mathbf{B}|$ are bounded above in a finite domain, this material invariance is incompatible with $\mathbf{A} \cdot \mathbf{B}$ being anywhere non-zero in a growing eigenfunction. Note that we are here referring to helicity density of the *complex* magnetic field $\mathbf{B}$. The helicity density $(\mathbf{A} + \mathbf{A}^*) \cdot (\mathbf{B} + \mathbf{B}^*)$ of the corresponding real magnetic field $\mathbf{B} + \mathbf{B}^*$ is also zero because it too is materially conserved, from the material conservation of $\mathbf{A} \cdot \mathbf{B}$ and $\mathbf{A}^* \cdot \mathbf{B}$; these properties result from the Cauchy solutions (1.2.17) and (4.2.6), which apply to complex as well as real fields, and the fact that $\mathbf{J}(\mathbf{a}, t)$ is real-valued.

4.3.2 The Frobenius Condition for Steady Flow

The vanishing of $\mathbf{a} \cdot \nabla \times \mathbf{a} = 0$ is a *Frobenius integrability condition* on $\mathbf{a}$ (see, e.g., Flanders 1963, Boothby 1986), which allows $\mathbf{a}$, and therefore $\mathbf{b}$, to be expressed in terms of scalar functions f and g:[35]

$$\mathbf{a} = f\nabla g, \qquad \mathbf{b} = \nabla f \times \nabla g. \tag{4.3.5}$$

This can be done, at least locally, at points where $\mathbf{a} \neq 0$. We therefore remove such points by working in a subvolume of $\mathcal{D}$:

$$\mathcal{D}' = \{\mathbf{x} : \mathbf{a}(\mathbf{x}) \neq 0\}. \tag{4.3.6}$$

Non-zero $\mathbf{a}$ implies that $\mathbf{u} \times \mathbf{b} \neq 0$ from (4.3.3); we thus exclude from $\mathcal{D}'$ zeros of $\mathbf{u}$, zeros of $\mathbf{b}$, and points where the vectors are parallel. Note that $\mathbf{u} \cdot \mathbf{n} = 0$ on the boundary $\partial \mathcal{D}'$ of $\mathcal{D}'$, for otherwise the flow would carry non-zero $\mathbf{b}$ out of $\mathcal{D}'$, yielding a contradiction.

We now consider the consequences of (4.3.5) for the case of steady flow. Substituting (4.3.5) into (4.3.3) we have

$$pf\nabla g = \mathbf{u} \times (\nabla f \times \nabla g) = (\nabla f)(\mathbf{u} \cdot \nabla g) - (\nabla g)(\mathbf{u} \cdot \nabla f). \tag{4.3.7}$$

Crossing this with ∇f yields $(pf + \mathbf{u} \cdot \nabla f)\mathbf{b} = 0$ and since $\mathbf{b} \neq 0$ in $\mathcal{D}'$,

$$pf + \mathbf{u} \cdot \nabla f = 0; \tag{4.3.8}$$

f is transported as a passive scalar. By a short calculation using the divergence theorem,

$$(p + p^*) \int_{\mathcal{D}'} |f|^2 \, d\mathbf{x} = -\int_{\partial \mathcal{D}'} \mathbf{u} \cdot \mathbf{n} |f|^2 \, dS = 0 \tag{4.3.9}$$

---

[35]The functions f and g are sometimes called *Clebsch potentials* or *Clebsch coordinates* from the form of vorticity used by Clebsch; see Lamb (1932), p. 248, and Roberts (1967).

(using $\mathbf{u} \cdot \mathbf{n} = 0$ on $\partial \mathcal{D}'$). For growth $p + p^* > 0$ and so $f = 0$ in $\mathcal{D}'$. Thus $\mathbf{a} = 0$ in $\mathcal{D}'$ and so in the whole of $\mathcal{D}$. Seeking a growing eigenfunction has led only to the zero function $\mathbf{a} = \mathbf{b} = 0$. The conclusion is that there are no smooth growing eigenfunctions for $\epsilon = 0$ (Moffatt & Proctor 1985).

4.3.3 Local Versus Global Integrability

In taking the integral (4.3.9) we are assuming that f is uniquely defined over the whole of $\mathcal{D}'$, whereas the condition (4.3.4) only guarantees existence of functions f and g in (4.3.5) locally.[36] Thus the deduction that there are no smooth eigenfunctions actually involves hidden conditions on the form of such eigenfunctions; this modifies the result to one denying the existence of eigenfunctions of a certain type. As it seems there is no general theorem indicating when (4.3.5) is valid globally, we investigate the construction of f and g in more detail.

Geometrically, the Frobenius condition (4.3.4) ensures the existence of a family of smooth surfaces throughout $\mathcal{D}'$ which are everywhere perpendicular to $\mathbf{a}$ (see, e.g., §2.9 of Arnold 1988). Such surfaces can be labelled locally by $g = \text{const.}$, and then f can be defined so that $\mathbf{a} = f\nabla g$. The fact that f and g are complex-valued functions (while most applications of (4.3.5) in MHD involve real functions) is inessential.

To give an example, one manner to define f and g locally is to consider a small surface element cutting $\mathbf{b}$ transversely (for both real and imaginary parts). For any solenoidal field the normal component of $\mathbf{b}$ on this surface can be used to define f and g as *flux functions* on this surface element. That is, we may define two intersecting families of smooth curves in the surface element, with labels (u, v), say, chosen such that $dS = du\, dv$ is the usual area element. We may then construct two complex-valued functions $f(u, v)$ and $g(u, v)$ on the surface such that

$$\frac{\partial f}{\partial u}\frac{\partial g}{\partial v} - \frac{\partial g}{\partial u}\frac{\partial f}{\partial v} = b_{\mathrm{n}}(u, v), \tag{4.3.10}$$

where b_{n} is the normal component of $\mathbf{b}$ evaluated on the surface. This can be done in many ways. This construction attaches labels (f, g) to each line of force, and since flux through the surface is given by $\int df\, dg$, we can extend (f, g) to functions of $\mathbf{x}$ in a neighborhood of the surface element. In this local neighborhood

$$\mathbf{b} = F(f, g)\nabla f \times \nabla g. \tag{4.3.11}$$

However, we may take $F = 1$, since we can otherwise replace g by

$$\tilde{g} = \int^{g} F(f, g')\, dg'. \tag{4.3.12}$$

---

[36]We are grateful to R. Kaiser for stressing this point to us.

To derive the first of (4.3.5) and finish the construction, we now uncurl to obtain

$$\mathbf{a} = f\nabla g + \nabla\phi(f,g), \tag{4.3.13}$$

where the form of ϕ reflects the constraint (4.3.4). We seek a representation of the form $f\nabla g + \nabla\phi(f,g) = F(f,g)\nabla G(f,g)$. It follows that

$$f + \phi_g = FG_g, \qquad \phi_f = FG_f, \tag{4.3.14a, b}$$

so that cross-differentiation to eliminate G yields $(f + \phi_g)F_f - \phi_f F_g = F$. This last equation may in principle be solved along (complex) characteristics,

$$\frac{df}{ds} = f + \phi_g(f,g), \qquad \frac{dg}{ds} = -\phi_f(f,g), \qquad \frac{dF}{ds} = F, \tag{4.3.15}$$

subject to the conditions $F = f$, $G = g$ on the line in (f,g) space defined by $\phi_f = \phi_g = 0$. Having thus obtained $F(f,g)$, G follows from (4.3.14b) by quadrature. Here, we have needed to assume $F \neq 0$, which follows from $\mathbf{a} \equiv f\nabla g + \nabla\phi \neq 0$. This completes one local construction leading to (4.3.5).[37]

Regardless of how the local construction is accomplished, there is no guarantee that the fields f and g can be defined globally throughout $\mathcal{D}'$. The problem is that $\mathcal{D}'$ may be horribly multiply-connected. For example, if the steady flow has a network of integrable islands, we expect $\mathbf{b}$ to be identically zero in the islands, where the field cannot be growing quickly. In such islands $\mathbf{a} = 0$ from (4.3.3) and so these islands have to be excluded from $\mathcal{D}'$. Suppose we define f and g in a small region of a multiply-connected $\mathcal{D}'$, and then extend the definition by following the surfaces $g = \text{const}$. If we go round a 'handle' in $\mathcal{D}'$ back to where we started, we may find the definitions disagree: f and g are multiple-valued. If f is multiple-valued, we conclude that the integral (4.3.9) does not make sense, or, equivalently, that the scalar advection equation (4.3.8) could carry in different values of f from round the handles of $\mathcal{D}'$ and allow the possibility of growth in f and $\mathbf{b}$.

To see potential difficulties, consider two examples of local magnetic field structure in cylindrical polar coordinates, (r, θ, z). First

$$\mathbf{a} = r\mathbf{e}_\theta, \quad \mathbf{b} = 2\mathbf{e}_z, \quad p = 2, \quad \mathbf{u} = -r\mathbf{e}_r + u_z\mathbf{e}_z \tag{4.3.16}$$

(Fig. 4.4a). We have exhibited a possible $\mathbf{u}$ and p to show that equation (4.3.3) does not exclude such an $\mathbf{a}$ locally; u_z may be chosen to make $\nabla \cdot \mathbf{u} = 0$. The line $r = 0$, where $\mathbf{a} = 0$, must be excluded from $\mathcal{D}'$. The field $\mathbf{a}$ can be represented by $f = r^2$, $g = \theta$. The scalar g is multiple-valued, and winds up

---

[37]Moffatt & Proctor (1985) point out that an alternative approach makes use of a construction of Ince (1926), provided (x, y, z) are taken to be complex-valued, with an eventual restriction to the real lines. The calculation relies on assumptions of analyticity which are significantly stronger than are actually needed in Frobenius' theorem.

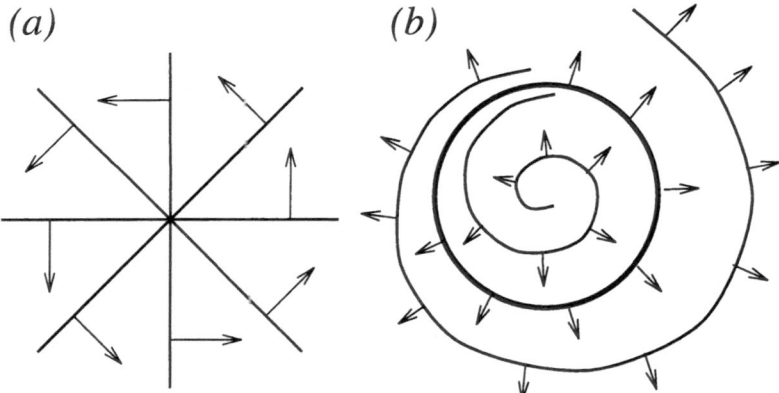

Fig. 4.4 Two examples of orthogonal surfaces; vectors show **a** and curves the corresponding surfaces. (a) See equation (4.3.16), and (b) equations (4.3.17, 18). Note that in (b) the closed surface $r = 2$ is shown bold.

by a constant every time we go round the origin. This is not problematic: since ∇g is single-valued, f is single-valued, and so can be used unambiguously in (4.3.9).

For a more troubling example in which f is multiple-valued and can only be made single-valued at the expense of singularities in g, consider the scalars defined in $r > 3/2$ by

$$f = re^{-\theta}, \qquad g = (r - 2)e^{\theta}, \tag{4.3.17}$$

corresponding to the fields

$$\mathbf{a} = (r, r - 2, 0), \qquad \mathbf{b} = (2r - 2)/r\, \mathbf{e}_z,$$
$$p = 1, \qquad \mathbf{u} = r/(2r - 2)\, (2 - r, r, u_z), \tag{4.3.18}$$

which are well-behaved in $r > 3/2$. The integral surfaces $g = $ const. are shown in Fig. 4.4(b); they comprise the circle $r = 2$ and, for $r > 2$ or $r < 2$, surfaces that spiral into this circle as θ is increased.

For this example f is multiple-valued, causing difficulties in the definition of the integral (4.3.9). One can try to avoid this by redefining g, which is not unique as the surfaces of constant g can always be relabelled by writing $\mathbf{a} = f\nabla g = f_1 \nabla g_1$, with

$$f_1(\mathbf{x}) = f(\mathbf{x})/h'(g(\mathbf{x})), \qquad g_1(\mathbf{x}) = h(g(\mathbf{x})), \tag{4.3.19}$$

for any scalar function $h(g)$.

To make f_1 single-valued in this example we must set $h'(g) = j'(\log g)g^{-1}$, where j' is any smooth periodic function, period 2π (for example, $j' = $ const., $j(g) \propto g$). This gives $h(g) = j(\log g)$ and

$$f_1 = r(r-2)/j'(\log g), \quad g_1 = j(\log g) \equiv j(\log(r-2) + \theta). \qquad (4.3.20)$$

We thus achieve a single-valued f_1, but only at a price, since now g_1 possesses a singularity at the radius $r = 2$ where $\log(r-2)$ diverges. At this radius the topology of the surfaces changes from an infinite accumulating spiral to closed circle. For this example we have a choice of multiple-valued f or singularities of g; the problem arises in this multiply-connected domain even though the underlying vector field $\mathbf{a}$ is entirely well-behaved.

Without further analysis to overcome these difficulties, it seems that the global definition of f and g is not ensured if $\mathbf{a}$ vanishes in the domain $\mathcal{D}$ (or if $\mathcal{D}$ itself is multiply-connected, say $\mathcal{D} = \mathbb{T}^3$). For $\epsilon = 0$ the conclusion is that there do not exist smooth, growing eigenfunctions satisfying the condition that $\mathbf{u} \times \mathbf{b} \neq 0$ everywhere. This is an awkward condition, since it applies to the eigenfunction $\mathbf{b}$ that we cannot specify and know little about.

It is nevertheless reasonable to take the non-existence of smooth eigenfunctions as a very plausible conjecture: it is unlikely that a typical stretching flow that is a non-diffusive fast dynamo would possess a well-behaved eigenfunction. A rough argument here is to assume a continuous eigenfunction with complex growth rate p, and to look at periodic orbits of the flow or map. Unless a periodic orbit $\mathbf{x}$, period t, has a Floquet multiplier $\lambda(\mathbf{x}, t)$ (that is, an eigenvalue of the Jacobian $\mathbf{J}^t(\mathbf{x})$) exactly equal to $\exp(pt)$, the magnetic field in a smooth eigenfunction must vanish on that orbit. Now in a chaotic flow periodic orbits are believed to be dense in a given chaotic region, and generally will have a range of distinct Floquet exponents, corresponding to uneven stretching. In such a case the field must vanish on a dense set and so must be zero everywhere, giving a contradiction.[38] This argument, although crude, appears to work well, since only the stacked baker's map (§3.1), cat map (§3.5) and cat flow (§6.4) are known to have smooth eigenfunctions, and all their periodic orbits obey this very special condition $\lambda(\mathbf{x}, t) = \exp(pt)$. For general flows and maps, however, our attitude is to regard the eigenvalue problem as classical only for positive ϵ.

---

[38]This idea is developed by Soward (1994a) who, in a study of pulsed Beltrami waves, discusses how one might define $\epsilon = 0$ eigenfunctions with singularities localized at each periodic orbit. The fast dynamo growth rate and the structure of the Floquet matrix on each orbit determine the form of the field locally; see also Núñez (1994).

5. Magnetic Structure in Steady Integrable Flows

From the examples of Chaps. 2 and 3 we have seen that fast dynamo activity is accompanied by the emergence of intense, small-scale magnetic structure. The most direct connection of this structure to the geometry of the flow is through the stretching of vectors as measured by the Liapunov exponent of the flow. There are, however, other ways to produce these effects in flows having zero Liapunov exponent. In the present chapter we shall consider processes of this kind for several simple *steady* flow fields. These examples will share many features with a classical problem of fluid mechanics, namely the formation and structure of *boundary layers* in viscous fluid flows at large Reynolds number $Re \equiv UL/\nu$, with ν the kinematic viscosity (see, e.g., Prandtl 1952, Batchelor 1967). Boundary-layer theory originated in Prandtl's observations of the flow adjacent to a rigid wall where the fluid adheres. The flow is arrested in a thin layer of thickness $O(Re^{-1/2})$. Mathematically, the Prandtl boundary-layer theory has been recognized to be a *singular perturbation* of the inviscid or Euler limit (see, e.g., Van Dyke 1975). In such a perturbation theory, the *ideal fluid* with zero viscosity is analogous to our perfectly-conducting limit. In the thin boundary layer, viscous forces are in balance with inertial forces and the ideal or Euler equations are not valid limits of the Navier–Stokes equations. Solutions of the Navier–Stokes equations which are valid in the limit of large Reynolds number, uniformly over the flow domain, must in general contain such boundary-layer structures.

In fast dynamo action, diffusion of the magnetic field similarly becomes important on small domains, where the equations of a perfect conductor are invalid. In general the boundaries of the domain will not play a significant role in the fast dynamo problem and there is no analog of the controlling influence of a rigid wall as in the Prandtl theory.[39] Rather, the flows we study below develop internal magnetic layers aligned with the streamlines of the flow, within which flux can be concentrated, and these layers have the structure envisaged in Prandtl's theory. As already indicated, we deal with *steady* velocity fields, since boundary layers can only form if the geometry of the flow changes on a time scale large compared to the characteristic time L/U. Our main object is to see how smooth steady flows on a scale L can concentrate flux into

---

[39]See, however, §9.5.3 for a discussion of boundary effects in maps.

thin sheets and tubes, and to study the role of diffusion in determining the structure of these magnetic features.

We have used the adjective *integrable* to describe the steady flows studied in the present chapter. The term is attached, somewhat vaguely, to a flow having no Lagrangian chaos, but it has a precise meaning in various special cases. The simplest class of flows of interest to us are steady two-dimensional incompressible flows of the form $\mathbf{u} = (u(x,y), v(x,y), w(x,y))$, introduced in §2.1. The *horizontal flow* $\mathbf{u}_H = (u, v, 0)$ has the form $\mathbf{u}_H = (\psi_y, -\psi_x, 0)$ in terms of the streamfunction $\psi(x,y)$. The *vertical flow* is given by the function $w(x,y)$. The Lagrangian coordinates of the horizontal motion are thus determined by solving

$$\frac{dx}{dt} = \psi_y, \qquad \frac{dy}{dt} = -\psi_x, \tag{5.0.1}$$

which, together with the vertical flow, determines the orbits of fluid particles as curves on the surfaces $\psi = \text{const}$. We call the latter *streamsurfaces* of the flow.

The equations (5.0.1) define an autonomous Hamiltonian system with two degrees of freedom and Hamiltonian $\psi(x,y)$. The Hamiltonian is then clearly an integral of the system,

$$\frac{d\psi}{dt} = \psi_x x_t + \psi_y y_t = 0, \tag{5.0.2}$$

whose values determine the streamlines. Any Hamiltonian system with two degrees of freedom having one integral is integrable in the technical sense of Hamiltonian systems (Arnold 1978). The z-component of velocity simply extends the Lagrangian orbits, still constrained to lie on a streamsurface, into the third dimension. We define this to be an integrable two-dimensional flow.[40]

When two-dimensional (or axial) symmetry is absent, the definition of integrability becomes more involved. An example of a solenoidal field that should be classified as integrable is one expressible in terms of the *Clebsch potentials* χ, θ, in the form $\mathbf{u} = \nabla\chi \times \nabla\theta$ (cf. §4.3.2). Here the streamlines are the lines of intersection of the two families of surfaces of constant potential. Boozer (1983) has exhibited a Hamiltonian for a significantly larger class of toroidal magnetic field configurations, which includes the Clebsch representation; see also Yoshida (1994). Applied to solenoidal velocity fields, Boozer's representation is in the form of a sum of two Clebsch representations,

---

[40]In general a Hamiltonian system having $2N$ degrees of freedom x_i, y_i for $i = 1, 2, \ldots, N$ is said to be integrable if there exist N distinct integrals $J_1, \ldots, J_N$ of the motion. These integrals must moreover *stand in involution*, that is, the Poisson brackets $[J_i, J_j] = \sum_{k=1}^{N} \{(\partial J_i/\partial y_k)(\partial J_j/\partial x_k) - (\partial J_i/\partial x_k)(\partial J_j/\partial y_k)\}$ all vanish. A natural $2N + 1$-dimensional generalization of the flows considered here would have $2N$ Lagrangian coordinates satisfying an integrable Hamiltonian system and a flow with velocity $w(x_1, y_1, \ldots, x_N, y_N)$ in the last coordinate direction.

$$\mathbf{u} = \nabla \psi \times \nabla \phi + \nabla \chi \times \nabla \theta. \tag{5.0.3}$$

From (5.0.3) we may extract the Hamiltonian system

$$\frac{d\chi}{d\phi} = \frac{\partial \psi}{\partial \theta}, \qquad \frac{d\theta}{d\phi} = -\frac{\partial \psi}{\partial \chi}. \tag{5.0.4}$$

Here ϕ may be identified with the time in a two-dimensional Hamiltonian $\psi(\chi, \theta, \phi)$.[41]

We may now define an integrable flow of the form (5.0.3) as one in which (5.0.4) is integrable in the Hamiltonian sense. Note that in toroidal geometry θ is the poloidal angle, ϕ the toroidal angle (on the axis of the torus), and ψ and χ are flux coordinates. This definition includes all flows discussed in this chapter, except possibly the 'skeletal' components of the generalized helical field discussed in §5.4.2 (Yoshida 1994). To obtain the two-dimensional flows we take $\phi = z$, $\theta = y$, $\psi = \psi(x, y)$, $\chi = \int w(x, y) \, dx$. Another example of an integrable form of (5.0.3) is the class of Euler flows considered by Arnold (1965). It is assumed that $\mathbf{u} \times (\nabla \times \mathbf{u}) = \nabla H$ where H is a smooth scalar function (the Bernoulli function) with non-zero gradient. Then $\mathbf{u}$ and $\nabla \times \mathbf{u}$ are tangent to the surfaces of constant H. These conditions can be shown to imply that $\psi = \psi(\chi)$, and so such a flow is integrable in the sense of (5.0.2); see Yoshida (1994) and references therein.

The simplest integrable flow with interesting dynamo properties is the *Ponomarenko flow*

$$\mathbf{u} = \begin{cases} (0, r\omega, W), & (r < a), \\ \mathbf{0}, & (r > a), \end{cases} \tag{5.0.5}$$

where ω and W are given constants. This is an integrable two-dimensional flow having helical streamlines in the circular cylinder $r < a$. It is not continuous on $r = a$. We shall consider the Ponomarenko model in §5.4.1. A natural generalization is the family of smooth flows of the form $\mathbf{u} = (0, r\Omega(r), W(r))$. These flows lack X-type stagnation points, which occur in spatially periodic arrangements of cellular flow with helical structure; X-points give rise to orbits with non-zero Liapunov exponent and a much richer boundary-layer structure. For this reason we focus our discussion in this chapter, not on circular helical flows, but rather on the simplest example of a *spatially-periodic* flow with helical structure.

---

[41]To prove (5.0.4) note that $d\chi/d\phi = (\mathbf{u} \cdot \nabla \chi)/(\mathbf{u} \cdot \nabla \phi)$. But $\mathbf{u} \cdot \nabla \chi = \nabla \psi \cdot (\nabla \phi \times \nabla \chi) = \nabla \theta \cdot (\nabla \phi \times \nabla \chi)(\partial \psi/\partial \theta)$ and $\mathbf{u} \cdot \nabla \phi = \nabla \theta \cdot (\nabla \phi \times \nabla \chi)$ yielding the first equation, and similarly for the second.

5.1 The Roberts Cell

The Roberts cell is an integrable two-dimensional flow which has been studied extensively in connection with the dynamo problem, beginning with Roberts (1972). The flow may be derived as the steady counterpart of the time-dependent Beltrami wave solutions; see §2.2. If we consider waves of equal amplitude, it is sufficient to take

$$\mathbf{u} \equiv (\mathbf{u_H}, w) = (\sin y, \sin x, \cos x - \cos y), \tag{5.1.1}$$

for waves of the same helicity. The horizontal component is determined by a streamfunction $\psi(x, y) = \cos x - \cos y$, in terms of which $\mathbf{u} = (\psi_y, -\psi_x, \psi)$. We call this the *wave form* of the Roberts cell. A more revealing expression for the streamfunction is obtained by a rotation of the coordinate system. Setting $x = (x' - y')/\sqrt{2}$ and $y = (x' + y')/\sqrt{2}$, there results

$$\psi = 2 \sin(x'/\sqrt{2}) \sin(y'/\sqrt{2}). \tag{5.1.2}$$

Thus

$$\mathbf{u'_H} = \sqrt{2}\left(\sin(x'/\sqrt{2})\cos(y'/\sqrt{2}), -\cos(x'/\sqrt{2})\sin(y'/\sqrt{2})\right). \tag{5.1.3}$$

Dividing primed quantities by $\sqrt{2}$ and dropping primes we thus obtain the *cell form* of the (rescaled) flow,

$$\mathbf{u} = (\psi_y, -\psi_x, \sqrt{2}w), \qquad \psi = \sin x \sin y. \tag{5.1.4}$$

While the wave form is often preferred for numerical computations, the cell form (5.1.4) reveals more clearly the flow geometry. In the cell form the cells are seen to be square and to have side π. The horizontal flow has hyperbolic, X-type stagnation points located at $(x, y) = (m\pi, n\pi)$ for integers m, n; see Fig. 5.1. The union of lines $x = m\pi$ and $y = n\pi$ for all integers m, n comprise the *separatrix* of the horizontal flow, dividing different regions of motion. The points $(x, y) = ((m - 1/2)\pi, (n + 1/2)\pi)$ are stagnation points which are *centers* of the horizontal flow, where the streamlines of the horizontal flow are approximately circular.

If we now add the vertical component, and consider the flow in three dimensions, the planes, $x = m\pi$ and $y = n\pi$, or equivalently $\psi = 0$, carry zero vertical velocity, and the lines $(x, y) = (m\pi, n\pi)$ have zero velocity.[42] The streamlines lie on the family of cylindrical streamsurfaces $\psi = \text{const.}$ The cell $0 \le x, y \le \pi$ will sometimes be called the *primary cell*.

---

[42]For waves of opposite helicity, $w = \cos x + \cos y$ in the wave form and $\sqrt{2}\cos x \cos y$ in the cell form; so the velocity does not vanish at the stagnation points of the horizontal flow. We define the Roberts cell with waves of like helicity and consider only this case in the present chapter.

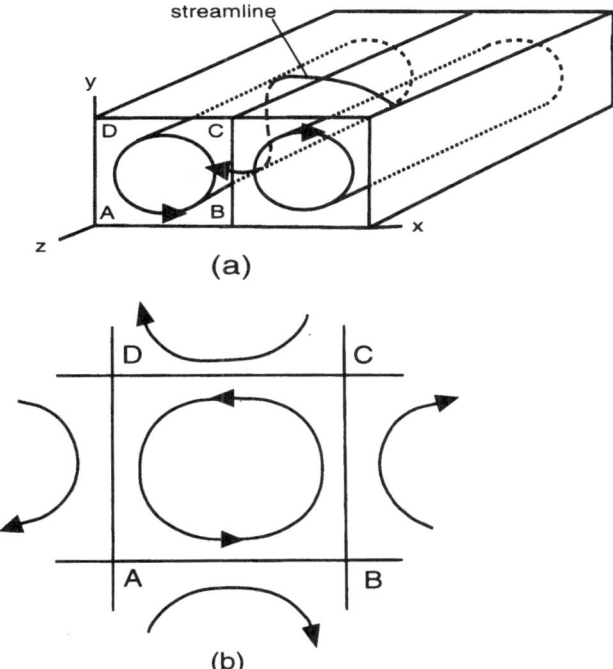

Fig. 5.1 The Roberts cell with $K = \sqrt{2}$: (a) helical structure, and (b) the primary cell.

Now adopting the cell form and (5.1.4), we see that in each cell the vorticity is $(\sqrt{2}\,\psi_y, -\sqrt{2}\,\psi_x, \omega)$ where $\omega = -\nabla^2\psi = 2\psi$. The Beltrami property of the flow, $\nabla \times \mathbf{u} = \sqrt{2}\,\mathbf{u}$, is evident. A generalized form of the Roberts cell, $\mathbf{u} = (\psi_x, -\psi_y, K\psi)$, with the same streamfunction but an arbitrary constant K, is sometimes used. Here the flow is Beltrami only if $K = \pm\sqrt{2}$. The cells alternate between positive and negative vorticity, and positive and negative vertical velocity, as the separatrix is crossed. The helicity density $\mathbf{u} \cdot \nabla \times \mathbf{u} = \sqrt{2}\,(\sin^2 x + \sin^2 y)$ is non-negative.

The hyperbolic points of the horizontal flow are crucial to the stretching properties of the Roberts cell. We shall study the appropriate stagnation-point flow in §5.2.1. The centers of the horizontal flow determine the axes around which the spiralling motions are developed. The planes $\psi = 0$, with horizontal projection onto the separatrix, mark the boundaries of these helical regions. The horizontal flow has closed streamlines in a helical region, and so tends to 'wind up' a horizontal magnetic field line which crosses it. Since the winding process must pull out paired lines of opposite direction (see Fig. 5.2), large shear is created in the field, which is then susceptible to smoothing by diffusion. The net effect is to displace field lines away from the centers, toward the boundaries of the helical regions, and it is termed 'flux expulsion'. We shall study this process in more detail, and as a time-dependent mechanism, in §5.2.2. The corresponding small-ϵ theory of steady magnetic field is discussed in §5.3.

The expulsion process produces magnetic layers near the separatrix of thickness $O(\epsilon^{1/2})$, where the field is intensified by a factor $O(\epsilon^{-1/2})$. It is at this stage that our subject makes contact with the classical boundary-layer of Prandtl. These magnetic boundary layers carry the flux associated with a finite mean field initially penetrating the cells. We shall study their structure, and their role in the dynamo process, in §5.5. Finally, in §5.6, we mention an analogous result for axisymmetric steady flow.

5.2 Basic Mechanisms

5.2.1 The Two-dimensional Hyperbolic Stagnation Point

The flow field in the immediate vicinity of a hyperbolic (or X-type) stagnation point of a steady two-dimensional incompressible flow has a velocity field of the form

$$\mathbf{u} = \beta(x, -y), \tag{5.2.1}$$

up to a translation and rotation, β being an arbitrary constant. For the Roberts cell in its cell form (5.1.4), the streamfunction satisfies $\psi \approx xy$ near (0,0), yielding the velocity field (5.2.1) locally with $\beta = 1$.

The flow has Lagrangian variables $x(a, b, t)$, $y(a, b, t)$ satisfying

$$\frac{\partial x}{\partial t} = \beta x, \quad \frac{\partial y}{\partial t} = -\beta y, \quad x(a, b, 0) = a, \quad y(a, b, 0) = b, \tag{5.2.2}$$

so that $(x, y) = (ae^{\beta t}, be^{-\beta t})$. For $\beta > 0$, most Lagrangian points are swept out in the $\pm x$-direction, and as $t \to \infty$ approach the x-axis asymptotically. Exceptions are points on the y-axis, which instead approach the origin asymptotically. If time is reversed, $t \to -\infty$, all points except those on the x-axis are swept close to the y-axis. For an initial vector $\mathbf{v}_0$ frozen in the flow,

$$\mathbf{v}(a, b, t) = (v_{0x}e^{\beta t}, v_{0y}e^{-\beta t}). \tag{5.2.3}$$

The v_x component is stretched exponentially as $t \to \infty$, while the v_y component is contracted, independently of position.

We consider now the time required for a Lagrangian point to move through the stagnation point flow. The point $(c/\beta L, L)$ on the streamline $\psi = \beta xy = c$ a distance L from the x-axis will take a time $\beta^{-1} \log |\beta L^2/c|$ to reach a point at $(L, c/\beta L)$ a distance L from the y-axis; this time grows logarithmically as

$c \to 0$. Applied to the hyperbolic points of the Roberts cell, we see that any streamline on the streamsurface $\psi = 0$ will tend to a line $(x, y) = (m\pi, n\pi)$ for some integers m and n, but will not attain this limit in finite time. On the other hand if a point moves on a streamsurface $\psi \neq 0$, the particle spends an increasing fraction of its time near the hyperbolic points in the limit $\psi \to 0$: the period diverges as $\log |\psi|$ and the limiting closed streamline containing stagnation points cannot be covered in a finite time by any Lagrangian point.

A basic measure of the stretching on the trajectory through a point a is the Liapunov exponent at a, $\Lambda_{\mathrm{Liap}}(a)$ (see §1.6). For the stagnation-point flow a positive Liapunov exponent β is obtained at all points, corresponding to the exponential growth of an initial vector parallel to the x-axis. However this is an unbounded flow, and the situation is quite different for stagnation points embedded in steady planar incompressible flows that are of finite extent, or are spatially periodic with zero mean flow, as for the horizontal part of the Roberts cell. There all streamlines that do not terminate at a stagnation point are closed. We then use an argument that depends on the fact that the Liapunov exponent $\Lambda_{\mathrm{Liap}}(a)$ is in fact the largest of a set of exponents (two for a planar flow[43]), measuring the stretching rates of all possible initial vectors (see §6.1). For an incompressible flow these exponents must sum to zero. On any closed streamline both Liapunov exponents vanish, since clearly a segment initially parallel to the streamline will not be stretched indefinitely. However, on a streamline terminating at a stagnation point one Liapunov exponent, $\Lambda_{\mathrm{Liap}}(a)$, is positive, the other negative. Indeed, let the stagnation point look locally like the one above. For a point a on the positive y-axis, an initial vector $\mathbf{v}_0 = \mathbf{e}_x$ stretches with exponent β, and the initial vector $\mathbf{v}_0 = \mathbf{e}_y$ with exponent $-\beta$: one or the other exponent is positive. Thus on curves that terminate at stagnation points the Liapunov exponent $\Lambda_{\mathrm{Liap}}(a) = |\beta|$ is positive, while on other streamlines it is zero. The Liapunov exponent of the flow Λ_{Liap} is zero, being an average over the plane. In the case of a periodic velocity field with non-zero mean flow, the same result applies when the tangent or cotangent of the mean flow inclination relative to the x-axis is a rational number; see §5.5.3.

Consider now the effect of the stagnation point flow on a planar magnetic field of the form $(\partial_y A, -\partial_x A)$. The evolution of A is, according to §1.2.6, governed by the dimensionless equation

$$A_t + xA_x - yA_y - \epsilon\nabla^2 A = 0, \tag{5.2.4}$$

with $\epsilon = \eta/(\beta L^2)$. If we change variables to $X = xe^{-t}$, $Y = ye^t$, we obtain

$$A_t - \epsilon(e^{-2t}A_{XX} + e^{2t}A_{YY}) = 0. \tag{5.2.5}$$

Let us first suppose that A is initially smooth and falls off rapidly with distance from the stagnation point, so that the field A may be written in terms of

---

[43]For the original two-dimensional flow, there is a third exponent corresponding to motion in z, which is zero.

its Fourier transform, $A(X, Y, t) = \int \hat{A}(k, l, t) \exp(ikX + ilY) \, dk \, dl$. We may then solve (5.2.5) in the form

$$A(x, y, t) = \int \hat{A}(k, l, 0) \, g(k, l, t) \, \exp(ikxe^{-t} + ilye^{t}) \, dk \, dl, \qquad (5.2.6a)$$

$$g(k, l, t) = \exp\left(\frac{\epsilon}{2} \left[k^2(e^{-2t} - 1) - l^2(e^{2t} - 1)\right]\right). \qquad (5.2.6b)$$

We can see from (5.2.6) the effect of the stagnation point on an initial distribution of A. The reduction of the effective x-wavenumber to zero reflects the stretching in x, which reduces all x-gradients to zero and in effect endows the field with the initial values of A near the y-axis. At the same time, for large t the compression in y causes y-gradients to diffuse away except from wavenumbers close to $l = 0$, and we find

$$A_y \sim -e^{-t}(2\pi/\epsilon^3)^{1/2} \, y \, e^{-y^2/2\epsilon} \int_{-\infty}^{+\infty} \hat{A}(k, 0, 0) \, e^{-\epsilon k^2/2} \, dk. \qquad (5.2.7)$$

Thus the field decays exponentially in time in the vicinity of the stagnation point.

If, on the other hand, there is initially a net flux of field through the y-axis, so that $A(0, +\infty) - A(0, -\infty) \neq 0$ and the Fourier transform does not exist, then the flux is compressed into a stationary flux sheet. For example, if $A(x, y, 0) = f(y) + A_1(x, y)$ where A_1 is localized as above, then the non-vanishing term in A may be obtained from the Poisson solution of the heat equation in Y. We then have

$$A_y \sim e^t \int_{-\infty}^{+\infty} f'(s) \frac{1}{\sqrt{4\pi\epsilon\tau}} \exp(-(ye^t - s)^2/4\epsilon\tau) \, ds, \qquad (5.2.8)$$

for large time, where $\tau = \frac{1}{2}(e^{2t} - 1)$. Thus

$$A_y \sim (f(+\infty) - f(-\infty)) \frac{1}{\sqrt{2\pi\epsilon}} e^{-y^2/2\epsilon}. \qquad (5.2.9)$$

This is a diffusive layer with Gaussian profile carrying finite flux. In the limit of large R, we thus see that a stagnation point creates an intense *flux sheet* on the stretching axis (the unstable manifold of the stagnation point), provided that the initial field has a net flux through the compression axis (the stable manifold of the stagnation point). The other field components decay locally, or are swept to infinity. Of course in bounded flows with closed streamlines these vanishing components will not be thus expelled, and then a different kind of boundary-layer structure, depending upon the recirculating flow, is possible (see §5.5.1).

5.2.2 Flux Expulsion in Planar Flow

In planar incompressible flows there are two kinds of stagnation points, the hyperbolic stagnation point already studied, and the *center*, of the form $\mathbf{u} = \beta(y, -x) = \beta r \mathbf{e}_\theta$ (in plane polar coordinates (r, θ)). In the Roberts cell, these centers have already been mentioned as the positions of the axes of the cells of helical motion, and there the horizontal flow locally has the above structure. For example near $(x, y) = (\pi/2, \pi/2)$ we have $\psi \approx 1 - (x^2 + y^2)/2$ and so $\beta = -1$.

At such a center, and indeed in any domain covered by a nested family of closed streamlines, a second basic mechanism affecting magnetic structure can occur. This mechanism is in a sense the complement to the creation of flux sheets and is usually referred to as *flux expulsion*. Flux expulsion can, among other possibilities, create voids of nearly constant potential, or small magnetic field. In a flow carrying finite flux, such voids can appear only if other, usually smaller, regions carry the flux. These regions of flux concentration tend to bound the closed-streamline 'eddies', as boundary layers or channels, in the manner already suggested. (Channels will be discussed in §5.3.2.) By this process flux expulsion supplies the flux which is trapped near the streamlines connecting hyperbolic points. As we shall see, the process of flux expulsion can also trap field as either closed loops on streamlines of the horizontal flow, or as spiralling lines on a streamsurface. In the absence of dynamo action these trapped fields decay slowly on a diffusion time scale, but given dynamo action they can create sites of excitation localized near a stream surface. We shall see examples of this in §5.4. The process is the basis for the states which can be constructed by the small-ϵ asymptotic theory discussed in §5.3.

We consider first the center defined above, but generalize the velocity to the circular eddy $\mathbf{u} = r\Omega(r)\mathbf{e}_\theta$, with $\Omega(r)$ a smooth function. We call this a *circular eddy*, the horizontal flow of the smooth extension of the Ponomarenko flow (5.0.5). We pass at once to the dimensionless equation for the magnetic potential in polar coordinates,

$$A_t + \Omega A_\theta - \epsilon\nabla^2 A = 0. \tag{5.2.10}$$

We look for a solution of (5.2.10) of the form

$$A(r, \theta, t) = f_m(r, t)e^{im(\theta - \Omega(r)t)}. \tag{5.2.11}$$

If $\epsilon = 0$, $f_m(r, t)$ is independent of time. For $\epsilon > 0$, substituting and arranging terms, we obtain

$$\frac{\partial f_m}{\partial t} - \epsilon\left(\frac{\partial^2 f_m}{\partial r^2} + \frac{1}{r}\frac{\partial f_m}{\partial r} - \frac{m^2}{r^2}f_m\right)$$
$$= -\epsilon\left(m^2 f_m \Omega_r^2 t^2 + 2im\frac{\partial f_m}{\partial r}\Omega_r t + im\frac{f_m}{r}\Omega_r t + imf_m\Omega_{rr}t\right). \tag{5.2.12}$$

If $m = 0$, (5.2.12) is the radial heat equation, so that the evolution of the field within the region of closed streamlines is on the diffusive time scale $1/\beta\epsilon = L^2/\eta$. Such a mode corresponds to the 'trapped' field lines mentioned earlier. The magnetic field has circular lines of force, and so cannot be associated with flux *through* the domain. The flux will appear in modes with $m \neq 0$. In this case the t^2 term on the right-hand side of (5.2.12) will eventually control the decay of the field. For small ϵ and large t we can approximate (5.2.12) by

$$\frac{\partial f_m}{\partial t} \approx -\epsilon m^2 f_m \Omega_r^2 t^2. \tag{5.2.13}$$

This leads to decay on the time-scale $\epsilon^{-1/3}/\beta$. We can understand the process of decay as consisting first of the creation of small-scale spiral structure, on the time scale β^{-1} of the eddy, by the 'winding up' of lines of force (Parker 1966, Weiss 1966, Moffatt & Kamkar 1983). When the radial scale is sufficiently small, decay of this spiral field ensues; see Fig. 5.2. The effect is to displace the flux toward the boundary of the region of closed streamlines.

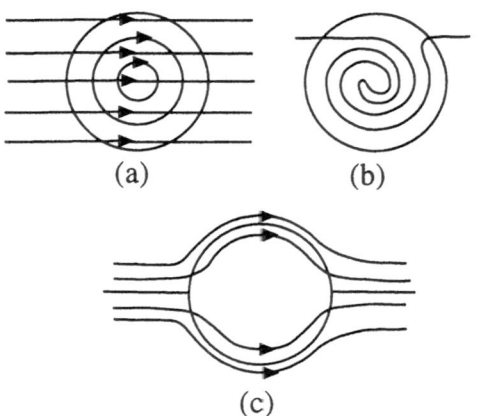

(a)   (b)

(c)

Fig. 5.2 Flux expulsion by a flow with closed streamlines. (a) Initial configuration: a uniform horizontal magnetic field penetrates a circular eddy. (b) A sample line of force is wound up on a time scale β^{-1}. (c) Decay of small-scale structure occurs on the time scale $R^{1/3}/\beta$, and no flux penetrates into the eddy.

Other aspects of this basic idea, including a final phase of adjustment of the field of circular magnetic lines, and flux expulsion from eddies that are not circular, are discussed by Rhines & Young (1983). For an arbitrary region of steady flow with closed streamlines $\psi = $ const., the problem is treated by writing the equation for A in the form

$$\frac{\partial A}{\partial t} + q\frac{\partial A}{\partial s} - \epsilon\nabla^2 A = 0, \qquad (q = |\mathbf{u}|), \tag{5.2.14}$$

where s is the arclength along a streamline, so that $\mathbf{u}\cdot\nabla = q\,\partial/\partial s$. We introduce a function $\Theta(s,\psi,t)$ defined by

$$\Theta(s,c,t) = \left(\theta - \frac{2\pi t}{T}\right)\bmod 2\pi, \qquad \theta = \frac{2\pi}{T}\int_0^s \frac{ds}{q}\bmod 2\pi, \tag{5.2.15}$$

where the integral is taken over the contour $\psi = c$ and

$$T(c) = \int_{\psi=c} \frac{ds}{q} \tag{5.2.16}$$

is the period of the orbit. Note that the right-hand side is constructed by analogy with the quantity $\theta - \Omega(r)t$ for the circular eddy. The θ in (5.2.15) is determined by first defining the point $s = 0$ on each streamline, as intersections of horizontal streamlines with a smooth arc extending from the center to the cell boundary, for example. θ is then an angle measurement based upon elapsed time on the orbit. Since the Ω in (5.2.11) is just 2π over the period on the circular orbit, we see the origin of the second term in the expression for Θ. The change of variables $(x,y,t) \to (\psi,\Theta,t)$ can be used to find an analog of (5.2.13) for the Roberts cell.

We remark that this change of variable $(x,y) \to (\psi,\Theta)$ is essentially equivalent to the introduction of action–angle variables in Hamiltonian mechanics. Both share the mapping of 'streamlines' into circles. The difference is that transformation to action–angle variables must be *canonical*, meaning here an area-preserving transformation (Arnold 1978). It turns out that the canonical action–angle variables are then $(\mathcal{A}/2\pi, \theta)$ where $\mathcal{A}(c)$ is the *area* enclosed within $\psi = c$. To check this, note first that

$$dx\,dy = ds\,dn = (1/q)\,ds\,d\psi, \tag{5.2.17}$$

from which we may deduce the well-known result for the period, $\mathcal{A}'(\psi) = T(\psi)$. From our definition of θ it then follows that

$$dx\,dy = (T/2\pi)\,d\theta\,d\psi = (1/2\pi)\,d\theta\,d\mathcal{A}, \tag{5.2.18}$$

which determines the action variable.

If now we substitute $A = f(\psi,\Theta,t)$ into (5.2.14), we obtain the equation analogous to (5.2.13):

$$\frac{\partial f}{\partial t} \approx \epsilon(2\pi t T_\psi q/T^2)^2 f_{\Theta\Theta} = \epsilon t^2 D(\Theta,\psi) f_{\Theta\Theta}. \tag{5.2.19}$$

This equation predicts flux expulsion by the same process as for the circular eddy. The trapped, long-lived flux must again be aligned with the streamlines of the eddy.

We remark that the mechanism of flux expulsion bears on an interesting question concerning the Roberts cell and other bounded or periodic two-dimensional flows: what is the value of h_{line}, the maximal growth rate of curve length, for these flows? A value of zero is suggested by the mechanism of flux expulsion, where the creation of spiral field lengthens lines linearly in time. The maximal rate of growth of line length near the separatrix is less obvious, since small oscillations of a line which repeatedly crosses the separatrix leads to flux loops which are pulled in opposite directions by the stagnation point flow. In fact Young (1977) shows that h_{top}, and so h_{line}, is zero in smooth steady two-dimensional flows on compact manifolds, which covers the cases here.

An interesting variant of flux expulsion occurs in the baker's maps studied in Chap. 3. In the folded baker's map flux is destroyed after N steps, where $\epsilon 4^N \approx 1$. For a three-layer baker's map which preserves total flux, however, actual expulsion occurs, leading presumably to the delta-function eigenfunction studied in §9.5.2.

5.3 Kinematic Prandtl–Batchelor Theory

In an early paper dealing with fluid motion at large Reynolds number, Prandtl (1904) observed that a steady planar Navier–Stokes flow must tend to a form having *constant vorticity* in a region of closed streamlines, in the limit of infinite Reynolds number. His argument is quite simple. Suppose that a family of closed streamlines $\psi = $ const. is given and has a well-behaved limit. Then, integrating the steady vorticity equation

$$\nabla \cdot (\mathbf{u}\omega - \nu \nabla \omega) = 0 \tag{5.3.1}$$

over a simple domain in the (x, y)-plane, bounded by a closed streamline C, and applying the divergence theorem, we find

$$\oint_C \frac{\partial \omega}{\partial n} \, ds = 0 \tag{5.3.2}$$

for any positive ν. Assuming that indeed viscous effects are small over the region, we may substitute for ω its limiting form $\omega_0(\psi)$, i.e., a solution of $\mathbf{u} \cdot \nabla \omega = 0$. Then (5.3.2) yields, since $\partial \psi / \partial n = q$,

$$\omega_0'(\psi) \oint_C q \, ds = 0, \qquad (q = |\mathbf{u}|). \tag{5.3.3}$$

This argument applies to any streamline of the family, so we deduce that $\omega = \omega_0 = $ const. over the interior of the region of closed streamlines. Thus this condition should characterize Euler flows of this form that are limits of Navier–Stokes flows for large Reynolds number.

Although in practice steady flows at high Reynolds number are usually unstable, this result is important as the first example of a very small number of asymptotic properties which can be established for steady inviscid limits of Navier–Stokes flows. The theorem was later discovered independently by Batchelor (1956), who also extended the result to axisymmetric flows. Further extensions have been discussed by Blennerhassett (1979) and Childress, Landman & Strauss (1990a), and there have been numerous applications in particular flow configurations (see, e.g., Lagerstrom 1975). A rigorous proof does not seem to have been given, nor is there a general extension to three dimensions.

The actual value of the constant ω_0 is not usually known *a priori*, and must be calculated from a complementary boundary-layer theory. When the boundary-layer pressure is constant, as for the case of a circular eddy, the constant can be deduced without knowing the boundary-layer structure (Batchelor 1956).

The title of this section, without the adjective *kinematic*, thus refers to the theory of inviscid limits of steady Navier–Stokes flows, in regions of small viscous effects. But the same ideas apply to kinematic theory in which the flow **u** is given, and ω is replaced by a passive scalar that is advected and diffused. This form of the theory is actually on firmer ground, since certain nonlinear effects, especially possible difficulties associated with a change in the geometry of the streamlines as the diffusivity tends to zero, are expelled from the problem.

5.3.1 Closed Streamlines

The magnetic problem which replaces Prandtl's vorticity theory is the steady equation for the planar potential $A(x, y)$,

$$\mathbf{u} \cdot \nabla A - \epsilon \nabla^2 A \equiv q \frac{\partial A}{\partial s} - \epsilon \nabla^2 A = 0. \tag{5.3.4}$$

For a domain of closed streamlines in two dimensions, the argument given above for vorticity may be repeated exactly, with the result that A has constant limit A_0 as $\epsilon \to 0$. Thus the magnetic field is zero. Prandtl–Batchelor theory is thus consistent with the transient process of flux expulsion studied in §5.2.2. The result of that process is a magnetic potential which is constant on closed streamlines. Constancy over the eddy is then developed over the diffusion time-scale and must involve diffusion from streamline to streamline.

5.3.2 Channels

A *channel* will be simply a bundle of streamlines that do not close, and along which A can have a finite variation (see Fig. 5.3). The change in A from arc C_1 (determined by a function $s = s_1(\psi)$) to arc C_2 is associated with a finite magnetic flux *across* streamlines of the channel.

Let us suppose that, along a given streamline, corresponding to a given value of c of ψ, A changes by $\Delta A(c)$ between the point $s = a$ on the arc C_1 and the point $s = b$ on the arc C_2. Then we must have, from (5.3.4),

$$\epsilon \int_a^b \frac{1}{q} \nabla^2 A \, ds = \Delta A(c). \tag{5.3.5}$$

If we now replace $A(\psi)$ in (5.3.5) by the expansion $\epsilon^{-1} A_0(\psi) + A_1(s, \psi) + O(\epsilon)$, appropriate to the limit of small diffusion, we see that

$$\Delta A_1(\psi) = \sigma(\psi) \frac{d^2 A_0}{d\psi^2} + \frac{dA_0}{d\psi} \int_a^b \frac{1}{q} \nabla^2 \psi \, ds, \tag{5.3.6}$$

where $\sigma(\psi) = \int_a^b q \, ds = \int_a^b (u \, dx + v \, dy)$ is the *circulation integral* for the channel. We now use the identity $dx \, dy = q^{-1} ds \, d\psi$ to obtain

$$\int_a^b \frac{1}{q} \nabla^2 \psi \, ds = \frac{d}{d\psi} \int_{D(\psi)} \nabla^2 \psi \, dx \, dy, \tag{5.3.7}$$

where $D(\psi)$ is the domain bounded by the given streamline, the edge streamline $\psi = \psi_1$ (see Fig. 5.3), and the arcs C_1 and C_2. A short calculation using the divergence theorem yields

$$\Delta A_1(\psi) = \frac{d}{d\psi} \left(\sigma(\psi) \frac{dA_0}{d\psi} \right) + S(\psi), \tag{5.3.8a}$$

where

$$S(\psi) = \frac{dA_0}{d\psi} (\tan \vartheta_2 - \tan \vartheta_1), \tag{5.3.8b}$$

and ϑ_i are the angles between the normal to the arcs C_i in the direction of the flow, and the local velocity vector.

From (5.3.8) we see that channels are *not* regions from which flux is expelled. On the contrary, fields of the order of $\epsilon^{-1} \Delta A$ are developed there. Physically, the process is one of stretching out along the channel any magnetic line which crosses it. A change in ΔA from a to b implies magnetic flux through the line joining a and b. In a steady flow at small ϵ, the field lines in the channel will then be stretched until the process is arrested by diffusion. This occurs only when the potential in the channel is $O(\epsilon^{-1} \Delta A)$. Thus any field which penetrates into the channel region will be stretched until its intensity is increased by a factor ϵ^{-1}. Only if ΔA and S vanish will these large fields not occur.

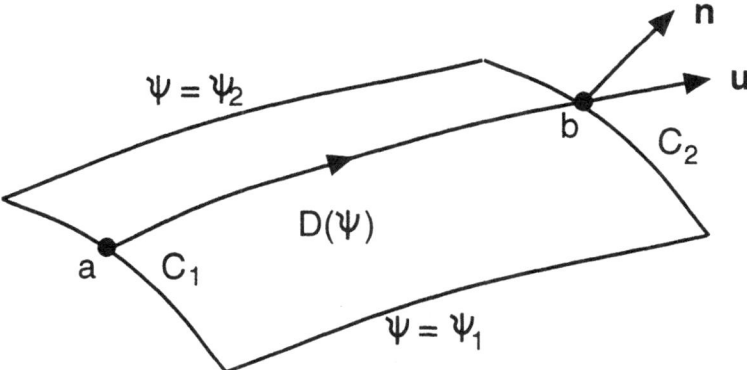

Fig. 5.3 A finite channel.

5.3.3 The Cat's-eye Flow

As an example of Prandtl–Batchelor theory in the scalar case, we take in (5.3.4) the planar flow with stream function

$$\psi = \sin x \sin y + \delta \cos x \cos y, \qquad (0 \le \delta \le 1). \qquad (5.3.9)$$

This is the horizontal, or planar, component of the flow studied by Childress & Soward (1989). The streamlines $\psi = \text{const.}$ are shown in Fig. 5.4.

We assume that the gradient of A has a non-zero spatial mean, corresponding to a unit mean magnetic field $(\overline{A}_y, -\overline{A}_x) = (\cos \phi, \sin \phi)$. Since A is arbitrary up to an additive constant, we may choose to set $A = 0$ in the limit $\epsilon \to 0$ within the *primary* eye, contained in the square $[0, \pi] \times [0, \pi]$ (see Fig. 5.4). In this case the limiting value of A will be $\pi(n \cos \phi - m \sin \phi)$ for the eye in the square $[m\pi, (m + 1)\pi] \times [n\pi, (n + 1)\pi]$. Now the limit of A should be continuous except at the separatrices, where boundary layers occur, and in the channels, where A diverges like ϵ^{-1}. To study the dominant channel we need the limit of ϵA, i.e., the function $A_0(\psi)$ of §5.3.2. This function must vanish on the separatrices which bound the channel, since the magnetic potential is $o(\epsilon^{-1})$ outside the channels.

We restrict attention to the primary channel, bounding the lower and right edges of the primary cell. The arcs C_1 and C_2 then coincide with the segment $(0,0)$ to $(\pi,0)$ and the segment (π, π) to $(\pi, 2\pi)$. Clearly $\Delta A = \Delta A_1 = \pi(\cos \phi - \sin \phi)$. Also $S(\psi) = 0$ since $\theta_1 = \theta_2$ (cf. Figs. 5.3, 4). The geometry of the eyes is introduced by the circulation integral $\sigma(\psi)$. To compute it from (5.3.9) note that $u^2 = \delta^2 - \psi^2 + (1 - \delta^2) \sin^2 x$ and similarly $v^2 = \delta^2 - \psi^2 + (1 - \delta^2) \sin^2 y$. It follows that the two terms $u\,dx + v\,dy$ make the same contribution; so

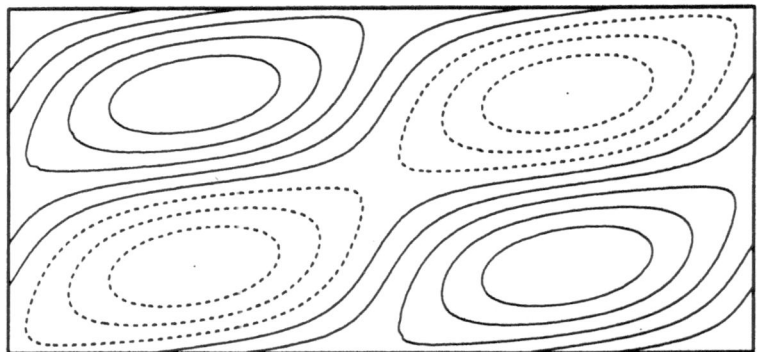

Fig. 5.4 The streamlines of the cat's-eye flow with $\delta = 0.3$. The domain shown is the $2\pi \times 2\pi$ cell. The dashed (solid) contours correspond to positive (negative) values of ψ. The channel is bounded by the streamlines $\psi = \pm 0.3$, which are also the stable and unstable manifolds of the hyperbolic stagnation points.

$$\sigma(\psi) = 2 \int_0^\pi \sqrt{\delta^2 - \psi^2 + (1 - \delta^2) \sin^2 z} \, dz. \tag{5.3.10}$$

As is to be expected in this doubly-periodic geometry, the integration gives us an elliptic integral, namely the complete elliptic integral of the second kind (Abramowitz & Stegun 1972),

$$\sigma(\psi) = 4\sqrt{1 - \psi^2} \, E(m), \qquad m = \frac{1 - \delta^2}{1 - \psi^2}, \tag{5.3.11a}$$

$$E(m) = \int_0^{\pi/2} (1 - m \sin^2 \theta)^{1/2} \, d\theta. \tag{5.3.11b}$$

We may now integrate (5.3.8) with the indicated values of ΔA_1 and S to obtain

$$\sigma(\psi) \frac{dA_0}{d\psi} = \pi (\cos\phi - \sin\phi)\psi, \tag{5.3.12}$$

where the additive constant is zero since the symmetry makes $dA/d\psi$ an odd function of ψ. Integrating again, we obtain

$$A_0 = \int_{-\delta}^{\psi} \pi (\cos\phi - \sin\phi)\psi/\sigma(\psi) \, d\psi, \tag{5.3.13}$$

where we have applied the boundary condition that $A_0(\pm\delta) = 0$. In Fig. 5.5 we show the profile of A_0 for several values of δ.

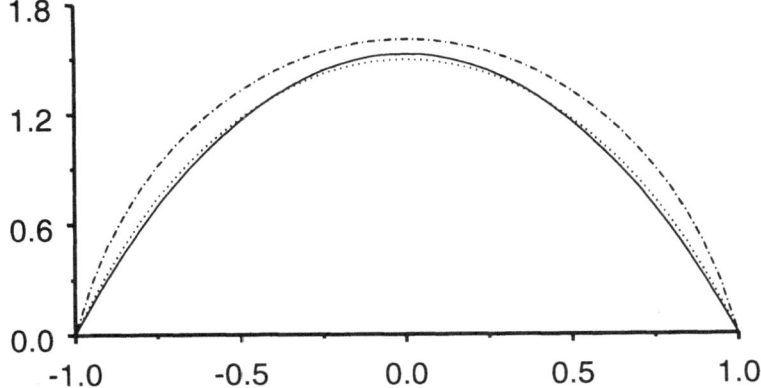

Fig. 5.5 The channel values of A_0/δ^2 for the cat's-eye flow as a function of ψ/δ for $\delta = 0.3$ (solid), 0.6 (dotted), 0.9 (dashed), and $\phi = 0$.

If $\phi = \pi/4$ or $5\pi/4$ we have from (5.3.13) that $A_0 = 0$. This signals a change of order, due to the alignment of the mean field with the channels. In this case we have $A \sim A_1 + \epsilon A_2 + \cdots$ and the function A_1 coincides at the edges of the channels with the constant values of A within the eyes. Thus in the primary channel we obtain from the equation analogous to (5.3.8a), namely

$$\frac{d}{d\psi}\left(\sigma(\psi)\frac{dA_1}{d\psi}\right) = 0, \qquad A_1 = A_1(\psi), \tag{5.3.14a}$$

the solution

$$A_1 = -\frac{\pi G(\psi)}{2G(0)}, \quad G(\psi) = \int_\psi^\delta \frac{d\psi}{\sigma(\psi)}, \quad (\phi = \pi/4, 5\pi/4). \tag{5.3.14b}$$

5.4 Helical Dynamos

5.4.1 Ponomarenko Models

The Ponomarenko flow (5.0.5) consists of solid-body rotation and uniform axial translation within a circular cylinder embedded in a rigid uniform conductor. Within this cylinder, fluid particles move on helices with speed $\sqrt{r^2\omega^2 + W^2}$ and helix pitch $\chi = W/r\omega$ depending upon r. We now summarize a few of the results which have been obtained for this simple geometry.

The analysis of this dynamo was first carried out by Ponomarenko (1973). The magnetic field is taken to have the separable form

$$\mathbf{B} = \mathbf{b}(r)\exp(im\theta + ikz + pt), \tag{5.4.1}$$

where $p(m, k, \epsilon)$ is the complex growth rate. Given the discontinuity in velocity on $r = a$, the appropriate jump conditions, in addition to the requirement that $\mathbf{B}$ vanish as $r \to \infty$, require the continuity of $\mathbf{B}$ and the tangential components of the electric field $\mathbf{E}$. Ponomarenko (1973) (see also Zeldovich et al. 1983) discussed this problem in the limit $\epsilon \to 0$ for fixed m, k of order unity. Roberts (1987) examines the problem using boundary-layer techniques near the cylinder boundary, and these methods are extended to the case m, $k = O(\epsilon^{-1/2})$ by Gilbert (1988).

Gilbert (1988) and Ruzmaikin, Sokoloff & Shukurov (1988) consider the case of smooth helical motions of the form $\mathbf{u} = (0, r\Omega(r), W(r))$, which generalize the Ponomarenko model but also serve to clarify the effect of discontinuities. A result of Gilbert (1988) is that the discontinuous flow (5.0.5) is a fast dynamo, albeit one of diffusive type in the classification of Chap. 4. The mechanism, which we discuss below, depends crucially on the discontinuity of the velocity field at the cylinder $r = a$.

The fact that the dynamo is diffusive follows from the vanishing of Γ_{sup}, as defined in §4.1.1. If $\epsilon = 0$, solutions of the initial value problem for $\mathbf{B}$ can be separated into those where B_r initially vanishes everywhere on $r = a$, and those where it does not. In the former case there is no growth at all, whereas in the latter case a small region of non-zero radial flux f through the bounding cylinder will generate a singular flux sheet as lines are sheared by the discontinuity. This creates a net flux Γ_S in the surface $r = a$ through a small orthogonal surface S, but its growth is bounded by a linear function of time. (In the case of a region of radial flux Φ, Γ_S will increase by Φ with each revolution of the cylinder.) It follows that $\Gamma_{\mathrm{sup}} = 0$ for the Ponomarenko flow. More formally, multiplying by a smooth test function $\mathbf{f}$ and integrating $|\mathbf{f} \cdot \mathbf{B}|$ over space, the limit for large time yields at most an $O(t)$ contribution from the axisymmetric component, and therefore $\Gamma(\mathbf{f}, 0) \leq 0$. This establishes the property mentioned in §4.1.1 that any weak measure of growth in a perfectly conducting Ponomarenko dynamo has zero or negative growth rate.

To summarize the calculations with diffusion, we note first that with (5.0.5) in the induction equation the dynamo problem reduces to a simple eigenvalue problem which can be solved explicitly in terms of Bessel functions. An especially simple result is obtained when $m\omega + kW = 0$. This condition ensures that in $r < a$ the components of the magnetic field are constant on the helical particle paths of the flow. In this case the natural scaling for m, k is $O(\epsilon^{-1/2})$, and the equation for p becomes asymptotically

$$p(m, k, \epsilon) \sim \pm\epsilon^{1/3}(im\omega/2a)^{2/3} - \epsilon(k^2 + m^2/a^2). \tag{5.4.2}$$

For fixed m, we see from (5.4.2) that $\mathrm{Re}\, p = O(\epsilon^{1/3})$ as $\epsilon \to 0$; dynamo action occurs but modes with $m = O(1)$ grow ever more slowly as $\epsilon \to 0$, suggesting slow dynamo action. However the growth rate (5.4.2) increases with m for a given ϵ and reaches a maximum at large wavenumbers when $m = O(\epsilon^{-1/2})$, with corresponding growth rates $p(m, k, \epsilon) = O(1)$. Thus the real part $\gamma(\epsilon)$, when maximized over all modes, is $O(1)$ and so the dynamo is fast, but it is a fast dynamo of diffusive type (cf. Fig. 4.1b).

The physical basis of the fast dynamo mechanism depends strongly upon the discontinuity of the flow. The resulting infinite shear pulls out small loops of field protruding radially through the cylinder $r = a$. This creates highly sheared field in the z, θ components, almost invariant along the helical streamlines. The θ component diffuses, and in cylindrical geometry generates radial field; this diffuses, however slightly, across the discontinuity at a rate which is sufficient to cause fast dynamo action.

The role of diffusion can be investigated most efficiently using boundary layer methods. There, we take m, k as $O(\epsilon^{-1/2})$ from the outset. The natural condition of alignment is then $m\omega + kW = O(1)$. This last combination becomes an independent parameter, measuring the slight deviation of the flow from the strict alignment enforced earlier. The boundary layer has a radial thickness $O(\epsilon^{1/2})$ and the field components may be developed in powers of this parameter. The diffusion of the $O(\epsilon^{1/2})$ radial field determines the equation for the maximal growth rate in terms of the helical pitch χ and the alignment parameter. Details may be found in Roberts (1987) and Gilbert (1988).

For the case of smooth helical flows $\mathbf{u} = (0, r\Omega(r), W(r))$, the natural class of dynamo modes at small ϵ occurs when m, $k = O(\epsilon^{-1/3})$. These modes turn out to be the most easily excited in general flows of this type (as was suggested by Ruzmaikin & Sokoloff 1980). The dynamo action is always slow,[44] the growth rate falling off as $\epsilon^{1/3}$, and the magnetic field is localized in a critical or resonant layer of thickness $O(\epsilon^{1/3})$. This critical layer is determined by the condition $m\omega'(r) + kW'(r) = 0$, which ensures that the induced field decays outside the layer. Similar considerations occur in classical boundary-layer stability analysis (Drazin & Reid 1981). A somewhat different ordering, leading to a critical layer of thickness $O(\epsilon^{1/4})$, occurs when we instead take m and k to be $O(1)$; see Ruzmaikin et al. (1988). We emphasize that in either case the resonant layers can exist on any closed stream surface compatible with the condition on m and k. The field trapped in regions of closed streamlines during flux 'expulsion' can therefore develop dynamo action near these critical layers, provided the necessary z-dependence of the field is present.

5.4.2 General Helical Flows

We shall attempt to give a brief summary of an extensive body of research dealing with what can be loosely described as helical dynamos, in the perfectly conducting limit. The first asymptotic theory of dynamo action at small ϵ, in flows which fall into our class of integrable flows, was formulated in the early 1960's by Braginsky. Braginsky (1964) considers velocity fields which are

---

[44]In fact dynamo action does not occur at all at small ϵ if the rate of change of the pitch with radius is too great.

close to being axisymmetric, and the various components are ordered in size relative to ϵ, in such a way that dynamo action can be achieved for small ϵ. The magnetic field is also close to axisymmetric. In view of Cowling's theorem (see §4.2), the non-axisymmetric parts of the fields are essential to the dynamo action, indicating that the near axial symmetry and small ϵ limit are closely linked. Since the theory allows a small-amplitude non-axisymmetric component of the velocity of a general kind, such a perturbation could introduce regions of Lagrangian chaos, and dynamos of this type could be in principle be fast. However this requires that the small perturbations survive the $\epsilon \to 0$ limit. In Braginsky's theory, however, the velocity perturbations are $O(\sqrt{\epsilon})$, and this leads to a growth rate of order ϵ^2 and a slow dynamo. Although Braginsky's geometry allows resonant dynamo action not unlike that of the Ponomarenko dynamo, his method emphasizes the distributed excitation obtained from the axisymmetric part of $\mathbf{u} \times \mathbf{B}$. This mean e.m.f. is produced by a process called the *alpha effect*. We discuss it in connection with the Roberts cell in §5.5.2. Thorough treatments of Braginsky's kinematic theory, and an extensive list of references, may be found in Roberts (1971) and Moffatt (1978).

Braginsky's asymptotic method was reexamined by Soward (1971, 1972), who greatly clarified the theory by casting it into a Lagrangian form. Since dynamo action is achieved in Braginsky's models by the addition of small $O(\epsilon^{1/2})$ velocity components to an axisymmetric flow, Soward exhibits perturbed Lagrangian variables which map the exact non-axisymmetric flow onto an axisymmetric flow. Expressed in these new variables, the induction equation contains additional terms. This leads to a modified axisymmetric induction equation containing new terms associated with the average effect of the non-axisymmetric velocity components, and the alpha effect is recovered.

Our main interest here in the Braginsky theory is the role of the axisymmetric velocity field as a kind of 'skeletal field' for achieving dynamo action. This skeletal field has a helical structure and thus axisymmetric fields of this general type are capable of slow dynamo action producing a *non-axisymmetric* magnetic field, by analogy with the Ponomarenko dynamo.

The role of helical symmetry in producing dynamo action was first elucidated by Lortz (1968). To exhibit the dynamos of Lortz, we first define a *helically symmetric field*. Consider cylindrical polar coordinates (r, θ, z) and set $\phi = m\theta + kz$. The vector field

$$\mathbf{h} = \frac{r\nabla r \times \nabla \phi}{m^2 + k^2 r^2} \tag{5.4.3}$$

determines the helical structure. Note that $\nabla \cdot \mathbf{h} = 0$ and $\nabla \times \mathbf{h} = -2mkh^2\mathbf{h}$; so $\mathbf{h}$ is itself a Beltrami field and a steady Euler flow. We shall say that a scalar field has *helical symmetry* if it is a function of (r, ϕ, t). Velocity and magnetic fields with helical symmetry have the form

$$\mathbf{u} = U\mathbf{h} + \nabla\psi \times \mathbf{h}, \tag{5.4.4a}$$

$$\mathbf{B} = B\mathbf{h} + \nabla\chi \times \mathbf{h}, \tag{5.4.4b}$$

where the four scalar fields U, ψ, B and χ have helical symmetry. Lortz shows that these functions may be chosen to satisfy the induction equation, thus obtaining a large class of steady dynamos. The key property is the non-vanishing of $\mathbf{h} \cdot \nabla \times \mathbf{h}$ in the helical flow. This scalar, which is proportional to mk and thus vanishes in the two-dimensional or axisymmetric limits, determines the alpha effect in the model. This theory was developed further by Benton (1979), who derived a class of time-dependent helical slow dynamos.

Soward (1990) has provided a unification of these various examples by allowing a general class of symmetries which include the helical symmetry. Soward considers an analogous field $\mathbf{h}$ of the form

$$\mathbf{h} = Q(\zeta,\xi)\nabla\zeta \times \nabla\xi, \tag{5.4.5}$$

where ζ, ξ = const. define two families of intersecting surfaces. The skeletal fields are then defined by formulas similar to (5.4.4). To these skeletal fields small perturbations can then be added, and are treated as in the Lagrangian form of Braginsky's analysis. It is perhaps the class of motions most typical of the the slow dynamo. Like the smooth Ponomarenko dynamo, for these flows the dominant excitation at small ϵ occurs on resonant or critical surfaces, and it is through small modifications near these resonant surfaces that the dynamos can be rendered fast. Presumably this can happen as in the Ponomarenko case, by using singular functions, or else through suitable perturbations which render the flows locally chaotic.

5.5 Analysis of the Roberts Cell and Extensions

5.5.1 Boundary-layer Structure

We turn now to a study of the steady boundary layers created by the Roberts cell when a uniform mean magnetic field permeates the plane. Assuming that the magnetic field is planar and of the form $(A_y, -A_x, 0)$, the equation of the vector potential $A(x,y)$ is

$$\mathbf{u_H} \cdot \nabla A - \epsilon\nabla^2 A = 0, \tag{5.5.1}$$

with $\epsilon \ll 1$ and $\mathbf{u_H} = (\sin x \cos y, -\cos x \sin y, 0)$. Because of flux expulsion, A must then be constant over the interior of any square cell. Thus any flux in the (x,y)-plane is expelled to the separatrix and will be found there in diffusive boundary layers, of thickness $O(\epsilon^{1/2})$ as $\epsilon \to 0$.

The structure of this boundary layer has been considered by Childress (1979), who gives references to earlier examples of periodic boundary-layer calculations in fluid mechanics, and by Anufriyev & Fishman (1982), Perkins & Zweibel (1987), Rosenbluth et al. (1987), Shraiman (1987), and Soward

(1987). To analyze the boundary layer we fix attention on the primary cell $0 < x, y < \pi$; see Fig. 5.1(b). We may assume that $A = 0$ over the interior of the cell, and further assume that the spatial average of the magnetic field is $(1, 0)$, i.e., that the mean field is in the direction of the positive x-axis. Then, the magnetic potential has the constant value $n\pi$ over the cell $m\pi < x < (m + 1)\pi$, $n\pi < y < (n + 1)\pi$.

The boundary-layer approximation is conveniently made using arclength s along the separatrix and the stream function ψ, with $(u, v) = (\psi_y, -\psi_x)$, as new coordinates.[45] For example, along the separatrix $0 < x < \pi$, $y = 0$, we have $s = x$ and so $\partial_x = \partial_s - v\partial_\psi$, $\partial_y = u\partial_\psi$. Thus (5.5.1) becomes

$$q\frac{\partial A}{\partial s} - \epsilon q^2 \frac{\partial^2 A}{\partial \psi^2} \simeq 0, \tag{5.5.2}$$

where $q \approx u$ is the speed of the fluid near the separatrix. We have used the fact that q varies little across the boundary layer, and have anticipated the scaling of ψ below, giving $q = q(s)$ approximately. Defining

$$\sigma = \int_0^s q\, ds, \qquad \xi = \epsilon^{-1/2}\psi, \tag{5.5.3}$$

we then obtain

$$\frac{\partial A}{\partial \sigma} - \frac{\partial^2 A}{\partial \xi^2} \simeq 0. \tag{5.5.4}$$

The magnetic potential thus satisfies the heat equation in these coordinates, in the boundary-layer approximation. The same equation applies on any segment of the separatrix, and we define σ to be continuous along any connected path along the separatrix starting from the point A in Fig. 5.1(b).

We next discuss the connections between various segments of the separatrix. The boundary-layer structure is periodic, in that as one goes around the primary cell A → B → C → D the boundary layer lying toward the interior feeds back onto itself. The parts of the boundary layer adjacent to the primary cell but lying outside it feed into neighboring cells. The magnetic potential has a symmetric structure which will allow connections to be made between these adjacent pieces. To study this we first note the following symmetries of the flow $\mathbf{u}$ in the (x, y)-plane:

$$T_1 : (x, y) \rightarrow (x + \pi, y + \pi), \qquad T_2 : (x, y) \rightarrow (x + \pi, y - \pi), \tag{5.5.5a}$$

$$R : (x, y) \rightarrow (\pi - y, x), \qquad S : (x, y) \rightarrow (\pi - x, y). \tag{5.5.5b}$$

The translations T_1, T_2 change A by an additive constant but do not change the magnetic field. This allows us to connect up diagonally adjacent cells. R is a rotation anti-clockwise through $\pi/2$ about the center $(\pi/2, \pi/2)$ of the

---

[45]In classical boundary-layer theory, these are known as *von Mises coordinates*. A particularly interesting application of these coordinates is made by Kaplun (1967).

primary cell. S is a translation and reflection. To obtain the effect on a vector field $\mathbf{v}$ think of a map of the plane with the field frozen in. For example consider the rotation R: $R\mathbf{v}(x,y) = \mathbf{J}_R\mathbf{v}(R^{-1}(x,y))$ where

$$\mathbf{J}_R = \begin{pmatrix} 0 & -1 \\ 1 & 0 \end{pmatrix} \tag{5.5.6}$$

is the Jacobian of the map R. The horizontal velocity field is invariant under R, for we have $\mathbf{u}_\mathrm{H}(R(x,y)) = (\cos x \sin y, \cos y \sin x) = \mathbf{J}_R\mathbf{u}_\mathrm{H}(x,y)$, or equivalently we observe that ψ is invariant under R.

To see how the magnetic field transforms, we need the boundary conditions which determine it. From the translational symmetries and the fact that $A = 0$ in the principal cell, we have

$$A(x,0) = -\pi/2, \qquad A(x,\pi) = \pi/2, \qquad (0 < x < \pi), \tag{5.5.7a}$$

$$\frac{\partial A}{\partial y} = 0, \qquad (0 < y < \pi, \ x = 0 \text{ or } \pi). \tag{5.5.7b}$$

Noting that σ increases by the integral of $\sin x$, or by 2, in going from point A to point B, we may proceed around the boundary of the primary cell, increasing σ by this amount on any leg of the separatrix. We then observe that (5.5.7) becomes

$$A(\sigma,0) = -\pi/2, \qquad (0 < \sigma < 2), \tag{5.5.8a}$$

$$\frac{\partial A}{\partial \xi}(\sigma,0) = 0, \qquad (2 < \sigma < 4), \tag{5.5.8b}$$

$$A(\sigma,0) = \pi/2, \qquad (4 < \sigma < 6), \tag{5.5.8c}$$

$$\frac{\partial A}{\partial \xi}(\sigma,0) = 0, \qquad (6 < \sigma < 8). \tag{5.5.8d}$$

The circulating nature of the boundary layer is expressed by

$$A(0,\xi) = A(8,\xi), \qquad (\xi > 0). \tag{5.5.9}$$

We now observe that the symmetry operation R^2 reproduces the primary cell but with the boundary conditions multiplied by -1. We thus obtain a useful symmetry (which results from the vanishing of the interior value of A)

$$A(\sigma+4,\xi) = -A(\sigma,\xi), \qquad (\xi > 0). \tag{5.5.10}$$

These conditions presuppose that a definite value of $A(\sigma,\xi)$ can be assigned when at the corners $\sigma = 0, 2, 4, 6$, and in particular that A is continuous there. This can be established, provided that $\xi > 0$, by examining the equation which is governing in the corner regions (Childress 1979). For example, near A we note that $\psi \approx xy$, and if we follow a streamline line $\xi = C$ into this corner region we have $xy\epsilon^{-1/2} \sim C$. Since x and y are comparable in a corner we have a square of size $O(\epsilon^{1/4})$ into which the boundary-layer flux is carried. The corner 'turning' of the boundary layers is thus accompanied by a thickening. If now we express (5.5.1) in the variables $\bar{x} = x\epsilon^{-1/4}, \bar{y} = y\epsilon^{-1/4}$, we have

$$\bar{x}\frac{\partial A}{\partial \bar{x}} - \bar{y}\frac{\partial A}{\partial \bar{y}} - \epsilon^{1/2}\overline{\nabla}^2 A \approx 0. \tag{5.5.11}$$

Thus diffusion is negligible in the corner when ϵ is small, which implies that A is constant on streamlines through the corner.

It should be emphasized that $A(\sigma, \xi = 0)$ is not, however, continuous at the corners $\sigma = 0$ and 4. The situation can be compared to a mathematically equivalent problem, a semi-infinite bar with initial temperature $T_0(\xi)$ for $\xi > 0$, σ being time. According to the conditions (5.5.8), the end $\xi = 0$ of the bar is held at temperature $-\pi/2$ for two time units, then insulated for 2 time units, then held a temperature $\pi/2$, then insulated. At times when the end is brought to a fixed temperature (that is, at $\sigma = 0$, 4), there is in general a discontinuity in temperature at $\xi = 0$. Applied to the magnetic potential, such a discontinuity is associated with a *flux sheet*, i.e., a delta-function concentration of magnetic field parallel to the separatrix. The associated diffusive *sub-layers*, which are needed to resolve the field there, were first treated explicitly by Anufriyev & Fishman (1982). These sub-layers are passive structures from the point of view of the main boundary layer. Apart from their presence near the corners, the magnetic field is of order $\epsilon^{-1/2}$ everywhere in the boundary layer.

We now turn to the analysis of the heat conduction problem with conditions (5.5.8). In the references already given, various techniques have been used. We outline here the method used by Soward (1987), since it results in a closed-form solution and is an elegant application of the powerful Wiener–Hopf technique to a heat conduction problem, a technique also used by Brown & Stewartson (1978) in a related problem in fluid mechanics. Introducing the complex-valued function

$$\tilde{A}(\sigma, \xi) = (2/\pi)[A(\sigma, \xi) + iA(\sigma + 2, \xi)], \tag{5.5.12a}$$

we have from (5.5.10)

$$\tilde{A}(\sigma + 2, \xi) = -i\tilde{A}(\sigma, \xi). \tag{5.5.12b}$$

To express the problem in terms of $\tilde{A}$, first solve the heat conduction problem in two parts to obtain,

$$A(2, \xi) = -\sqrt{\pi}\int_{\xi/2\sqrt{2}}^{\infty} e^{-t^2}\,dt$$
$$+ \frac{1}{\sqrt{8\pi}}\int_0^{\infty} A(0, \xi')\left[e^{-(\xi-\xi')^2/8} - e^{-(\xi+\xi')^2/8}\right]d\xi', \tag{5.5.13a}$$

$$A(4, \xi) = \frac{1}{\sqrt{8\pi}}\int_0^{\infty} A(2, \xi')\left[e^{-(\xi-\xi')^2/8} + e^{-(\xi+\xi')^2/8}\right]d\xi'. \tag{5.5.13b}$$

The first term on the right of (5.5.13a) accounts for the Dirichlet boundary condition (5.5.8a), the second mapping $A(0,\xi)$ to $A(2,\xi)$ with a homogeneous Dirichlet boundary condition. (5.5.13b) maps A on the second leg with the homogeneous Neumann condition (5.5.8b). We now differentiate (5.5.13) and integrate by parts. Letting $\xi = \sqrt{8}\,x$ and defining

$$f(x) = \frac{\partial \tilde{A}}{\partial \xi}(2, \sqrt{8}\,x), \qquad (5.5.14)$$

the resulting equations can be combined into the complex integral equation

$$
\begin{aligned}
f(x) =& (\lambda/\sqrt{2\pi})e^{-x^2} \\
&+ \frac{i}{\sqrt{\pi}} \int_0^\infty \left[f(x')e^{-(x-x')^2} - f^*(x')e^{-(x+x')^2} \right] dx',
\end{aligned}
\qquad (5.5.15a)
$$

where

$$\lambda = 1 + A(0-, 0+) = 1 + i + i\tilde{A}(2-, 0+) = 1 + i - i\sqrt{8} \int_0^\infty f(x)\, dx, \quad (5.5.15b)$$

using (5.5.8, 12) and (5.5.14). λ is an unknown constant which shall be determined along with the solution by (5.5.15a).

The solution of the integral equation (5.5.15), following Soward (1987), will be sought in Fourier space, using the half-transforms

$$g_+(k) = \int_0^\infty f(x)\, e^{ikx}\, dx, \qquad (5.5.16a)$$

$$g_-(k) = \int_{-\infty}^0 f(x)\, e^{ikx}\, dx. \qquad (5.5.16b)$$

To do this, we use the fact that the real part of $\tilde{A}_{\xi\xi}(\sigma, 0) = \tilde{A}_\sigma(\sigma, 0)$, as well as the imaginary part of $\tilde{A}_\xi(\sigma, 0)$, both vanish on $0 < \sigma < 2$, allowing analytic extension of f into $x < 0$ by

$$f(-x) = f^*(x). \qquad (5.5.17)$$

In Fourier space, after multiplying (5.5.15a) by e^{ikx} and integrating from $-\infty$ to ∞, the integral equation then takes the simpler form

$$g_+ + g_- = [\lambda/\sqrt{2} + i(g_+ - g_-)]e^{-k^2/4}. \qquad (5.5.18)$$

To implement the Wiener–Hopf method, Soward notes from (5.5.17) that $g_-(k) = g_+^*(k)$ when k is real, which suggests rewriting (5.5.18) as

$$
\begin{aligned}
&e^{-i\pi/4}\left[g_+ - i\lambda/\sqrt{8}\right] \sinh\left(k^2/8 - i\pi/4\right) \\
&= e^{i\pi/4}\left[g_- + i\lambda/\sqrt{8}\right] \sinh\left(k^2/8 + i\pi/4\right).
\end{aligned}
\qquad (5.5.19)
$$

The object is to arrange (5.5.19) to have the form $F_+(k) = F_-(k)$ where $F_\pm$ are analytic functions in the upper and lower halves of the complex k-plane and hence in a common strip containing the real axis. This then implies $F_+ = F_- = F$ is analytic everywhere and hence constant. The constant can be fixed by the condition $g_+ \to 0$ as $\operatorname{Im} k \to +\infty$. A key step for achieving this decomposition is the use of the factorial representation of sinh (see, e.g., Abramowitz & Stegun 1972, p. 85),

$$\sinh z = z \prod_{k=1}^{\infty} (1 - (z/k\pi)^2). \tag{5.5.20}$$

After factoring and rearranging the terms of the product on the right of (5.5.20), this procedure yields

$$g_+(k) = \frac{\lambda}{\sqrt{8}} \left[i - e^{i\pi/4} \frac{k + (-1+i)\sqrt{\pi}}{k + (1+i)\sqrt{\pi}} \prod_{n=1}^{\infty} R_n(k) \right], \tag{5.5.21a}$$

where

$$R_n(k) = \frac{[k + (-1+i)\sqrt{(4n+1)\pi}] \, [k + (1+i)\sqrt{(4n-1)\pi}]}{[k + (1-i)\sqrt{(4n+1)\pi}] \, [k + (-1+i)\sqrt{(4n-1)\pi}]}. \tag{5.5.21b}$$

The constant λ then follows from (5.5.15b),

$$\lambda = 1 + i - i\sqrt{8}\, g_+(0) = 1 + i + \lambda(1 - e^{i\pi/4}), \tag{5.5.22}$$

which yields $\lambda = \sqrt{2}$.

We note a physical quantity of interest when this solution is interpreted as an example of convective diffusion of a passive scalar. A mean magnetic field $(1,0)$ corresponds to a mean gradient in A of $(0,1)$. If we compute the mean convective transport of A over the primary cell, this can be reduced to an integral over the segment $2 < \sigma < 4$:

$$\frac{1}{\pi^2} \int vA \, dx \, dy \sim \frac{2\epsilon^{1/2}}{\pi} \int_0^{\infty} A \, d\xi, \qquad (2 < \sigma < 4) \tag{5.5.23}$$

$$= -\epsilon^{1/2} \operatorname{Im} \left[\int_0^{\infty} \xi \frac{\partial \tilde{A}}{\partial \xi} \, d\xi \right], \qquad (0 < \sigma < 2).$$

This quantity is the effective diffusivity of A, i.e., the spatial mean of vA in a unit mean gradient Calling this $\epsilon^{1/2}G$, we have

$$G = 8 \operatorname{Re} \left[\frac{dg_+}{dk} \right]_{k=0}, \tag{5.5.24a}$$

and using (5.5.21) there results

$$G = \sqrt{2/\pi} \sum_{n=1}^{\infty} \frac{(-1)^{n+1}}{\sqrt{2n+1}} = 1.0655. \tag{5.5.24b}$$

The ratio of the dimensional effective diffusivity to the molecular diffusivity is $G\epsilon^{-1/2}$. The alternating series reflects the $1/\sqrt{\sigma}$ decay of the fundamental solution of the heat equation, when the more awkward Neumann series solution of the integral equation is used (Childress 1979).

5.5.2 Dynamo Action

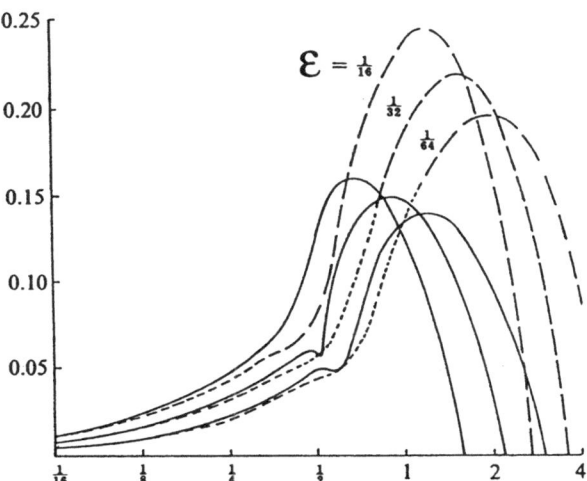

Fig. 5.6 Growth rates $p(k)$ plotted against wavenumber k as computed by Roberts (1972) (solid curves), compared with Soward's (1987) asymptotic theory for large R (dashed curves). The values of $\epsilon = 1/R$ are indicated. (Fig. 5 of Soward (1987), reproduced with permission of Cambridge University Press.)

The original studies of Roberts (1972) dealt with slow dynamo action and magnetic field of the form $\mathbf{B}(x, y, z, t) = e^{pt+ikz}\,\mathbf{b}(x, y)$. His results are summarized in Fig. 5.6. These results are typical of a slow dynamo — a maximal (over k) growth rate which decreases with decreasing ϵ. The physical mechanism involved in the dynamo was discussed heuristically by Roberts (1972) and in more detail by Soward (1987). Since it leads naturally to models such as the stretch–fold–shear map we summarize the description here. Consider a mean field $(1, 0)e^{ikz}$. In this case the boundary layers on separatrix segments parallel to the y-axis are seats of dynamo action, and we shall show how this dynamo action creates a field component of the form $(0, 1)e^{ikz}$. But the same argument applies, by the symmetry R (see 5.5.5b), to the latter component, so that the original mean field can be amplified.

The mechanism for producing mean y-field from mean x-field is shown in Fig. 5.7. The flux tongues are produced by the action of the X-type corner regions and extend into the boundary layers. The process, while here occur-

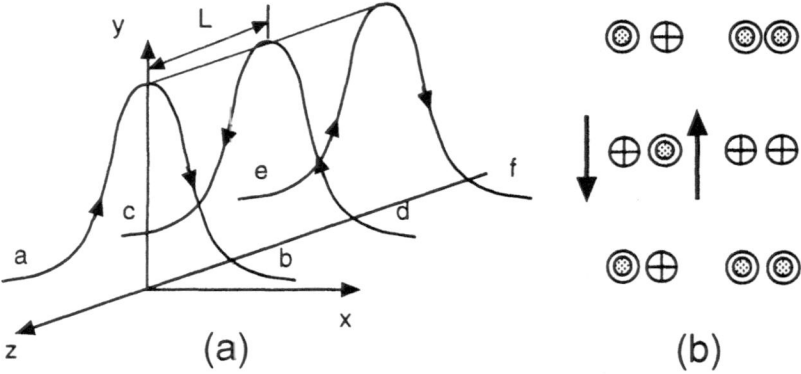

Fig. 5.7 The production of mean y-field from mean x-field in the Roberts cell. (a) The stretch mechanism, producing flux tongues, and (b) the twist or shear mechanism, which brings field of like sign into coincidence and simultaneously produces vertical field. L is the half wavelength π/k.

ring in a diffusive layer, is similar to the intensification of field in channels; see §5.3.2. To the extent that lines of force lie in a given horizontal plane, there is no *mean* field produced. The twist or shear mechanism shown in Fig. 5.7(b) is therefore essential. The vertical shear of the flow twists the flux tongues, tending to bring y-field of like sign into proximity. This produces a mean field containing the component $(0,1)e^{ikz}$. The process of reinforcement thus results from the phase shift produced by the effect of the vertical motion on the z-dependent mean field. The SFS map of §3.2 is constructed to apply this same basic mechanism to the entire domain, rather than just to the boundary layers of the cell.

Turning to analysis of the dynamo for small ϵ, the boundary-layer structure discussed in §5.5.1 can be developed further, allowing fields of the form $e^{pt+ikz}\,\mathbf{b}(x,y)$. Given a magnetic field with non-zero (horizontal) mean, as in the planar case, the resulting boundary-layer solution can then be used to compute the average induction created in the layers, by integration of $\mathbf{u} \times \mathbf{B}$ over the cell. In this way the average electromotive force can be related to the average field penetrating the cell, and a dynamo process identified in the mean components. Since the induction problem is linear in $\mathbf{B}$, the resulting relation is linear and of the form

$$\overline{\mathbf{u} \times \mathbf{B}} \approx \alpha \cdot \overline{\mathbf{B}}, \tag{5.5.25}$$

where α is a symmetric matrix; the electromagnetic force is said to create an *alpha effect*. The same result emerges from the asymptotic method of Braginsky (1964) noted above, and the mean field analysis of Krause & Rädler

(1980). The idea goes back to the seminal paper of Parker (1955). The usefulness of the concept to fast dynamo theory depends upon the accessibility of a relation of the form (5.5.25) in the perfectly conducting limit; see Moffatt (1978). Childress (1979) studied the scalar α obtained in the generalized Roberts cell, $\mathbf{u} = (\psi_y, -\psi_x, K\psi)$, at wavenumber $k = 0$, in the limit of large R. The magnetic field then reduces to $(A_y(x, y), -A_x(x, y), KB(x, y))$, where A satisfies (5.5.1) and

$$\mathbf{u_H} \cdot \nabla B - \epsilon \nabla^2 B = -\mathbf{u_H} \cdot \nabla A. \tag{5.5.26}$$

In the boundary layer limit the relevant solution of (5.5.26), compatible with the vanishing of B over the interior of the cell, is $B = \xi A_\xi / 2$. The alpha matrix can then be evaluated with the result that $\alpha = \alpha \mathbf{I}$, where the dimensionless form is $\alpha = -KG\epsilon^{1/2}/2$ with G given by (5.5.24). Thus, at small ϵ, molecular diffusion of magnetic field is nominally smaller than the alpha effect by a factor $\epsilon^{1/2}$. On the other hand the alpha effect still vanishes at zero ϵ.

The effect of non-zero vertical wave number k is however essential for dynamo action, as is clear from Fig. 5.6. The physical mechanism discussed above suggests that the maximum growth rate at small ϵ occurs when the vertical wave length is of the order of the boundary-layer thickness. The possibility of fast dynamo action in the Roberts cell was therefore examined by Soward (1987) in the limit where $\beta = k\epsilon^{1/2}$ is $O(1)$. The magnetic field $\mathbf{B} = \mathbf{b}e^{ikz}$ now takes the form $\mathbf{b} = (\mathbf{b_H}, KB)$ where $\mathbf{b_H}$ is the horizontal component (now not divergence-free). The equation for this is

$$(P + \mathbf{u_H} \cdot \nabla)\mathbf{b_H} = \mathbf{b_H} \cdot \nabla \mathbf{u_H} + \epsilon \nabla_H^2 \mathbf{b_H}, \tag{5.5.27}$$

where $\nabla_H = (\partial_x, \partial_y, 0)$ is the horizontal gradient and

$$P = p + \beta^2 + i\epsilon^{-1/2} K \beta \psi. \tag{5.5.28}$$

The vertical magnetic field is then recovered from the solenoidal condition $\nabla_H \cdot \mathbf{b_H} = -i\epsilon^{-1/2} K \beta B$.

The important magnetic structure is still confined to boundary layers, so that in many respects the analysis, though more involved, is similar in spirit to that for small k discussed above. An essential difference is in the behavior of the corner regions. The presence of the quantity P means that fields are not carried unchanged along streamlines in the corners, but rather, behave as $\exp(-P \int q^{-1} ds)$, where $q = |\mathbf{u_H}|$. Now, from §5.2.1, the time taken by a fluid particle to pass through a corner region is $\tau \sim -\log \psi \sim \log \epsilon^{-1/2}$ for fixed ξ. In term of this time variable, Soward finds an expression for the growth rate, by a method similar to the matching of the boundary layers through corner regions, of the form

$$p = (2/\pi)^{1/2} \beta K e^{(-p+\beta^2)\tau} - \beta^2. \tag{5.5.29}$$

There follows a maximal growth rate, written in terms of $R = 1/\epsilon$:

$$p_{\text{max}} \sim \frac{\log \tau}{2\tau} \sim \frac{\log \log R}{\log R}. \tag{5.5.30}$$

The dynamo is thus slow, but barely so. In fact, as Soward notes, the Roberts cell can be made into a fast dynamo by changing the flow locally, in the corner regions, so that the time of transit of fluid particles becomes independent of ψ, instead of diverging logarithmically. For example, given $\delta \ll 1$ set $\psi = a_\delta \sin x \sin y$ where

$$a_\delta = 1 + \log^2(r/\delta) \tag{5.5.31}$$

when r, the radial distance to the X-point, is less than δ, and otherwise $a_\delta = 1$. The modification, which introduces logarithmic singularities into the vorticity field, yields $\tau \sim -2 \log \delta$, and hence a fast dynamo, which we refer to as the 'modified Roberts flow'.

It should be emphasized that the modification introduces what may be called 'tearing' of a fluid element, in the same way as occurs at the cylinder boundary in the Ponomarenko flow (5.0.5). For fluid particles having the same initial y-position, $y = -\delta$, but lying on streamlines $\psi = \pm c$ for small c, will be separated by a distance 2δ in fixed finite time, irrespective of how small c may be. This tearing property must account for the fast dynamo action, although the process is somewhat different from the tearing that occurs in the Ponomarenko model.

This result is important because it underscores the delicate role of smoothness of the velocity field on the realizability of fast dynamo action. The Ponomarenko and modified Roberts motions are similar in that they are both fast dynamos of diffusive type with singular flow fields, and both become slow dynamos under smoothing.

5.5.3 Cat's Eyes and Mean Flow

As final examples of two-dimensional integrable flows we consider two flows which might be regarded as 'transitional', lying between the geometry of the Roberts cell and chaotic flows such as the ABC flow with $A = B = 1$, $C = \epsilon$.

The first is the cat's-eye flow introduced in §5.3.3, and defined by (5.1.4) and (5.3.9). For this flow, Childress & Soward (1989) consider the effective diffusion of a scalar field in the horizontal velocity, as well as the alpha effect for the dynamo at zero vertical wave number. The primary motivation is to determine the role of the channels. In the context of boundary-layer theory, the channel of width δ becomes part of the layer structure when $\beta \equiv \delta\epsilon^{-1/2} = O(1)$. When $\beta \gg 1$ the channel should be regarded as bounded by separate layers, each a separatrix of the flow (see Fig. 5.4).

If $\delta = O(1)$, so the channels are of finite width, we found in §5.3.2 that at small ϵ there can be a dramatic effect on the horizontal mean field. The result is the enhancement of both effective diffusion and the alpha effect. The study of this phenomenon in the boundary-layer limit assumes $\beta = O(1)$ and must be carried out numerically. A direct attack using the Wiener–Hopf approach of §5.5.1 does not appear possible, but the procedure can be used to study the large-β asymptotics of the boundary-layer limit, under the assumption that the layers on either side of the channel do not influence one another. The result is that the effective diffusion is determined by the 2×2 matrix

$$\mathbf{D}_{\text{eff}} = D_{\text{molecular}}(\mathbf{I} + \epsilon^{-1/2}\mathbf{I}^+\mu^+ + \epsilon^{-1/2}\mathbf{I}^-\mu^-), \tag{5.5.32a}$$

where

$$\mathbf{I}^\pm = \frac{1}{2}\begin{pmatrix} 1 & \pm 1 \\ \pm 1 & 1 \end{pmatrix} \tag{5.5.32b}$$

and $\mu^\pm$ are functions of β. Note that $\mu^\pm(0) = \mu = 1.065$, the value for the Roberts cell. For large β, the asymptotic solutions give

$$\mu^+ \sim \beta^3/3, \tag{5.5.33a}$$

an estimate which agrees with the leading terms computed from Prandtl–Batchelor theory for small δ. Indeed, from (5.3.10) and (5.3.13), and with $\phi = 0$ say, we obtain $A_0 \approx \pi(\psi^2 - \delta^2)/8$ for small δ. To compute advective flux in the y-direction, we average $vA_0 = -\psi_x A_0$ over the cell. This yields $\delta^3/6$. Since the comparison mean gradient is unity and $A \approx \epsilon^{-1}A_0$, this gives the dominant contribution (5.5.33a) in (5.5.32).

The cross-channel diffusion is much smaller and, while unobtainable from Prandtl–Batchelor theory, is accessible from the Wiener–Hopf asymptotics:

$$\mu^- \sim 1/\beta. \tag{5.5.33b}$$

Similar results are available for the 2×2 alpha effect matrix.

A helpful physical argument leading to diffusion rates of the form (5.5.33) is given by Crisanti et al. (1990). Consider a particle executing a random walk. The molecular diffusion, with diffusivity $D_{\text{molecular}} = D$ say, will cause the particle to escape the channel after time $T_c \sim L^2\delta^2/D$ where $UL/D = 1/\epsilon$ (ϵ and δ being dimensionless). During this time it moves a distance $L_c \sim UT_c \sim UL^2\delta^2/D$. If we suppose T_c to be large compared to the time scale L/U of the flow, then we have the condition $\delta\epsilon^{-1/2} \gg 1$. Since the probability of finding a particle in a channel is proportional to δ, we obtain $D_{\text{eff}} \sim \delta L_c^2/T_c \sim L^2U^2\delta^3/D \sim D\epsilon^{-1/2}\beta^3$ as given in (5.5.32, 33a). On the other hand cross-channel transport is associated with time T_c but length L, so $D_{\text{eff}} \sim \delta L^2/T_c = D/\delta \sim D\epsilon^{-1/2}/\beta$, as in (5.5.32, 33b) These estimates also apply to Lagrangian markers on field lines and so can be used to deduce field strengths as well.

The cat's-eye flow very likely remains a slow dynamo, although the dynamo study analogous to Soward (1987) for the Roberts cell has not been carried out. The main reason for considering the cat's eye flow is to emphasize the effective transport of the channels. Fast dynamo activity, and turbulent diffusion, actually involve *smaller* rates of transport. Exponential stretching and chaos go hand in hand, and the coherent channel structure is destroyed. On the other hand the mechanism by which field lines are stretched remains the same in the small, even in random channels.

A second example, which modifies the Roberts cell so as to create more involved boundary-layer structure including channels, involves the addition of a small horizontal mean flow $\delta(\cos\theta, \sin\theta, 0) = \delta\mathbf{U}$ to the velocity given by (5.1.4). We now denote the latter by $\mathbf{u}_0$ so $\mathbf{u} = \mathbf{u}_0 + \delta\mathbf{U}$. To understand what happens for $\delta > 0$, consider a fluid particle initially on the separatrix of $\mathbf{u}_0$, e.g., on the interval $0 < x < \pi$, $y = 0$. Since the streamfunction associated with $\mathbf{u}_0$ vanishes on this separatrix, the streamfunction for $\mathbf{u}$ there has the value $\delta(y\cos\theta - x\sin\theta)$. If the particle is now released in the flow, it will, under the action of the mean flow, eventually find itself again on the separatrix. But this point carries the same stream function value. Thus the intersections of the particle orbit with the separatrix are also the intersections with a straight line at angle θ through the initial point; see Fig. 5.8.

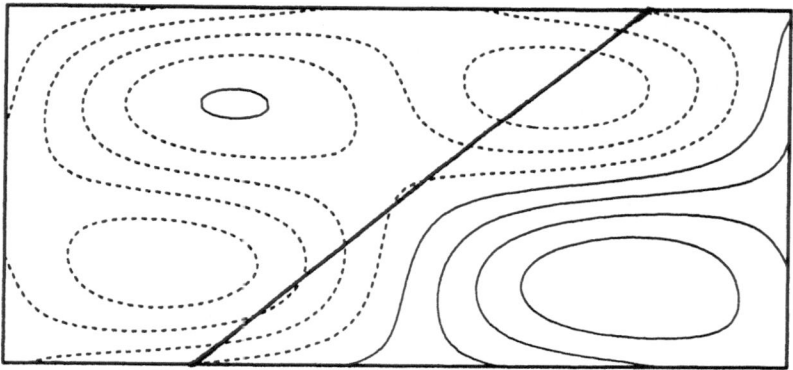

Fig. 5.8 The Roberts cell with mean flow, $\delta = 0.3$ and $\theta = \pi/6$. The domain shown is the $2\pi \times 2\pi$ cell and the straight line is aligned with the mean flow.

It follows that if $\tan\theta$ is a rational number then the particle orbit, viewed as an orbit on a two-torus rather than the periodic plane, will eventually close. If, however, $\tan\theta$ is irrational, the orbit cannot close, and will fill a channel densely. In the latter case any boundary layer structure is unresolvable; so the asymptotic methods for the flows with zero mean will not work. The only hope for carrying out calculations analogous to the large-β theory of the cat's-eye flow is thus to assume a rational tangent and then make ϵ sufficiently small to resolve the boundary layers bounding the channels. It turns out to be possible to extract estimates analogous to the $\beta^3/3$ asymptotic term (5.5.33a) for the cat's-eye flow, for a mean flow with arbitrary rational tangent (Soward

& Childress 1990). This involves Prandtl–Batchelor theory and a delicate matching of channels of arbitrarily long spatial period. The 2×2 effective diffusion matrix takes the form

$$\mathbf{D}_{\text{eff}}/D_{\text{molecular}} \sim \epsilon^{-1/2} \mu(\beta, \theta) \mathbf{UU}, \tag{5.5.34a}$$

where again $\beta = \delta \epsilon^{-1/2}$. If $\tan \theta = p/q$, where p, q are relatively prime integers,

$$\mu = \frac{\beta^3 \pi^3}{48(p+q)(p^2+q^2)^{1/2}\Lambda}, \tag{5.5.34b}$$

with $\Lambda = 1$ for $p+q$ odd and $\Lambda = 1 - 1/(p+q)^2$ for $p+q$ even. (The topology of the channels divides into two classes, depending upon whether $p+q$ is even or odd.) From (5.5.34) it follows that the leading term of $\mathbf{D}_{\text{eff}}$, which diverges like $1/\epsilon$ as $\epsilon \to 0$ just as for the cat's-eye flow, vanishes in the limit $p, q \to \infty$, when the tangent becomes irrational.

Now at an irrational tangent the separatrices are almost periodic and dense in the channel (on the two-torus), and (5.5.34b) shows that this configuration removes the divergence of $\mathbf{D}_{\text{eff}}$ at small ϵ. This is important as the first instance of the effect of complexity of channel structure on the transport. These results raise the possibility of enhanced transport which could remain in the limit $\epsilon \to 0$. However, we now know that in the present example the enhanced diffusion is in fact stopped completely in the irrational limit, i.e., the effective diffusivity is comparable to the molecular value, presumably a result of the almost-periodic channel structure (see the mathematical results discussed below).

Soward & Childress (1990) also carried out numerical calculations for $\beta = \delta \epsilon^{-1/2}$ moderately large. These results for $\alpha(\theta)$ and $D_{\text{eff}}(\theta)$ were then compared with particular estimates at rational tangents. These calculations suggest a 'spiky' variation with θ at very small ϵ, with perfectly conducting limits being differentiable nowhere.

Recently there have been several important mathematical studies of enhanced transport which have put these formal calculations on firmer ground. Avellaneda & Majda (1991) give an abstract setting for determining effective diffusion for a large class of steady periodic flows without mean motion. A closely related theory of diffusion in a porous medium with possible mean motion is given by Koch et al. (1989). In particular in the theory of Koch et al., our case of irrational tangent would correspond to diffusion which is in the limit of the order of the molecular diffusion, an ordering they predict whenever the streamlines are dense modulo 2π. More recently, Majda & McLaughlin (1993) and Fannjiang & Papanicolaou (1994) have analyzed using various mathematical tools a number of steady periodic flows, some of which include mean motion. A large body of work on the related problem of transport in random flows is reviewed by Isichenko (1992). Majda & McLaughlin (1993) apply the theory of Avellaneda & Majda (1991) to flows with mean motion. They establish rigorously that the rational case has the divergence with $\epsilon \to 0$ already noted. They can also confirm the Koch *et al.* (1989) result that the irrational case is one of *minimal* enhanced diffusion, comparable to the molecular value.

As we indicated at the beginning of this section, the flows considered here are variants of the Roberts cell exhibiting enhanced dynamo action (as measured by the alpha effect), of a kind that would suggest transition to fast dynamos. (Note that as yet no calculations of dynamo action in the cats-eye and mean flow models, analogous to those of §5.5.2 for the Roberts cell, have been carried out). However it would be misleading to characterize them as transitional toward chaotic fast dynamo action. The appropriate transitional flows are the nearly integrable cellular flows of Chap. 8, and the ABC flows with $C \ll 1$ (Gilbert 1992). Slow dynamo action in generalized helical flows as developed by Soward (1990) would occur simultaneously with fast dynamo action in small regions of chaos, with non-diffusive fast dynamo action emerging once the region of chaos became sufficiently large.

5.6 Axisymmetric Boundary Layers

These last results concerning two-dimensional periodic flows with and without mean motion suggest that order one α and D_{eff} are inaccessible in laminar flows — either there is a divergence with ϵ associated with finite channels, or else transport is reduced to its molecular value. We give an example in the present section which shows that in three dimensions this need not be the case, even under the constraint of axial symmetry. The simplest axisymmetric velocity field having the form of the Roberts cell (5.1.4) is, in cylindrical polar coordinates,

$$(u_z, u_r, u_\phi) = r^{-1}(\partial\psi/\partial r, -\partial\psi/\partial z, K\psi). \tag{5.6.1}$$

Here the domain of interest is the cylinder $0 < z < 1$, $0 \leq r < r_0$, and while we do not specify $\psi(z,r)$ we shall want it to vanish on the boundary of the cylinder and have the property that the surfaces $\psi = \text{const.}$ have the form of a single family of nested toroidal domains. Also we assume that

$$\psi \sim r^2 f(z) \tag{5.6.2}$$

for small r. For such a flow Childress (1979) (see also Soward 1988) showed that the alpha coefficient relating a mean magnetic field $(B_0, 0, 0)$ to a mean z-current tends to a finite non-zero value as $\epsilon \to 0$. The magnetic field developed by the flow is assumed to have the same form as the velocity,

$$(B_z, B_r, B_\phi) = r^{-1}(\partial A/\partial r, -\partial A/\partial z, r^2 B). \tag{5.6.3}$$

Studies of magnetohydrodynamic convection in flows of this kind, by Galloway, Proctor & Weiss (1978), and more recently by Jones & Galloway

(1993a, b), show the development of magnetic fields of this form, where the dominant field occurs at an intense flux rope along the axis of the cylinder. This is the axisymmetric form of flux expulsion, and since it tends to concentrate a flux $\pi r_0^2 B_0$ into a rope of thickness $O(\epsilon^{-1/2})$, the field in the rope is of order $1/\epsilon$. We assume a periodic extension of the field in z, so that B_z and B_ϕ are even with respect to $z = 0, 1$, and set $B_\phi = 0$ on $r = r_0$. Then a boundary-layer theory may be developed along the lines of the boundary-layer theory of the Roberts cell summarized in §5.5.1. It has a somewhat more complicated structure since the 'corner' flows at the axis are quite different from those at the edges of the cylinder. At the axis and at $z = 1$, given $f(z) \geq 0$ in ψ, the flux rope fans out into a flux sheet along the upper wall of the cylinder. Since the local corner flow at the axis has stream function $r^2 f'(1)(z - 1)$, which is $O(\epsilon)$ in the flux rope, we see that this sheet has thickness $O(\epsilon)$ as it develops along the upper wall, and so is thin compared to the nominal boundary layer thickness. As a result, the boundary layer structure must also involve a flux tube of thickness $O(\epsilon^{1/4})$ at the axis, formed as the boundary layer on the *lower* wall $z = 0$ of the cylinder *converges* into its corner flow at the axis. The structure thus consists of two nested flux tubes with associated corner flows at the axis, a boundary layer of thickness $O(\epsilon^{1/2})$, and two-dimensional corner flows at the edges.

It might be thought that the dominant alpha effect should occur on the structure of maximum field intensity, namely the inner flux rope. This is not, however, the case. The z-component of the average of $\mathbf{u} \times \mathbf{B}$ over the cylinder acquires the contribution of order unity from the upper wall of the cylinder, in the boundary layer. The mechanism is as follows. In that boundary layer, the field $\mathbf{B}$ is pulled out into a spiralling field with ϕ-components of order $\epsilon^{-1/2}$. The radial velocity of the fluid through these lines of force is of order unity, and the resulting e.m.f. $\mathbf{u} \times \mathbf{B}$ is of order $\epsilon^{-1/2}$. When integrated over the boundary layer, an α of order unity is produced.

Childress & Soward (1985) considered this mechanism in the context of steady three-dimensional flows, and applied it to the '111' ABC flow. The physical process should create intense induction at unstable two-dimensional manifolds in Beltrami flows, but as yet the complexity of these manifolds has prevented this from being a useful technique for computing a growth rate in a steady fast dynamo. Soward (1989) has reviewed various results indicated above for steady flow at large R, with special emphasis on the implications for possible fast dynamo action.

6. Upper Bounds

The broad theme of this chapter is the derivation of upper bounds for fast dynamo growth rates. We have already met the line-stretching exponent h_{line} as a plausible upper bound in §2.4 and §3.3; this is discussed further in §6.1. In §6.2 we consider a conjecture of Oseledets (1993), which is related to methods of obtaining fast dynamo growth rates using Fredholm determinants (§6.3). In §6.4 we discuss Arnold's (1972) suspension of the cat map as a flow on a manifold; this is an example of an Anosov flow, and in §6.5 we develop Bayly's (1986) analysis of fast dynamo action in such flows. Section 6.6 follows Vishik's (1988, 1989) construction of an approximate Green's function, which gives an upper bound on fast dynamo growth rates. In §§6.7, 8 we consider the implementation of diffusion in terms of noisy trajectories, sketch Klapper & Young's (1995) proof that h_{top} bounds fast dynamo growth rates, and discuss the flux conjecture.

6.1 Field Growth and Topological Entropy

In Chaps. 2 and 3 we have seen how chaotic stretching of magnetic field, together with constructive folding, can lead to growth of average fields and fast dynamo action. Chaotic flows are associated with positive Liapunov exponents, which measure the asymptotic growth of vectors. Given a starting point $\mathbf{a}$ and a vector $\mathbf{v}$ in a flow let

$$\chi(\mathbf{a}, \mathbf{v}) = \limsup_{t \to \infty} \frac{1}{t} \log |\mathbf{J}^t(\mathbf{a})\mathbf{v}|. \tag{6.1.1}$$

By Oseledets' (1968) multiplicative ergodic theorem $\chi(\mathbf{a}, \mathbf{v})$ exists for almost all $\mathbf{v}$ and $\mathbf{a}$ in a given chaotic region; we call this value Λ_{Liap} the (leading) Liapunov exponent. In fact Λ_{Liap} is the first in a family of exponents. Given a point $\mathbf{a}$ define the positive-definite, symmetric matrix

$$\Lambda^t(\mathbf{a}) = (\mathbf{J}^t(\mathbf{a}))^{\mathrm{T}} \mathbf{J}^t(\mathbf{a}), \tag{6.1.2}$$

with the property that $|\mathbf{J}^t(\mathbf{a})\mathbf{v}|^2 = \mathbf{v} \cdot \Lambda^t(\mathbf{a})\mathbf{v}$, and set

$$\Lambda_{\mathbf{a}} = \lim_{t \to \infty} (\Lambda^t(\mathbf{a}))^{1/2t}; \tag{6.1.3}$$

this matrix exists for almost all $\mathbf{a}$.[46] The logarithms of the eigenvalues of $\Lambda_\mathbf{a}$ are the same for almost all $\mathbf{a}$, and are called Liapunov exponents, $\Lambda_{\text{Liap}} \equiv \Lambda_{\text{Liap}}^{(1)} \geq \Lambda_{\text{Liap}}^{(2)} \geq \cdots \geq \Lambda_{\text{Liap}}^{(d)}$; they give the asymptotic rate of growth of vectors in the corresponding subspaces of $\Lambda_\mathbf{a}$ (Oseledets 1968). Here d is the dimension of space and for a volume-preserving flow the sum of the Liapunov exponents is zero. Since the definition (6.1.3) is sensitive to the asymptotic behavior of trajectories, the Liapunov exponents are not usually continuous functions of $\mathbf{a}$.

As the Liapunov exponent gives the rate at which almost all vectors stretch, one might conjecture that this is an upper bound on the fast dynamo growth rate, and would be achieved in the absence of cancellations. However as we saw in §§2.4, 3.3.1, magnetic field can grow faster than the largest Liapunov exponent (Bayly 1986). A more plausible upper bound for field growth without cancellation is the line-stretching exponent h_{line} (Finn & Ott 1988a, b), given in equation (1.6.3) as the rate of growth of smooth material curves.

The line-stretching exponent is generally larger than the Liapunov exponent; we give a heuristic argument (Klapper 1991) to explain this. Since the Liapunov exponent is the same at almost all points, we can average (6.1.1) over area (and over $\mathbf{v}$ in a single chaotic region:

$$\Lambda_{\text{Liap}} = \lim_{t \to \infty} \frac{1}{t} <\log |\mathbf{J}^t(\mathbf{a})\mathbf{v}|>. \tag{6.1.4}$$

Now consider the rate of stretching of a finite curve in this chaotic region; after a time, the curve will be spread all over the region. (The argument depends strongly on this mixing assumption.) Its subsequent length will be given by the average of $\mathbf{J}^t(\mathbf{a})\mathbf{v}$ over the region; therefore

$$h_{\text{line}} = \lim_{t \to \infty} \frac{1}{t} \log <|\mathbf{J}^t(\mathbf{a})\mathbf{v}|>. \tag{6.1.5}$$

This formula is easily verified for the generalized baker's maps of §3.3.1. Jensen's inequality (which follows from the convexity of log),

$$\log <\cdot> \geq <\log \cdot>, \tag{6.1.6}$$

implies that $h_{\text{line}} \geq \Lambda_{\text{Liap}}$, with equality only when the stretching is completely uniform, as in the cat map.

The line-stretching exponent is closely related to the topological entropy h_{top} of the flow by theorems of Newhouse (1987) and Yomdin (1987), who show that the topological entropy gives the maximum growth rate of finite k-dimensional volumes in a smooth (that is, C^∞) flow, maximized over $k =$

---

[46]We shall avoid technicalities in the definition of Liapunov exponents (see the lucid discussion in Ruelle 1989a). Let us just say that 'almost all' means with respect to an *ergodic invariant measure*, which for our purposes is volume restricted to a single connected region of chaos.

$1, \ldots, d$. Thus for smooth planar area-preserving flows $h_{\text{top}} = h_{\text{line}}$, while in three dimensions $h_{\text{line}} \leq h_{\text{top}}$.

There are a number of equivalent definitions of h_{top} following the seminal paper of Adler, Konheim & McAndrew (1965); see, for example, Walters (1982). Here we give a definition of Dinaburg (1970) and Bowen (1971). Given a flow M^t on a compact metric space, $t > 0$ and $\epsilon > 0$, we say two points $\mathbf{x}$ and $\mathbf{y}$ are (t, ϵ)-separated if

$$d(M^\tau \mathbf{x}, M^\tau \mathbf{y}) > \epsilon, \quad \text{for some } \tau \text{ with } 0 < \tau < t; \tag{6.1.7}$$

here $d(\mathbf{x}, \mathbf{y})$ is the distance between $\mathbf{x}$ and $\mathbf{y}$. We then define $N_{\text{sep}}(t, \epsilon)$ to be the maximum number of points $\mathbf{x}_i$ that can be mutually (t, ϵ)-separated. In other words if we observe the system for a time t and if we can only resolve points more than ϵ apart, then we can observe at most $N_{\text{sep}}(t, \epsilon)$ distinct orbits. Finally we define the *topological entropy* of the flow $\mathbf{u}$ by

$$h_{\text{top}} = \lim_{\epsilon \to 0+} \limsup_{t \to \infty} \frac{1}{t} \log N_{\text{sep}}(t, \epsilon), \tag{6.1.8}$$

which is the rate at which the system reveals information about its structure to us as t increases. This explains the choice of the word 'entropy' as used in the mathematical theory of communication (see, e.g., Shannon & Weaver 1964). Positive topological entropy is a hallmark of chaos in a flow, as it indicates sensitive dependence on initial conditions. It has the property that it is invariant under time-reversal, $h_{\text{top}}(\mathbf{u}(\mathbf{x}, t)) = h_{\text{top}}(-\mathbf{u}(\mathbf{x}, -t))$.[47]

Vishik (1992) and Klapper & Young (1995) have shown that topological entropy is an upper bound on the fast dynamo growth rate, a result first conjectured by Finn & Ott (1988 a, b). Let us give the result, quoted from Klapper and Young but specialized to the case of smooth flows and smooth fields. Define first the *supremum fast dynamo growth rate* $\overline{\gamma}_0 \equiv \limsup_{\epsilon \to 0} \gamma(\epsilon)$; note that in this monograph we use an infimum definition for the fast dynamo growth rate γ_0 and that $\gamma_0 \leq \overline{\gamma}_0$. Their result implies that $\overline{\gamma}_0 \leq h_{\text{top}}$.

Theorem 6.1 *(Vishik 1992, Klapper & Young 1995) Let $\mathbf{u}$ and $\mathbf{B}_0$ be smooth divergence-free vector fields supported on a compact domain $\mathcal{D} \subset \mathbb{R}^d$. Let $\mathbf{f}$ be the time-one map of the steady flow generated by $\mathbf{u}$ and let $\mathbf{B}(\mathbf{x}, t; \epsilon)$ be the solution of the induction equation with initial condition $\mathbf{B}_0(\mathbf{x})$. Then*

$$\limsup_{\epsilon \to 0} \limsup_{n \to \infty} \frac{1}{n} \log \int_{\mathcal{D}} |\mathbf{B}(\mathbf{x}, n; \epsilon)| \, d\mathbf{x} \leq h_{\text{top}}(\mathbf{f}), \tag{6.1.9}$$

where $h_{\text{top}}(\mathbf{f})$ is the topological entropy of $\mathbf{f}$ (restricted to $\mathcal{D}$). This upper bound is also valid for $\epsilon = 0$ and in this case h_{top} may be replaced by h_{line} (I. Klapper and L.S. Young, personal communication).

---

[47]Note also that Λ_{Liap} can show complicated behavior when a system parameter is varied, while h_{top} tends to be better behaved; see §9.5 of Ruelle (1989a). It is also easier to measure h_{top} numerically.

As stated here this theorem is for steady flows; however within the more general framework of Klapper and Young, it applies equally to time-periodic flows. The result immediately implies an anti-fast-dynamo theorem: a flow must have chaos in the sense $h_{\text{top}} > 0$ to be a fast dynamo. Note that a bound is not given in terms of h_{line} for positive ϵ, only for $\epsilon = 0$. The result was first proved by Vishik (1992) for smooth flows using methods from geometric measure theory; it was generalized by Klapper & Young (1995) to give bounds for non-smooth flows, using methods based on noisy trajectories, which we discuss below in §6.7.

For smooth two-dimensional flows $\mathbf{u} = (\mathbf{u}_{\text{H}}, u_z)$, the quantities h_{line} and h_{top} may be calculated from the planar horizontal flow $\mathbf{u}_{\text{H}}$ rather than the full flow, as there is no exponential stretching in the z-direction, only linear shearing. Thus for these flows $h_{\text{top}} = h_{\text{line}}$ and we retrieve h_{line} as an upper bound on γ_0. As we have noted in §5.2.2, for steady, smooth, two-dimensional flows, h_{top} and h_{line} are zero (Young 1977), and so these are eliminated from the class of fast dynamos; for example the result excludes fast dynamo action in Roberts' cellular flow of §5.5.2, a result that deserves more than passing attention since the actual decay of $\gamma(\epsilon)$ with ϵ is extremely weak. For smooth three-dimensional flows h_{line} can be smaller than h_{top}. However if the flow is steady $h_{\text{line}} = h_{\text{top}}$; this result follows from constructions in Klapper & Young (1995). Thus it is only for unsteady three-dimensional flows that the quantities h_{line} and h_{top} may be distinct.

Curiously, while $h_{\text{line}}(\mathbf{u}(\mathbf{x}, t))$ is a natural bound on the growth of length of magnetic field lines for $\epsilon = 0$, and so on the flux growth rate Γ_{sup}, the numerical technique of Chap. 2 suggested an upper bound of $h_{\text{line}}(-\mathbf{u}(\mathbf{x}, -t))$, i.e., the line-stretching exponent of the time-reversed flow: there we tracked the flux through a surface S by following the boundary of S in the time-reversed flow. The flows of Chap. 2 have time-reversing symmetries and so $h_{\text{line}}(\mathbf{u}(\mathbf{x}, t)) = h_{\text{line}}(-\mathbf{u}(\mathbf{x}, -t))$. In any case for two-dimensional flows and steady three-dimensional flows $h_{\text{line}}(\mathbf{u}(\mathbf{x}, t)) = h_{\text{top}} = h_{\text{line}}(-\mathbf{u}(\mathbf{x}, -t))$. For unsteady three-dimensional flows $h_{\text{line}}(\mathbf{u}(\mathbf{x}, t)) \neq h_{\text{line}}(-\mathbf{u}(\mathbf{x}, -t))$ is possible, but in any case neither of these can exceed h_{top}. We remark that reversed flows occur in the formulation of the adjoint induction problem; see §9.3.1.

All the discussion above is for smooth flows; there is a similar version of the theorem for smooth maps and pulsed diffusion. If the map or flow $\mathbf{u}$ is not smooth, but only C^k for $k \geq 1$, Klapper and Young give an upper bound of the form h_{top} plus a correction depending on the smoothness of the flow or map. The correction accounts for the fact that while topological entropy bounds the length of curves 'in the large', lack of smoothness in the flow can cause local crinkling of lines and enhance growth. Because of the smoothness requirement, these results do not apply to the Ponomarenko or modified Roberts diffusive fast dynamos, which are only piecewise continuous. Nor do they apply to generalized baker's maps, SFS or SFSl, although h_{top} is an upper bound on Γ_{sup} in these cases, and numerically appears to be an upper bound on γ_0. For $\epsilon = 0$ Klapper & Young (1995) also give a lower bound on the growth rate of the integral in equation (6.1.9), maximized over all initial conditions, in terms of the topological entropy.

6.2 Oseledets' Conjecture

The definitions and properties of h_{line} and h_{top} are the same for a map M when time t is replaced by number of iterations n. Let us now consider a map and let $N_{\text{orb}}(n)$ be the number of periodic points $\mathbf{x}$ with period n: $M^n\mathbf{x} = \mathbf{x}$. Then for certain classes of chaotic maps the topological entropy is given by the growth in the number of periodic orbits with n:

$$h_{\text{top}} = \limsup_{n \to \infty} \frac{1}{n} \log N_{\text{orb}}(n) \qquad (6.2.1)$$

(see §8.5 of Walters 1982). For example, this applies to the cat map of §3.5, whose periodic orbits are given by $(M^n - I)\mathbf{x} = \mathbf{0}$ mod 1. Since $M^n - I$ maps any unit square into a parallelogram of area $|\det(M^n - I)| = \lambda^n + \lambda^{-n} - 2$, it may be seen that the number of distinct periodic orbits on $\mathbb{T}^2$ is precisely this area: $N_{\text{orb}}(n) = \lambda^n + \lambda^{-n} - 2$ and so $h_{\text{top}} = h_{\text{line}} = \log \lambda$.

This property of h_{top} is connected with a conjecture of Oseledets (1993), who associates an index ± 1 with each unstable periodic orbit $\mathbf{x} = M^n\mathbf{x}$ by

$$\text{ind}(\mathbf{x}, n) = \text{sign} \det (\mathbf{J}^n(\mathbf{x}) - \mathbf{I}) = \text{sign} \prod_j (\lambda_j(\mathbf{x}, n) - 1), \qquad (6.2.2)$$

where $\lambda_j(\mathbf{x}, n)$ are the Floquet multipliers of the orbit, i.e., the eigenvalues of $\mathbf{J}^n(\mathbf{x})$. If the index is negative the orbit is called *untwisted*, otherwise it is *twisted*.[48] To interpret this consider an unstable periodic orbit of an area-preserving two-dimensional map having Floquet multipliers $\lambda(\mathbf{x}, n)^{\pm 1}$ with $|\lambda(\mathbf{x}, n)| > 1$, and take a field vector aligned with the direction corresponding to $\lambda(\mathbf{x}, n)$. The index is

$$\text{ind}(\mathbf{x}, n) = \text{sign}[(\lambda(\mathbf{x}, n) - 1)(\lambda^{-1}(\mathbf{x}, n) - 1)] = \text{sign}(-\lambda(\mathbf{x}, n)). \qquad (6.2.3)$$

If $\lambda > 1$ the orbit is untwisted, and after one traversal of the orbit the field vector returns pointing in the same direction. If $\lambda < -1$, the orbit is twisted and the vector returns with its direction reversed.

Let $N_{\text{tw}}(n)$ be the sum of the indices of periodic points, period n:

$$N_{\text{tw}}(n) = \sum_{\mathbf{x} \in \text{per}(n)} \text{ind}(\mathbf{x}, n), \qquad (6.2.4a)$$

---

[48] Note that Oseledets (1993) has the opposite definition of twisted and untwisted periodic orbits.

where $\mathrm{per}(n)$ is the set of all fixed points of M^n; in two dimensions,

$$N_{\mathrm{tw}}(n) = \sum_{\mathbf{x} \in \mathrm{per}(n)} \mathrm{sign}(-\lambda(\mathbf{x}, n)). \qquad (6.2.4b)$$

Define

$$h_{\mathrm{tw}} = \limsup_{n \to \infty} \frac{1}{n} \log |N_{\mathrm{tw}}(n)|. \qquad (6.2.5)$$

Oseledets (1993) conjectures that for a 'typical' dynamical system the fast dynamo growth rate is h_{tw}. More precisely, here is the conjecture quoted from his paper:

Conjecture 5 (Oseledets 1993) For a 'typical' dynamical system the fast dynamo exponent is the growth rate of the number of twisted and untwisted periodic orbits whose lengths are bounded by t on the support of the invariant measure with the largest Kolmogorov Sinai (KS) entropy.

Since this is a conjecture and words such as typical remain to be elucidated, we will not attempt to explain the conditions given nor the term KS entropy (see, e.g., Ruelle 1989a); we take it that only periodic orbits in the 'most chaotic' region are to be included. This conjecture has close connections with the topological entropy and with methods based on Fredholm determinants and zeta functions, to be discussed in the next section. In table 4.1 of §4.1 we listed the systems for which the conjecture is known to hold in the column C5.

Oseledets' conjecture is easily verified for cat maps and baker's maps. For the cat map all orbits are untwisted and $h_{\mathrm{tw}} = h_{\mathrm{top}} = \log \lambda = \gamma_0$. Consider a generalized baker's map with N strips and orientation factors ϵ_i, as in §3.3.1. Start with unit x-directed field in the square (see Fig. 3.10) and apply the map n times to obtain N^n strips of field. It can be shown, for example using symbolic dynamics (§3.3.2, Guckenheimer & Holmes 1983), that there are N^n periodic points of period n, one in each strip. For each periodic point $\mathbf{x}$, the Floquet multiplier of largest modulus, $\lambda(\mathbf{x}, n)$, is positive if the field in the strip points in the x-direction, giving a negative index and an untwisted orbit. If $\lambda(\mathbf{x}, n) < 0$ the field points in the $-x$-direction, and the orbit is twisted. In fact

$$N_{\mathrm{orb}}(n) = N^n, \qquad N_{\mathrm{tw}}(n) = -\left(\sum_{i=1}^{N} \epsilon_i\right)^n; \qquad (6.2.6)$$

so $h_{\mathrm{tw}} = \log |\sum_{i=1}^{N} \epsilon_i|$ and gives the correct fast dynamo growth rate. The indices of the periodic points keep track of the sign flips in the magnetic field, and so N_{tw} gives a correct evaluation of the evolution of flux.

If M has the property that the whole phase space is chaotic, and so all periodic orbits are included in the sum (6.2.4), $N_{\mathrm{tw}}(n)$ is an invariant of the map M^n called the Lefshetz number $L(M^n)$ (Lefshetz 1930). This can be calculated from the action of M^n on the space of cohomologies,

$$L(M^n) = \sum_p (-1)^p \operatorname{trace} M_p^n, \tag{6.2.7}$$

where M_p^n is the linear map induced by M^n in the p-th cohomology of the space. Thus in this case Oseledets' conjecture equates the fast dynamo growth rate with the *asymptotic Lefshetz number of M*:

$$h_{\text{tw}} = \limsup_{n \to \infty} \frac{1}{n} \log |L(M^n)| \tag{6.2.8}$$

(Oseledets 1993). We have neither the space nor the expertise to go into the machinery of cohomology, but the result can be checked for diffeomorphisms of $\mathbb{T}^2$ (§10.1).

6.3 Fredholm Determinants, Periodic Orbits and Zeta Functions

The conjecture of Oseledets is closely connected with general methods of calculating properties of a chaotic system from sums over the periodic orbits of the system. These orbits form a 'skeleton' within the chaos, and by considering orbits of increasing length it is often possible to construct convergent approximations to quantities that describe the chaotic system. These methods were pioneered in the field of quantum chaos (see, for example, Gutzwiller 1990), and are applied to the study of dynamical systems (see Kadanoff & Tang 1984, Grebogi, Ott & Yorke 1988, Artuso, Aurell & Cvitanović 1990 a, b) and anomalous diffusion (Artuso 1991).

In fast dynamo theory the important parameter of a chaotic system is the fast dynamo exponent γ_0. Work relating this to periodic orbits has been published by Vishik (1989), Oseledets (1993), Aurell & Gilbert (1993), and Balmforth *et al.* (1993); see also de la Llave (1993), Núñez (1993, 1994). The general idea is to manipulate the dynamo operator so that the leading eigenvalue is expressed in terms of sums over periodic orbits; an approximation eventually leads to equations (6.2.4) and Oseledets' conjecture emerges.

With these methods, the fast dynamo problem is reformulated as a question about the periodic orbit structure of a flow or map. This throws the difficulty of calculating γ_0 by taking the limit $\epsilon \to 0$ onto the difficulty of classifying periodic orbits of the system, something that appears possible analytically only for special, *hyperbolic* systems, such as cat maps and baker's maps. One can draw a parallel with the flux conjecture: while this conjecture transforms the fast dynamo problem into one about the behavior of functionals or fluxes of the field for $\epsilon = 0$, it is not a simple matter to verify that

$\Gamma_{\sup} > 0$ for other than the simplest maps or flows. Furthermore both the flux conjecture and the methods using periodic orbit sums have only been justified for dynamos based on hyperbolic systems, and not for more realistic systems. Nevertheless either reformulation, even if not yet established mathematically, gives useful information on how the limit $\epsilon \to 0$ relates to setting $\epsilon = 0$, and on the properties of a dynamical system that assist fast dynamo action.

6.3.1 Fredholm Theory

We consider a pulsed model in which we apply a smooth, volume-preserving map M on some bounded or periodic region of space $\mathcal{D}$, and then apply diffusion with a heat kernel H_ϵ, for example a Gaussian. The diffusive induction operator is $\mathbf{T}_\epsilon = H_\epsilon \mathbf{T}_0$, and we may represent the perfect induction operator $\mathbf{T} \equiv \mathbf{T}_0$ by the action of a singular integral kernel $\mathbf{T}(\mathbf{x}, \mathbf{y})$:

$$\mathbf{T}\mathbf{B}(\mathbf{x}) = \int \mathbf{T}(\mathbf{x}, \mathbf{y})\mathbf{B}(\mathbf{y})\, dy, \tag{6.3.1}$$

$$\mathbf{T}(\mathbf{x}, \mathbf{y}) = \phi(\mathbf{x})\delta(M^{-1}\mathbf{x} - \mathbf{y})\mathbf{J}(\mathbf{y}). \tag{6.3.2}$$

It is convenient to use the same symbol for $\mathbf{T}$ and the corresponding kernel, without much danger of confusion. Here $\phi(\mathbf{x})$ is an optional complex phase shift, allowing us to include models such as SFSI and cat maps with shear. This $\epsilon = 0$ kernel (6.3.2) is singular, but with diffusion the corresponding kernel $\mathbf{T}_\epsilon(\mathbf{x}, \mathbf{y})$ for $\mathbf{T}_\epsilon$ is everywhere bounded, as the heat kernel H_ϵ smooths out the delta function in (6.3.2).

We shall perform a series of manipulations on the induction operator $\mathbf{T}_\epsilon$. These are valid for $\epsilon > 0$, but are difficult to justify when $\epsilon = 0$. It is helpful as an example to think of replacing $\mathbf{T}_\epsilon$ by a 2×2 matrix:

$$\mathbf{T}_\epsilon = \begin{pmatrix} a & 0 \\ 0 & b \end{pmatrix}, \tag{6.3.3}$$

with eigenvalues $\mu = a$, $b \in \mathbb{C}$. These eigenvalues can be found from zeros z of the characteristic equation:

$$F_\epsilon(z) \equiv \det(\mathbf{I} - z\mathbf{T}_\epsilon) = 0, \qquad (\mu = 1/z). \tag{6.3.4}$$

Here $\mathbf{I}$ is the identity operator. We can also expand $F_\epsilon(z)$ in terms of the traces $c_{n\epsilon}$ of the iterated operators $\mathbf{T}_\epsilon^n$:

$$F_\epsilon(z) = \exp \operatorname{tr} \log(\mathbf{I} - z\mathbf{T}_\epsilon) \tag{6.3.5a}$$

$$= \exp \left(\sum_{n=1}^{\infty} -c_{n\epsilon} \frac{z^n}{n} \right), \qquad (c_{n\epsilon} = \operatorname{tr} \mathbf{T}_\epsilon^n). \tag{6.3.5b}$$

Equation (6.3.5b) results from expanding the logarithm in (6.3.5a) as a power series, which will generally be valid for z sufficiently small. For our example, $F_\epsilon(z) = (1 - az)(1 - bz)$ and $c_{n\epsilon} = a^n + b^n$.

We also define a logarithmic derivative of F_ϵ:

$$\Omega_\epsilon(z) = -z \frac{d}{dz} \log F_\epsilon(z) \equiv -z \frac{F'_\epsilon(z)}{F_\epsilon(z)} = \sum_{n=1}^{\infty} c_{n\epsilon} z^n, \qquad (6.3.6)$$

which has a simple pole at each zero of $F_\epsilon(z)$. The eigenvalue of $\mathbf{T}_\epsilon$ of largest modulus, $\mu_{\max}$, corresponds to the pole of $\Omega_\epsilon(z)$ closest to the origin. The distance of this pole from the origin is the radius of convergence R_ϵ of Ω_ϵ; thus

$$|\mu_{\max}| = R_\epsilon^{-1} = \limsup_{n \to \infty} |c_{n\epsilon}|^{1/n}. \qquad (6.3.7)$$

For the example, this yields correctly $|\mu_{\max}| = \max(|a|, |b|)$.

As discussed in §9.1.2, for $\epsilon > 0$ the operator $\mathbf{T}_\epsilon$ is a compact operator in the space L^2 of square-integrable functions, having a spectrum made up only of point eigenvalues μ with no accumulation point different from zero. Since the kernel $\mathbf{T}_\epsilon(\mathbf{x}, \mathbf{y})$ is bounded for $\epsilon > 0$ it follows from the classical theory of Fredholm (see Chap. 3 of Courant & Hilbert 1953), that equations (6.3.4–7) hold for $\mathbf{T}_\epsilon$, and yield information about the eigenvalues just as in the finite-dimensional case. Here the traces $c_{n\epsilon} = \operatorname{tr} \mathbf{T}_\epsilon^n$ are defined by:

$$c_{n\epsilon} \equiv \int \operatorname{tr}[\mathbf{T}_\epsilon(\mathbf{x}_1, \mathbf{x}_2)\mathbf{T}_\epsilon(\mathbf{x}_2, \mathbf{x}_3) \ldots \mathbf{T}_\epsilon(\mathbf{x}_n, \mathbf{x}_1)] \, d\mathbf{x}_1 \, d\mathbf{x}_2 \ldots d\mathbf{x}_n; \qquad (6.3.8)$$

in this equation $\operatorname{tr}[\,\cdot\,]$ is the trace of the product of matrices $[\,\cdot\,]$. The determinant $F_\epsilon(z)$ in (6.3.4) is called a *Fredholm determinant* and is *entire*, that is, analytic everywhere.

In the context of dynamo theory we can write from (6.3.7):

$$\gamma(\epsilon) = \log R_\epsilon^{-1} = \log \limsup_{n \to \infty} |c_{n\epsilon}|^{1/n}. \qquad (6.3.9)$$

If the limit as $\epsilon \to 0$ may be taken we obtain

$$\gamma_0 = \log R_0^{-1} = \log \limsup_{n \to \infty} |c_{n0}|^{1/n}. \qquad (6.3.10)$$

However the theory of Fredholm fails for the singular $\epsilon = 0$ kernel (6.3.2). We shall nevertheless assume the limit can be taken and set $\epsilon = 0$, taking (6.3.10) as a working hypothesis we shall check for various models. Going from (6.3.9) to (6.3.10) is decidedly non-trivial and not always justifiable, as we shall see.

With $\epsilon = 0$, the traces c_{n0} may be determined fairly easily. For example:

$$c_{10} = \int \phi(\mathbf{x})\delta(M^{-1}\mathbf{x} - \mathbf{x}) \operatorname{tr} \mathbf{J}(\mathbf{x}) \, d\mathbf{x}. \qquad (6.3.11)$$

The integral receives a contribution only from points where $M^{-1}\mathbf{x} = \mathbf{x}$, i.e., from all fixed points, $\mathbf{x} \in \operatorname{per}(1)$, of M. Near to a fixed point $\mathbf{x}_0$, set $\mathbf{x} = \mathbf{x}_0 + \mathbf{y}$; then

$$\delta(M^{-1}\mathbf{x} - \mathbf{x}) = \delta\left([\mathbf{J}(\mathbf{x}_0)^{-1} - \mathbf{I}]\mathbf{y}\right) = \frac{\delta(\mathbf{y})}{|\det(\mathbf{I} - \mathbf{J}(\mathbf{x}_0)^{-1})|} \tag{6.3.12}$$

and so

$$c_{10} = \sum_{\mathbf{x}\in\text{per}(1)} \frac{\phi(\mathbf{x})\,\text{tr}\,\mathbf{J}(\mathbf{x})}{|\det(\mathbf{I} - \mathbf{J}(\mathbf{x})^{-1})|}. \tag{6.3.13}$$

This generalizes straightforwardly to the nth trace:

$$c_{n0} = \sum_{\mathbf{x}\in\text{per}(n)} \frac{\phi_n(\mathbf{x})\,\text{tr}\,\mathbf{J}^n(\mathbf{x})}{|\det(\mathbf{I} - \mathbf{J}^n(\mathbf{x})^{-1})|}, \tag{6.3.14a}$$

where $\phi_n(\mathbf{x}) = \phi(\mathbf{x})\phi(M\mathbf{x})\cdots\phi(M^{n-1}\mathbf{x})$ is the product of the phase factors round the orbit and $\mathbf{J}^n(\mathbf{x}) = \partial M^n \mathbf{x}/\partial \mathbf{x}$ is the Floquet matrix for the orbit. This formula is written for any number of dimensions; the version for two-dimensional area-preserving maps (with possible phase shifts) is:

$$c_{n0} = \sum_{\mathbf{x}\in\text{per}(n)} \phi_n(\mathbf{x}) \frac{\lambda(\mathbf{x},n) + \lambda(\mathbf{x},n)^{-1}}{|\lambda(\mathbf{x},n) + \lambda(\mathbf{x},n)^{-1} - 2|}, \tag{6.3.14b}$$

where $\lambda(\mathbf{x},n)^{\pm 1}$ are the eigenvalues of the matrix $\mathbf{J}^n(\mathbf{x})$ with $|\lambda(\mathbf{x},n)| \geq 1$ always assumed. We have in mind maps such as baker's maps, cat maps, SFS and SFSl, and maps obtained from periodic two-dimensional flows such as MW+, with the phase shifts coding the z-motion.

The simplest application of this is to the cat map of §3.5, with no shear, $\phi(\mathbf{x}) \equiv 1$. Here $\lambda(\mathbf{x},n) = \lambda^n$, and there are $N_{\text{orb}}(n) = \lambda^n + \lambda^{-n} - 2$ periodic orbits of length n. Therefore $c_{n0} = \lambda^n + \lambda^{-n}$ from (6.3.14b) and (6.3.10) gives the correct fast dynamo growth rate $\gamma_0 = \log\lambda$. Furthermore from (6.3.5b) we obtain explicitly, $F_0(z) = (1 - z\lambda)(1 - z/\lambda)$, which correctly gives the eigenvalues $\mu = \lambda$ and $\mu = 1/\lambda$.

Non-trivial phase shifts, $\phi \neq \text{const.}$, can be included to obtain the model of 'cat map with shear' of §10.4 but, except for the simplest cases, the traces c_{n0} and $F_0(z)$ must be evaluated numerically; results confirm formula (6.3.10). Weak diffusion can be included perturbatively in (6.3.14) and gives the correct growth rate for $0 < \epsilon \ll 1$ using (6.3.9). The numerical evidence is that one can pass from (6.3.9) to (6.3.10) for cat maps with shear (Aurell & Gilbert 1993).

We return to consider (6.3.14), as these expressions are plainly not valid for all maps. For example equation (6.3.14a) does not make sense for the time-one map of a steady flow: in this case each Floquet matrix $\mathbf{J}^n(\mathbf{x})$ has an eigenvalue of unity, corresponding to the direction along streamlines, and the denominators in (6.3.14a) vanish. Steady flows are considered separately in §6.3.3 below.

For definiteness let us consider a two-dimensional map (always assumed to be area-preserving): if it possesses a periodic point $\mathbf{x}$ for which the corresponding Floquet matrix is a pure rotation, the point is called *elliptic*. If the rotation angle is rational then $\mathbf{J}^n(\mathbf{x}) = \mathbf{I}$ for some n and the denominator of (6.3.14b) vanishes. If the rotation angle is irrational, then the denominator never vanishes, but attains arbitrarily small values for large n. When a zero or small denominator arises for a periodic orbit in (6.3.14), this is a signal that for $0 < \epsilon \ll 1$ diffusion is important in determining the contribution to the sum from the orbit. To calculate traces in this case one would have to reinstate diffusion, $\epsilon > 0$, calculate the trace $c_{n\epsilon}$ and then let $\epsilon \to 0$. This remains largely unexplored territory for dynamo theory, but techniques have been developed in quantum mechanics: see Chap. 9 of Ozorio de Almeida (1988) and references therein.

The interpretation of (6.3.14) is clear when all orbits are hyperbolic, that is, none of the eigenvalues of $\mathbf{J}^n(\mathbf{x})$ have modulus unity. In fact we shall adopt a stronger condition: still working in two dimensions for simplicity, we demand that there are positive constants C and ν such that $|\lambda(\mathbf{x}, n)| \geq C \exp(\nu n)$ for all $\mathbf{x}$ and n. This means that we are considering an *Anosov* or *hyperbolic* map M. These terms are defined in §6.5 below; examples are baker's maps, cat maps and perturbed cat maps. For *smooth* hyperbolic systems it can be shown that $c_{n\epsilon} \to c_{n0}$ as $\epsilon \to 0$: diffusion has a small effect on each trace as $\epsilon \to 0$ (Aurell & Gilbert 1993).[49]

Let us therefore work with the simplest case of a two-dimensional hyperbolic system (with possible phase shifts); each quotient in the sum (6.3.14b) can be approximated by

$$\frac{\lambda(\mathbf{x}, n) + \lambda(\mathbf{x}, n)^{-1}}{|\lambda(\mathbf{x}, n) + \lambda(\mathbf{x}, n)^{-1} - 2|} = \text{sign}(\lambda(\mathbf{x}, n)) + O(\lambda(\mathbf{x}, n)^{-1}) \qquad (6.3.15)$$

to obtain

$$c_{n0} = -N_{\text{tw}}(n) + O\left(N_{\text{orb}}(n) \Big/ \min_{\mathbf{x} \in \text{per}(n)} |\lambda(\mathbf{x}, n)|\right), \qquad (6.3.16)$$

where

$$N_{\text{tw}}(n) = \sum_{\mathbf{x} \in \text{per}(n)} \phi_n(\mathbf{x}) \, \text{sign}(-\lambda(\mathbf{x}, n)). \qquad (6.3.17)$$

This is a natural generalization of definition (6.2.4b) to include phase shifts.

Now provided the error term in (6.3.16) grows more slowly than N_{tw}, we recover from (6.3.10) Oseledets' conjecture:

$$\gamma_0 = h_{\text{tw}} \equiv \limsup_{n \to \infty} \frac{1}{n} \log |N_{\text{tw}}(n)| \qquad (6.3.18)$$

for two-dimensional maps, generalized to include complex phase shifts. A straightforward condition to ensure that the error term grows more slowly than $N_{\text{tw}}(n)$ in (6.3.16) is

---

[49]Note that the limit $c_{n\epsilon} \to c_{n0}$ may not be uniform in n, and so this is not a proof that (6.3.10) follows from (6.3.9).

$$h_{\text{tw}} > h_{\text{top}} - \Lambda_{\text{Liap}}^{\min}. \tag{6.3.19}$$

Here h_{top} is an upper bound on the growth rate of $N_{\text{orb}}(n)$ with n for a hyperbolic system (see §8.5 of Walters 1982), and h_{tw} gives the growth rate of $N_{\text{tw}}(n)$. We have bounded the growth rate of $\min_{\mathbf{x} \in \text{per}(n)} |\lambda(\mathbf{x}, n)|$ from below by the minimum Liapunov exponent $\Lambda_{\text{Liap}}^{\min} = \min_{\mathbf{x}} \Lambda_{\text{Liap}}(\mathbf{x}) \geq \nu$ in this hyperbolic system; this is the largest stretching rate on an orbit, minimized over all orbits.

Finally, before we study examples, note that if the approximation $c_{n0} \simeq -N_{\text{tw}}(n)$ from (6.3.16) is made and substituted into (6.3.5b), we obtain in place of the Fredholm determinant $F_0(z)$ the function:

$$1/\zeta_{\text{dyn}}(z) \equiv \exp \sum_{n=1}^{\infty} -\frac{z^n}{n} \left(\sum_{\mathbf{x} \in \text{per}(n)} \phi_n(\mathbf{x}) \operatorname{sign} \lambda(\mathbf{x}, n) \right) \tag{6.3.20a}$$

(V.I. Oseledets, personal communication, Aurell & Gilbert 1993). The function ζ_{dyn} is called a Ruelle zeta function (Ruelle 1978, 1989b, c, Artuso et al. 1990a, b). It can also be written as

$$1/\zeta_{\text{dyn}}(z) = \prod_{p=1}^{\infty} \prod_{\mathbf{x} \in \text{prim}(p)} \left(1 - z^p \phi_p(\mathbf{x}) \operatorname{sign} \lambda(\mathbf{x}, p) \right) \tag{6.3.20b}$$

(Ruelle 1978); $\text{prim}(p)$ is the set of *primitive* periodic orbits of length p, that is, periodic orbits, each traversed only once, and counted only once modulo circular permutation.

When the approximation (6.3.16) is valid, the function $1/\zeta_{\text{dyn}}$ has a zero at the leading zero of $F_0(z)$, giving the fast dynamo growth rate, and Oseledets' conjecture holds. More generally, if the expansion in (6.3.15) is extended as a series, and then substituted into (6.3.5b), we obtain $F_0(z)$ as a product of functions, the first term of which is $1/\zeta_{\text{dyn}}(z)$. This will be seen explicitly for SFS1 in (6.3.25) below.

6.3.2 Generalized Baker's Maps and SFS1

These points are best illustrated by the examples of generalized baker's maps and SFS1. For a generalized baker's map with N strips, widths p_i and orientation factors ϵ_i, we shall not give $F_0(z)$ as it is quite complicated. From (6.2.6) Oseledets' conjecture is verified, with h_{tw} as the fast dynamo growth rate. However the condition (6.3.19) amounts to

$$\log \left| \sum_i \epsilon_i \right| > \log N - \min_i \log p_i^{-1}, \tag{6.3.21}$$

which could fail for strong cancellation or very uneven stretching.

We now explore the SFS1 model of §3.4 in somewhat more detail. With $k = 1$ and $\epsilon = 0$, recall from equation (3.4.5) that the leading growth rate is either $\Gamma_0 = \log|2\sin\pi\alpha|$ or $\Gamma_1 = \log|\cos\pi\alpha|$, whichever is larger. For the folded baker's map there are 2^n points in $per(n)$. Of these $n!/m!(n-m)!$ make m visits to the interval $0 \leq y < 1/2$ in one orbit, and $n-m$ visits to the interval $1/2 \leq y < 1$; for the SFSl model their phase factor is $\phi_n(\mathbf{x}) = e^{i\pi\alpha m}e^{-i\pi\alpha(n-m)}$ and the Jacobian matrix has leading eigenvalue $\lambda(\mathbf{x}, n) = 2^m(-2)^{n-m}$. Summing the terms in (6.3.17) yields

$$N_{\text{tw}}(n) = -(2i\sin\pi\alpha)^n, \qquad h_{\text{tw}} = \log|2\sin\pi\alpha|. \tag{6.3.22}$$

This gives correctly the flux growth rate Γ_0; whenever the flow is a perfect dynamo this is the fastest flux growth rate, and condition (6.3.19), which amounts here to $h_{\text{tw}} > 0$, is satisfied. We have not captured the second eigenvalue Γ_1, which can be dominant for some values of α, although only when the flow is not a perfect dynamo. This is not surprising, since in this case $h_{\text{tw}} \leq 0$ and condition (6.3.19) is violated. To obtain the second and higher eigenvalues we need to keep track of the exact values of c_{n0}.

For this purpose we use the parallel map $T \equiv T^{\text{SFSl}}$ (3.4.2). This is not measure-preserving, but a calculation analogous to that in §6.3.1 yields:

$$c_{n0} = \sum_{\mathbf{x}\in per(n)} \phi_n(\mathbf{x}) \frac{\lambda(\mathbf{x}, n)}{|\lambda(\mathbf{x}, n) - 1|}$$

$$= \sum_{m=0}^{n} \binom{n}{m} \frac{(-1)^{n-m}e^{i\pi\alpha m}e^{-i\pi\alpha(n-m)}}{1 - 2^{-n}(-1)^{n-m}}. \tag{6.3.23}$$

Expand the denominator as a power series and sum over m to give

$$c_{n0} = (2i\sin\pi\alpha)^n + (\cos\pi\alpha)^n + ((i/2)\sin\pi\alpha)^n + \cdots; \tag{6.3.24}$$

the first term is $-N_{\text{tw}}(n)$ in agreement with (6.3.16). Substituting (6.3.24) in (6.3.5b) yields finally:

$$F_0(z) = (1 - 2iz\sin\pi\alpha)(1 - z\cos\pi\alpha)(1 - (iz/2)\sin\pi\alpha)\cdots$$

$$= \prod_{k=0}^{\infty} \left[1 - z\,2^{-k}(e^{i\pi\alpha} - (-1)^k e^{-i\pi\alpha})\right]. \tag{6.3.25}$$

We recover all the eigenvalues (3.4.4) of SFSl for $\epsilon = 0$, and have expressed F_0 as a product, whose first term is $1/\zeta_{\text{dyn}}(z) = 1 - 2iz\sin\pi\alpha$.

Note that we do not obtain any information about the diffusive dynamo action seen in §9.6 when $\Gamma_0, \Gamma_1 < 0$ and yet $\gamma_0 > 0$. This is not picked up by the periodic orbit sums and so for this system one cannot pass from (6.3.9) to (6.3.10) in general. This appears to be because of lack of smoothness of the underlying baker's map; see §6.3.4 below.

The picture for SFS appears to be similar, but unfortunately the sums over periodic orbits are only available numerically. It has been verified numerically that the Fredholm determinant and zeta function (and hence Oseledets' conjecture) all give good approximations to the fast dynamo growth rate for $\alpha = 1.0$, 1.5 and 2.0 (Aurell & Gilbert 1993).

6.3.3 Steady Flows

We now consider three-dimensional steady fluid flows; these are studied by Vishik (1989) and Balmforth *et al.* (1993). We use the induction operator $\mathbf{T}_\epsilon(t)$ acting on solenoidal magnetic fields. The trace of $\mathbf{T}_\epsilon(t)$ involves sums over the set 'prim' of primitive periodic orbits, that is, orbits traversed just once. We choose any point $\mathbf{x}$ on each primitive orbit $\xi \in$ prim, and let $T(\xi)$ be the period of the orbit. If we place a small section S across the orbit at $\mathbf{x}$, we obtain a Poincaré map from S to S whose Jacobian has eigenvalues $\lambda^{\pm 1}(\xi)$; these are *transverse Floquet multipliers*, and are independent of the choice of $\mathbf{x}$ on the orbit. From Vishik (1989) the trace $\vartheta(t) \equiv \mathrm{tr}\, \mathbf{T}_\epsilon(t)$ for a steady flow with no fixed points and $\epsilon > 0$ is given (in a weak sense) by

$$\vartheta(t) = \sum_{\xi \in \mathrm{prim}} T(\xi) \sum_{m=1}^{\infty} \frac{\lambda(\xi)^m + \lambda(\xi)^{-m}}{|\lambda(\xi)^m + \lambda(\xi)^{-m} - 2|} \delta(t - mT(\xi)) + O(\sqrt{\epsilon}). \quad (6.3.26)$$

A further term appears if fixed points are present. For $\epsilon > 0$, $\vartheta(t)$ is related to the complex growth rates p_i of eigenfunctions of $\mathbf{T}_\epsilon$ by (Vishik 1989)

$$\vartheta(t) = \sum_i \exp(p_i t). \quad (6.3.27)$$

We set $\epsilon = 0$ in (6.3.26) and follow Balmforth *et al.* (1993), who give further discussion. Setting $\epsilon = 0$ begs many questions, particularly about the decoupling of the field along streamlines (see §7.3). Our approach here is to obtain expressions that can be tested for models, in the absence of firm analysis. The analog of the $\epsilon = 0$ Fredholm determinant (6.3.5, 14) is given by

$$F_0(s) = \exp\left(-\sum_{\mathrm{prim}} \sum_{m=1}^{\infty} \frac{1}{m} \frac{\lambda^m + \lambda^{-m}}{|\lambda^m + \lambda^{-m} - 2|} e^{smT}\right). \quad (6.3.28)$$

The set 'prim' is again the set of primitive periodic orbits, that is, orbits traversed once only. The zeros of this function are $-p_i$, where p_i are the dynamo eigenvalues (Balmforth *et al.* 1993), under certain, as yet unverified, assumptions about the properties of $F_0(s)$ as a complex function.

The quotients in (6.3.28) can be approximated to yield a zeta function approximation to $F_0(s)$:

$$1/\zeta_{\mathrm{dyn}}(s) = \exp\left(-\sum_{\mathrm{prim}} \sum_{m=1}^{\infty} m^{-1} \mathrm{sign}(\lambda^m)\, e^{smT}\right) \quad (6.3.29a)$$

$$= \prod_{\mathrm{prim}} \left(1 - \mathrm{sign}(\lambda)\, e^{sT}\right); \quad (6.3.29b)$$

(6.3.29b) is obtained by recognizing that for each primitive orbit the sum over m in (6.3.29a) gives a logarithm. This equation is analogous to (6.3.20) if e^s is replaced by z. In fact this is the first term in an expansion of $F_0(s)$ as an infinite product (Balmforth et al. 1993).

We would now like to retrieve, albeit roughly, a version of Oseledets' conjecture for three-dimensional steady flows. Here we have to ignore the Floquet multiplier of unity along streamlines, and interpret the index of a periodic orbit ξ as $\mathrm{ind}(\xi) = -\mathrm{sign}(\lambda(\xi))$, in other words as the index based only on the transverse Floquet multipliers (cf. 6.2.3).

Consider the function $\Omega(z)$ given by a sum over the set 'per' of all periodic orbits, including multiple transversals of each orbit,

$$\Omega(z) = \sum_{\xi \in \mathrm{per}} \mathrm{sign}(\lambda(\xi))\, z^{T(\xi)} \equiv \sum_{\xi \in \mathrm{prim}} \sum_{m=1}^{\infty} \mathrm{sign}(\lambda(\xi)^m)\, z^{mT(\xi)}. \qquad (6.3.30)$$

This is the same as $\log \zeta_{\mathrm{dyn}}$ if we substitute $e^s = z$ and omit the factor m^{-1} in (6.3.29a). This factor is likely to be insignificant for periodic orbits of long period, since in a chaotic system the exponential proliferation of number of orbits with increasing period implies that most long periodic orbits are primitive. However we do not have a proof of this key assumption. A zero of $1/\zeta_{\mathrm{dyn}}(s)$ corresponds to $s = -p_i$ for an eigenvalue p_i; in this case $\Omega(z)$ will diverge for $z = \exp(-p_i)$. The radius of convergence of $\Omega(z)$ is the same as that of

$$\Xi(z) = \sum_{n=1}^{\infty} \sum_{\xi \in \mathrm{per}(n)} \mathrm{sign}(\lambda(\xi))\, z^n, \qquad (6.3.31)$$

where $\mathrm{per}(n)$ is the set of periodic orbits ξ with $n - 1 < T(\xi) \le n$. By considering the radius of convergence of (6.3.31) we then recover Oseledets' conjecture for steady flows, $\gamma_0 = h_{\mathrm{tw}}$ with[50]

$$h_{\mathrm{tw}} \equiv \limsup_{n \to \infty} \frac{1}{n} \log |N_{\mathrm{tw}}(n)|, \qquad (6.3.32a)$$

$$N_{\mathrm{tw}}(n) = -\sum_{\xi \in \mathrm{per}(n)} \mathrm{sign}(\lambda(\xi)). \qquad (6.3.32b)$$

There are few examples of steady flows for which this can be tested; we consider a flow of Finn et al. (1989, 1991) in §7.3, and another example may be found in Balmforth et al. (1993).

---

[50]This assumes that the growth rate of the sum of indices of the periodic orbits with $T \le n$ is the same as the growth rate of N_{tw} defined here. This would only fail for the case of zero or negative growth rate.

6.3.4 Discussion

For a two-dimensional map M with possible phase shifts, $N_{tw}(n)$ in (6.2.4b, 6.3.17) has the property that it is the same under time-reversal, $M \to M^{-1}$. This agrees with the result that dynamo growth rates $\gamma(\epsilon)$ are the same for a pulsed model using either M or its inverse M^{-1} (see §9.3.1). For unsteady two-dimensional flows, although Oseledets' conjecture suggests that γ_0 is invariant under time-reversal, for $\epsilon > 0$ there is no obvious relation between the dynamo eigenvalues of a flow and its time-reversed counterpart, although this is an interesting line of enquiry in particular cases. (As elsewhere, given a flow $\mathbf{u}(\mathbf{x}, t)$ the time-reversed flow is defined to be $-\mathbf{u}(\mathbf{x}, -t)$.)

For more than two dimensions, the analog of (6.3.17) is

$$N_{tw}(n) = \sum_{\mathbf{x} \in per(n)} \left(\phi_n(\mathbf{x}) \max_i \lambda_i(\mathbf{x}, n) \prod_{|\lambda_i(\mathbf{x},n)|<1} |\lambda_i(\mathbf{x}, n)| \right); \qquad (6.3.33)$$

recall that $\lambda_i(\mathbf{x}, n)$ are the eigenvalues of $\mathbf{J}^n(\mathbf{x})$. The primary application of this would be to three-dimensional unsteady flows. This expression is not explicitly invariant under time-reversal, and does not seem to be related to the definition (6.2.4a) for N_{tw} in any obvious way.

We have not given any formal justification for the manipulations above in the case $\epsilon = 0$. For a mapping, the Fredholm determinant $F_\epsilon(z)$ for $\epsilon > 0$ is known to be an entire complex function (having no poles), by classical theory; we also know that for a smooth hyperbolic system $c_{n\epsilon} \to c_{n0}$ as $\epsilon \to 0$. Unfortunately this does not guarantee that $F_0(z)$ is also entire: it could have poles in the complex plane. However if one can show that $F_0(z)$ is also entire, then the effect of diffusion in a smooth hyperbolic system is a regular perturbation, and one can pass from (6.3.9) to (6.3.10) in the limit $\epsilon \to 0$ (Aurell & Gilbert 1993). Aurell (1992) has used methods of Rugh (1994) to show that for systems based on hyperbolic maps, $F_0(z)$ is indeed entire, and so (6.3.10) is valid for smooth hyperbolic maps.

This means that the above techniques are valid for cat maps with shear and related models (see §10.4). Baker's maps are not smooth, and so (6.3.10) is not generally valid for SFS and SFSl; the diffusive dynamo action, produced by the subtle interaction of discontinuities in the map and weak diffusion, is not captured by the periodic orbit sums for $\epsilon = 0$. No proof of (6.3.10) is yet available for smooth steady flows, or for more general classes of maps. Another interesting problem is to characterise which part of the spectrum of the diffusionless operator $\mathbf{T}_0$ is given by F_0. This information might suggest how to relate the spectrum of $\mathbf{T}_0$ to the limiting eigenvalues of $\mathbf{T}_\epsilon$ for $\epsilon > 0$ in more general maps and flows.

A flow such as MW+ is not hyperbolic and the strong conditions above on periodic orbits will not be satisfied. There are also troublesome elliptic orbits present, as is evident from the Poincaré section in Fig. 2.4. A study of the periodic orbits and Oseledets' conjecture would be valuable but difficult, since periodic orbits of realistic flows would need to be found numerically, or perhaps in the limiting cases of near-integrability or strong chaos (see Soward 1994a).

6.4 Dynamo Action in the Cat Flow

In §3.5 we discussed fast dynamo action for the cat map $M : \mathbf{x} \to \mathbf{Mx}$. As discussed there the cat map has the drawback (from the point of view of astrophysical applications) that it cannot be achieved by a smooth flow on $\mathbb{T}^2$: the mean magnetic field over $\mathbb{T}^2$ is multiplied by $\mathbf{M}$ at each iteration of the map, whereas for a smooth flow on $\mathbb{T}^2$ the mean field is constant.[51] The situation is reminiscent of the stacked baker's map, which is a fast dynamo but does not model a planar flow; instead one has to appeal to a three-dimensional flow such as STF to achieve the doubling of flux. These arguments carry across to any diffeomorphism M of $\mathbb{T}^2$: if M is a fast dynamo, it cannot be achieved by a smooth flow on $\mathbb{T}^2$; see §10.1.

Mappings of $\mathbb{T}^2$ such as the cat map can be achieved by steady flows, but only on peculiar three-dimensional manifolds: chaos in the map is thrown onto complications in the structure of the manifold. The construction of such a flow for the cat map is given by Arnold (1972), and Arnold et al. (1981) show that this flow is a fast dynamo. Consider again the cat map

$$M : \mathbf{x} \to \mathbf{Mx}, \qquad \mathbf{M} = \begin{pmatrix} 2 & 1 \\ 1 & 1 \end{pmatrix}, \tag{6.4.1}$$

with eigenvectors $\mathbf{e}_{\pm}$ and eigenvalues $\lambda^{\pm 1}$ such that $\lambda > 1$. To construct the manifold on which the steady flow will live, start with $\mathbb{R}^3$ and first identify points according to

$$\mathbf{x} = (x, y, z) \sim (x + 1, y, z) \sim (x, y + 1, z). \tag{6.4.2}$$

These two identifications fold each (x, y)-plane into a two-torus. Now fold up the z-direction using the matrix $\mathbf{M}$ and identifying

$$(2x + y, x + y, z) \equiv (\mathbf{M}(x, y), z) \sim (x, y, z + 1). \tag{6.4.3}$$

We denote the compact manifold which results from these three identifications by $\mathcal{M}$. This quotient construction can be generalized to other diffeomorphisms of $\mathbb{T}^2$.

For the time being let us think of $\mathcal{M}$ as the box $[0, 1]^3$ with the planes of its surface identified by (6.4.2, 3). We define a flow $\mathbf{u} = (0, 0, 1)$ on $\mathcal{M}$.

---

[51] Alternatively, the cat map M has a non-trivial action on the fundamental group (the group of all directed curves identified under continuous deformation) of the torus, and so cannot be achieved by a flow on $\mathbb{T}^2$ — this formalizes the argument about how curves are mapped under M given in §3.5.

A Lagrangian particle at $\mathbf{a} = (a, b, c)$ when $t = 0$ is transported by the flow to $\mathbf{x}(t)$ with

$$\mathbf{x}(1) = (a, b, c + 1) = (\mathbf{M}(a, b) \bmod 1, c), \tag{6.4.4}$$

then to

$$\mathbf{x}(2) = (\mathbf{M}(a, b) \bmod 1, c + 1) = (\mathbf{M}^2(a, b) \bmod 1, c), \tag{6.4.5}$$

and so forth. Thus the time-one map of the flow $\mathbf{u}$ maps particles by the cat map in the (x, y)-coordinates; the flow is a *suspension* of the cat map on $\mathcal{M}$, which we call the *cat flow* for brevity.

If we attach a vector $\mathbf{v} = (v_x, v_y, v_z)$ to the point $\mathbf{x}$ and freeze it in the flow, the vector is mapped to $(\mathbf{M}^n(v_x, v_y), v_z)$ after an integer time n and generally stretched by the cat map. The vector is stretched instantaneously as the particle moves from $z = 1$ to $z = 0$ using the identification (6.4.3). This unphysical stretching arises because we are using the usual metric

$$ds_{\text{usual}}^2 = dx^2 + dy^2 + dz^2, \qquad |\mathbf{v}|_{\text{usual}}^2 = v_x^2 + v_y^2 + v_z^2 \tag{6.4.6}$$

in $[0, 1]^3$ to measure the length of a vector, and this metric is not invariant under the identification (6.4.3).

We can define a new metric for which this stretching happens smoothly (Arnold *et al.* 1981). Change coordinates from (x, y, z) to (x_+, x_-, z) with

$$\mathbf{x} = x_+ \mathbf{e}_+ + x_- \mathbf{e}_- + z \mathbf{e}_z, \tag{6.4.7}$$

and define the new metric in $\mathbb{R}^3$ by

$$ds^2 = \lambda^{2z} dx_+^2 + \lambda^{-2z} dx_-^2 + dz^2. \tag{6.4.8}$$

A right-handed orthonormal set of vectors in the new metric is

$$\mathbf{f}_+ = \lambda^{-z} \mathbf{e}_+, \quad \mathbf{f}_- = \lambda^z \mathbf{e}_-, \quad \mathbf{f}_z = \mathbf{e}_z, \tag{6.4.9}$$

and we define the components of a vector in this basis by

$$\mathbf{v} = v_+ \mathbf{f}_+ + v_- \mathbf{f}_- + v_z \mathbf{f}_z; \tag{6.4.10}$$

the length of $\mathbf{v}$ under the new metric is

$$|\mathbf{v}|^2 = v_+^2 + v_-^2 + v_z^2. \tag{6.4.11}$$

Note that, despite appearances, there is no inconsistency between definition (6.4.9, 10) for the components of vectors and (6.4.7), which gives a coordinate transformation.

It may be checked that the basis $\{\mathbf{f}_+, \mathbf{f}_-, \mathbf{f}_z\}$ and the metric (6.4.8) are invariant under the identifications (6.4.2, 3) and so vary smoothly on $\mathcal{M}$. When a general vector $\mathbf{v}$ is 'identified' from z to $z - 1$ using (6.4.3) its components (v_+, v_-, v_z) and length $|\mathbf{v}|$ are unchanged. A general vector carried in the flow $\mathbf{u}$ on $\mathcal{M}$ is stretched smoothly. Of course one can instead follow a vector in $\mathbb{R}^3$ under the flow and obtain the same answer — the length in the new metric grows as the vectors are translated to larger and larger z-values.

With the metric (6.4.8) defined we can set up equations for the magnetic field; in the (x_+, x_-, z) coordinate system (6.4.7) with the orthonormal basis (6.4.9), and writing $\partial_\pm = \partial/\partial x_\pm$, we have

$$\nabla = (\lambda^{-z}\partial_+, \lambda^z\partial_-, \partial_z),$$
$$\nabla \cdot \mathbf{B} = \lambda^{-z}\partial_+ B_+ + \lambda^z\partial_- B_- + \partial_z B_z,$$
$$\nabla \times \mathbf{B} = (\lambda^z[\partial_- B_z - \partial_z(\lambda^{-z}B_-)], \tag{6.4.12}$$
$$\lambda^{-z}[\partial_z(\lambda^z B_+) - \partial_+ B_z], \lambda^{-z}\partial_+ B_- - \lambda^z\partial_- B_+),$$
$$\nabla^2\phi = \lambda^{-2z}\partial_+^2\phi + \lambda^{2z}\partial_-^2\phi + \partial_z^2\phi.$$

We use the induction equation in the form

$$\partial_t\mathbf{B} + (\mathbf{u}\cdot\nabla)\mathbf{B} - (\mathbf{B}\cdot\nabla)\mathbf{u} = \epsilon\nabla^2\mathbf{B}, \tag{6.4.13}$$

with $\nabla\cdot\mathbf{B} = 0$. Note that $(\mathbf{B}\cdot\nabla)\mathbf{u} = (\mathbf{B}\cdot\nabla)\mathbf{e}_z = 0$ and, writing $\mu = \log\lambda$,

$$(\mathbf{u}\cdot\nabla)\mathbf{B} = \partial_z(B_+ e^{-\mu z}\mathbf{e}_+ + B_- e^{\mu z}\mathbf{e}_- + B_z\mathbf{e}_z)$$
$$= (\partial_z B_+ - \mu B_+)\mathbf{f}_+ + (\partial_z B_- + \mu B_-)\mathbf{f}_- + \partial_z B_z\mathbf{f}_z. \tag{6.4.14}$$

Since $\nabla^2\mathbf{B} = \nabla\nabla\cdot\mathbf{B} - \nabla\times\nabla\times\mathbf{B}$, the induction equation becomes[52]

$$\partial_t B_+ + \partial_z B_+ - \mu B_+ = \epsilon\left((\nabla^2 - \mu^2)B_+ + 2\mu\lambda^{-z}\partial_+ B_z\right), \tag{6.4.15a}$$
$$\partial_t B_- + \partial_z B_- + \mu B_- = \epsilon\left((\nabla^2 - \mu^2)B_- - 2\mu\lambda^z\partial_- B_z\right), \tag{6.4.15b}$$
$$\partial_t B_z + \partial_z B_z = \epsilon\left(\nabla^2 B_z + 2\mu(-\lambda^{-z}\partial_+ B_+ + \lambda^z\partial_- B_-)\right). \tag{6.4.15c}$$

Consider the case $\epsilon = 0$: the solution to equations (6.4.15) is simply

$$(B_+, B_-, B_z)(x, y, z, t) = (\lambda^t B_{0+}, \lambda^{-t} B_{0-}, B_{0z})(x, y, z - t), \tag{6.4.16}$$

with the initial condition $\mathbf{B}_0(x, y, z)$. Field is carried in the flow, stretched in the $\mathbf{f}_+$ direction and compressed in the $\mathbf{f}_-$ direction.

For $\epsilon \geq 0$, two families of eigenfunctions are given by

$$\mathbf{B}_\pm = \mathbf{f}_\pm e^{2\pi imz+pt}, \quad p(\epsilon) = \pm\mu - 2\pi im - \epsilon(4\pi^2 m^2 + \mu^2), \tag{6.4.17}$$

for integers m. The growth rate of the plus modes tends to $\mu = \log\lambda > 0$ in the limit $\epsilon \to 0$; thus the flow is a fast dynamo with $\gamma_0 = \log\lambda$, as for the cat map (Arnold et al. 1981). Plainly from (6.4.16) weak measures of growth yield $\Gamma_{\sup} = \gamma_0$ and so the cat dynamo is also a perfect dynamo satisfying conjectures 1–4 of §4.1.1. Oseledets' conjecture is also satisfied.

---

[52] Arnold et al. (1981) obtain a separable equation for B_z with the incorrect right-hand side $\epsilon(\nabla^2 - 2\mu\partial_z)B_z$. Note that these terms are equivalent to ours if the sign of the term $-\lambda^{-z}\partial_+ B_+$ in (6.4.15c) is changed and $\nabla\cdot\mathbf{B} = 0$ is used.

Note that there is degeneracy in (6.4.17) when $\epsilon = 0$; growing plus eigenfunctions for different m have different frequencies but the same (real) growth rate. When weak diffusion is introduced this degeneracy is destroyed and the growth rates are spaced out by steps of $O(\epsilon)$. In a numerical simulation with $0 < \epsilon \ll 1$ from general initial conditions one would soon see the correct growth rate, but it would take a long time, of order $1/\epsilon$, for the dominant eigenfunction to emerge. While the growth rate is given by the stretching and folding in the dynamical system, the weak effect of diffusion selects the fastest growing mode, that with least structure along streamlines. Following flux with $\epsilon = 0$, one would see the correct growth rate, but the frequency of oscillation would depend on the initial condition. These qualitative phenomena appear generally in steady flows (see §7.3).

The above eigenmodes (6.4.17) are independent of x_+ and x_-. Modes that do have structure in these directions decay rapidly, as $\exp(-t/O(\epsilon))$. The dramatic decay arises because the mapping $\mathbf{M}$ tends to introduce fine structure as the field is advected around $\mathcal{M}$, and so the effect of weak diffusion is very potent. We refer the reader to Arnold *et al.* (1981) for further information, and only mention here that in the original cat map the equivalent modes decay super-exponentially with diffusion (Vishik 1992, §10.4.1): the suspension of the map as a flow turns super-exponential decay into rapid exponential decay.

6.5 Anosov Flows

The cat map is an example of an *Anosov map*, and its suspension as a flow on $\mathcal{M}$ is an example of an *Anosov flow* (sometimes called a C-flow or U-flow). These were introduced by Anosov (1962); see Arnold (1988) for a clear discussion and examples. These systems, like the cat map, do not capture the complexity of fast dynamo action in realistic flows, but are natural systems to analyse mathematically. Bayly (1986) was the first to study Anosov flows more general than the cat flow, and made the discovery that the growth rate of a fast dynamo could exceed the Liapunov exponent, a behavior also observed in the smooth, physically reasonable flows of Chap. 2.

We discuss Bayly's work below, but first let us define Anosov flows. Suppose $\mathbf{u}$ is a steady flow on a compact, smooth manifold $\mathcal{M}$ with integral curves $M^t\mathbf{x}$. Vectors frozen in the flow can be transported using the corresponding Jacobian: a vector $\mathbf{v}$ at $\mathbf{x}$ is carried to $\mathbf{J}^t(\mathbf{x})\mathbf{v}$. We call the flow an *Anosov flow* if the following properties hold (Anosov 1962).

A   The space of vectors $V_\mathbf{x}$ at each point $\mathbf{x}$ can be written as the direct sum of three non-trivial subspaces

$$V_\mathbf{x} = E_\mathbf{x}^\mathrm{u} \oplus E_\mathbf{x}^\mathrm{s} \oplus E_\mathbf{x}^\mathrm{n}. \tag{6.5.1}$$

B   The one-dimensional subspace E_x^n comprises all real multiples of the flow velocity $u(x)$: $E_x^n = \lin\{u\}$, where $\lin S$ is the *linear span* of the vectors in a set S.

C   The spaces E_x^u and E_x^s depend continuously on x and are invariant under the flow, that is:

$$J^t(x)E_x^u = E_{M^tx}^u, \qquad J^t(x)E_x^s = E_{M^tx}^s. \tag{6.5.2}$$

D   There exist positive constants C and ν such that

$$\begin{aligned} v \in E_x^u &\Longrightarrow |J^t(x)v| \le C|v|\exp(\nu t), &(t < 0), \\ v \in E_x^s &\Longrightarrow |J^t(x)v| \le C|v|\exp(-\nu t), &(t > 0). \end{aligned} \tag{6.5.3}$$

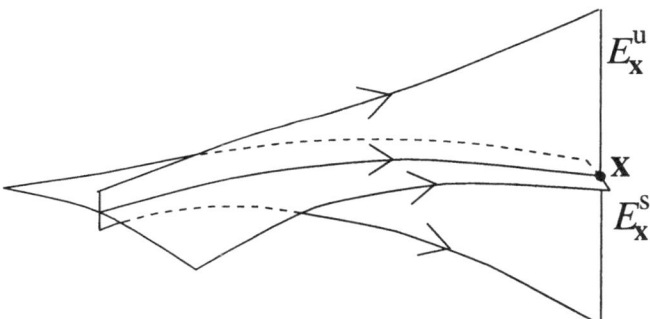

Fig. 6.1 Local behavior in an Anosov flow.

The space E_x^s is called the *stable* (or *contracting*) *direction* at x: vectors in this space are contracted exponentially as $t \to \infty$. Similarly vectors in the *unstable* (or *expanding*) *direction* E_x^u are contracted exponentially as $t \to -\infty$ (Fig. 6.1). For a flow in three dimensions these must be one-dimensional subspaces. The cat flow is perhaps the simplest example of an Anosov flow with $E_x^u = \lin\{e_+\}$, $E_x^s = \lin\{e_-\}$ and $E_x^n = \lin\{e_z\}$. In fact equality holds in (6.5.3) with $\nu = \log\lambda$ and $C = 1$.

In a similar way one can define *Anosov maps*, of which the cat map is the prime example; at each point V_x is the direct sum of two non-trivial subspaces, E_x^u and E_x^s, satisfying conditions (C) and (D). More generally if a flow or map satisfies these conditions only on a subset $\Lambda \subset \mathcal{M}$, Λ is called a *hyperbolic set* or is said to have *hyperbolic structure*; for an example, see the Smale horseshoe in §7.3.1. For an Anosov flow the whole manifold defines a hyperbolic set $\Lambda = \mathcal{M}$.

Anosov maps and flows are important in the theory of dynamical systems because they are *structurally stable* (Anosov 1962). Smooth perturbations to an Anosov map M, such as a cat map, give another Anosov map N, which is *topologically equivalent* to the original map, i.e., $M = F^{-1}NF$ where F is some homeomorphism. The same is true for Anosov flows: a perturbed Anosov flow is *topologically orbitally equivalent* to the original flow under some homeomorphism. The word orbital is present because the homeomorphism maps streamlines to streamlines: the parameterization by time along the streamlines may be different in the two cases.

Magnetic field evolution is in fact sensitive to this parameterization, since field perpendicular to streamlines can be sheared by velocity differences to give field along streamlines. In the cat flow the z-motion is uniform, and so there is no conversion of B_x or B_y to B_z by shearing. However if the flow is perturbed so that the z-motion is non-uniform, such shear will take place and new terms appear in the induction equation (6.4.15). These terms depend on the parameterization of time along streamlines and so are not invariant under topological orbital equivalence; however it seems unlikely that these terms would play much of a role in determining the fast dynamo growth rate.

We sketch Bayly's (1986) analysis of magnetic field growth in a three-dimensional Anosov flow; see also Vishik (1989). Although the cat flow is the simplest Anosov flow, it gives a misleading impression of just how pleasant Anosov flows are: for a general three-dimensional Anosov flow, the directions E_x^u and E_x^s are C^1 — continuously differentiable functions of space — but are generally not twice differentiable (Arnold & Sinai 1962, Anosov 1963).

Let us define $e_+(x) \in E_x^u$ as a C^1 field of unit vectors in the unstable direction; this assumes that the field of directions E_x^u is *orientable*. This is certainly correct for the cat flow, and remains true after sufficiently small perturbations, but does not hold for general Anosov flows; for a counterexample follow the discussion of §6.4, but replace (6.4.3) by

$$(-\mathbf{M}(x,y), z) \sim (x, y, z + 1) \tag{6.5.4}$$

(see Klapper 1991).

If we carry $e_+(x)$ in the flow for a short time δt to a nearby point x' we obtain a multiple of $e_+(x')$ by condition (C). Thus

$$e_+(x) + \delta t(e_+ \cdot \nabla u - u \cdot \nabla e_+) \simeq (1 + \delta t c_+)e_+(x'), \tag{6.5.5}$$

where $c_+(x)$ is the rate of stretching of vectors in the expanding direction given by

$$c_+ e_+ = e_+ \cdot \nabla u - u \cdot \nabla e_+, \quad c_+ = e_+ \cdot (e_+ \cdot \nabla)u. \tag{6.5.6}$$

Now integrating along a trajectory yields

$$\mathbf{J}^t(x)e_+(x) = \exp\left(\int_0^t c_+(M^\tau x)\, d\tau\right) e_+(M^t x), \tag{6.5.7}$$

from which (6.1.1) gives:

$$\chi(\mathbf{x}, \mathbf{e}_+) = \lim_{t \to \infty} \frac{1}{t} \int_0^t c_+(M^\tau \mathbf{x}) \, d\tau. \tag{6.5.8}$$

Using the fact that Anosov flows are ergodic (Anosov 1963) we can for almost all $\mathbf{x}$ replace the time-average in (6.5.8) by a space-average $<\cdot>$ over the manifold, to give the Liapunov exponent of the flow:

$$\Lambda_{\text{Liap}} = <c_+(\mathbf{x})>. \tag{6.5.9}$$

Since we need stretching of field lines, it makes sense to choose a magnetic field aligned with $\mathbf{e}_+$; however this vector field is only C^1 and so does not allow calculation of $\nabla^2 \mathbf{e}_+$. Therefore Bayly first smooths $\mathbf{e}_+$ over diffusive scales of order $\epsilon^{1/2}$, setting

$$\mathbf{e}(\mathbf{x}) = \int \epsilon^{-3/2} s'(\epsilon^{-1/2} |\mathbf{x} - \mathbf{y}|) \, \mathbf{e}_+(\mathbf{y}) \, d\mathbf{y}, \tag{6.5.10}$$

where $s(x)$ is a smooth function that is positive for $0 \le x < 1$, zero for $x \ge 1$ and is normalised by $4\pi \int_0^1 x^2 s(x) \, dx = 1$. From (6.5.10)

$$|\mathbf{e}(\mathbf{x}) - \mathbf{e}_+(\mathbf{x})| = O(\epsilon^{1/2}), \tag{6.5.11}$$

and from this we can set

$$c(\mathbf{x}) \equiv \mathbf{e} \cdot (\mathbf{e} \cdot \nabla)\mathbf{u} = c_+(\mathbf{x}) + O(\epsilon^{1/2}) \tag{6.5.12a}$$

with

$$c\mathbf{e} = \mathbf{e} \cdot \nabla \mathbf{u} - \mathbf{u} \cdot \nabla \mathbf{e} + O(\epsilon^{1/2}). \tag{6.5.12b}$$

Now Bayly seeks an eigenfunction of the form

$$\mathbf{B}(\mathbf{x}, t) = [\beta(\mathbf{x})\mathbf{e}(\mathbf{x}) + O(\epsilon^{1/2})] \exp[(p + O(\epsilon^{1/2}))t]. \tag{6.5.13}$$

Here p, $\beta = O(1)$ and β generally has structure on scales of order $\epsilon^{1/2}$. Substituting this in the induction equation and retaining only leading order terms yields a scalar equation for β:

$$(p + \mathbf{u} \cdot \nabla)\beta = c\beta + \epsilon\nabla^2\beta. \tag{6.5.14}$$

This uses the results $\nabla \mathbf{e} = O(1)$ and $\nabla^2 \mathbf{e} = O(\epsilon^{-1/2})$, which follow from (6.5.10) and $\mathbf{e}_+$ being C^1; Bayly uses somewhat weaker conditions.

Take $\beta(\mathbf{x})$ to be everywhere positive and divide through by β:

$$p + \mathbf{u} \cdot \nabla \log \beta = c + \epsilon\beta^{-1}\nabla^2\beta. \tag{6.5.15}$$

Average over space and use the divergence theorem to obtain:

$$p = <c> + \epsilon <|\nabla \log \beta|^2> \ge <c> = \Lambda_{\text{Liap}} + O(\epsilon^{1/2}) \tag{6.5.16}$$

from (6.5.9, 12a). The growth rate $\gamma(\epsilon) = \mathrm{Re}\,p$ is, up to corrections of order $\epsilon^{1/2}$, greater than or equal to the Liapunov exponent. In the limit $\epsilon \to 0$ the dynamo is fast with $\gamma_0 \geq \Lambda_{\mathrm{Liap}}$.

For general Anosov flows the inequality is strict (P. Collet, personal communication), confirming that γ_0 is generally larger than the Liapunov exponent in the absence of cancellations. Equality holds when the expanding direction, contracting direction and the expansion rate $c_+(\mathbf{x})$ depend smoothly on space (Vishik 1989), as for the cat flow. For general Anosov flows, however, these quantities do not depend smoothly on space. In the general case P. Collet (personal communication) has conjectured that the growth rate is the topological entropy (see Childress *et al.* 1990b), but a proof has not yet been obtained. Rigorous results on spectra for Anosov flows with $\epsilon = 0$ are given by Mather (1968) and de la Llave (1993); see also Thompson (1990).

When it comes to modelling astrophysical fast dynamos, Anosov flows have severe limitations, since they live on manifolds which are not easily embedded in real space. Margulis (1967) and Plante & Thurston (1972) have shown that a manifold $\mathcal{M}$ supporting an Anosov flow has a fundamental group in which the number of distinct loops increases exponentially with length. For $\mathbb{T}^3$ the number goes up algebraically, while for $\mathbb{R}^3$ the fundamental group is completely trivial. Thus the manifold $\mathcal{M}$ is more complicated than $\mathbb{R}^3$ or $\mathbb{T}^3$, and it seems unlikely that an Anosov flow can be embedded in real space. It seems difficult to extend the above analysis to study dynamo action in more realistic flows. Some related models, called 'scalar dynamos', can be studied by these techniques (Bayly 1993); a passive scalar is advected and amplified in a general flow by an equation similar to (6.5.15).

From the practical point of view, Anosov flows lack the properties of very uneven stretching and folding. For $\epsilon = 0$ and with the initial condition $\mathbf{B}_0(\mathbf{x}) = \mathbf{e}_+(\mathbf{x})$, the strength of $\mathbf{B}$ at every point is bounded below by a growing exponential; this is a very strong condition. We also assumed that the field of stretching directions $E_{\mathbf{x}}^{\mathrm{u}}$ is orientable, which means that with a suitable choice of initial conditions there is no cancellation — nearby field vectors are always aligned. This is a far cry from realistic flows, such as MW+ of Chap. 2, for which with $\epsilon = 0$ cancellations seem endemic.

6.6 An Approximate Green's Function

In this section we describe Vishik's (1988, 1989) construction of an asymptotic Green's function for the evolution of magnetic field in the limit of weak diffusion (for further developments see Dobrokhotov *et al.* 1993). The Green's function can also be calculated by Weiner methods (§6.7, Dittrich *et al.* 1984). The idea is to work in a fixed time interval $t \in [0, T]$ in the limit $\epsilon \to 0$. The $\epsilon = 0$ Green's function is given by the Cauchy solution and contains a delta function. For small positive ϵ, diffusion smears out the field and the delta function becomes a Gaussian of width $O(\sqrt{\epsilon})$ at leading order. As t is bounded while $\epsilon \to 0$, this is *not* the fast dynamo limit, $t \to \infty$ *then* $\epsilon \to 0$. Nevertheless we shall see that construction of an approximate Green's function gives an upper bound on fast dynamo growth rates, and allows one to exclude certain classes of flows from being fast dynamos. The construction is also interesting because it justifies, to some extent, the use of pulsed diffusion and is related to our discussion of noisy trajectories and shadowing theory in §§6.7, 8.

Following Vishik (1989) and Soward (1972) closely, the setting is a steady, smooth flow $\mathbf{u}(\mathbf{x})$ in ordinary or periodic space. The diffusivity is ϵ everywhere, although the flow may be localised in some region of space. Consider the evolution of magnetic field by the induction equation

$$\partial_t \mathbf{B} + \mathbf{u} \cdot \nabla \mathbf{B} - \mathbf{B} \cdot \nabla \mathbf{u} = \epsilon \nabla^2 \mathbf{B}, \tag{6.6.1}$$

with $\mathbf{B}(\mathbf{x}, 0) = \mathbf{B}_0(\mathbf{x})$ and $\nabla \cdot \mathbf{u} = 0$. Let $\mathbf{T}_\epsilon(t)$ be the induction operator and $\mathbf{T}_\epsilon(\mathbf{x}, \mathbf{y}, t)$ be the matrix-valued Green's function of (6.6.1) for $\epsilon \geq 0$:

$$\mathbf{B}(\mathbf{x}, t) = (\mathbf{T}_\epsilon(t)\mathbf{B}_0)(\mathbf{x}) = \int \mathbf{T}_\epsilon(\mathbf{x}, \mathbf{y}, t)\mathbf{B}_0(\mathbf{y}) \, d\mathbf{y}. \tag{6.6.2}$$

If there is no diffusion, $\epsilon = 0$, the Cauchy solution applies and yields the singular perfectly conducting Green's function

$$\mathbf{T}_0(\mathbf{x}, \mathbf{y}, t) = \delta(M^{-t}\mathbf{x} - \mathbf{y})\mathbf{J}^t(\mathbf{y}). \tag{6.6.3}$$

Weak diffusion will smear out this delta function to a narrow Gaussian hump. To quantify this we use a field $\mathbf{C}(\mathbf{a}, t)$ that is constant for $\epsilon = 0$:

$$\mathbf{C}(\mathbf{a}, t) = \mathbf{J}^{-t}(\mathbf{a})\mathbf{B}(\mathbf{x}, t), \qquad \mathbf{J}^t(\mathbf{a})\mathbf{C}(\mathbf{a}, t) = \mathbf{B}(\mathbf{x}, t), \tag{6.6.4a, b}$$

where here and elsewhere, $\mathbf{x} = M^t \mathbf{a}$. The new field $\mathbf{C}(\mathbf{a}, t)$ is obtained by *pulling back* the field $\mathbf{B}(\mathbf{x}, t)$ along the Lagrangian trajectories (Soward 1972). We can think of evolving $\mathbf{B}_0(\mathbf{x})$ for a time t using the induction equation with $\epsilon \geq 0$ to give us $\mathbf{B}(\mathbf{x}, t)$. We then turn off diffusion and reverse the flow field for a time t to give us the field $\mathbf{C}(\mathbf{a}, t)$ related to $\mathbf{B}(\mathbf{x}, t)$ by frozen field evolution (1.2.18).

If $\epsilon = 0$ the field $\mathbf{C}$ is constant, $\mathbf{C}(\mathbf{a}, t) = \mathbf{B}_0(\mathbf{a})$. To obtain an equation for $\mathbf{C}$ when $\epsilon \neq 0$ first differentiate (6.6.4b) at fixed $\mathbf{a}$ and rearrange as

$$\mathbf{J}\partial_t \mathbf{C} = \partial_t \mathbf{B} + \mathbf{u} \cdot \nabla \mathbf{B} - \mathbf{B} \cdot \nabla \mathbf{u}. \tag{6.6.5}$$

We have omitted the arguments from $\mathbf{J}^t(\mathbf{a})$, $\mathbf{B}(\mathbf{x}, t)$ and $\mathbf{C}(\mathbf{a}, t)$. This deals with the advection terms of (6.6.1); if the diffusion terms are written in terms of the Lagrangian coordinate $\mathbf{a}$ we obtain an equation for $\mathbf{C}$,

$$\partial_t C_i = \epsilon J_{ij}^{-1} J_{mk}^{-1} \frac{\partial}{\partial a_m} J_{nk}^{-1} \frac{\partial}{\partial a_n} J_{jl} C_l; \tag{6.6.6}$$

here the partial derivatives act on everything to their right. Any explicit reference to the original flow field has disappeared, although the Jacobians and their derivatives contain information about the local stretching and folding in the flow. This equation was obtained by Soward (1972) in his study of Braginsky's dynamo; it can also be written in an elegant form (Vishik 1989):

$$\partial_t \mathbf{C} = -\epsilon \nabla \times \mathbf{\Lambda} \nabla \times \mathbf{\Lambda} \mathbf{C} + \epsilon \mathbf{\Lambda}^{-1} \nabla \nabla \cdot \mathbf{C}. \tag{6.6.7}$$

Here ∇ means gradient with respect to $\mathbf{a}$ and acts on everything to its right. The positive-definite, symmetric matrix $\mathbf{\Lambda}^t(\mathbf{a})$ is defined in (6.1.2). Equations (6.6.6, 7) are a recasting of the induction equation, with diffusion, in terms of Lagrangian coordinates.

Equation (6.6.6) for $\mathbf{C}$ can be rearranged in the suggestive form

$$\partial_t C_i = \epsilon \partial_m \Lambda_{mn}^{-1} \partial_n C_i + \text{other terms}, \tag{6.6.8}$$

where the 'other terms' are terms linear in $\mathbf{C}$ or $\partial \mathbf{C}/\partial \mathbf{a}$ and do not involve second order derivatives of $\mathbf{C}$. Since the matrix function $\mathbf{\Lambda}^t(\mathbf{a})$ is always positive definite and symmetric the terms retained give diffusion, albeit anisotropic, non-stationary and inhomogeneous. Because of this the Green's function $\mathbf{S}_\epsilon(\mathbf{a}, \mathbf{b}, t)$ for $\mathbf{C}$ in the full equation (6.6.6) can be expanded in an asymptotic series in powers of $\sqrt{\epsilon}$, the leading term of which is pure diffusion:

$$\mathbf{S}_\epsilon^0(\mathbf{a}, \mathbf{b}, t) = \frac{\mathbf{I}}{(4\pi\epsilon)^{3/2}(\det \mathbf{Q})^{1/2}} \exp\left(\frac{-(\mathbf{a} - \mathbf{b}) \cdot \mathbf{Q}^{-1}(\mathbf{a} - \mathbf{b})}{4\epsilon}\right). \tag{6.6.9}$$

The positive definite symmetric matrix

$$\mathbf{Q}(\mathbf{a}, t) = \int_0^t [\mathbf{\Lambda}^\tau(\mathbf{a})]^{-1} d\tau \tag{6.6.10}$$

contains information about the stretching near the trajectory through $\mathbf{a}$.

The construction of this leading-order approximate Green's function (6.6.9) gives some justification to the technique of pulsing diffusion. For definiteness suppose we have a magnetic field $\mathbf{B}$ evolving with continuous weak diffusion in a flow comprising Beltrami waves of duration T. From equations (6.6.4) we can write without approximation

$$\mathbf{T}_\epsilon(T) = \mathbf{T}_0(T)\mathbf{S}_\epsilon(T), \tag{6.6.11}$$

where $\mathbf{S}_\epsilon(t)$ is the operator for evolution of $\mathbf{C}$-field under (6.6.6). We again use the same symbol for an operator and its kernel. In terms of Green's functions:

$$\begin{aligned} \mathbf{T}_\epsilon(\mathbf{x}, \mathbf{b}, T) &= \int \mathbf{T}_0(\mathbf{x}, \mathbf{a}, T)\mathbf{S}_\epsilon(\mathbf{a}, \mathbf{b}, T) \, d\mathbf{a} \\ &= \mathbf{J}(M^{-T}\mathbf{x})\mathbf{S}_\epsilon(M^{-T}\mathbf{x}, \mathbf{b}, T). \end{aligned} \tag{6.6.12}$$

Now if $\epsilon \ll 1$ we may be able to use the leading order approximate Green's function $\mathbf{S}_\epsilon^0$ (6.6.9) to evolve $\mathbf{C}$ instead of the exact one. In this case (6.6.11) or (6.6.12) simply amount to pulsing diffusion, i.e., applying diffusion with the kernel (6.6.9) instantaneously, followed by advection without diffusion using M^T. From continuous diffusion the pull-back gives us a pulsed model, but with anisotropic and inhomogeneous diffusion. This gives some justification to the use of pulsed diffusion: if a fast dynamo process is insensitive to the specific type of diffusion for $\epsilon \ll 1$, similar results should be obtained for continuous or pulsed diffusion. One could in principle write down the leading-order Green's function for a Beltrami wave, and insert this as diffusion in a pulsed code to obtain a better approximation to continuous diffusion.[53]

The approximate Green's function is used by Vishik (1988, 1989) in a proof of an upper bound on fast dynamo growth rates and an anti-fast-dynamo theorem. This result was first proved by Oseledets (1984), using rather different methods; this paper is difficult to obtain, but his result is mentioned in Latushkin & Stepin (1991).

Recall first the *supremum fast dynamo growth rate* $\overline{\gamma}_0 \equiv \limsup_{\epsilon \to 0} \gamma(\epsilon)$, with $\gamma_0 \leq \overline{\gamma}_0$, and the definition of $\Lambda_{\text{Liap}}^{\max}$ from §1.6; in terms of (6.1.1, 2),

$$\Lambda_{\text{Liap}}^{\max} = \sup_{\mathbf{a},\mathbf{v}} \chi(\mathbf{a},\mathbf{v}), \qquad \chi(\mathbf{a},\mathbf{v}) = \limsup_{t \to \infty} \frac{1}{2t} \log |\mathbf{v} \cdot \Lambda^t(\mathbf{a})\mathbf{v}|. \qquad (6.6.13)$$

The theorem is:

Theorem 6.2 *(Oseledets 1984, Vishik 1988, 1989) For a smooth, steady flow $\mathbf{u}$ the supremum growth rate $\overline{\gamma}_0$ is no greater than the maximum Liapunov exponent $\Lambda_{\text{Liap}}^{\max}$. Therefore if there is no exponential stretching in the flow, $\Lambda_{\text{Liap}}^{\max} = 0$, then only slow dynamo action is possible.*

Intuitively, at least one vector frozen in the flow must stretch exponentially for a fast dynamo.

We discuss this result following Vishik (1988, 1989): the reader should consult these papers for more detail. There are two possible settings; in the first $\mathbf{u}(\mathbf{x})$ is a steady, smooth, space-periodic flow, and the magnetic field has the same periodicity; in this case we use integrals over one periodicity box. In the second $\mathbf{u}(\mathbf{x})$ is a steady, smooth flow in ordinary space which is zero outside a bounded region, $\mathcal{D}$; here we use integrals over all space. In each case the diffusivity ϵ is uniform in space and the magnetic field has finite energy. We follow growth in the L^2 norm of the magnetic field $\mathbf{B}(\mathbf{x},t)$,

$$\|\mathbf{B}(\mathbf{x},t)\|_2 = \left(\int |\mathbf{B}(\mathbf{x},t)|^2 \, d\mathbf{x} \right)^{1/2}. \qquad (6.6.14)$$

Take $\epsilon \ll 1$ and let $\mathbf{B}_\epsilon$ be the fastest growing dynamo eigenfunction with complex growth rate $p(\epsilon) = \gamma(\epsilon) + i\omega(\epsilon)$. Then

---

[53]There is one caveat, namely that the leading-order Green's function does not exactly preserve the solenoidality of a vector field, but will generally introduce a weak divergence of magnitude $O(\sqrt{\epsilon})$.

$$\mathbf{T}_\epsilon(T)\mathbf{B}_\epsilon = \exp(p(\epsilon)T)\mathbf{B}_\epsilon \tag{6.6.15}$$

and so

$$\|\mathbf{T}_\epsilon(T)\mathbf{B}_\epsilon\|_2 = \exp(\gamma(\epsilon)T)\,\|\mathbf{B}_\epsilon\|_2, \quad \|\mathbf{T}_\epsilon(T)\|_2 \geq \exp(\gamma(\epsilon)T). \tag{6.6.16a,b}$$

The factorization (6.6.11) implies that

$$\|\mathbf{T}_\epsilon(T)\|_2 \leq \|\mathbf{T}_0(T)\|_2\,\|\mathbf{S}_\epsilon(T)\|_2. \tag{6.6.17}$$

The operator $\mathbf{S}_\epsilon(T)$ is diffusion at leading order from (6.6.9) and so cannot increase the magnetic energy much when $\epsilon \ll 1$; the following estimate holds:

$$\|\mathbf{S}_\epsilon(T)\|_2 \leq 1 + O(\sqrt{\epsilon}), \quad (\epsilon \to 0). \tag{6.6.18}$$

Any fast energy increase must arise from stretching by $\mathbf{T}_0(T)$. Under frozen field evolution

$$\|\mathbf{T}_0(t)\mathbf{C}\|_2^2 = \int \mathbf{C}^*(\mathbf{a}) \cdot \Lambda^t(\mathbf{a})\mathbf{C}(\mathbf{a})\,d\mathbf{a} \tag{6.6.19}$$

by (6.1.2, 6.3) and so

$$\|\mathbf{T}_0(t)\|_2^2 \leq \sup_{\mathbf{a},\mathbf{v}} \frac{\mathbf{v} \cdot \Lambda^t(\mathbf{a})\mathbf{v}}{|\mathbf{v}|^2}. \tag{6.6.20}$$

From (6.6.13) this implies that

$$\limsup_{t \to \infty} \frac{1}{t} \log \|\mathbf{T}_0(t)\|_2 \leq \Lambda_{\text{Liap}}^{\max} \equiv \sup_{\mathbf{a},\mathbf{v}} \chi(\mathbf{a},\mathbf{v}), \tag{6.6.21}$$

using a compactness argument detailed in Vishik (1988) that allows the limit of large t and the supremum over $\mathbf{a}$ and $\mathbf{v}$ to be interchanged.

We can now prove theorem 6.2 by *reductio ad absurdum*. First let $\overline{\gamma}_0$ be the supremum fast dynamo growth rate and assume that $\Lambda_{\text{Liap}}^{\max} < \overline{\gamma}_0$. From (6.6.21) we can fix a time T such that

$$\|\mathbf{T}_0(T)\|_2 \leq \exp(\Lambda_{\text{Liap}}^{\max}T + \delta T/3) = \exp(\overline{\gamma}_0 T - 2\delta T/3), \tag{6.6.22}$$

where $\delta = \overline{\gamma}_0 - \Lambda_{\text{Liap}}^{\max} > 0$. With T fixed consider the limit $\epsilon \to 0$. From the definition of $\overline{\gamma}_0$ there exists a sequence $\epsilon_i \to 0$ for which the growth rates of the fastest growing eigenfunctions satisfy $\gamma(\epsilon_i) \geq \overline{\gamma}_0 - \delta/3$; therefore applying $\mathbf{T}_{\epsilon_i}(T)$ to these eigenfunctions gives

$$\|\mathbf{T}_{\epsilon_i}(T)\|_2 \geq \exp(\overline{\gamma}_0 T - \delta T/3) \tag{6.6.23}$$

from (6.6.16b). But from (6.6.17, 18, 22)

$$\|\mathbf{T}_{\epsilon_i}(T)\|_2 \leq \exp(\overline{\gamma}_0 T - 2\delta T/3)\,(1 + O(\sqrt{\epsilon_i})) \tag{6.6.24}$$

and when ϵ_i is sufficiently small this contradicts (6.6.23). Thus $\overline{\gamma}_0 \leq \Lambda_{\text{Liap}}^{\max}$ and if $\Lambda_{\text{Liap}}^{\max} = 0$, so that there is no stretching of vectors in the flow, a fast dynamo is impossible.

This anti-fast-dynamo result is less powerful than the topological entropy bound. It does not exclude fast dynamo action in smooth laminar flows with isolated stagnation points, which have no chaos and zero topological entropy, for example the Roberts' dynamo. The supremum in (6.6.20, 21) is sensitive to stretching on sets of zero measure, such as points, lines and surfaces. This sensitivity to stretching on a 'small' set is not surprising, since a fast-growing magnetic field could become localised very close to the set as $\epsilon \to 0$. This phenomenon occurs in the boundary layer dynamos of Chap. 5, and even in chaotic dynamos the flux tends to be concentrated on sets of small measure as $\epsilon \to 0$ (see §7.1).

Note that the condition on the flow being smooth is essential for the theory of the previous section, and so the theorem does not apply to fast dynamos with discontinuities. Consider Ponomarenko's dynamo which by any sensible definition would have $\Lambda_{\text{Liap}}^{\max} = 0$, yet $\gamma_0 > 0$. However Vishik's (1988) compactness argument for (6.6.21) requires differentiability of the flow. Also (6.6.20) does not give useful information, since at the discontinuity the matrix Λ^t is infinite. In fact the key property of this flow and the modified Roberts cell fast dynamo (§5.5.2) is that the operator $\mathbf{T}_0(T)$ has infinite L^2 norm and so no useful bounds are obtained from diffusionless evolution. If we place a loop of field across the discontinuity and allow the flow to operate for a time, we obtain a delta-function sheet of field at the discontinuity, which has finite flux, but infinite energy.

6.7 Noisy Trajectories and Diffusion

One method of dealing with diffusion in fast dynamos is to replace deterministic diffusion $\epsilon \nabla^2 \mathbf{B}$ by an average over noisy trajectories (see, e.g., Zeldovich et al. 1988). This device allows another view of the effects of diffusion and throws some light on the flux conjecture. It lies at the heart of the proof of Klapper & Young (1995) that the topological entropy h_{top} is an upper bound on fast dynamo growth rates. The method can also be used to obtain approximate Green's functions. In the limit $\epsilon \to 0$ for a finite time-interval, the noisy trajectories form a cloud close to the deterministic $\epsilon = 0$ path, with an approximately Gaussian distribution. This distribution can be calculated perturbatively (Dittrich et al. 1984) to recover the approximate Green's function of §6.6. The description in terms of noisy paths has been used by Molchanov, Ruzmaikin & Sokoloff (1983, 1984, 1985) and Zeldovich et al. (1987, 1988) in the discussion of random dynamos and intermittency.

Consider first the advection and diffusion of a passive scalar. At a microscopic level diffusion of a scalar arises from the random walking of scalar molecules, which undergo Brownian motion relative to the flow. In obtaining a macroscopic advection–diffusion equation an average over these paths is taken and this yields the usual Laplacian term. One can, however, still view the evolution at a microscopic level; for example, numerically one can obtain effective diffusivities by following random-walking particles (Drummond, Duane & Horgan 1984).

The microscopic physics of magnetic field diffusion is more complicated, but nevertheless one can write down a prescription for moving field vectors in a given flow $\mathbf{u}(\mathbf{x}, t)$, which after averaging yields the usual diffusion term. The notation becomes a little cumbersome, because the noisy flows are neither steady nor periodic in time. Follow a particle path from position $\mathbf{a}$ at time s to position $M_s^t \mathbf{a}$ at time t according to

$$\partial_t M_s^t \mathbf{a} \, dt = \mathbf{u}(M_s^t \mathbf{a}, t) \, dt + \sqrt{2\epsilon} \, d\mathbf{w}(t), \qquad M_s^s \mathbf{a} = \mathbf{a}. \tag{6.7.1}$$

We have added to the flow a term $\mathbf{w}(t)$ which depends on time but not on position, this being a three-dimensional *Brownian motion* or *Weiner process*; see McKean (1969) for mathematical background. Each component $w_i(t)$, $t > 0$, is an independent random process satisfying

$$<w_i(t)> = 0, \qquad <w_i(s)w_j(t)> = \delta_{ij} \min(s, t). \tag{6.7.2}$$

For any $s \geq 0$, $\Delta w_i(t) = w_i(t + s) - w_i(s)$ is again a Brownian motion independent of $w_i(t)$ for $t \leq s$; the process has 'independent increments'. This allows us to express (6.7.2) for infinitesimal time intervals dt by

$$<dw_i> = 0, \qquad <dw_i dw_j> = \delta_{ij} \, dt, \tag{6.7.3}$$

showing that we should think of $d\mathbf{w}$ as being of order $\sqrt{dt}$.

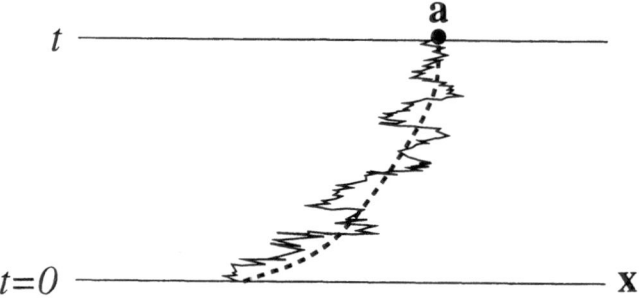

Fig. 6.2 A Weiner path in a flow (solid). For small ϵ this approximates the deterministic $\epsilon = 0$ path (dashed). (After Molchanov *et al.* 1983.)

The term $\mathbf{w}(t)$ represents the microscopic random-walking of particles that leads to macroscopic diffusion. With $\mathbf{u} = 0$ trajectories

$$\mathbf{x} = M_0^t \mathbf{a} = \mathbf{a} + \sqrt{2\epsilon}\,\mathbf{w}(t) \tag{6.7.4}$$

are pure Brownian motion, and a point source at $\mathbf{a}$ when $t = 0$ spreads out into a Gaussian distribution at time t with mean $<\mathbf{x}> = \mathbf{a}$ and isotropic variance matrix: $<(\mathbf{x}-\mathbf{a})(\mathbf{x}-\mathbf{a})> = 2\epsilon t\mathbf{I}$, which increases linearly with time. Most trajectories have $\mathbf{x} - \mathbf{a} = O(\sqrt{\epsilon t})$: a few will wander further than this but this occurrence is extremely rare, because of the rapid decay of a Gaussian distribution. When $\mathbf{u}$ is non-zero the Brownian motion and the flow field are superposed. Most of these *Weiner paths* remain a distance of order $\sqrt{\epsilon t}$ from the $\epsilon = 0$ deterministic path (see Fig. 6.2); for small ϵ and finite t these have an approximately Gaussian distribution, which is generally anisotropic because of the action of the flow in stretching out certain directions.

In a given realization of $\mathbf{w}(t)$ we can evolve a frozen field by the Cauchy solution from some initial condition,

$$\mathbf{B_w}(\mathbf{x}, t) = \mathbf{J}_0^t(M_t^0\mathbf{x})\,\mathbf{B}(M_t^0\mathbf{x}, 0); \tag{6.7.5}$$

differentiating (6.7.1) gives equations for the Jacobian $\mathbf{J}_s^t(\mathbf{a}) = \partial M_s^t\mathbf{a}/\partial\mathbf{a}$:

$$\partial_t \mathbf{J}_s^t(\mathbf{a}) = [\nabla\mathbf{u}(M_s^t\mathbf{a}, t)]^{\mathrm{T}}\mathbf{J}_s^t(\mathbf{a}), \qquad \mathbf{J}_s^s(\mathbf{a}) = \mathbf{I}. \tag{6.7.6}$$

Note that the term $\mathbf{w}(t)$ has zero gradient and so does not stretch vectors; it thus does not appear explicitly in the equation for $\mathbf{J}$, but enters implicitly as it affects the trajectory $M_s^t\mathbf{a}$.

Now ensemble average the Cauchy solution (6.7.5) over all realizations of the Brownian motion $\mathbf{w}(t)$:

$$\mathbf{B}(\mathbf{x}, t) \equiv <\mathbf{B_w}(\mathbf{x}, t)> = <\mathbf{J}_0^t(M_t^0\mathbf{x})\mathbf{B}(M_t^0\mathbf{x}, 0)>; \tag{6.7.7}$$

here $\mathbf{x}$ is held fixed and so the initial position and the Jacobian depend on the realization of $\mathbf{w}(t)$. Field is picked up from all random trajectories that arrive at $\mathbf{x}$ at time t and then averaged.

The ensemble-averaged field $\mathbf{B}(\mathbf{x}, t)$ satisfies the induction equation for the flow $\mathbf{u}(\mathbf{x}, t)$ and diffusivity ϵ. Let us make this plausible, following, for example, Klapper (1992b), by fixing $\mathbf{x}$ and considering an infinitesimal time interval from t to $t + dt$. From (6.7.7),

$$\begin{aligned}\mathbf{B}(\mathbf{x}, t+dt) &= <\mathbf{J}_0^{t+dt}(M_{t+dt}^0\mathbf{x})\mathbf{B}(M_{t+dt}^0\mathbf{x}, 0)> \\ &= <\mathbf{J}_t^{t+dt}(M_{t+dt}^t\mathbf{x})\mathbf{J}_0^t(M_t^0 M_{t+dt}^t\mathbf{x})\mathbf{B}(M_t^0 M_{t+dt}^t\mathbf{x}, 0)>.\end{aligned} \tag{6.7.8}$$

The increments $d\mathbf{w}$ are independent in the time intervals $(0, t)$ and $(t, t + dt)$; so we can average over $(0, t)$ first and use (6.7.7) to obtain an equation

$$\mathbf{B}(\mathbf{x}, t+dt) = <\mathbf{J}_t^{t+dt}(M_{t+dt}^t\mathbf{x})\mathbf{B}(M_{t+dt}^t\mathbf{x}, t)> \tag{6.7.9}$$

for evolution from t to $t + dt$. From (6.7.1)

$$M_t^{t+dt}(\mathbf{x}) = \mathbf{x} + \mathbf{u}\, dt + \sqrt{2\epsilon}\, d\mathbf{w}, \quad M_{t+dt}^t(\mathbf{x}) = \mathbf{x} - \mathbf{u}\, dt - \sqrt{2\epsilon}\, d\mathbf{w} \quad (6.7.10)$$

(quantities without arguments are evaluated at $(\mathbf{x}, t)$) and from (6.7.6)

$$\mathbf{J}_t^{t+dt}(M_{t+dt}^t\mathbf{x}) = \mathbf{I} + (\nabla\mathbf{u})^{\mathrm{T}} dt. \quad (6.7.11)$$

Using (6.7.10) we can Taylor expand

$$\mathbf{B}(M_{t+dt}^t\mathbf{x}, t) = \mathbf{B} - \mathbf{u} \cdot \nabla\mathbf{B}\, dt - \sqrt{2\epsilon}\, d\mathbf{w} \cdot \nabla\mathbf{B} + \epsilon\, dw_i dw_j\, \partial_i\partial_j\mathbf{B} \quad (6.7.12)$$

up to order dt or $(d\mathbf{w})^2$. Substituting (6.7.11, 12) in the average (6.7.9) yields

$$
\begin{aligned}
\mathbf{B}(\mathbf{x}, t + dt) &= \; <\mathbf{B} + \mathbf{B} \cdot \nabla\mathbf{u}\, dt - \mathbf{u} \cdot \nabla\mathbf{B}\, dt \\
&\quad - \sqrt{2\epsilon}\, d\mathbf{w} \cdot \nabla\mathbf{B} + \epsilon\, dw_i dw_j\, \partial_i\partial_j\mathbf{B}> \quad (6.7.13) \\
&= \mathbf{B} + (\mathbf{B} \cdot \nabla\mathbf{u} - \mathbf{u} \cdot \nabla\mathbf{B} + \epsilon\nabla^2\mathbf{B})\, dt,
\end{aligned}
$$

using (6.7.3). This is the induction equation with diffusivity ϵ.

The idea of averaging over noisy paths can form the basis of numerical methods (see Drummond, Duane & Horgan 1983, Drummond & Horgan 1986); it changes a partial differential equation into an ensemble of noisy ordinary differential equations. One chooses a point $\mathbf{x}$ and integrates noisy particle paths back in time to calculate $\mathbf{B}(\mathbf{x}, t)$ from equation (6.7.7); the more paths one uses the smaller is the statistical error in the calculation. This is used by Klapper (1992b) in a numerical study of fast dynamo action in a patched, chaotic two-dimensional flow. The numerical method has the advantages that $\mathbf{B}(\mathbf{x}, t)$ need only be calculated at a single point $\mathbf{x}$ to obtain a growth rate and that the magnetic Reynolds number $R = 1/\epsilon$ can be set formally to very large values; however there are limits on the magnetic Reynolds numbers that can be used reliably in fast dynamo studies. There is an initial transient before clear exponential growth of field emerges, whose length increases with R, and there is cancellation in calculating the average (6.7.7); as R increases the amount of computer power needed to overcome these increases rapidly.

We now discuss briefly the proof of Klapper & Young (1995) that $h_{\mathrm{top}} \geq \gamma_0$, that is, theorem 6.1 of §6.1, without technical details. Klapper and Young first prove that the result for $\epsilon = 0$, which is reasonable in terms of the picture of magnetic field lines as material curves in the fluid. The problem then is to extend the result to $\epsilon > 0$. From the above discussion one can include diffusion by averaging the evolution over a set of noisy flows (6.7.1). Now for $\epsilon \ll 1$, most of these noisy flows are, most of the time, close to $\mathbf{u}$, and so stretch material lines by at most about h_{top}. Occasionally the noise is large, the noisy flow is far from $\mathbf{u}$, and the line-stretching rate is hard to control. However these events are so rare for small ϵ that they have a small effect on the ensemble-averaged growth rate of material lines; in the limit $\epsilon \to 0$ the quantity h_{top} becomes an upper bound.

6.8 Shadowing and the Flux Conjecture

The description in terms of noisy trajectories can be used to give some support to the flux conjecture (Klapper 1991, 1992a, b, 1993). The key idea comes from dynamical systems theory and is known as 'shadowing'. Consider following a chaotic orbit of a map M numerically; the presence of numerical errors means that a computed trajectory $\{\mathbf{x}_j\}$ satisfies only $|\mathbf{x}_{j+1} - M\mathbf{x}_j| < \alpha$ for some $0 < \alpha \ll 1$; this is called an α-*pseudo-orbit*. An important question is whether this bears any resemblance to a true orbit of the map: is there a true orbit $\mathbf{y}_{j+1} = M\mathbf{y}_j$ which is always close to $\{\mathbf{x}_j\}$? We seek a true orbit $\{\mathbf{y}_j\}$ that β-*shadows* $\{\mathbf{x}_j\}$, i.e., for which $|\mathbf{y}_j - \mathbf{x}_j| < \beta$ for all j. This is possible for hyperbolic systems such as Anosov maps, which have the *shadowing property*: given $\beta > 0$, there is an $\alpha > 0$ such that every α-pseudo-orbit is β-shadowed by a true orbit (Anosov 1967, Bowen 1978, Newhouse 1980). Remarkably this is true without any restriction on the range of j, and this reflects the special properties of hyperbolic systems.

In the case of a dynamo with very weak diffusion, we have seen that diffusion can be treated by averaging over noisy orbits. The key observation of Klapper is that most noisy orbits can be replaced by nearby true orbits in systems for which shadowing holds. The average of magnetic field over noisy orbits then becomes approximately an average over true orbits. The diffusive solution for $0 < \epsilon \ll 1$ can therefore be replaced approximately by an average over the Cauchy solution for $\epsilon = 0$; this is a form of the flux conjecture, but is only valid for a finite time.

We follow Klapper (1993) and then give further discussion. Let M be an area- and orientation-preserving Anosov map on a compact two-dimensional manifold. It is helpful to have in mind a perturbed cat map on $\mathbb{T}^2$; see §10.1. In a pulsed dynamo let us first apply the map M, at times $t = n+$, and then diffuse the magnetic field for a unit time with diffusivity $0 < \epsilon \ll 1$. For $\epsilon \geq 0$, let $\mathbf{B}_\epsilon(\mathbf{x}, n)$ be the magnetic field after n iterations at time $t = n$ resulting from some given initial condition. According to §6.7, we can obtain field $\mathbf{B}_\epsilon(\mathbf{X}, n)$ at some given point $\mathbf{X}$ by following noisy trajectories back from $\mathbf{X}$ to $t = 0$, bringing field forwards, and averaging. The first step is to find the probability distribution $\rho_0(\mathbf{x})$ of noisy trajectories at time $t = 0$. Note that the analysis is approximate; errors arise in numerous places and are generally $O(\sqrt{\epsilon})$, but not uniformly in time.

Consider how the distribution of noisy paths evolves through one pulse going *backwards* in time. During a diffusion phase, the noisy paths spread out and a probability distribution $\rho(\mathbf{x})$ is convoluted with the heat kernel

$$H_\epsilon(\mathbf{x}) = h_\epsilon(x)h_\epsilon(y), \qquad h_\epsilon(x) = (4\pi\epsilon)^{-1/2}\exp(-x^2/4\epsilon). \tag{6.8.1}$$

For the map phase, a given probability distribution of particles $\rho(\mathbf{x})$ *after* the application of M is the result of a probability distribution $M^*\rho$ *before* the mapping, where M^* is the *pull-back* defined by

$$(M^*\rho)(\mathbf{x}) = \rho(M\mathbf{x})\det(\partial M\mathbf{x}/\partial\mathbf{x}) = \rho(M\mathbf{x}). \tag{6.8.2}$$

Combining (6.8.1, 2), the probability density function $\rho_j(\mathbf{x})$ at time $t = j$ for the noisy trajectories that arrive at $\mathbf{X}$ when $t = n$ satisfies

$$\rho_{j-1} = M^*(H_\epsilon * \rho_j), \qquad \rho_n(\mathbf{x}) = \delta(\mathbf{x} - \mathbf{X}), \tag{6.8.3}$$

where the operation $*$ is convolution. For simplicity we shall assume that M has a fixed point, and take $\mathbf{X}$ to be this fixed point, $M\mathbf{X} = \mathbf{X}$; this assumption simplifies discussion, but can easily be relaxed, as we indicate below.

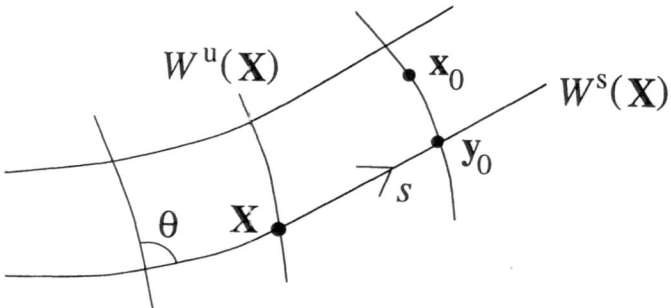

Fig. 6.3 The local grid structure of stable and unstable manifolds for the Anosov map M.

The map M is an Anosov map, which means that at $\mathbf{X}$ there are well-defined stable and unstable directions $E_{\mathbf{X}}^s$ and $E_{\mathbf{X}}^u$. These can be extended into stable and unstable manifolds $W^s(\mathbf{X})$ and $W^u(\mathbf{X})$ passing through $\mathbf{X}$:

$$W^s(\mathbf{X}) = \{\mathbf{x} : \lim_{n\to\infty} |M^n\mathbf{x} - M^n\mathbf{X}| = 0\},$$
$$W^u(\mathbf{X}) = \{\mathbf{x} : \lim_{n\to-\infty} |M^n\mathbf{x} - M^n\mathbf{X}| = 0\} \tag{6.8.4}$$

(see, e.g., Guckenheimer & Holmes 1983). Locally the stable and unstable manifolds of $\mathbf{X}$ and nearby points $\mathbf{x}$ form a grid-like structure (Fig. 6.3). The grid is curved over $O(1)$ scales and is not orthogonal, but the angle $\theta(\mathbf{x})$ between stable and unstable manifolds is bounded away from zero. The action of M is to give exponential contraction of points along $W^s(\mathbf{X})$ and exponential separation of points along $W^u(\mathbf{X})$.

Without detailed calculations it is clear how the distribution of noisy trajectories will evolve backwards in time from the fixed point $\mathbf{X}$ by (6.8.3). Diffusion tends to spread out trajectories isotropically while the pull-back tends to stretch the distribution exponentially along the *stable* manifold $W^s(\mathbf{X})$ and compress it along the *unstable* manifold $W^u(\mathbf{X})$. The result for long

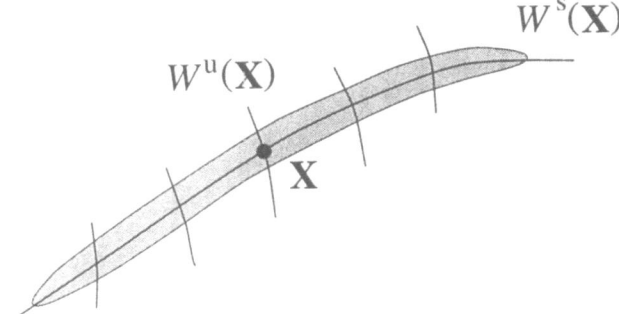

Fig. 6.4 The cloud of noisy particles spreads into a long thin distribution along $W^s(\mathbf{X})$.

backwards time is a distribution taking the form of a long thin sausage along $W^s(\mathbf{X})$, depicted in Fig. 6.4. Its width is of order $\sqrt{\epsilon}$ characteristic of the balance between exponential compression and diffusion.

We have the distribution of noisy orbits at $t = 0$. For each noisy orbit $\{\mathbf{x}_j\}$ in the distribution, a true orbit that shadows it for $0 \le t \le n$ is obtained by projecting $\mathbf{x}_0$ along the grid of unstable manifolds to a point $\mathbf{y}_0$ on the stable manifold $W^s(\mathbf{X})$ (Fig. 6.3). The orbit $\{\mathbf{y}_j\}$ from this point remains close to $\{\mathbf{x}_j\}$ for $0 \le t \le n$, and arrives at $\mathbf{y}_n$, close to $\mathbf{X}$, when $t = n$. Since the noisy orbit $\{\mathbf{x}_j\}$ and the true orbit $\{\mathbf{y}_j\}$ are close, so are the Jacobians used to carry field from time $t = 0$ to time $t = n$. Thus the value of the field $\mathbf{B}_\epsilon(\mathbf{X}, n)$, being an average of the field carried along the noisy paths, is approximately the average of the Cauchy solution $\mathbf{B}_0(\mathbf{X}, n)$ over the distribution ν_n of arrival points, $\mathbf{y}_n$, of the shadowing orbits. This distribution will be approximately Gaussian, lie on $W^s(\mathbf{X})$ and be localised within distance $O(\sqrt{\epsilon})$ from $\mathbf{X}$.

The distribution of final arrival points is obtained by projecting the distribution $\rho_j(\mathbf{x})$ of noisy orbits onto $W^s(\mathbf{X})$, that is, by integrating $\rho_j(\mathbf{x})$ over the unstable manifolds to obtain a distribution $\bar{\rho}_j(s)$, where s is arclength along $W^s(\mathbf{X})$ measured from $\mathbf{X}$ (Fig. 6.3). Then an equation similar to (6.8.3) is obtained for the evolution of $\bar{\rho}_j(s)$ under mapping and diffusion. The distribution $\bar{\rho}_0(s)$ at time zero is then brought forward under the mapping and no diffusion to obtain the distribution $\nu_n(s)$ of arrival times, given approximately for large n and small ϵ by the heat kernel (see 6.8.1)

$$\nu_n(s) = h_{\epsilon_0}(s), \qquad \epsilon_0 = \epsilon \sum_{j=0}^{\infty} \left[(g^j)'(0) \sin \theta \right]^{-2} \tag{6.8.5}$$

(Klapper 1993). The field with diffusion ϵ at $\mathbf{X}$ after some time $t = n \gg 1$ is given approximately by averaging the Cauchy solution over the stable manifold $W^s(\mathbf{X})$ using a Gaussian distribution with parameter ϵ_0, independent of n.

In (6.8.5) $g(s)$ is the map from $W^s(\mathbf{X})$ to itself induced by M^{-1}; its presence keeps track of the interaction of diffusion and stretching along the stable manifold in backwards time. The angle θ is the angle between stable and unstable directions at $\mathbf{X}$ and arises from the interaction of diffusion and projection onto $W^s(\mathbf{X})$. The angle $\theta(s)$ is bounded away from zero and $(g^j)'$ increases exponentially for an Anosov system; so this series converges geometrically. Since $\epsilon_0 = O(\epsilon) \ll 1$ this amounts to averaging over $E^s(\mathbf{X})$.[54]

Although we took $\mathbf{X}$ to be a fixed point in the above, there is no difficulty in extending this to a general point $\mathbf{X}$ (Klapper 1993); in this case we use a sequence of maps g_j, which map points on $W^s(M^{-j+1}\mathbf{X})$ to points on $W^s(M^{-j}\mathbf{X})$. The result analogous to (6.8.5) is then

$$\nu_n(s) = h_{\epsilon_0}(s), \qquad \epsilon_0 = \epsilon \sum_{j=0}^{\infty} [(g_j \circ g_{j-1} \circ \cdots \circ g_1)'(0) \sin \theta_j]^{-2}, \quad (6.8.6)$$

where θ_j is the angle between stable and unstable manifolds at $M^{-j}(\mathbf{X})$.

The result (6.8.6) is a finite-time form of the flux conjecture, relating the evolution of the solution with diffusion to an average of the Cauchy solution. It uses a Gaussian average over the Cauchy solution, rather than a flux or arbitrary smooth functional. The result is not uniform in time, however, because of the $O(\sqrt{\epsilon})$ errors in its derivation. The conclusion then is that the flux conjecture works for a time, but has not been proved for infinite time, and this remains an interesting problem; see §5 of Klapper (1993). Note that the theory goes through for an Anosov map in (x, y) with shear in a third direction z giving complex phase shifts as in the models of §10.4.

Another problem is to extend the analysis to maps that are not Anosov and have non-trivial folding of field. Generally shadowing does not hold for such systems: this is clear if there are islands present as a true orbit cannot escape an island whereas a noisy one can. However even if there are no islands, shadowing can break down at the folds of the map where the stretching is weak, and the angle between stable and unstable directions is small.

Nevertheless shadowing in non-Anosov systems can be achieved for quite long times before breakdown occurs (Grebogi et al. 1990). Furthermore Klapper (1992b) finds that the Gaussian average (6.8.6) gives results consistent with the true diffusive solution for a patched, chaotic cellular flow. The reason is probably that the noisy trajectories that cannot be shadowed have only weak magnetic field stretching, and so can probably be ignored in the final average; the magnetic field growth in a fast dynamo is dominated by atypical orbits with strong stretching and these are likely to be shadowable even in non-Anosov systems (Klapper 1992b). Justifying this idea mathematically however remains a challenge.

---

[54]Note that the stable direction $E^s(\mathbf{X})$ is not generally perpendicular to the unstable direction $E^u(\mathbf{X})$, the latter being easily identified from pictures of field evolution.

7. Magnetic Structure in Chaotic Flows

In this chapter we discuss the structure of magnetic fields in chaotic fast dynamos. The results contrast with those in Chap. 5 for laminar flows: in a chaotic flow the field is less organized and chaotic streamlines spread field over finite volumes of space. Furthermore a chaotic flow generally stretches some field vectors more than others, leading to the phenomenon of intermittency — the concentration of very strong fields in very small regions of space (§7.1). Another feature of magnetic fields in chaotic flows is the generation of fields that change sign repeatedly on small scales; this can be quantified by means of a *cancellation exponent*, introduced by Ott *et al.* (1992), which can also be related to growth rates (§7.2). In §7.3 we consider properties of magnetic field and fast amplification mechanisms in steady three-dimensional flows. A 'cut–paste' operation is described: this can operate in steady flows with stagnation points to supplement the stretching, folding and shearing of field.

7.1 Intermittency

To study intermittency and the distribution of field in fast dynamos, we return to some of the simplest fast dynamo models, those based on generalized baker's maps. These, despite their simplicity, appear to capture many qualitative features of magnetic fields in chaotic flows; related discussion for random flows may be found in §§11.1–3. We follow closely the analysis of Finn & Ott (1988a, b).

Consider a stacked, generalized baker's map, with variable stretching. For simplicity take a map having two strips and, in the notation of §3.3.1, set $N = 2$, $p_1 = \alpha$, $p_2 = \beta$ and $\epsilon_1 = \epsilon_2 = +1$. In each iteration the field is stretched by factors $1/\alpha$ and $1/\beta$ and then reassembled; α and β are taken to satisfy $\alpha + \beta = 1$ and $0 < \alpha, \beta < 1$. The case $\alpha = \beta = 1/2$ is special, since all vectors are stretched by the same amount: rather, we have in mind the general case $\alpha \neq 1/2$, and take $\alpha > \beta$ without loss of generality.

To fix ideas, start with a constant, unit magnetic field in the x-direction (Fig 3.10) and zero diffusion, $\epsilon = 0$. After n iterations of the map there are 2^n strips of field, each carrying unit flux. These strips have a range of widths: for each $m = 0, 1, \ldots, n$ there are $N_n(m)$ strips of width $W_n(m)$ in which the field has strength $B_r(m)$, where

$$N_n(m) = n!/(m!(n-m)!), \tag{7.1.1a}$$
$$W_n(m) = \alpha^{n-m}\beta^m, \tag{7.1.1b}$$
$$B_n(m) = 1/W_n(m) \tag{7.1.1c}$$

— call these 'm-strips' for brevity.

For large n the distribution $N_n(m)$ tends to a Gaussian form by the central limit theorem. Think of tossing a fair coin at each iteration to see whether one increments m by 0 or by 1: the mean increment is $1/2$ and the variance is $1/4$. After n iterations, the mean of m is $\mu = n/2$ and the variance is $\sigma^2 = n/4$. Set $s = (m-\mu)/\sigma$; then for large n the number of strips with $|s| < r$ is approximately

$$S_n(r) = 2^n \int_{-r}^{r} \exp\left(-s^2/2\right) \frac{ds}{\sqrt{2\pi}} \equiv 2^n \operatorname{erf}\left(r/\sqrt{2}\right), \tag{7.1.2}$$

where $\operatorname{erf}(z) = \pi^{-1/2} \int_{-z}^{z} \exp(-t^2)\, dt$. This approximation becomes good as $n \to \infty$ for fixed r. It means that most m-strips have $m = n/2 + O(\sqrt{n})$ when n is large.

For definiteness set $r = 2$, for which $\operatorname{erf}\left(r/\sqrt{2}\right) \simeq 0.95$. Equation (7.1.2) means that 95% of the flux is contained in the 95% of the m-strips with $|s| < 2$, i.e., $|m - \mu| < 2\sigma$. How much area do these strips actually occupy? The simplest approximation is to say that all these strips have a width of approximately $W_n(n/2) = (\alpha\beta)^{n/2}$, so that the amount of area taken up is roughly

$$A_n(|s| < r) \approx (\alpha/\beta)^{n/2}\, 2^n \operatorname{erf}\left(r/\sqrt{2}\right) \simeq 0.95(4\alpha\beta)^{n/2}. \tag{7.1.3}$$

In the special case $\alpha = 1/2$ this is 95% of the area; however for $\alpha \neq 1/2$, it is easily verified that $4\alpha\beta < 1$ and so the area tends to zero as n tends to infinity. After large numbers of iterates most of the flux is concentrated in a very small region of space and the magnetic field is called *spatially intermittent*; the flux of the magnetic field can be described as concentrating on a *fractal set* whose dimension can be calculated (Finn & Ott 1988a, b).

The area (7.1.3) is in fact an underestimate; to improve on it we take into account the variation of $W_n(m)$ with m in (7.1.1b). In terms of s:

$$W_n(m) = \alpha^{\mu-\sigma s}\beta^{\mu+\sigma s} = (\alpha\beta)^{\mu}(\beta/\alpha)^{\sigma s}. \tag{7.1.4}$$

From (7.1.2) the total area occupied by strips with $|s| < r$ is

$$A_n(|s| < r) \simeq (4\alpha\beta)^{\mu} \int_{-r}^{r} (\beta/\alpha)^{\sigma s} \exp\left(-s^2/2\right) \frac{ds}{\sqrt{2\pi}}. \tag{7.1.5}$$

Since $\sigma \gg 1$ for large n, the integral may be approximated using Laplace's method to yield

$$A_n(|s| < r) \simeq (4\alpha\beta)^\mu \left[\frac{(\alpha/\beta)^{\sigma r} \exp(-r^2/2)}{\sigma \log(\alpha/\beta)\sqrt{2\pi}} (1 + O(\sigma^{-1})) \right]. \qquad (7.1.6)$$

The term $[\cdot]$ in (7.1.6) represents a multiplicative correction to the approximation (7.1.3) that grows large as $n \to \infty$. However the quantity A_n still tends to zero as $n \to \infty$ and the field becomes ever more intermittent.

Consider also the distribution of field after n iterates; if we pick a point at random in the unit square, the probability of it lying on an m-strip is the total area of m-strips, $P_n(m) = N_n(m)W_n(m)$. Using the central limit theorem again for large n we see that the distribution $P_n(m)$ of values of m tends to a Gaussian (except at the tails of the distribution). In terms of coin-tossing, we toss a biased coin to decide whether to increment m by 0 (probability α) or 1 (probability β); the mean increment is $0 \cdot \alpha + 1 \cdot \beta = \beta$ and the variance is $\alpha\beta$. Thus for large n, the distribution of m values is Gaussian, mean $n\beta$, variance $n\alpha\beta$. Now the value of $\log|\mathbf{B}|$ in an m-strip is given by

$$\log B_n(m) = -n\log\alpha + m\log(\alpha/\beta) \qquad (7.1.7)$$

and is linearly related to m. Thus $\log|\mathbf{B}|$ also has a Gaussian distribution, with mean $-n(\alpha\log\alpha + \beta\log\beta)$, related to the Liapunov exponent in (7.1.11) below, and variance $n\alpha\beta(\log(\alpha/\beta))^2$. The field $|\mathbf{B}|$ itself takes a *log-normal distribution* (Finn & Ott 1988a, b). The reason is that $|\mathbf{B}|$ is given by the product of a large number of independent random quantities. This argument is correct for most of the field but breaks down for the very weak and very strong field corresponding to the tails of the distribution.

Another means of characterizing the intermittency in the magnetic field, and its tendency to concentrate on a sparse set, is to follow the growth of *moments* $<|\mathbf{B}(\mathbf{x}, n)|^p>$ of the magnetic field distribution as $n \to \infty$ (Falcioni, Paladin & Vulpiani 1989, Finn & Ott 1990). Here $<\cdot>$ is an average over space, the unit square in our case. To use quantities that are easily compared, we follow $<|\mathbf{B}|^p>^{1/p}$ and define its growth rate as

$$\Lambda_p = \lim_{n\to\infty} (np)^{-1} \log <|\mathbf{B}(\mathbf{x}, n)|^p>, \qquad (p > 0); \qquad (7.1.8)$$

thus Λ_p gives the growth rate of the L^p norm of the field. Some of these exponents are familiar: Λ_1 is same as the growth rate h_{line} of unsigned flux or of material lines, and $2\Lambda_2$ gives the growth rate of magnetic energy. If the limit $p \to 0$ is taken using l'Hôpital's rule, we obtain

$$\Lambda_0 \equiv \lim_{p\to 0} \Lambda_p = \lim_{n\to\infty} n^{-1} <\log|\mathbf{B}(\mathbf{x}, n)|>, \qquad (7.1.9)$$

which may be identified with the Liapunov exponent Λ_{Liap}. The exponents Λ_p are useful in the study of random flows, and are revisited in §§11.2, 3.

For the two-strip generalized baker's map,

$$<|\mathbf{B}(\mathbf{x}, n)|^p> = \sum_{m=0}^{n} N_n(m)W_n(m)B_n^p(m) = (\alpha^{1-p} + \beta^{1-p})^n \qquad (7.1.10)$$

and so

$$\Lambda_p = p^{-1}\log(\alpha^{1-p} + \beta^{1-p}), \quad \Lambda_0 = \alpha\log(1/\alpha) + \beta\log(1/\beta) \qquad (7.1.11)$$

(Finn & Ott 1990). This gives correctly $h_{\text{line}} = \Lambda_1 = \log 2$ and $\Lambda_{\text{Liap}} = \Lambda_0$. Except in the special case of uniform stretching, $\alpha = 1/2$, the exponents Λ_p vary with p; they increase with p (see §11.3), but are bounded above by $\lim_{p \to \infty} \Lambda_p = \log(1/\beta)$ (as $\alpha > \beta$). For large p the Λ_p are dominated by the fastest stretching vectors, which stretch by $1/\beta$ per iteration of the map.

To interpret these results first consider a flow or map with $\epsilon > 0$. As $n \to \infty$ the magnetic field is dominated by a finite number of eigenfunctions with greatest growth rate $\gamma(\epsilon)$. Since the spatial form of the eigenfunctions is fixed, independent of n, it follows that all the exponents Λ_p are equal to $\gamma(\epsilon)$. Thus in the above calculations with $\epsilon = 0$, the fact that these exponents are different (for $\alpha \neq 1/2$) shows that the field is not settling down to some simple spatial form, but is acquiring finer and finer structure as n increases. The growth rate depends on the L^p norm used, as discussed in §4.1.5. The larger p is, the more sensitive $<|\mathbf{B}|^p>^{1/p}$ is to the peak values of $\mathbf{B}$; since Λ_p increases with p, in the limit $n \to \infty$ the field gains ever stronger magnetic field in ever smaller regions of space and becomes spatially intermittent.[55]

The above qualitative features of magnetic field evolution with $\epsilon = 0$ have been verified numerically for maps based on pulsed waves. For field in a three-dimensional ABC *map*, based on the ABC flow, Finn & Ott (1988a, b) illustrate graphically the vanishingly small region of space that contains much of the flux, and Falcioni et al. (1989) show that different moments (7.1.8) of the field grow at different rates. For a related *web map* (Finn et al. 1991) and the 111 ABC flow (Lau & Finn 1993a), it has been verified that field vectors adopt a log-normal distribution (except in the tails of the distribution).

The situation for $0 < \epsilon \ll 1$ in a fluid flow is illustrated in Fig. 7.1, which shows the evolution of magnetic moments for Otani's MW+ flow. The magnetic field is $\mathbf{B} = \mathbf{b}(x, y, t)\exp(ikz)$ with $k = 0.8$; the diffusivity is $\epsilon = 10^{-4}$. Figure 7.1 shows the evolution with time of the quantities

$$M_0(t) = <\log(|b_x|^2 + |b_y|^2)^{1/2}>, \qquad (7.1.12a)$$

$$M_p(t) = \log <(|b_x|^2 + |b_y|^2)^{p/2}>^{1/p}, \qquad (7.1.12b)$$

for $p = 0$–4. During the initial transient before diffusion becomes effective, the quantities $M_p(t)$ grow at different rates. However after a short time $t_{\text{diff}} = O(\log(1/\epsilon))$, the reduction in scale is stopped by diffusive spreading and the field settles down into an eigenfunction, albeit with complicated small-scale structure (cf. Fig 4.3). The $M_p(t)$ then increase at the same rate, $\gamma(\epsilon)$. Only

---

[55]For our purposes the definition of intermittency is that Λ_p is not constant as a function of p. Other definitions are used; for example Baxendale & Rozovskii (1993) require that $\Lambda_r \to \infty$ as $p \to \infty$. By this definition only the delta-correlated random flows mentioned in §11.3 qualify as intermittent.

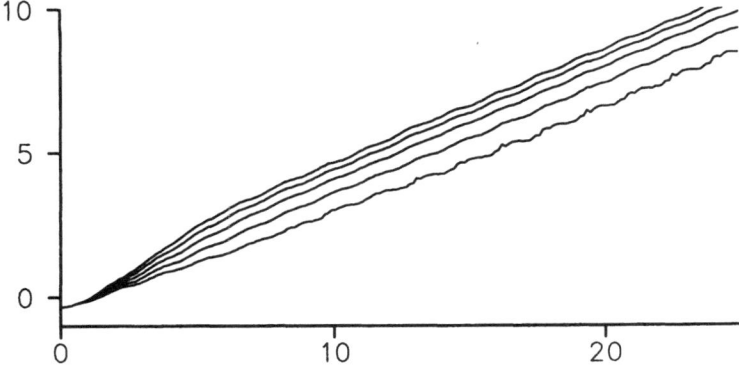

Fig. 7.1 Growth in moments for the MW+ flow with $k = 0.8$ and $\epsilon = 10^{-4}$. The quantities $M_p(t)$ are plotted against time t for $p = 0\text{--}4$, reading the curves from bottom to top.

a flux should, by the flux conjecture, show virtually no change in slope when diffusion becomes effective at t_{diff}; this is seen in baker's map models (Finn & Ott 1990).

Although much of the field tends to concentrate in a small region of space for $\epsilon = 0$, it is awkward to exploit this fact analytically to prove fast dynamo action. One cannot easily neglect weak field and concentrate on strong field in a limited region of space with simpler geometry, because a fast dynamo involves a subdominant large-scale field in a sea of fluctuations. The errors made in neglecting fields, however weak, may swamp the large-scale field one needs to track. A way to avoid this is to work in unbounded space and send unwanted field to infinity, concentrating on generation in a finite region. Dynamos of this type are studied by Finn *et al.* (1989, 1991), Lau & Finn (1993*a, b*) and Balmforth *et al.* (1993), and are discussed in §7.3 below.

Here and elsewhere, the discussion has focussed on baker's maps, which have agreeable properties characteristic of hyperbolic and Anosov systems, such as clear separation of stretching and contracting directions. However a realistic fast dynamo, such as the flow MW+, has a complicated mixture of strong stretching and weak stretching, near islands, KAM surfaces and folds. Even for parameter values for which the phase space appears entirely chaotic, some fluid elements (those near the folds) are only stretched weakly. This non-hyperbolic mixture of weak and strong stretching is absent from baker's maps and cat maps. Despite this, the qualitative behavior of field in baker's maps appears to be a good guide to the qualitative behavior in more realistic flows and maps. When a fast dynamo in a smooth flow, say MW+, is modelled by a baker's map, regions of weak stretching are poorly modelled. However, fluid elements in such regions carry only weak magnetic field. In effect they belong to the long tail of the distribution of field vectors, and are negligible when an average of the field is calculated. This natural weighting of strongly stretched fluid elements emphasizes the 'most' chaotic regions and appears to be why baker's maps are good at modelling qualitatively the amplification and field structure in fast dynamos. The picture is very different for scalar transport; here there is no amplification through stretching and all fluid elements are equally weighted. Integrable islands can play a crucial role in capturing and releasing scalar (Mackay, Meiss & Percival 1984).

Further analysis of the properties of the fractal set on which the magnetic flux concentrates has been given by Finn & Ott (1988a, b), Ott & Antonsen (1988, 1989), Falcioni et $al.$ (1989), Antonsen & Ott (1991), Városi, Antonsen & Ott (1991), Du & Ott (1993a), Du, Tél & Ott (1994b) and Cattaneo et $al.$ (1995). In these studies $measures$ are associated with the distribution of magnetic flux or, in a parallel development, the gradient of a passive scalar. Once this is done it is possible to describe these measures in terms of $parti$-$tion$ $functions$ and a $spectrum$ of $fractal$ $dimensions$, using the machinery of $multifractals$. We refer the reader to the above papers for discussion.

7.2 The Cancellation Exponent

A striking feature of magnetic fields in fast dynamo simulations is rapid changes of sign in space; see, for example, Fig 2.9. Sign changes are an important aspect of the fast dynamo process, since they indicate how well the fluid flow folds field constructively, and so generates significant fluxes. For the STF picture or the cat map the field has no changes of sign, corresponding to perfectly constructive folding, strong fluxes and no cancellation of field. Folding in a planar flow, however, leads to field with alternating sign, no growth in overall flux and near-complete cancellation. The existence of a connection between sign changes and growth rates is particularly clear in the generalized baker's map models of §3.3: it is the flips in sign that reduce the flux growth rate from the theoretical maximum, the growth rate of material lines in the flow, h_{line}.

To quantify the sign changes and cancellations in a complicated magnetic field, and ultimately relate these to fast dynamo growth rates, Ott et $al.$ (1992) and Du & Ott (1993a, b) introduce the idea of a $cancellation$ $exponent$. They give two equivalent formulations; we follow that in appendix A of Du & Ott (1993b). Suppose we have a magnetic field $\mathbf{B}$ which is a fast dynamo eigenfunction for $0 < \epsilon \ll 1$. Take a cube V of side δ (much larger than the order $\sqrt{\epsilon}$ diffusive length scale) and define

$$\xi(V, l) = \int_V |H_{l^2}\mathbf{B}| \, d\mathbf{x}. \tag{7.2.1}$$

Here H_{l^2} is a heat operator which takes a convolution, $H_{l^2}\mathbf{B} = h_{l^2} * \mathbf{B}$, of $\mathbf{B}$ with a normalized Gaussian, $h_{l^2}(\mathbf{x}) \propto \exp(-\mathbf{x}^2/4l^2)$, width l, smoothing the field over scales of order l.

Consider decreasing l from δ; as we do this $H_{l^2}\mathbf{B}$ will reveal more of the fine structure present in $\mathbf{B}$. In particular more changes of sign and peaks of greater intensity will appear. Thus as l is decreased the integral increases and we might expect to see a range with scaling behavior

$$\xi(V,l) \sim Kl^{-\kappa}, \qquad (\sqrt{\epsilon} \ll l \ll \delta), \tag{7.2.2}$$

where K is a constant. The cancellation exponent κ characterizes the sign changes in the field induced by the fast dynamo process, and is expected to be the same for all cubes V in the same chaotic region. It is also a measure of the amount of field in the fluctuations, a topic taken up in Chap. 12, and quantifies an aspect of the self-similarity of fast dynamo eigenfunctions.

The scaling behavior (7.2.2) is observed for magnetic fields of fast dynamos and for vorticity in high Reynolds number flow (Ott et $al.$ 1992, Du & Ott 1993a, b, Du, Ott & Finn 1994a, Brandenburg 1995). Bertozzi & Chhabra (1994) have related κ to the classical idea of a Hölder exponent: the Hölder exponent of a function is one minus the cancellation exponent of its derivative.

7.2.1 Cancellation Exponents and Growth Rates

We now summarize the derivation of Du & Ott (1993b) of a formula for the growth rate of a fast dynamo in terms of κ and the stretching behavior of the flow, quantifying how cancellations reduce the growth rate from a value determined simply by h_{line}. We shall keep track of

$$\xi(\mathcal{D},l,n) \equiv \int_{\mathcal{D}} |H_{l^2}\mathbf{B}(\mathbf{x},n)| \, d\mathbf{x}, \qquad (\sqrt{\epsilon} \ll l \ll l_0), \tag{7.2.3}$$

for some choice of l_0, and relate it at discrete times $n = 0$ and $n \gg 1$ for a map M (or periodic flow) in, say, three-dimensional space. Here $\mathcal{D}$ is a single connected chaotic region and $\mathbf{B}(\mathbf{x},n) = \exp(pn)\mathbf{b}(\mathbf{x})$ the fastest growing eigenfunction, with complex growth rate $p = \gamma + i\omega$; the eigenfunction is taken to be unique, and localized in $\mathcal{D}$. We work with a complex magnetic field, whereas in the development of Du & Ott (1993b) the corresponding real field is used.

To do this we need to split $\mathcal{D}$ into small pieces. First consider a sphere S of small radius δ about some point $\mathbf{a}$. After n iterations of M, the sphere is mapped approximately into an ellipsoid $M^n S$, whose shape is given by the Jacobian $\mathbf{J}^n(\mathbf{a})$. The lengths of the principal axes of the ellipsoid are δ times the quantities $L_1(\mathbf{a},n)$, $L_2(\mathbf{a},n)$ and $L_3(\mathbf{a},n)$, which are defined as the square roots of the eigenvalues of the symmetric matrix $\Lambda^n(\mathbf{a}) = (\mathbf{J}^n(\mathbf{a}))^T\mathbf{J}^n(\mathbf{a})$ (cf. 6.1.2). We denote the corresponding orthogonal eigenvectors $\mathbf{e}_i = \mathbf{e}_i(\mathbf{a})$.

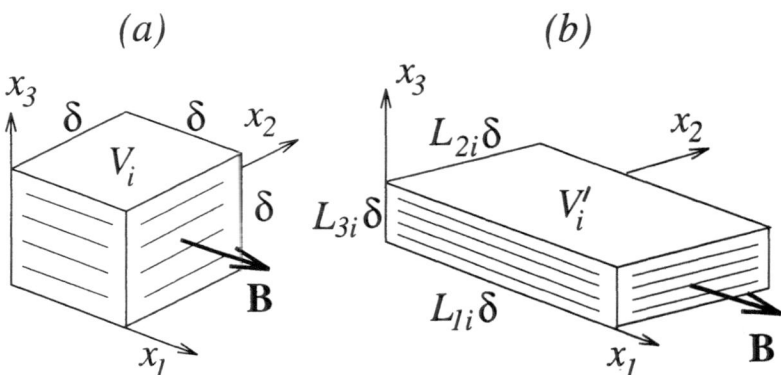

Fig. 7.2 Evolution of a small cube in a chaotic map: (a) the cube V_i, side δ, is stretched into (b) a rectangular slab V_i', with sides $L_{1i}\delta$, $L_{2i}\delta$ and $L_{3i}\delta$.

The quantities L_k are called *finite time Liapunov numbers*; they satisfy $L_1L_2L_3 = 1$ and can be ordered to satisfy $L_1 \geq L_2 \geq L_3$. These are related to the Liapunov exponents $\Lambda^{(k)}_{\text{Liap}}$ (see §6.1), since

$$\lim_{n\to\infty} \frac{1}{n}\log L_k(\mathbf{a}, n) = \Lambda^{(k)}_{\text{Liap}} \qquad (7.2.4)$$

for almost all $\mathbf{a}$.

At time $t = 0$ let us divide the chaotic region into small cubes V_i of side δ centered at points $\mathbf{a}_i$, with sides aligned locally with $\mathbf{e}_1$, $\mathbf{e}_2$ and $\mathbf{e}_3$ (Fig. 7.2a).[56] We take δ to be sufficiently small that after n iterations the image of each cube, $V_i' = M^n V_i$, is approximately a rectangular slab with sides $L_{1i}\delta$, $L_{2i}\delta$ and $L_{3i}\delta$, where $L_{ki} \equiv L_k(\mathbf{a}_i, n)$ (Fig 7.2b). Later we will take the limit $n \to \infty$, and so have in mind $L_{1i} \gg 1 \gg L_{3i}$.

In order to proceed further, Du and Ott restrict the class of maps under consideration to those that generate magnetic field having *sheet-like structure*, much like field in the generalized baker's maps: after a time, the field takes the form of a pile of thin sheets, with rapid variation through the sheets and slow variation across. A crude condition for sheet-like structure is that the second Liapunov exponent $\Lambda^{(2)}_{\text{Liap}} \geq 0$, or that for typical $\mathbf{a}$, $L_2(\mathbf{a}, n) \geq 1$. More precise conditions are given in the appendix of Du & Ott (1993a). The other possibility, which is excluded, is *rope-like structure*, $\Lambda^{(2)}_{\text{Liap}} < 0$, in which the field takes the form of a bundle of ropes, as for the STF picture; see also the diffusive rope-like structures discussed in §5.6.

---

[56]These directions do not vary rapidly in space; see §5A of Finn & Ott (1988b).

Let us focus on one cube V_i. Since the map generates sheet-like structure, the field $\mathbf{B}(\mathbf{x}, 0)$ in V_i takes the approximate one-dimensional form

$$\mathbf{B}(\mathbf{x}, 0) \simeq f(x_3)\mathbf{e}_1, \tag{7.2.5}$$

using local axes (x_1, x_2, x_3) shown in Fig 7.2(a). From (7.2.5) and the scaling law (7.2.1, 2) we can write

$$\xi(V_i, l, 0) \equiv \int_{V_i} |H_{l^2}\mathbf{B}(\mathbf{x}, 0)| \, d\mathbf{x}$$
$$\simeq \delta^2 \int_0^\delta |H_{l^2} f(x_3)| \, dx_3 \sim Kl^{-\kappa}, \quad (\sqrt{\epsilon} \ll l \ll \delta). \tag{7.2.6}$$

$\xi(V_i, l, 0)$ is given by smoothing f over scales l and integrating its modulus; we use the same symbol, H_{l^2}, for both one-dimensional and three-dimensional heat operators.

Under evolution, the resulting field in V_i' will be approximately

$$\mathbf{B}(\mathbf{x}, n) \simeq H_{\epsilon_0} g(x_3) L_{1i}\mathbf{e}_1, \quad g(x_3) = f(x_3/L_{3i}) \tag{7.2.7}$$

with respect to local axes (x_1, x_2, x_3) in V_i' (Fig 7.2b). The field is stretched by L_{1i}, its scale reduced by L_{3i} and it is smoothed by the kernel H_{ϵ_0}, which represents the effect of weak diffusion during the n iterations and could be calculated approximately as explained in §6.6. Provided there is sufficient stretching, $\epsilon_0 = O(\epsilon)$; its precise value is not important.

To obtain $\xi(V_i', l, n)$ for $\sqrt{\epsilon} \ll l \ll \delta L_{3i}$, we smooth the field in V_i' over the scale l with H_{l^2} and integrate its modulus. Now $H_{l^2} H_{\epsilon_0} = H_{l^2+\epsilon_0} \simeq H_{l^2}$. Furthermore since g is obtained from f by a linear compression in (7.2.7), smoothing the function g with H_{l^2} is the same as smoothing f with the broader heat kernel $H_{(l/L_{3i})^2}$ and then reducing its scale by the factor L_{3i}. It may thus be seen from (7.2.6, 7) that

$$\xi(V_i', l, n) \sim L_{1i} K (l/L_{3i})^{-\kappa}, \quad (\sqrt{\epsilon} \ll l \ll \delta L_{3i}). \tag{7.2.8}$$

The scaling of the quantity ξ is the same in V_i and V_i', up to a constant factor. Eliminating $Kl^{-\kappa}$ between (7.2.6, 8) gives

$$\int_{V_i'} |H_{l^2}\mathbf{B}(\mathbf{x}, n)| \, d\mathbf{x} \simeq L_{1i} L_{3i}^\kappa \int_{V_i} |H_{l^2}\mathbf{B}(\mathbf{x}, 0)| \, d\mathbf{x}, \quad (\sqrt{\epsilon} \ll l \ll \delta L_{3i}). \tag{7.2.9}$$

The factor L_{1i} comes from the stretching, and the factor L_{3i}^κ from the cascade of fluctuations past the scale l.

Sum over all cubes V_i in $\mathcal{D}$:

$$\int_{\mathcal{D}} |H_{l^2}\mathbf{B}(\mathbf{x}, n)| \, d\mathbf{x} \simeq \sum_i L_{1i} L_{3i}^\kappa \int_{V_i} |H_{l^2}\mathbf{B}(\mathbf{x}, 0)| \, d\mathbf{x}. \tag{7.2.10}$$

Now if n is large, the field in V_i, which is determined by the history of stretching up to time $t = 0$, and the quantity $L_{1i} L_{3i}^\kappa$, determined by stretching from

$t = 0$ to $t = n$, will be effectively uncorrelated (Du & Ott 1993b). Given this independence assumption one can write

$$\int_{\mathcal{D}} |H_{l^2}\mathbf{B}(\mathbf{x}, n)|\, d\mathbf{x} \simeq <L_1(\mathbf{a}, n)[L_3(\mathbf{a}, n)]^{\kappa}> \sum_i \int_{V_i} |H_{l^2}\mathbf{B}(\mathbf{x}, 0)|\, d\mathbf{x}$$

$$= <L_1(\mathbf{a}, n)[L_3(\mathbf{a}, n)]^{\kappa}> \int_{\mathcal{D}} |H_{l^2}\mathbf{B}(\mathbf{x}, 0)|\, d\mathbf{x}, \qquad (7.2.11)$$

where $<\cdot>$ is an average over points $\mathbf{a}$ in the chaotic region. Finally, since $\mathbf{B}$ is an eigenfunction, we have

$$\exp(n\gamma) \simeq <L_1(\mathbf{a}, n)[L_3(\mathbf{a}, n)]^{\kappa}>, \qquad (7.2.12)$$

which relates the growth rate to the cancellation exponent and the finite time Liapunov numbers. This approximation should improve as n is increased; taking also $\epsilon \to 0$, we obtain

$$\gamma_0 = \lim_{n \to \infty} \frac{1}{n} \log <L_1(\mathbf{a}, n)[L_3(\mathbf{a}, n)]^{\kappa}>, \qquad (7.2.13)$$

relating the fast dynamo growth rate to κ and the finite time Liapunov numbers $L_i(\mathbf{a}, n)$ (Du & Ott 1993b). If there are no cancellations, $\kappa = 0$, and the growth rate reduces to the rate of stretching of finite lines, h_{line}.

The above argument accounts for fluctuations of field as they are cascaded beyond the scale l; one could fix l as the dissipation scale, but this is not essential. There is little mention of dissipation in the above arguments. However use of the limit $n \to \infty$ in (7.2.13) requires that $\epsilon \to 0$ implicitly, since we need $\sqrt{\epsilon} \ll \delta L_{3i}$ to use (7.2.9); in the limit $n \to \infty$, $L_{3i} \to 0$, and we also require $\delta \to 0$ (so the rectangular slabs are not too distorted).

If diffusion is switched off and the field evolved with $\epsilon = 0$ for n iterations a cancellation exponent κ' may be defined using (7.2.1, 2), and will describe the field in the range $l_* \ll l \ll \delta$, where l_* is the smallest scale in the field. It is likely that $\kappa = \kappa'$, as discussed by Du & Ott (1993a), this following from the idea that a diffusive eigenfunction can be obtained by smoothing field evolved with $\epsilon = 0$; see §4.1.3. Up to normalisation, one might expect $H_{l^2}\mathbf{B}$, $\mathbf{B}$ being an eigenfunction with $0 < \sqrt{\epsilon} \ll l$, to be the same as for a field evolved with $\epsilon = 0$ for sufficiently many iterations that the smallest scales l_* satisfy $l_* \ll l$. Numerical evidence for this is presented by Du & Ott (1993a).

In the calculations above we use integrals of the modulus of the field over small regions V_i and V_i'; this means that the previous analysis goes through if we include complex phase shifts encoding motion in an ignorable direction, as for SFS. These change the phase of the field, but are approximately constant over small regions.

Finally note that we have assumed tacitly that the image V_i' of V_i is a rectangular block (Fig 7.2b). This will hold for some systems (for example, baker's maps) but for others V_i' will actually be a thin parallelepiped of thickness $\delta L_{3i} \sin \theta_i$ where θ_i is the angle between the x_3-direction and the

(x_1, x_2)-plane. The factor of $\sin \theta_i$ may be traced through the argument and modifies (7.2.13) to

$$\gamma_0 = \lim_{n \to \infty} \frac{1}{n} \log <L_1(\mathbf{a}, n)[L_3(\mathbf{a}, n) \sin \theta(\mathbf{a}, n)]^\kappa >. \qquad (7.2.13')$$

For baker's maps and SFS, $\sin \theta = 1$ and (7.2.13, 13′) are the same. For hyperbolic or Anosov systems $\sin \theta$ is bounded away from zero, and so (7.2.13) and (7.2.13′) will agree. For non-hyperbolic systems, $\sin \theta$ is not bounded below and will be small near folds in the field. While (7.2.13′) suggests that in this case (7.2.13) may overestimate the growth rate, this is unlikely since L_{1i} will generally also be small near folds, and the average in (7.2.13, 13′) will probably be dominated by trajectories for which $\sin \theta = O(1)$.

It is not at all straightforward to extend the analysis to flows generating rope-like structure, such as STF. The functions f and g in (7.2.5, 7) become two-dimensional, and are related by an anisotropic linear scaling transformation. To make progress, one would need to modify (7.2.1) to allow anisotropic heat kernels, with attendant complications. An indirect approach is to consider the time-reversed flow, which generates sheet-like structure. For maps with pulsed diffusion and steady flows $\gamma(\epsilon)$ is invariant under time-reversal. This does not hold for unsteady flows with continuous diffusion, but in this case it is likely that γ_0 is invariant; see discussion in §9.3.1.

7.2.2 Cancellation Exponents for Baker's Maps and SFS

Let us calculate the cancellation exponent for a generalized, stacked baker's map (§3.3.1) with four strips, widths α, β, μ and ν, and verify (7.2.13) for the growth rate. At each iteration let us reverse the third strip only; so the flux growth rate is $\Gamma = \log 2$. Let $B_1(y)\mathbf{e}_x$ be the eigenfunction with diffusion ϵ and V be the unit square. Normalize B_1 by dividing by the total flux Φ and insert this in (7.2.1, 2):

$$\xi(V, l) = |\Phi|^{-1} \int_V |H_{l^2} B_1| \, dy \sim K l^{-\kappa}, \qquad (\sqrt{\epsilon} \ll l \ll 1). \qquad (7.2.14)$$

Now perform another iteration of the generalized baker's map and diffusion to obtain a field $B_2(y)$ from $B_1(y)$, with flux 2Φ. The field $B_2/2\Phi$ is the same as B_1/Φ and so we also have

$$\xi(V, l) = |2\Phi|^{-1} \int_V |H_{l^2} B_2| \, dy \sim K l^{-\kappa}. \qquad (7.2.15)$$

This can be written in terms of the field B_1 using the baker's map;[57] consider the portion $0 \leq y \leq \alpha$ of V:

---

[57] We have in mind pulsing diffusion ϵ during the unit time between iterations; however the effect of diffusion can be neglected in these calculations since $H_{l^2} H_\epsilon = H_{\epsilon + l^2} \simeq H_{l^2}$ as $l^2 \gg \epsilon$. However at the lower limit of this range, $l^2 \simeq \epsilon$, this argument fails and ϵ must be taken into account.

$$\int_0^\alpha |H_{l^2} B_2(y)|\, dy = \int_0^\alpha \alpha^{-1} |H_{l^2}[B_1(y/\alpha)]|\, dy$$

$$= \int_0^1 |H_{(l/\alpha)^2} B_1(y)|\, dy = |\Phi| K (l/\alpha)^{-\kappa}, \tag{7.2.16}$$

from (7.2.14). Here we have ignored the effect of diffusion in carrying field across $y = \alpha$, a negligible process for small l; the width of the heat kernel is changed because of the scaling that relates B_1 and B_2 in this strip.

Applying the same argument for the other three strips:

$$\int_V |H_{l^2} B_2|\, dy = |\Phi| K l^{-\kappa} (\alpha^\kappa + \beta^\kappa + \mu^\kappa + \nu^\kappa). \tag{7.2.17}$$

Substituting this in (7.2.15) yields an implicit equation for κ:

$$2 = \alpha^\kappa + \beta^\kappa + \mu^\kappa + \nu^\kappa \tag{7.2.18}$$

(Du & Ott 1993a). We check that this gives the correct growth rate in (7.2.13); for a baker's map, $L_3 = 1/L_1$ and so (7.2.13) reduces to

$$\gamma_0 = \lim_{n\to\infty} \frac{1}{n} \log <L_1(\mathbf{a}, n)^{1-\kappa}>. \tag{7.2.19}$$

$<L_1(\mathbf{a}, n)^{1-\kappa}>$ can be written as a sum over 4^n strips:

$$<L_1(\mathbf{a}, n)^{1-\kappa}> = \sum_{\text{types of strips}} [\text{no. of strips}]\,[\text{strip width}]\,[L_1 \text{ for strip}]^{1-\kappa}$$

$$= \sum_{\substack{k,l,m,p \\ k+l+m+p=n}} \frac{n!}{k!\, l!\, m!\, p!} [\alpha^k \beta^l \mu^m \nu^p]\, [\alpha^{-k} \beta^{-l} \mu^{-m} \nu^{-p}]^{1-\kappa}$$

$$= (\alpha^\kappa + \beta^\kappa + \mu^\kappa + \nu^\kappa)^n = 2^n. \tag{7.2.20}$$

This gives the correct growth rate $\gamma_0 = \log 2$.

The formula (7.2.18) can be generalized to an SFSI model based on an N-strip generalized baker's map with orientation factors $\epsilon_i = \pm 1$, strip widths p_i and complex phase shifts ϕ_i with $|\phi_i| = 1$:

$$\exp \Gamma \equiv \left| \sum_{i=1}^N \epsilon_i \phi_i \right| = \sum_{i=1}^N p_i^\kappa \tag{7.2.21}$$

(Du & Ott 1993b). While the argument in (7.2.19, 20) will go through in this case, clearly the diffusive fast dynamo action of SFSI (and SFS) cannot be detected by this approach. Perhaps in this case the (unproven) scaling *ansatz* (7.2.2) does not hold.

The cancellation exponent κ can be thought of as measuring the efficiency of a fast dynamo: how much flux is 'wasted' by cancellation. The idealized STF picture, in which the field is always brought into alignment, may be modelled by a two-strip baker's map, given by stretching factors $1/\alpha$ and $1/\beta$

with $\alpha + \beta = 1$ (Finn & Ott 1988b). If the strips are brought into alignment at each mapping, the equation for κ is $\alpha^{\kappa} + \beta^{\kappa} = 2$, implying $\kappa = 0$ and no cancellations. In the case where one of the strips is flipped, the equation becomes $\alpha^{\kappa} + \beta^{\kappa} = 0$, corresponding to $\kappa = \infty$, complete cancellation and decay of field. A proof that $\kappa = \infty$ for general smooth planar flows would be valuable.

For the SFS map the field adopts sheet-like structure as there is no stretching in the z-direction, $L_2 \equiv 1$. Bayly & Rado (1993) calculate κ and γ_0 numerically and obtain good agreement with (7.2.19); in this case $L_1 = 2^n$ and so $\gamma_0 = (1 - \kappa) \log 2$. Figure 7.3 shows their results for the SFS growth rate calculated numerically (solid), and the growth rate calculated from (7.2.19) (dashed), plotted as a function of αk. The agreement is good, including the mode-crossings. In three dimensions Du & Ott (1993b) consider a mapping based on pulsed waves, for which the magnetic field adopts sheet-like structure. They find the cancellation exponent κ numerically and calculate the average in (7.2.13) numerically, by following trajectories. The resulting growth rate agrees with the growth of flux in the flow. Cattaneo $et\ al.$ (1995) calculate κ and growth rates for several two-dimensional flows. For the CP flow (2.5.5) they report that (7.2.13) is incorrect; the reason for this remains to be understood.

There exist now two formulae relating the fast dynamo growth rate to the stretching and folding properties of the flow; that of Du and Ott (7.2.13) involving finite time Liapunov numbers and the cancellation exponent, and that of Oseledets (§6.2) in terms of twisted and untwisted periodic orbits. Both reduce to h_{line} in the absence of sign flips in the field and both quantify the reduction in growth rate caused by cancellations in the field. It is an open problem as to how the two might be related; for example, can one calculated κ from sums over periodic orbits?

7.3 Steady Flows

Relatively little is known about fast dynamo mechanisms and magnetic field structure for three-dimensional steady flows. Numerical studies for $\epsilon > 0$ and flux growth calculations for $\epsilon = 0$ were described for the ABC and Kolmogorov flows in §2.6. In this section we discuss features specific to steady flows and work on modelling fast dynamo action in such flows.

To begin with set $\epsilon = 0$; it is clear geometrically that if the field vector and flow vector are parallel at a point, they remain so under advection and stretching. Thus we can decompose a magnetic field into components parallel and perpendicular to streamlines:

$$\mathbf{B} = \phi \mathbf{u} + \mathbf{B}_{\perp}, \qquad \mathbf{u} \cdot \mathbf{B}_{\perp} = 0. \tag{7.3.1}$$

Projecting the $\epsilon = 0$ induction equation along and perpendicular to $\mathbf{u}$ yields equations for ϕ and $\mathbf{B}_{\perp}$:

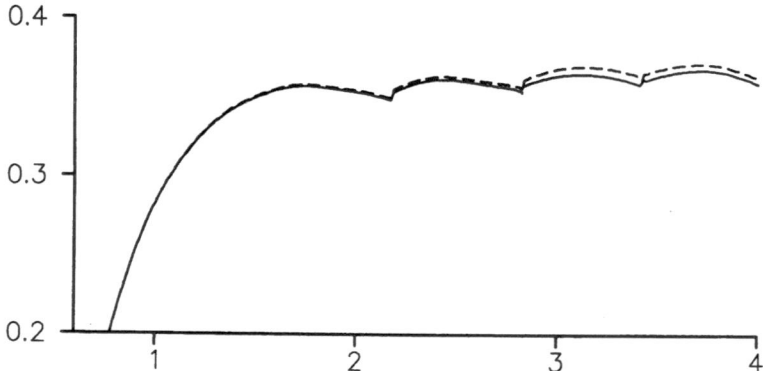

Fig. 7.3 Fast dynamo growth rate γ_0 as a function of αk for the SFS model. The solid line shows growth rates obtained numerically; the dashed line shows those obtained from calculating κ numerically and then using the formula (7.2.19). (After Bayly & Rado 1993, numerical results courtesy B.J. Bayly.)

$$(\partial_t + \mathbf{u} \cdot \nabla)\phi = u^{-2}\mathbf{u} \cdot (-\mathbf{u} \cdot \nabla \mathbf{B}_\perp + \mathbf{B}_\perp \cdot \nabla \mathbf{u}), \tag{7.3.2}$$

$$(\partial_t + \mathbf{P}_\perp \mathbf{u} \cdot \nabla)\mathbf{B}_\perp = \mathbf{P}_\perp \mathbf{B}_\perp \cdot \nabla \mathbf{u}, \tag{7.3.3}$$

where $\mathbf{P}_\perp = \mathbf{I} - u^{-2}\mathbf{u}\mathbf{u}^{\mathrm{T}}$ is a projection perpendicular to streamlines. Notice that the equation for ϕ is that for a passive scalar with a source term depending on $\mathbf{B}_\perp$. If the transverse field $\mathbf{B}_\perp$ is everywhere zero ϕ is materially conserved, while for $\mathbf{B}_\perp \neq 0$, shearing of transverse field by variations in $\mathbf{u}$ generates ϕ. Furthermore ϕ does not appear in the equation for $\mathbf{B}_\perp$. The conclusion is that for $\epsilon = 0$ the scalar ϕ is slaved to $\mathbf{B}_\perp$, which can be considered independently (Finn *et al.* 1991, Gilbert 1992).

Consider first the case $\mathbf{B}_\perp \equiv 0$; so ϕ has no source term. If $\mathbf{u}$ has stagnation points $\mathbf{u} = 0$, the field ϕ will generally have singularities, diverging as $1/r$ close to zeros of $\mathbf{u}$, since if $\mathbf{B}_0$ and ϕ_0 are fields at $t = 0$ then $\phi_0 = u^{-2}\mathbf{u} \cdot \mathbf{B}_0$. As the flow operates these large values of ϕ can be carried materially into regions where $\mathbf{u} = O(1)$ and appear as large fields emerging from an X-type stagnation point. Although the field produced in this way can grow exponentially in time, the flux that can be generated is bounded since the singularities in ϕ are integrable in two or three dimensions.

If $\mathbf{B}_\perp \neq 0$ there is generation of ϕ-field from transverse field by shearing. This process is very effective near X-points, where the last term of (7.3.2) is large. For a two-dimensional X-point flow $\mathbf{u} = (x, -y, 0)$ in $-1 \leq x, y \leq 1$, the process is shown in Fig 7.4: an initial field vector $\mathbf{e}_x$ is decomposed into parallel and transverse components. After traversing the stagnation point this becomes a large x-directed vector. When decomposed, the parallel part is large, but the length of the transverse part is unchanged.

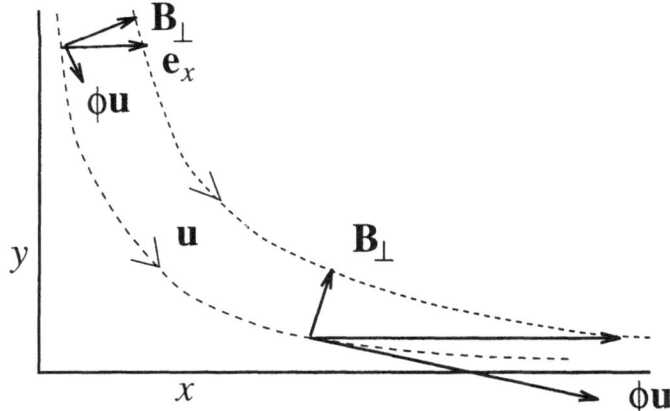

Fig. 7.4 The conversion of transverse field $\mathbf{B}_\perp$ to parallel field $\phi\mathbf{u}$ at a two-dimensional stagnation point.

Thus although the parallel field $\phi\mathbf{u}$ is slaved to $\mathbf{B}_\perp$, large values are generated in small regions near two-dimensional stagnation points, and this can make flux of the parallel field difficult to measure. This was noted by Gilbert & Childress (1990) and Gilbert (1992), who constructed a three-dimensional steady flow by 'patching' together simple flows, so as to conserve volume and keep the stretching of fluid elements bounded (during finite time intervals). The resulting patched flow contains (non-generic) lines of two-dimensional stagnation points. Field was decomposed into parallel and transverse fields using (7.3.1), and clear flux growth in $\mathbf{B}_\perp$ over a long period was measured, suggestive of fast dynamo action in view of the flux conjecture. Flux of the $\phi\mathbf{u}$ component was, however, impossible to measure beyond moderate times. The situation for three-dimensional stagnation points is more complicated; these can amplify or suppress $\mathbf{B}_\perp$, and so the separation into $\phi\mathbf{u}$ and $\mathbf{B}_\perp$ has fewer numerical advantages.

In view of the slaving of parallel field to $\mathbf{B}_\perp$, it appears likely that it is sufficient to ignore the parallel field in seeking flux growth for $\epsilon = 0$ in steady flows. One might conjecture that no smooth linear functional of parallel field can grow exponentially without some linear functional of the transverse field also growing exponentially. However we know of no results along these lines.

If the fluid is not perfectly conducting, $0 < \epsilon \ll 1$, parallel and transverse fields become coupled. In particular diffusive terms involving $\nabla\phi$ appear in the equation for $\mathbf{B}_\perp$. These diffusive effects are essential in the operation of the discontinuous fast dynamos discussed in Chap. 5. If the flux conjecture is correct, and fast dynamo action can be studied by setting $\epsilon = 0$ and following fluxes, then the diffusive coupling of parallel and transverse fields should become asymptotically irrelevant for fast-growing fields in the limit $\epsilon \to 0$.

This has, however, not been proved, and a proof is likely to be delicate in view of 'nearly fast' dynamo action in the Roberts cell (§5.5.2), in which this coupling is important and yet growth rates fall off extremely slowly in the limit $\epsilon \to 0$. The possibility of diffusive eigenfunctions whose growth rate falls off very slowly should be born in mind for steady flows and might be the reason for the discrepancies between flux growth rates and low-ϵ growth rates for the 111 ABC flow seen in §2.6.

7.3.1 The SFS Mechanism in Steady Flows

We move on to discuss fast dynamo mechanisms in steady flows. Two mechanisms have been analyzed and to describe these it is helpful to begin by considering a Poincaré section and Poincaré map. We take a fixed surface S immersed in the flow, and follow the evolution of field on that surface, as in §2.6. For definiteness let the section S be given by $z = 0$ and describe points on S by coordinates (x, y). The Poincaré map P relates the subsequent crossings of S by an advected particle. If the particle crosses S at $(x, y, 0, t)$ then its next crossing is at $(x', y', 0, t')$ where $(x', y', t') = P(x, y, t)$ (Fig. 7.5). For a steady flow $t' - t$ is a function of (x, y) only and (x', y') are functions of (x, y) alone.[58]

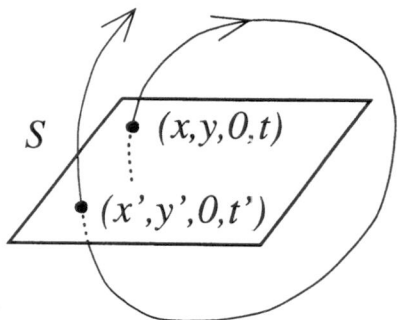

Fig. 7.5 A Poincaré section S.

The map P can also be used to move vectors in the flow from S to S provided the component u_z normal to S is nowhere zero. Given this we write a vector at (x, y, t) in the form $\mathbf{B} = B_x \mathbf{e}_x + B_y \mathbf{e}_y + B_u \mathbf{u}$. This vector can

---

[58]Note that the Poincaré map for a volume-preserving flow generally does not preserve the usual area $dx \, dy$, but instead a weighted area $u_z \, dx \, dy$. This is because volume conservation requires that $u_z \, dt \, dx \, dy = u'_z \, dt' \, dx' \, dy'$ and for steady flow $dt = dt'$.

thought of as joining two infinitesimally close particles that pass through the section at (x, y, t) and $(x+dx, y+dy, t+dt)$, where $(dx, dy, dt) \propto (B_x, B_y, B_u)$. The pair of particles next cross S at (x', y', t') and $(x' + dx', y' + dy', t' + dt')$ where

$$(dx', dy', dt') = \partial P(x, y, t)/\partial(x, y, t)\,(dx, dy, dt), \tag{7.3.4}$$

and so $\mathbf{B}$ is carried to $\mathbf{B}' = B'_x \mathbf{e}'_x + B'_y \mathbf{e}'_y + B'_u \mathbf{u}'$, where

$$(B'_x, B'_y, B'_u) = \partial P(x, y, t)/\partial(x, y, t)\,(B_x, B_y, B_u). \tag{7.3.5}$$

Thus the Poincaré map contains sufficient information to allow field vectors to be carried in the flow from S to S.

If a flow is constructed so that the Poincaré map is available analytically, field vectors can be tracked numerically very efficiently for $\epsilon = 0$. To find the field at $\mathbf{x}$ on S at time $t = T$ one follows individual trajectories from points on S back in time by iterating P^{-1}. At $t = 0$ the trajectory will not generally lie on the section; the initial point $\mathbf{a}$ has to be found and the initial field $\mathbf{B}_0(\mathbf{a})$ calculated. While following the trajectory back in time the Jacobians of P in (7.3.5) may be calculated and multiplied together. The resulting product can be used to find the field at $\mathbf{x}$ at time T using the Cauchy solution. For steady flow, only a single trajectory need be followed for each $\mathbf{x}$ on S to find field for a range of final times T. For more information see Gilbert & Childress (1990), Finn *et al.* (1991), Gilbert (1992) and Klapper (1992*b*).

What fast dynamo mechanisms might exist in steady three-dimensional flows? Two have been proposed. If we ignore time, a map from the Poincaré section S to S is a two-dimensional map, which at first sight is not conducive to dynamo action. However one way to avoid anti-dynamo theorems is to use time-delays to shear field structure along the streamlines and realize a familiar stretch–fold–shear mechanism (Finn *et al.* 1989, 1991, Klapper 1991). The second mechanism makes use of possible discontinuities in the Poincaré map and is the subject of the next subsection.

An SFS mechanism in a steady flow is shown schematically in Fig. 7.6, which shows a block of field, varying in the direction along streamlines, being processed by the flow. The field is first stretched and folded in the steady flow; this part of the flow is called the *map region*. It is then sheared along streamlines in the *delay region*: the flow is arranged so that the upper streamlines move more slowly than the lower ones. This can be achieved in an incompressible flow provided the streamlines are allowed to spread out where the flow is slow (Finn *et al.* 1989, 1991). The process effects the SFS moves in a steady flow and allows amplification of field. The separation of the flow into map and delay regions is artificial, and in a general flow the folding and shearing processes would occur simultaneously.

This SFS process certainly occurs in steady flows. Finn *et al.* (1989, 1991) study a number of examples. In the map region they consider models with: (*a*) a folded baker's map, (*b*) a horseshoe map, and (*c*) smooth spatially

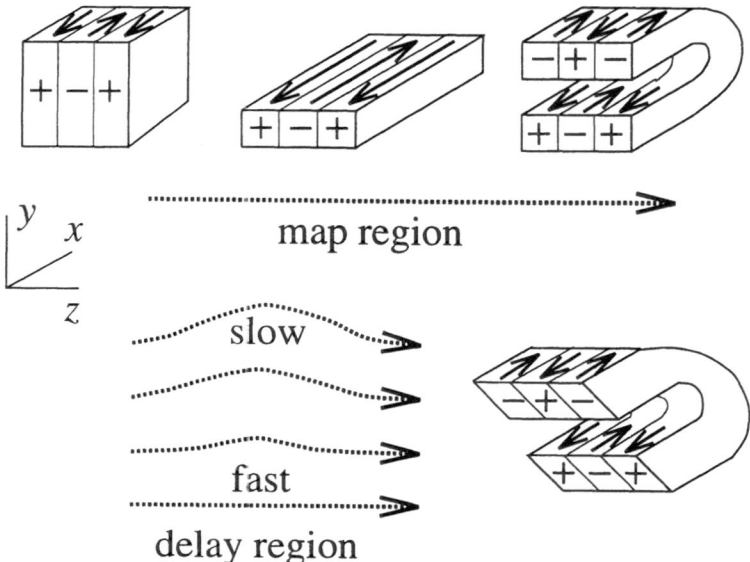

Fig. 7.6 Stretch–fold–shear in a steady flow. Field varies along the direction of the streamlines. As the field is advected it is stretched, folded, and then sheared, allowing like-signed field to be brought together.

periodic maps. The time delays applied to the field are either (1) piecewise constant or (2) continuous. Models (a1) and (b1) can be analyzed and are essentially stretch–fold–slide models, although the dispersion relation takes a different form. Numerical simulations for types (a2), (b2) and (c2) reveal growing fluxes suggestive of fast dynamo action by the flux conjecture.

Rather than discuss types (a1, 2), based on the familiar, discontinuous folded baker's map, we focus on type (b1), for which the mapping can be chosen to be smooth (i.e., infinitely differentiable) but at the expense of working in unbounded space. The model is based on the *Smale horseshoe* (see, e.g., Guckenheimer & Holmes 1983) and is taken from Finn *et al.* (1989, 1991). Consider the smooth map depicted in Fig 7.7. A square, lying in the plane, is stretched linearly by a factor three and folded to give a horseshoe shape. Two thirds of the field is carried back into the square in the two arms of the horseshoe; the rest remains in the fold. Without loss of smoothness, the map can be arranged so that the region $0 \leq y \leq 1/3$ of the original square maps linearly into the first arm $0 \leq x \leq 1/3$ of the horseshoe, and similarly for the second arm:

$$M(x,y) = \begin{cases} (x/3, 3y), & (0 \leq y \leq 1/3), \\ (1 - x/3, 3(1 - y)), & (2/3 \leq y \leq 1). \end{cases} \tag{7.3.6}$$

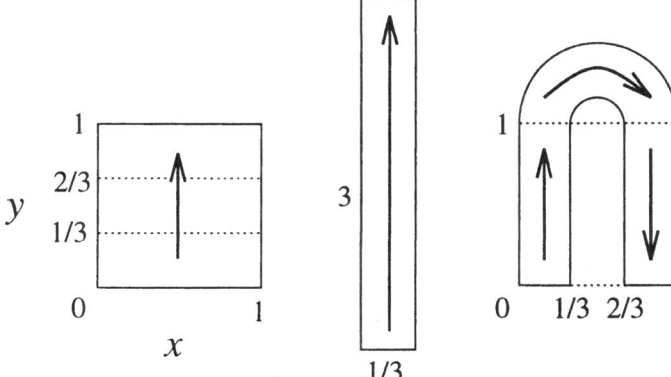

Fig. 7.7 The horseshoe map. (After Finn *et al.* 1991.)

The middle third $1/3 \leq y \leq 2/3$ is mapped smoothly into the fold.

To analyze this configuration with $\epsilon = 0$, we follow field in the unit square, and so need to be able to ignore any field in the fold or between the arms of the horseshoe. The fold is labelled 0 in Fig. 7.8(a) and its images under subsequent iterations are labelled 1, 2, 3 . . .: the map can be arranged so that the field in the fold piles up against the $-y$-axis and is stretched off to infinity. Thus in unbounded space we may ignore legitimately the field in the folds for $\epsilon = 0$. In bounded space these folds could not be ignored; see the discussion of STF in §1.4.

The other concern is whether field can arrive in between the arms of the horseshoe. Figure 7.8(b) shows the relevant region, labelled 0, together with its pre-images, -1, -2, -3 . . ., which pile up against the $-x$-axis. With a suitable initial condition, shown in Fig 7.8(c), Finn *et al.* (1991) arrange that no field ever enters in between the arms of the horseshoe. These choices allow us to consider only field in the unit square $\mathcal{D}$ and the horseshoe map restricted to $\mathcal{D}$. To do this, of course, the unboundedness of $\mathbb{R}^2$ is used in an essential way, and this limits the applicability of this model for natural dynamos, when we always have in mind finite regions of fluid.

Note that under M^n, the image of $\mathcal{D}$ comprises 2^n vertical rectangles, with total area $(2/3)^n$. The field concentrates asymptotically on a *Cantor set* of lines, which has area zero! (Incidentally this furnishes an example of a hyperbolic set as defined in §6.5.) It appears paradoxical that we can hope to achieve a dynamo, but this is possible as long as the flux on the diminishing area grows. Similar dynamos in which the flux concentrates on a chaotic *repeller* may be found in Lau & Finn (1993*a, b*), who consider a flow with stagnation points, and Balmforth *et al.* (1993), who study a system

(a)   *(b)*

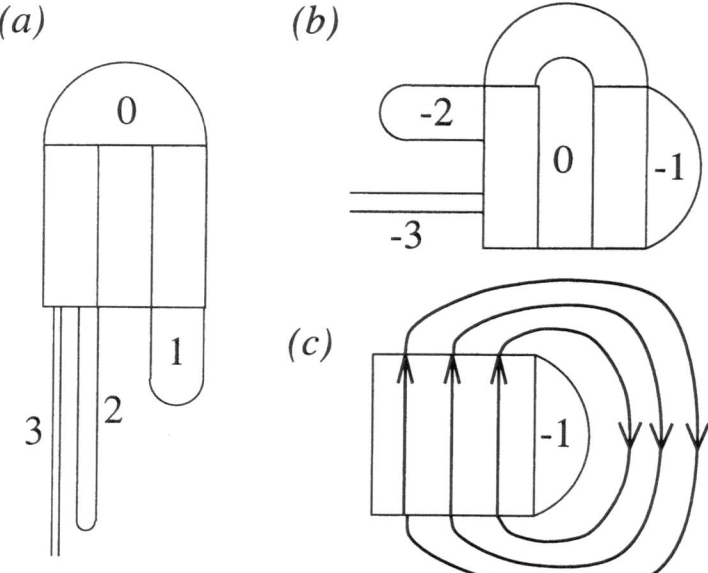

(c)

Fig. 7.8 (*a*) Images of the fold under iteration of M. (*b*) Pre-images of the region between the arms of the horseshoe. (*c*) Initial condition. (After Finn *et al.* 1991.)

with Shilnikov-type chaos near a homoclinic orbit (see, e.g., Guckenheimer & Holmes 1983).

To achieve flux amplification we embed M in a smooth flow which carries field in the z-direction. Using the schematic Fig 7.6, we suppose that in the map region of the flow the field is stretched and folded in (x, y)-planes so as to effect the horseshoe map M. To account for the delay region, in which Lagrangian particles travel at different speeds, we suppose that it takes a constant time T_1 for particles in the first arm of the horseshoe to travel from one map region to the next and time T_2 for particles in the second arm. Again this can be arranged with a smooth flow in the delay region, because of the gap between the arms of the horseshoe. The flow and field are taken to be periodic in z; so the map–delay cycle is repeated.

Thus we have achieved a version of stretch–fold–shear or, more precisely, stretch–fold–slide. To find the flux growth rate Γ for $\epsilon = 0$, note that the total flux $\Phi(t)$ across $\mathcal{D}$ at time t is the flux $\Phi(t - T_1)$ that arrives in the first arm plus the flux $-\Phi(t - T_2)$ that arrives in the second arm; the minus sign arises from the folding. We thus obtain a delay equation for the total flux:

$$\Phi(t) = \Phi(t - T_1) - \Phi(t - T_2). \tag{7.3.7}$$

If the flux has complex growth rate $p = \Gamma + i\omega$ then

$$1 = \exp(-pT_1) - \exp(-pT_2). \tag{7.3.8}$$

This dispersion relation for the complex growth rate is awkward to deal with, and depends on whether T_1/T_2 is rational or irrational. Similar time-delay dispersion relations arise in the study of nearly-integrable flows in Chap. 8. Much of the discussion below is taken from Finn *et al.* (1991), which contains further useful information, including analysis of eigenfunctions; see §9.6.4. For simplicity of discussion we assume the flux conjecture holds.

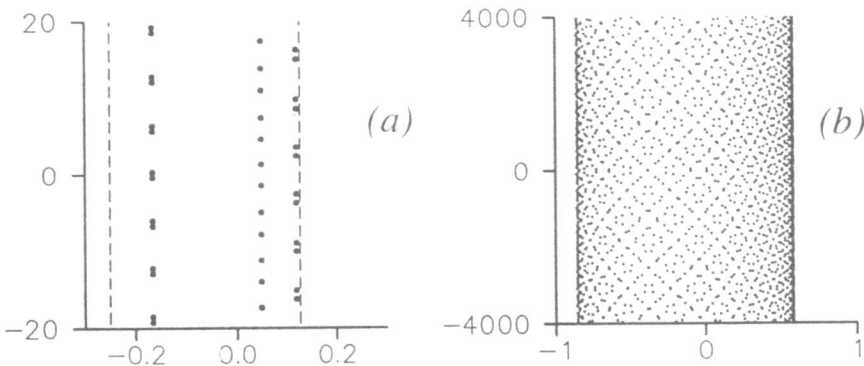

Fig. 7.9 Complex growth rates $p = \Gamma + i\omega$ (dots) for the SFSl steady fast dynamo with $\epsilon = 0$. In (a) $T_1 = 5$, $T_2 = 6$, and (b) $T_1 = 1$, $T_2 = \sqrt{2}$. In both cases the vertical dashed lines give the bounds p_+ and p_-.

(A) Suppose T_1/T_2 is rational; then by rescaling time, we can take T_1 and T_2 to be integers. The equation for the complex growth factor $\mu = \exp p$ is

$$1 = \mu^{-T_1} - \mu^{-T_2}, \tag{7.3.9}$$

and is in this case equivalent to a polynomial in μ. Unfortunately the simple cases $T_1/T_2 = 1$, 2 and 1/2 do not give growth. It appears that for growth the degree of the polynomial must be at least three, and must be solved numerically. As an example take $T_1 = 5$ and $T_2 = 6$, for which (7.3.9) becomes $\mu^6 - \mu + 1 = 0$, with six complex roots corresponding to growth rates

$$p - 2in\pi \simeq 0.118 \pm 2.56i, \ 0.0486 \pm 1.42i, \ -0.167 \pm 0.363i, \tag{7.3.10}$$

for any integer n; the higher the value of n, the higher the frequency and the more structure the eigenfunction has along streamlines. Figure 7.9(a) shows these roots as dots on the complex p plane. Growth occurs when $\mathrm{Re}\, p > 0$. The dashed vertical lines are given by $\mathrm{Re}\, p = p_\pm$ where

$$e^{-T_1 p_+} + e^{-T_2 p_+} = 1, \qquad \left| e^{-T_1 p_-} - e^{-T_2 p_-} \right| = 1. \qquad (7.3.11)$$

It can be shown that these are upper[59] and lower bounds on real growth rates given by (7.3.8) (Finn et al. 1991). For this case $p_- \simeq -0.2509$ and $p_+ \simeq 0.1264$, so that the upper bound is nearly attained by the fastest growth rates of (7.3.10).

Each real flux growth rate Γ is degenerate, being shared by many modes with differing frequencies ω. In a computer simulation with $\epsilon = 0$ the maximum real growth rate of flux will be robust, but the frequency will depend on which modes are excited by the initial conditions. The dynamo has a strong *memory* of initial conditions. With weak diffusion the degeneracy will be broken; in particular diffusion along streamlines will tend to suppress modes with very high ω.[60] In a simulation with $0 < \epsilon \ll 1$ a robust real growth rate will emerge quickly. However the precise dynamo mode and corresponding frequency will take a long time to emerge; the appropriate time-scale for this is the reciprocal of the difference in real parts of the fastest and next-fastest growing eigenfunctions. Because this splitting is determined by diffusion it will presumably be small and scale as some positive power of ϵ; the time-scale will be correspondingly long. This phenomenon is clear for the cat flow of §6.4, for which the splitting is $O(\epsilon)$ and the time-scale an impressive $O(\epsilon^{-1})$.

(B) If T_1/T_2 is irrational, equation (7.3.9) is not equivalent to a polynomial. Figure 7.9(b) shows growth rates (dots) found numerically in the case $T_1 = 1$ and $T_2 = \sqrt{2}$. The bounds (7.3.11) still apply and are $p_- = -0.8553$ and $p_+ = 0.5802$, shown as (scarcely visible) dashed lines bounding the roots. The complex growth rates are not periodic in frequency ω, and have flux growth rates Γ which fill the interval $[p_-, p_+]$ densely, without attaining the limits in general; in this case $\Gamma_{\mathrm{sup}} = p_+$.

Consider following flux from smooth initial conditions for case (B) with $\epsilon = 0$. As $t \to \infty$, modes with real growth rates approaching p_+ will dominate, although generally the growth rate p_+ will not be attained by any one mode. Rather, the closer the real growth rate of a mode is to p_+ the larger will be the corresponding frequency ω. These high-frequency modes will however only be weakly excited from smooth initial conditions. Thus, rather bizarrely, in a simulation the real flux growth rate would rapidly tend to p_+, but the frequency of the field would increase indefinitely as $t \to \infty$. Correspondingly, for $\epsilon > 0$ the fastest growing mode would change as $\epsilon \to 0$ in a series of mode crossings. As ϵ decreases the dominant mode will possess higher frequencies, more structure along streamlines, and growth rates approaching p_+.

---

[59]Finn et al. (1991) identify p_+ with the h_{line} and topological entropy; p_+ is indeed the rate of increase of line length *within* the square, but since additional line-length is trapped in the folds, the topological entropy of the flow may be larger than p_+.

[60]Just which of the nearly degenerate modes will be favored for small ϵ is unknown, but given that diffusion can sometimes help amplification, we know of no reason why the fastest growing modes must have $\omega = O(1)$: small-scale modes with $\omega = O(\epsilon^{-\alpha})$ for some α with $0 < \alpha < 1/2$ could dominate for small ϵ.

In this particular SFS1 model, clearly possibility (B) is typical, given that 'most' numbers T_2/T_1 are irrational. Unfortunately there are few results available for more general classes of steady flows. For $\epsilon = 0$, Gilbert (1991 a, 1992) follows flux in the ABC 111 flow and a patched flow based on an ABC flow with $A = B = 1$, $C \ll 1$. For each symmetry class of magnetic fields, the growth rate is robust, but the frequency depends on the initial condition: the more complicated the initial field the greater the frequency. There is no evidence of the frequency increasing with time, and so case (A) for T_1/T_2 rational seems most appropriate. On the other hand it is not clear that SFS is the dominant amplification mechanism in these flows; see §7.3.2 below. Finn et al. (1991) and Schmidt, Chernikov & Rogalsky (1993) obtain flux growth for $\epsilon = 0$ for smooth, steady flows, and obtain results consistent with (A).

Thus on the extremely limited evidence available, it appears that possibility (B) is special to the steady SFS1 model and a consequence of the piecewise constant time delay; possibility (A) appears to be more likely for steady smooth flows. Numerical investigation of mode degeneracy and frequency selection in the limit $\epsilon \to 0$ for steady flows appears to be impossible with current computers. As well as the simulations of ABC flows and Kolmogorov flows of §2.6, studies for $0 < \epsilon \ll 1$ have been carried out by Zheligovsky (1993) for a Beltrami flow in a sphere surrounded by an insulator, and by Lau & Finn (1993a, b) for a chaotic scattering flow involving stagnation points. Some rigorous results on spectra in the limit $\epsilon \to 0$ are given by Vishik (1989).

Note that for the SFS1 map with $\epsilon = 0$ there is exact degeneracy of modes with different values of αk and the same real growth rate (Fig 3.12), akin to case (A); for the SFS map there appears to be similar degeneracy (Figs. 3.7, 7.3), but whether this is exact (A) or approximate (B) is unknown. These degeneracies have similar implications for perfect evolution and mode selection in the limit $\epsilon \to 0$ (when k is not fixed) both for the maps and for the two-dimensional flows they model. For example suppose that in, say, MW+, the magnetic field is free to adopt the optimal k, instead of it being fixed as is usual in simulations but artificial physically. In the limit $\epsilon \to 0$ the time taken for the most unstable mode to emerge is likely to increase, and we cannot rule out a sequence of mode crossings, with the optimal value of k increasing indefinitely.

The dispersion relation (7.3.9) can also be recovered from the zeta function methods of §6.3.3. We can classify periodic orbits by the number n of times they undergo the map–delay sequence. Given n, for $m = 0, 1, \ldots, n$ there are $n!/m!(n-m)!$ orbits ξ with period $T(\xi) = mT_1 + (n-m)T_2$ and transverse Floquet multiplier $\lambda(\xi) = 3^m(-3)^{n-m}$. From (6.3.30) we obtain

$$\Omega(z) = \sum_{n=1}^{\infty} (z^{T_1} - z^{T_2})^n = \frac{z^{T_1} - z^{T_2}}{1 - z^{T_1} + z^{T_2}}; \qquad (7.3.12)$$

poles of $\Omega(z)$ give eigenvalues $p = -\log z$ and this condition recovers the dispersion relation (7.3.9). It appears rather more difficult to evaluate $N_{\mathrm{tw}}(n)$ in (6.3.32b) directly and so confirm Oseledets' conjecture for this flow.

7.3.2 The Stretch–Fold–Cut–Paste Mechanism

Another mechanism for fast dynamo action in a steady flow may be motivated by again considering the Poincaré map P from S to S, defined in the previous subsection. If the flow contains X-type stagnation points the map P will be discontinuous, even when the flow is perfectly smooth. There will be a set C of curves and/or points that are drawn into stagnation points and never return to the section (see Fig 7.10). The set C contains a point when a one-dimensional stable manifold of a stagnation point first intersects S, and a curve for a two-dimensional stable manifold. The period of return to the section, $t' - t$, diverges logarithmically as C is approached on S, and P is discontinuous across C. A discontinuous Poincaré map enables one to evade anti-dynamo theorems applicable to smooth two-dimensional flows, and allows the possibility of growth of flux on S; as a familiar example think of the stacked baker's map.

Let us temporarily disregard the time delays, and consider a Poincaré map P acting on the two-dimensional section, $P : (x, y) \to (x', y')$. We set $\epsilon = 0$, ignore the slaved B_u component of field along streamlines, and evolve the (B_x, B_y) components using the 2×2 Jacobian $\partial(x', y')/\partial(x, y)$; see (7.3.5). We thus have a field lying in the plane S and a map P from S to S.

We can then imagine a fast dynamo in a steady flow that involves two processes: the stretching and folding of field by the chaotic streamlines in the flow, and the cutting up and reassembly of field because of stagnation points — a 'stretch–fold–cut–paste' dynamo mechanism (Gilbert 1992). These processes would be encoded in P as a smooth distortion of S, causing stretching and folding of field, but not constructively, followed by a cutting up of the section along curves in C and reassembly. This cutting and reconnection of field lines can then lead to constructive reinforcement of field, as for the stacked baker's map. We suspect that given a 'general' distortion of S, say by a standard map or application of a sine-wave flow, and a 'general' cutting up and reassembly, fast flux growth would occur, though this is at present no more than a conjecture.

A more concrete example of a fast dynamo that uses this mechanism is depicted in Fig. 7.11, which shows schematically how one can achieve the basic stacked baker's map (Fig 3.2) using lines of two-dimensional stagnation points to cut and reassemble transverse field. A unit square of field A is carried up to the horizontal stagnation line B (dotted) and divided into two rectangles, which are carried to C and D; at the same time the rectangles are stretched so as to double the field. The dynamo is spatially periodic and the field alternates as shown; so the rectangle at C meets another rectangle D'

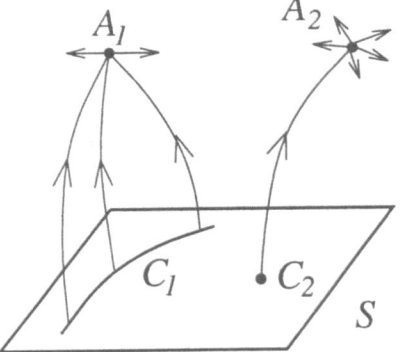

Fig. 7.10 Schematic Poincaré section S. Points on the curve C_1 are drawn into the stagnation point A_1, and the point C_2 is drawn into A_2. At C_1 and C_2 the Poincaré map is discontinuous.

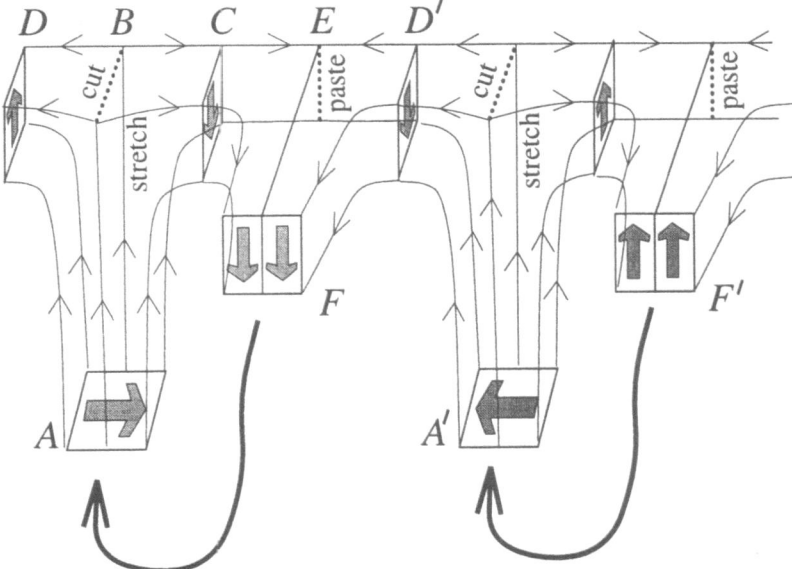

Fig. 7.11 Schematic stretch–fold–cut–paste dynamo, described in the text.

oriented as shown, and these are combined at the vertical stagnation line E (dotted) to give a square F. Using the spatial periodicity and the stagnation lines we have achieved a stacked baker's map $A \to F$ and doubled the flux of the transverse field. The resulting fields at F and F' can now be fed by the steady flow back into A and A' (with the correct orientation), and the cycle repeated.

The principal omission from this picture is the time delay associated with the stagnation lines. The time of travel from A to F varies, and diverges for particles that pass close to the stagnation lines B and E. How can we be sure that fast flux growth will persist when these delays are included?

Fortunately, for the flow shown in Fig 7.11 we can choose an initial condition so that the folding of transverse field is always constructive; transverse field vectors are always aligned, despite delays. For example take the initial transverse field at A to be a square of field aligned with the arrow shown, and extend this upwards a little to give a rectangular slab of constant transverse field. Take a similar initial field at A' aligned with the arrow there. In the subsequent evolution, the direction of the transverse field will always be that depicted by the arrows in Fig 7.11; the stretching, cutting and pasting is such as never to bring opposed transverse field into contact. Of course this lack of cancellation is partially a result of our factoring out the parallel field; the sign of this component will probably vary rapidly.

Therefore in this instance, if we simply neglect the field that is most delayed by the stagnation lines in, say, calculating the average field on the square A at a given time, we bound this average *from below*. This is an unusual occurrence — usually we cannot neglect field in calculating a flux because of the possibility of cancellation. However for our choice of initial condition there are no cancellations! Let us arrange the flow so that field vectors take a unit time to travel from A round the circuit back to A, except vectors which come within a given distance of the stagnation lines B and E: these suffer varying (unbounded) delays. This can be done without compromising the smoothness of the flow.

With these choices, over a unit time the flow achieves the stacked baker's map shown in Fig 7.12, where the shaded regions represent field vectors that are delayed near stagnation points. We arrange the flow so that these regions have width δ and 2δ, as shown; the shaded region has area $3\delta - 2\delta^2$. We start with unit field $\mathbf{B} = \mathbf{e}_y$, apply the stacked baker's map (say M) and then discard the delayed field by setting the field to zero in the shaded region; let us call the resulting map M_*. The process is then repeated, and after n iterations of M_* the field is 2^n or 0 at each point. Call the area of the *support* of the magnetic field (i.e., region where the field is non-zero) $A_n(\delta)$; the average field after n iterations of M_* is $2^n A_n(\delta)$, and this bounds the average field in the real delay system from below. Although $A_n(\delta) \to 0$ as $n \to \infty$, growth of the average field $2^n A_n$ is possible provided the decay of area is slow enough.

We do not know $A_n(\delta)$ explicitly, but if we take $\delta = 2^{-m}$ for some fixed m, we can use a *Markov partition* (see, e.g., Guckenheimer & Holmes 1983) to calculate the asymptotic behavior of $A_n(\delta)$.[61] Divide the unit square into

---

[61]Rigorous results concerning loss of area in a phase space with a 'hole' in it may be found in Pitskel (1991).

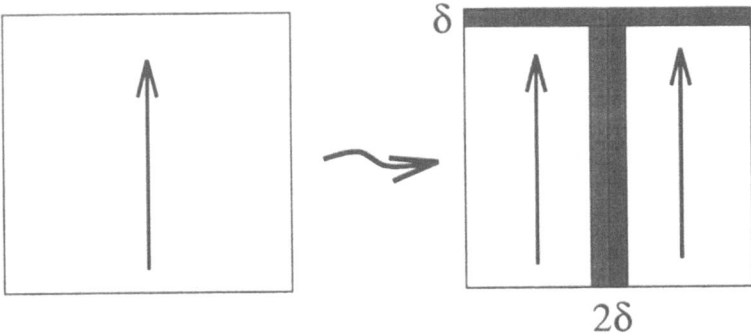

Fig. 7.12 Baker's map: field is to be set to zero in the shaded regions.

smaller squares of side δ, labelled S_i, for $i = 1, \ldots, 2^{2m}$. Under the stacked baker's map M the interior of each square is mapped into exactly two others, and we can define a $2^{2m} \times 2^{2m}$ transition matrix $\mathbf{P}$ by

$$P_{ij} = \begin{cases} 1/2, & (S_i \to S_j), \\ 0, & (\text{otherwise}). \end{cases} \tag{7.3.13}$$

The matrix $\mathbf{P}$ describes how M moves areas or, equivalently, the support of a magnetic field from square to square.

Now the shaded region comprises $3 \times 2^m - 2$ of the squares S_i, and we can define another matrix Q_{ij} by setting to zero all entries of P_{ij} for which the square S_i is in the shaded region. The matrix $\mathbf{Q}$ gives the effect of M_* on the support of the field, that is, the effect of applying the baker's map and then setting to zero field in the shaded regions. Finally

$$A_n(\delta) \sim \text{const.} \times [\lambda(\delta)]^n, \qquad (n \to \infty), \tag{7.3.14}$$

where $\lambda(\delta)$ is the largest eigenvalue of $\mathbf{Q}$. This bounds the growth rate $\Gamma(\delta)$ of average field in the true delay system by

$$\Gamma(\delta) \geq \Gamma_*(\delta) \equiv \log(2\lambda(\delta)). \tag{7.3.15}$$

It may be checked that $\lambda(1/4) = \lambda(1/8) = 1/2$, corresponding to a bound of zero and so no proof of flux growth. However

$$\lambda(1/16) = (1 + \sqrt{5})/4, \quad \Gamma_*(1/16) = \log((1 + \sqrt{5})/2) > 0; \tag{7.3.16}$$

the eigenvalue λ here is computed numerically, but is found to be half the golden mean. Also

$$\lambda(1/32) \simeq 0.9196, \quad \Gamma_*(1/32) \simeq 0.6094. \tag{7.3.17}$$

For both of these values of δ, there is therefore growth of flux in the time-delay system of Fig 7.11. Thus it is possible to have fast flux growth despite the effect of stagnation points in delaying field vectors; this suggests fast dynamo action by the flux conjecture.

Note that strong fields parallel to the streamlines are developed near the stagnation line B (Fig 7.11) and these intersect the square in the horizontal shaded region (Fig. 7.12) and its images. These fields take into account the apparent cutting of transverse field lines and ensure that the total field remains divergence-free. In other words the field lines that appear to be cut are not really cut, but are connected by lengths of parallel field. These parallel fields can be neglected for $\epsilon = 0$ but are coupled for $\epsilon > 0$; it remains for their effect to be assessed and bounded, as the flux conjecture would suggest.

A stretch–fold–cut–paste mechanism appears to operate in the cellular patched flow of Gilbert & Childress (1990) and Gilbert (1992), which contains similar non-generic stagnation lines. This study uses a number of Poincaré sections, and the corresponding Poincaré maps are discontinuous. Using the Cauchy solution (with the correct time delays derived from the flow field) flux growth in the transverse field is observed. The flux growth continues (at approximately the correct rate) when the time delay in the map is set to a constant, independent of position; this turns the Poincaré maps into two-dimensional maps, and shows the importance of the cutting and pasting at discontinuities, needed to avoid the anti-dynamo theorems. It thus appears that it is a combination of stretching and folding, and cutting and pasting that is the mechanism at work in the original flow. However note that generally if a stretch–fold–cut–paste mechanism is present in a flow, the SFS mechanism is also likely to be present, and the two could both contribute to a fast dynamo; they could be difficult to disentangle, and could interact in a complicated way.

8. Nearly Integrable Flows

The idea of studying fast dynamo action near the onset of chaos, for a flow which is 'close' to being integrable, is an appealing one. If we neglect the effects of stretching and equate integrability with slow dynamo action, this puts us at the boundary of fast dynamo action, where one might hope to understand the geometry of slow and fast mechanisms. At the same time, any such study faces a basic obstacle, namely that one expects that then the fast dynamo growth rate will be *small*, with mean flux a small component of the total field, and therefore possibly difficult to evaluate either numerically or analytically.

The main analytical advantage of nearly-integrable systems is that one can use the method of Melnikov (1963). Applied to two-dimensional, time-dependent incompressible flow, Melnikov's theory deals with perturbed stream functions of the form $\psi(x, y, t; \varepsilon) = \psi_0(x, y) + \varepsilon\psi_1(x, y, t)$ where $\varepsilon \ll 1$, i.e., a two-dimensional Hamiltonian with time-dependent perturbation.[62] The chaos introduced by the perturbation is localized near *separatrices* of the steady component defined by ψ_0. A typical separatrix is a streamline of the steady component which connects two X-type stagnation points. These are the hyperbolic critical points of ψ_0, where the velocity vanishes and the streamline structure is locally a saddle. A fast dynamo model will thus depend upon the form of the unperturbed, integrable flow, and upon the nature of the small perturbation. A distinct, and in some respects less natural, variant of Melnikov's method applies to steady flows which are nearly two-dimensional (see Gilbert 1992). In the present chapter we restrict attention to two-dimensional unsteady flows.

The theory of perturbed Hamiltonian systems, of which the above two-dimensional flow problem is an example, thus provides the theoretical bedrock for the studies of the present chapter. Among many treatments of the theory and its applications we mention Arnold (1978), Guckenheimer & Holmes (1983), Lichtenberg & Lieberman (1983) and Zaslavsky *et al.* (1991). The present chapter will be concerned, not so much with this underlying theory, as with the implications it has for the structure of the magnetic field in a chaotic region. Our object is really to exploit in one special case techniques which

---

[62]In this chapter we use ε to denote a small parameter in the flow, retaining ϵ for the inverse magnetic Reynolds number R.

allow a microscopic focus on magnetic structure in a chaotic system. This goal stands in contrast to the 'anti-integrable' limit considered in Chap. 10, where another microscopic picture (based upon self-similarity) is appropriate. We shall show below that an approximate description of dynamo action in a perfect conductor is accessible in the nearly-integrable setting, although explicit calculations of growth rate remain difficult and there are few analytical results.

The fast dynamo problem shares certain features with the problem of turbulent transport of a scalar or vector field. Slow dynamo action can be compared with enhanced diffusion in eddying flows (Isichenko 1992), while fast dynamo action corresponds to turbulent transport in the limit of small dissipation. For the case of nearly-integrable flows, the study of transport in chaotic systems has been pioneered by Wiggins and his co-workers (Wiggins 1992), who have developed a geometry of the lobe structures associated with material transport. The transport of a scalar field is linked (in the case of planar flows) to the areas of lobe structures in the heteroclinic tangle. In the model to be developed here, it is rather the *boundaries* of the lobes which carry flux sheets. The transport of a phase of the magnetic field, together with the stretching and folding of the lobes, will then provide the basis for fast dynamo action.

8.1 The Onset of Chaos

8.1.1 Time-dependent Boundary Layers

The first question which might be asked is how the well-defined magnetic structures of steady, small-ϵ dynamo theory change by the addition of small unsteadiness of the flow.[63] A simple example of this is the advection–diffusion problem associated with the horizontal flow in a Roberts cell. The velocity field is $\mathbf{u} = (\psi_y, -\psi_x, 0)$ with $\psi = \psi_0 + \varepsilon\psi_1$, and the unperturbed streamfunction is $\psi_0 = \sin x \sin y$. We consider the equation for the magnetic potential,

$$A_t + \mathbf{u} \cdot \nabla A - \epsilon\nabla^2 A = 0. \tag{8.1.1}$$

The boundary-layer analysis of the unperturbed, steady case given in §5.5 applies, with the results described there, and our task is to extend this boundary-layer theory to account for small unsteadiness. Our interest is of course small ϵ, and it is reasonable to ask whether the *perturbed* boundary layer might maintain its integrity as a thin concentration of magnetic lines even though its location depends upon time. If this happens, the dynamo theory becomes closely linked to the Melnikov theory, as we discuss below.

---

[63]The material in this section includes some unpublished work contributed by I. Klapper.

To investigate this we retain the (σ, ξ) formulation of §5.5.1 We now stipulate, however, that σ and ξ are both based upon the *unperturbed* flow,

$$\sigma = \int_0^s q_0 \, ds, \qquad \xi = R^{1/2} \psi_0, \tag{8.1.2}$$

where $q_0 = |\nabla \psi_0|$ and s is arclength along the unperturbed separatrix. In our example the separatrix is a line connecting any two adjacent stagnation points, which are the points $(m\pi, n\pi)$, with m, n arbitrary integers.

In terms of these variables (8.1.1) has the boundary-layer limit

$$A_t + q_0^2(A_\sigma - A_{\xi\xi}) + \beta q_0 v_1 A_\xi = 0, \tag{8.1.3}$$

where $v_1 = -\partial \psi_1 / \partial x$ and $\beta = \varepsilon \sqrt{R}$ is a parameter which measures the effect of the time-dependent perturbation on a boundary layer. If it is assumed that a time scale of the perturbation can be based upon the unperturbed velocity and cell size, then β measures the lateral deflection of the unstable manifold in units of the boundary-layer thickness. Since v_1 is here evaluated at the separatrix, it may be regarded as a function of s or σ and t alone. We may therefore introduce the function $Q(\sigma, t)$ defined by

$$Q_t + q_0^2 Q_\sigma + q_0 v_1 = 0, \qquad Q(0, -\infty) = 0. \tag{8.1.4}$$

A formal solution of this problem can be expressed in terms of a Lagrangian variable $\sigma(\sigma_0, t)$ defined by

$$\frac{d\sigma}{dt} = q_0^2(\sigma), \qquad \sigma(\sigma_0, 0) = \sigma_0. \tag{8.1.5}$$

Then

$$Q(\sigma(\sigma_0, t), t) = -\int_{-\infty}^t q_0(\sigma(\sigma_0, t)) \, v_1(\sigma(\sigma_0, t), t) \, dt. \tag{8.1.6}$$

Note that the integral on the right is performed at fixed σ_0, yielding a function of σ_0 and t. To determine the function $Q(\sigma, t)$ on the left we must substitute σ_0 as a function of (σ, t). We observe that the function Q is closely related to the Melnikov function discussed in the next section (cf. 8.1.19).

In terms of Q we have the following basic result: suppose that $\tilde{A}(\sigma, \xi, t)$ is a solution of the time-dependent *unperturbed* boundary-layer equation,

$$\tilde{A}_t + q_0^2(\tilde{A}_\sigma - \tilde{A}_{\xi\xi}) = 0. \tag{8.1.7}$$

Then $A(\sigma, \xi, t) = \tilde{A}(\sigma, \xi + \beta Q(\sigma, t), t)$ is a solution of (8.1.3). This is easily proved by direct substitution. We may think of Q as a *shift function*. It determines the shift of the streamlines of the unperturbed boundary layer in the direction normal to the separatrix, which is needed to make the boundary-layer conform to the perturbed flow. In effect it tells us that the boundary layer is *fixed* to the curve determined by Q, at least in the limit $\varepsilon \to 0$ for fixed β.

The solution of (8.1.3) comes down to evaluating (8.1.6), generally not a simple task, and then solving for the unperturbed time-dependent boundary layer from appropriate initial and boundary conditions. The boundary conditions involve the analysis of the corner regions of size $O(R^{-1/4})$ as discussed in §5.5.1. There A is a material scalar to first order. Indeed, in these regions $|\mathbf{u}_0| = O(R^{-1/4})$, while $|\mathbf{u}_1| = O(\varepsilon) = O(\beta R^{-1/2})$, and therefore $\mathbf{u}_1$ may be neglected. Here it is important to distinguish between the $O(R^{-1/2})$ thickness of the layer and the $O(R^{-1/4})$ scale at the corner when computing $\mathbf{u}_0 \cdot \nabla A$ and $\mathbf{u}_1 \cdot \nabla A$. We thus have the approximate equation

$$A_t + \mathbf{u}_0 \cdot \nabla A \approx 0. \tag{8.1.8}$$

The conditions at $\sigma = 0$ for our unperturbed boundary-layer problem are determined by transport though a corner region of A as determined on adjacent separatrix segments. These provide the 'input' data to be propagated to the next stagnation point by the unperturbed boundary-layer solution, with Q telling us how this straight boundary layer must actually be shifted to account for the perturbation. The shifted layer then provides, when evaluated through the terminal corner region, input to the next layer, and so on. This is a difficult program to carry out in detail, although it is similar to the steady boundary-layer solutions. We are not aware of any computed results, although the problem is of considerable importance as the proper place to study the transition process. The parameter β provides control over the chaotic regions. For β small, the boundary layer is essentially laminar but there will presumably be some small, weak chaotic zones buried within it. As β becomes large, the boundary layer will follow the time-dependent curve determined by $Q(\sigma, t)$, the thickness then being small compared to the lateral excursions. This suggests that the relevant perfectly conducting limit is that of a time-dependent *flux sheet*. This observation is the basis for a model discussed in §8.3.

8.1.2 Melnikov's Method

The curve $\xi = Q(\sigma, t)$ which figured in the boundary-layer analysis of §8.1.1 is a one-dimensional manifold (smoothly deformable into a straight line) which emerges from the stagnation point at $\sigma = 0$. It is equivalent to a structure arising in the well-known theory of Melnikov (1963) dealing with two-dimensional systems with small time-periodic perturbations. Helpful discussions of the Melnikov method may be found in Guckenheimer & Holmes (1983), Bertozzi (1988), and Wiggins (1992). We shall develop it here only in the context of the unperturbed streamfunction $\psi_0 = \sin x \sin y$ considered above. We show this configuration in Fig. 8.1. We indicate two hyperbolic stagnation points of $\mathbf{u}_0$, here $(0,0)$ and $(\pi, 0)$, and corresponding *stable and unstable manifolds*. The latter are time-dependent curves defined as follows: the unstable manifold of A, $W^{\mathrm{u}}(A)$, at a given time is the set of points whose Lagrangian orbits $\mathbf{x}_{\mathrm{u}}(t)$ in the perturbed flow $\mathbf{u}$ have the property that $\mathbf{x}_{\mathrm{u}}(t) \to (0,0)$ as $t \to -\infty$. The stable manifold of A, $W^{\mathrm{s}}(A)$, is similarly the set of points $\mathbf{x}_{\mathrm{s}}(t)$ such that $\mathbf{x}_{\mathrm{s}}(t) \to (0,0)$ as $t \to +\infty$. These manifolds, which connect to the critical points of the unperturbed flow, are limiting cases of orbits which sweep past the critical points.[64] These manifolds replace, in the perturbed problem, the lines $x = m\pi$, $y = n\pi$ which defined the separatrix of the unperturbed flow.

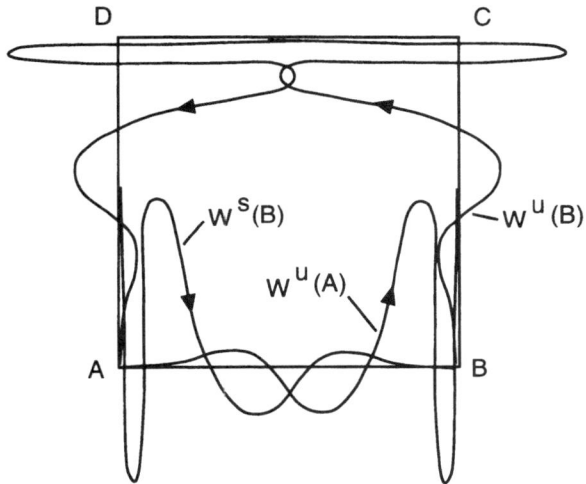

Fig. 8.1 Perturbation of a heteroclinic orbit by a time-periodic disturbance. The unstable manifold at $A = (0,0)$, $W^{\mathrm{u}}(A)$, and the corresponding stable and unstable manifolds at $B = (\pi,0)$ are indicated.

Melnikov's method uses general results from the theory of invariant manifolds and a useful computational criterion to determine the asymptotic form of the stable and unstable manifolds in the limit of small ε. The procedure also provides a condition ensuring that stable and unstable manifolds intersect *transversely*, that is, with non-coincident tangents. Such an intersection is known to lead to a sensitive dependence upon initial conditions associated with chaos (Guckenheimer & Holmes 1983), and so may be taken as the condition for the existence of chaotic motion in the neighborhood of the separatrix.

---

[64]In the presence of time-dependent perturbations, critical points of $\mathbf{u}$ may or may not depend on time. They are fixed in MW+ε (see (8.2.4) below), but are helical periodic orbits for CP+ε (8.2.6). This motion of points is a higher order effect in the Melnikov formulation, a result of the spreading of streamlines near a critical point.

We shall describe the method as it applies to our example. Let $x(t, x_0)$ be the solution of

$$\frac{dx}{dt} = q_0(x, 0) = \frac{\partial \psi_0}{\partial y}(x, 0) = \sin x, \qquad x(0) = x_0, \tag{8.1.9}$$

where we continue to fix upon the heteroclinic connection from $(0,0)$ to $(\pi, 0)$. This Lagrangian variable represents *simultaneously* all points on the hetero-clinic orbit (of the unperturbed system) at a given time t. It will be useful, however, to consider a label different from the initial position, and rather use the arrival time at a fixed line $x = x_0$ where $0 < x_0 < \pi$. Any point can be chosen, since all points cross the line at *some* time. If this arrival time is t_0, then the desired Lagrangian variable is, in terms of the solution of (8.1.9),

$$x(t, t_0) \equiv x(t - t_0, x_0). \tag{8.1.10}$$

Consider now the manifold $W^u(A)$. Taking y as $O(\varepsilon)$, we may compute its location to the order we need from the expressions

$$\frac{dx}{dt} = \sin x + O(\varepsilon), \tag{8.1.11a}$$

$$\frac{dy}{dt} = -y \cos x + \varepsilon v_1(x, 0, t) + O(\varepsilon^2). \tag{8.1.11b}$$

These equations determine the manifolds $W^u(A)$ and $W^s(B)$ when we add the conditions $(x, y) \rightarrow (0, 0)$ as $t \rightarrow -\infty$ and $(x, y) \rightarrow (\pi, 0)$ as $t \rightarrow \infty$, respectively. The quantity we seek is the lateral displacement Δ_u of the point arriving at time t_0 at the line $x = x_0$. Using (8.1.11a) we may write (8.1.11b) in the form

$$\frac{d(y \sin x)}{dt} = \varepsilon v_1 \sin x + O(\varepsilon^2), \tag{8.1.12}$$

and then integrate to obtain

$$\Delta_u \sin x_0 = \varepsilon \int_{-\infty}^{t_0} \sin\left(x(t, t_0)\right) v_1(x(t, t_0), 0, t)\, dt + O(\varepsilon^2). \tag{8.1.13a}$$

Similarly, for the displacement Δ_s of $W^s(B)$ at the same location we obtain

$$-\Delta_s \sin x_0 = \varepsilon \int_{t_0}^{+\infty} \sin\left(x(t, t_0)\right) v_1(x(t, t_0), 0, t)\, dt + O(\varepsilon^2). \tag{8.1.13b}$$

Adding, we obtain to leading order the *Melnikov function* $M(t_0)$,

$$M(t_0) = \varepsilon \int_{-\infty}^{+\infty} \sin\left(x(t, t_0)\right) v_1(x(t, t_0), 0, t)\, dt, \tag{8.1.14}$$

in terms of which the separation distance $\Delta_u - \Delta_s$ may be computed to order ε:

$$\Delta_{\mathrm{u}} - \Delta_{\mathrm{s}} = M(t_0)/\sin x_0 + O(\varepsilon^2). \tag{8.1.15}$$

The procedure indicated here in a special, but typical, case leads easily to the general formulation for an arbitrary smooth heteroclinic connection, which includes the homoclinic case when A and B coincide.

It follows that, as t_0 varies, any simple zero of $M(t_0)$ is indicative of a transversal or non-tangential intersection of the manifolds $W^{\mathrm{u}}(A)$ and $W^{\mathrm{s}}(B)$, and in fact such an intersection can be established rigorously. To take an example of interest to us later, let $\psi_1 = \cos x \cos y \cos \omega t$. Taking $x_0 = \pi/2$, we compute $\sin x(t, t_0) = \mathrm{sech}(t - t_0)$ from (8.1.11a) and

$$
\begin{aligned}
M(t_0) &= \varepsilon \int_{-\infty}^{+\infty} \sin^2 x(t, t_0) \cos \omega t \, dt \\
&= \varepsilon \int_{-\infty}^{+\infty} \mathrm{sech}^2 (t - t_0) \cos \omega t \, dt.
\end{aligned}
\tag{8.1.16}
$$

The integral may be evaluated using residues (see Bertozzi 1988) and there results

$$M(t_0) = \varepsilon \frac{\pi\omega}{\sinh \pi\omega/2} \cos \omega t_0 \equiv \varepsilon F(\omega) \cos \omega t_0. \tag{8.1.17}$$

There thus exist simple zeros of the Melnikov function and a 'heteroclinic tangle' results, as is shown in Fig. 8.1.

As Ponty et al. (1993) and Ponty, Pouquet & Sulem (1995) have noted, the largest values of $F(\omega)$ are usually identified with the most extensive regions of chaos (Ottino 1989). We plot the function $F(\omega)$ in (8.1.17) in Fig. 8.2. We thus expect the domain of chaos to be largest when ω is small. We can also estimate the thickness of the web as $O(\varepsilon)$ in that case. This follows from the fact that we are dealing with a slowly-varying cat's-eye flow; see Fig. 5.4. Whenever $\cos \omega t$ is positive the channels are oriented as in Fig. 5.4, but when the sign is negative the x-component of velocity is reversed and the mean slope of the channel reverses sign. A fluid particle can make long excursions but nevertheless remains in the region of width ε occupied by the channels.

When ω is large the chaos is only a small part of the channel, localized near the separatrix and with broader portions at the hyperbolic points. The reason for this is the high frequency of the lateral oscillations of a fluid particle, which therefore remains on average near the same unperturbed streamline. As a hyperbolic point is approached, however, the particle will move one way or the other and this depends upon the phase of the oscillation. Thus sensitive dependence on initial conditions occurs only very close to the unperturbed separatrix.

We now compare the Melnikov formulation with the unsteady boundary layer analysis of §8.1.1. Continuing with the example just discussed, so that $q_0(x, 0) = \sin x$, we see that the Lagrangian problem (8.1.5) has the solution $\sigma = 2(1 + Ce^{-2t})^{-1}$, the arbitrary positive constant being determined by the

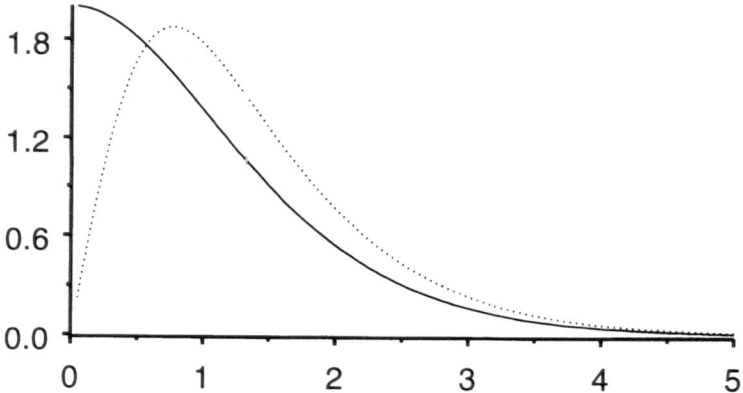

Fig. 8.2 The amplitude of the Melnikov function $F(\omega)$ for the flow MW+ε, as defined by (8.1.17) (solid), and the Melnikov function $G(\omega)$ for the flow CP+ε, as defined by (8.2.10) (dotted).

initial condition $\sigma(0) = \sigma_0$. To make a comparison with the Melnikov function we set $C = e^{2t_0}$, or $\sigma_0 = 2(1 + e^{2t_0})^{-1}$, in which case the function Q defined by (8.1.4) becomes (with $v_1(x, 0) = \varepsilon \sin x \cos \omega t$)

$$
\begin{aligned}
Q(\sigma(\sigma_0, t), t) &= -\varepsilon \int_{-\infty}^{t} \frac{d\sigma}{dt} \cos \omega t \, dt \\
&= -\varepsilon \int_{-\infty}^{t} \operatorname{sech}^2 (t - t_0) \cos \omega t \, dt,
\end{aligned}
\tag{8.1.18}
$$

using (8.1.5, 6, 9). Thus

$$
\lim_{t \to \infty} Q(\sigma(\sigma_0(t_0), t), t) = -M(t_0).
\tag{8.1.19}
$$

This identification of the shift function and the Melnikov function is natural and it underscores the origins of chaos in dissipative boundary-layer theory. The zeros of the Melnikov function are associated with lobes developing in the heteroclinic structure. In the boundary layer, and for small or moderate β, the lobes of $W^u(A)$ are folded and stretched against the unstable manifold at B, $W^u(B)$, but can be drawn out only slightly before they are smoothed by diffusion. We would thus expect small regions of stretching and chaos to exist near the hyperbolic points, and enlarge with increasing β, ultimately spreading out to form the chaotic web associated with small ε and large β. These boundary-layer structures would represent the likely seat of fast dynamo action in its earliest stages, that is, the first appearance of stretching over a domain of finite size.

8.2 Dynamo Action in the Chaotic Web

The calculation of dynamo action in nearly-integrable time-dependent flows involves not only the geometry of the stable and unstable manifolds, but also, for the class of models which concern us, the connections between separatrix regions associated with adjacent cells. These connections, which were alluded to in the preceding section, depend very much upon the nature of the perturbing streamfunction ψ_1 and the symmetries of the flow as a whole. In addition, the dynamo problem brings in the third dimension through the essential z-component of velocity and the z-dependence of the magnetic field, $\mathbf{B}(x, y, z, t) = e^{ikz}\mathbf{b}(x, y, t)$. We turn now to the investigation of the likely magnetic structure in a perfect conductor as revealed by the geometry of the chaotic web.

8.2.1 The Flows Defined

In the search for fast dynamo action, there is considerable freedom in the choice of the unperturbed steady flow, as well as of the form of the perturbation. We shall focus here on two flows. The first is the nearly-integrable counterpart of the MW flow of Otani (1988, 1993); see §2.2. The symbol MW± indicates Beltrami waves of equal amplitude and having helicity of the same (+) or opposite (−) sign. The second is the CP flow of Galloway & Proctor (1992). The latter flow has been studied in its nearly-integrable form by Ponty et $al.$ (1993, 1995). Recall that MW± is defined by

$$\mathbf{u}(x, y, t) = \alpha(t)(0, \sin x, \cos x) + \beta(t)(\sin y, 0, \mp \cos y), \tag{8.2.1}$$

where, after a translation of time (cf. 2.2.1), we have $\alpha(t) = 2\sin^2 t$ and $\beta(t) = 2\cos^2 t$. To introduce flows near integrability, we replace $2\cos^2 t$ by $1 + \varepsilon \cos \omega t$ and $2\sin^2 t$ by $1 - \varepsilon \cos \omega t$, yielding for MW+

$$\mathbf{u} = (\psi_y, -\psi_x, \psi) \tag{8.2.2}$$

where

$$\psi = \cos x - \cos y + \varepsilon(\cos x + \cos y)\cos \omega t. \tag{8.2.3}$$

This is the wave form of the flow we shall denote by MW+ε, defined in analogy with the Roberts cell of §5.1. The cell form is obtained as in the Roberts flow, by the change of variables above (5.1.2). The cell form of MW+ε is then

$$\mathbf{u} = (\psi_y, -\psi_x, \sqrt{2}\,\psi), \quad \psi = \sin x \sin y + \varepsilon \cos x \cos y \cos \omega t. \tag{8.2.4}$$

We recall that the velocities of the wave form are larger than those of the cell form by a factor $\sqrt{2}$. At the same time the size of the helical cells is reduced by a factor $1/\sqrt{2}$ in the cell form. From the induction equation it then follows that the growth rate for a magnetic field of the form $\mathbf{B}(x, y, z, t) = e^{ikz}\mathbf{b}(x, y, t)$, computed in the wave form at a magnetic Reynolds number R, must equal that computed in the corresponding cell form at Reynolds number $2R$ and wavenumber $\sqrt{2}k$. We shall make use of the correspondence below. Numerical calculations are often facilitated in the wave form but we prefer the cell form for analysis.

Thus the cell form of the flow MW$+\varepsilon$ is defined by (8.2.4). The analogous cell form with Beltrami waves of opposite helicity, MW$-\varepsilon$, is defined by (8.2.4) but the z-component is no longer $w = \sqrt{2}\psi$, but rather

$$w = \sqrt{2}\,(\cos x \cos y + \varepsilon \sin x \sin y \cos \omega t). \tag{8.2.5}$$

Note that a more general form of these flows can be used, in which $\alpha = A(1 + \varepsilon \cos \omega t)$ and $\beta = B(1 + \varepsilon \cos \omega t)$. The symmetric case $A = B = 1$ is natural since this makes the two waves equivalent in action. If $A \neq B$ the unperturbed flow is the cat's-eye flow, up to a translation, with separatrices bounding both sides of the channels; see §5.5.3 and Fig. 5.4. The resulting Melnikov theory is more involved, but there is no indication that any new features appear.

For the CP flow we focus on the nearly-integrable variant studied by Ponty et al. (1993, 1995). The more general case of distinct wave amplitudes is treated in Ponty et al. (1995). The wave form of the flow we shall denote by CP$\pm\varepsilon$ is

$$\mathbf{u} = (\cos(y + \varepsilon \sin \omega t), \sin(x + \varepsilon \cos \omega t), w), \tag{8.2.6a}$$

$$w = \sin(y + \varepsilon \sin \omega t) \pm \cos(x + \varepsilon \cos \omega t), \tag{8.2.6b}$$

the $+$ sign again corresponding to waves of the same helicity (the case studied by Ponty et al. 1993, 1995). For applying the Melnikov theory, we first introduce the change of variables

$$\xi = x + \varepsilon \cos \omega t, \qquad \eta = y + \pi/2 + \varepsilon \sin \omega t. \tag{8.2.7}$$

Relative to coordinates (ξ, η, z), the horizontal fluid velocity is (u', v') with

$$u' = \sin \eta - \varepsilon \omega \sin \omega t, \quad v' = \sin \xi + \varepsilon \omega \cos \omega t. \tag{8.2.8}$$

The corresponding cell form of the CP$\pm\varepsilon$ flow is then obtained by the same change of variables as before. It is defined by the streamfunction

$$\psi = \sin x \sin y + \varepsilon(\omega/\sqrt{2})(-x \sin(\omega t + \pi/4) + y \cos(\omega t + \pi/4)), \tag{8.2.9a}$$

and the vertical velocity

$$w = \begin{cases} \sqrt{2} \sin x \sin y, & (\text{CP}+\varepsilon), \\ -\sqrt{2} \cos x \cos y, & (\text{CP}-\varepsilon). \end{cases} \tag{8.2.9b}$$

This flow therefore stands in the same relation to the cell flow with mean motion discussed in §5.5.3 as the MW$\pm\varepsilon$ flow stands relative to the cat's-eye flow, the angle of the mean motion here rotating with angular velocity ω.

The Melnikov function for the flow (8.2.9) and the heteroclinic connection from $(0,0)$ to $(\pi,0)$ is easily computed as in (8.1.16), and there results (Ponty et al. 1993, 1995):

$$M(t_0) = 2^{-1/2}\,\varepsilon\omega\pi\,\mathrm{sech}(\omega\pi/2)\sin(\omega t_0 + \pi/4)$$
$$\equiv \varepsilon G(\omega)\sin(\omega t_0 + \pi/4). \tag{8.2.10}$$

We show the function $G(\omega)$ in Fig. 8.2. Figure 8.3 contains some representative Poincaré sections, which confirm the role of $G(\omega)$ as an indicator of the size of the web.

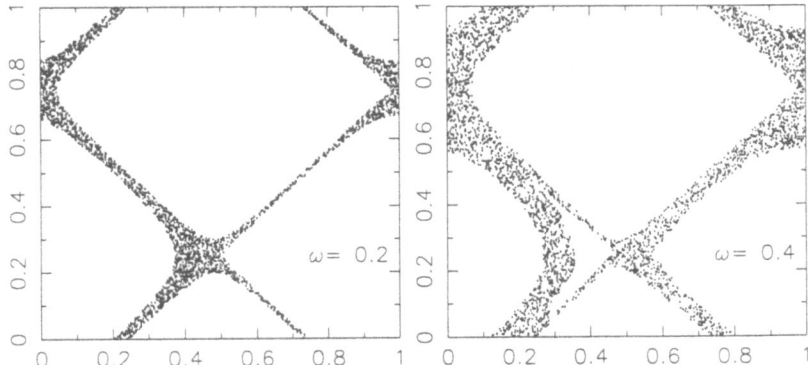

Fig. 8.3 Poincaré sections in time for the perturbed circularly polarized flow, CP+ε with $\varepsilon = 0.1$ and $\omega = 0.2$ (left), $\omega = 0.4$ (right). (From Ponty et al. (1993), reproduced with permission of Cambridge University Press.)

8.2.2 Asymptotics of the Manifolds

For the flows just presented fast dynamo action can be assumed to occur only within the chaotic web. Although, as we have suggested in §8.1.1, the geometry of the stable and unstable manifolds emanating from the critical points will partially determine the structure of the magnetic field, we will need more detailed asymptotic structure than is reflected in the Melnikov function. Also, to compute the phase of the magnetic field we must have an expression for the vertical drift of a fluid particle lying on a manifold. For these we must examine more closely the local Lagrangian structure. In the present section we consider the integration of the horizontal flow for the asymptotic expansion in ε of the Lagrangian coordinates of W^u and W^s, and then the calculation of vertical motion. In these calculations we shall always use the cell-based form of the equations, scaled so that the unperturbed stream function is $\sin x \sin y$.

From (8.1.11, 12) we have that, to the order retained in (8.1.11), $W^u(A)$ is given by

$$x(t, t_0) = \sin^{-1}(\text{sech}(t - t_0)), \tag{8.2.11a}$$

$$y(t, t_0) = \varepsilon \cosh(t - t_0) \int_{-\infty}^{t} \sin x(s, t_0)\, v_1(x(s, t_0), 0, s)\, ds. \tag{8.2.11b}$$

Here t_0 is the arbitrary constant which equals the time of arrival of particles at $x = \pi/2$, and v_1 is the $O(\varepsilon)$ term of v. The branch of $\sin^{-1}$ is such that $x \to 0$ as $t \to -\infty$, $x \to \pi$ as $t \to +\infty$. The endpoint of integration in (8.2.11b) is dictated by the vanishing of y as $t \to -\infty$. For large $t - t_0$ we have from (8.1.14) and (8.2.11b),

$$y = M(t_0)\cosh(t - t_0) \tag{8.2.12}$$

to this order, with M given by (8.1.17) for MW$\pm\varepsilon$ (for which $v_1 = \cos\omega t \sin x$) and by (8.2.10) for CP$\pm\varepsilon$ (with $v_1 = (\omega/\sqrt{2})\sin(\omega t + \pi/4)$). This expression describes the structure of the folds of $W^u(A)$ as they are stretched out by the hyperbolic flow at B. Field frozen into this flow will thus be folded and stretched as it is carried into the hyperbolic points. As the folds are carried away from B, they are simultaneously compressed against the unstable manifold emanating from B (see Fig. 8.1). The dynamo action, should it occur, could be the result of constructive folding through the phase factors developed by the vertical flow. Thus the asymptotics of dynamo action will depend crucially on the nature of this vertical flow, the underlying horizontal Melnikov structure being much the same for any unsteady perturbation.

We consider now the calculation of vertical drift. For MW$+\varepsilon$ the third Lagrangian equation is, from (8.2.4, 11),

$$\frac{1}{\sqrt{2}}\frac{dz}{dt} = \varepsilon \int_{-\infty}^{t} \frac{\cos\omega s}{\cosh^2(s - t_0)}\, ds - \varepsilon \cos\omega t \tanh(t - t_0) + O(\varepsilon^2)$$
$$= W(t, t_0)/\sqrt{2} + O(\varepsilon^2). \tag{8.2.13}$$

Integrating (8.2.13) over some given interval $[t_1, t_2]$ we have

$$\Delta z \equiv \int_{t_1}^{t_2} W(t, t_0)\, dt. \tag{8.2.14}$$

The second term in (8.2.13) may be integrated by parts, the term which integrates out yielding the contribution

$$\Delta z_1 = -\frac{\varepsilon\sqrt{2}}{\omega} \sin\omega t \tanh(t - t_0)\Big|_{t_1}^{t_2}. \tag{8.2.15a}$$

The remaining term can be combined with the double integral coming from the first term in (8.2.13). If the order of integration is reversed, one integration is possible. The resulting single integrals yield the second contribution to Δz:

$$\Delta z_2 = \frac{\varepsilon\sqrt{2}}{\omega} \int_{t_1}^{t_2} \frac{\sin\omega t}{\cosh^2(t-t_0)}\, dt + \varepsilon\sqrt{2}\,(t_2 - t_1) \int_{-\infty}^{t_1} \frac{\cos\omega t}{\cosh^2(t-t_0)}\, dt$$

$$+ \varepsilon\sqrt{2} \int_{t_1}^{t_2} \frac{(t_2 - t)\cos\omega t}{\cosh^2(t-t_0)}\, dt. \tag{8.2.15b}$$

We now impose the conditions $t_2 - t_0 \gg 1$ and $t_0 - t_1 \gg 1$. This ordering will follow from the fact that t_1 and t_2 will refer to times when the fluid particle is near the stagnation points of the flow, where it lingers for a long time compared to the nominal 'turnover' time of the cell.

Applying these conditions we may evaluate Δz_2 approximately as

$$\Delta z_2 \approx (\varepsilon\pi^2/\sqrt{2})\omega \cosh(\omega\pi/2)\, \mathrm{cosech}^2(\omega\pi/2) \sin\omega t_0$$

$$+ \varepsilon\sqrt{2}\pi(t_2 - t_0)\omega\, \mathrm{cosech}(\omega\pi/2)\cos\omega t_0, \quad (\mathrm{MW}+\varepsilon). \tag{8.2.15c}$$

We note that when $\omega \gg 1$ the largest contribution to Δz is Δz_1:

$$\Delta z = \Delta z_1 + \Delta z_2 \approx \Delta z_1 \approx \frac{-\varepsilon\sqrt{2}}{\omega}(\sin\omega t_1 + \sin\omega t_2). \tag{8.2.16a}$$

When ω is large the phase of the magnetic field is then dominated by a modulated oscillation, of amplitude ε/ω, as a function of t_2. Also the phase changes are small, since $w = \sqrt{2}\psi$ for MW$+\varepsilon$ and is therefore $O(\varepsilon)$ in the separatrix. For MW$-\varepsilon$ we see from (8.2.5) that $dz/dt = \sqrt{2}\cos x + O(\varepsilon^2)$ and therefore

$$\Delta z \approx \sqrt{2}\log\left(\frac{\cosh(t_2 - t_0)}{\cosh(t_1 - t_0)}\right) \approx \sqrt{2}(t_1 + t_2 - 2t_0), \tag{8.2.16b}$$

for $t_2 - t_0 \gg 1$ and $t_0 - t_1 \gg 1$. This linear drift with amplitude $O(1)$ occurs because w is $O(1)$ and of opposite signs near the two critical points A and B.

For the CP$\pm\varepsilon$ flows analogous results are obtained. We focus on CP$+\varepsilon$. In this case there is no leading contribution of the form (8.2.15a) and the approximate form for the vertical drift may be obtained from a term corresponding to the last term on the right of (8.2.15b),

$$\Delta z \approx \varepsilon\omega\pi(t_2 - t_0)\,\mathrm{sech}(\omega\pi/2)\sin(\omega t_0 + \pi/4) \tag{8.2.17}$$

$$- (\varepsilon\omega\pi^2/2)\sinh(\omega\pi/2)\,\mathrm{sech}^2(\omega\pi/2)\cos(\omega t_0 + \pi/4), \quad (\mathrm{CP}+\varepsilon).$$

We shall show in §8.4.1 that the approximation (8.2.16a) for MW$+\varepsilon$ and the expression (8.2.16b) for MW$-\varepsilon$ both fail to give dynamo action in the model to be developed in §8.3. This means that Δz_2 as given by (8.2.15c) is the only possible source of phase shifting yielding dynamo action for MW$+\varepsilon$ at this level of approximation. We shall make use of (8.2.15c) for MW$+\varepsilon$ and (8.2.17) for CP$+\varepsilon$ in developing our model.

Table 8.1 Symmetries of MW$\pm\varepsilon$.

Operator	Image of (x, y, t)	ψ	$w\ (+)$	$w\ (-)$
I	(x, y, t)	$+1$	$+1$	$+1$
R	$(\pi - y, x, t + \pi/\omega)$	$+1$	$+1$	-1
R^2	$(-x, -y, t)$	$+1$	$+1$	$+1$
R^3	$(y, \pi - x, t + \pi/\omega)$	$+1$	$+1$	-1
S	$(\pi + y, x, t)$	-1	-1	-1
RS	$(-x, y, t + \pi/\omega)$	-1	-1	$+1$
$R^2 S$	$(\pi - y, -x, t)$	-1	-1	-1
$R^3 S$	$(x, -y, t + \pi/\omega)$	-1	-1	$+1$

8.2.3 Symmetry and the Cell to Cell Connections

The dynamo mechanism just described treats the separatrix region as analogous to a magnetic boundary layer. Just as the advection–diffusion problem for the magnetic field in the various cellular flows considered in §5.5 involves connections between the boundary layers in adjacent cells, the calculation of the magnetic structure in the separatrix region will use relations between cells derived from the symmetry of the unperturbed flow and the time-periodic perturbation. The present discussion parallels that of the symmetry of pulsed Beltrami waves in §10.2.1.

We shall consider first the flows MW$\pm\varepsilon$, defined by (8.2.4) and (8.2.5), for which the streamfunction is spatially periodic. We consider only magnetic fields with the same space-periodicity, i.e., fields on the two-torus given by identifying $(x, y) \sim (x + \pi, y + \pi) \sim (x - \pi, y + \pi)$. Also instead of considering growing fields, it is convenient to use the Floquet representation to write $\mathbf{b}(\mathbf{x}_{\mathrm{H}}, t) = \mathbf{c}(\mathbf{x}_{\mathrm{H}}, t)e^{pt}$ where $\mathbf{c}$ is periodic with period $2\pi/\omega$ in time. The symmetries of fields of this type may be expressed in terms of the two transformations R and S defined by

$$R: \quad (x, y, t) \to (\pi - y, x, t + \pi/\omega), \tag{8.2.18a}$$

$$S: \quad (x, y, t) \to (\pi + y, x, t). \tag{8.2.18b}$$

The horizontal velocity field $(u, v) = \mathbf{u}_{\mathrm{H}}$ transforms under R according to

$$R\mathbf{u}_{\mathrm{H}}(R(\mathbf{x}, t)) = \frac{\partial R\mathbf{x}}{\partial \mathbf{x}} \cdot \mathbf{u}_{\mathrm{H}}(\mathbf{x}, t), \tag{8.2.19}$$

and similarly for S. R is a rotation of $\pi/2$ about the center of the cell with a time shift, while RS is a reflection in the y-axis and a time shift. We may then compile the following table 8.1 of symmetries which leave $\mathbf{u}_{\mathrm{H}}$ invariant and multiply w by ± 1. The indicated symmetry operations form a group equivalent to the dihedral group D_4, i.e., the symmetry group of the square; see, for example, Hamermesh (1962)

Here the change of sign of ψ occurs by a reversal of orientation of the coordinate system, ψ being a pseudo-scalar, and is allowed. The reversal of sign of w corresponds to changing z to minus z, or to taking the complex conjugate of the field when we use the usual substitution $\mathbf{B}(\mathbf{x}, t) = e^{ikz}\mathbf{b}(\mathbf{x}_\mathrm{H}, t)$. However this also requires taking the complex conjugate of the growth rate p of an eigenfunction. It follows that we can only use w-reversing symmetries, for example to superpose fields as in (8.2.20) below, if p is real. In the absence of any guarantee of the latter condition, we disallow these symmetries. This leaves I, R, R^2 and R^3 in table 8.1 for MW$+\varepsilon$, and I, R^2, RS and $R^3 S$ for MW$-\varepsilon$. These are subgroups of D_4: a cyclic group and the four-group. Focusing first on MW$+\varepsilon$, we show the manifold structure of a typical cell in Fig. 8.4. Here we indicate how the magnetic structure on different sections of the separatrix web matches up at times t and $t + \pi/\omega$.

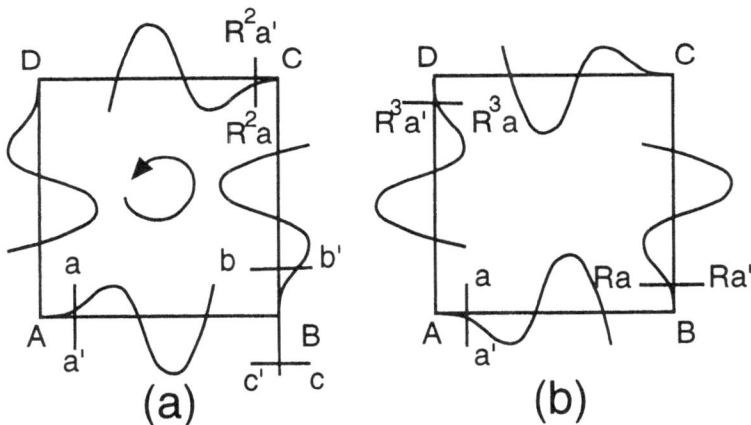

Fig. 8.4 Manifold structure for MW$+\varepsilon$ at times (a) t, and (b) $t + \pi/\omega$. We indicate the mapping of half-sections a and a' of the separatrix.

Using the representation theory of finite groups (see Hamermesh 1962), we may then find certain fields with simple symmetry properties, namely which are eigenfunctions of R. Given any admissible field $\mathbf{c}$, we may define the following fields $\mathbf{c}_k$ for $k = 1, 2, 3, 4$:

$$\mathbf{c}_1 = (I - iR - R^2 + iR^3)\mathbf{c}, \tag{8.2.20a}$$
$$\mathbf{c}_2 = (I - R + R^2 - R^3)\mathbf{c}, \tag{8.2.20b}$$
$$\mathbf{c}_3 = (I + iR - R^2 - iR^3)\mathbf{c}, \tag{8.2.20c}$$
$$\mathbf{c}_4 = (I + R + R^2 + R^3)\mathbf{c}. \tag{8.2.20d}$$

We then have $R\mathbf{c}_k = i^k \mathbf{c}_k$. It is thus sufficient to consider eigenfunctions that satisfy the condition

$$Rc = \rho c, \tag{8.2.21}$$

where ρ is any fourth root of unity.[65]

We introduce the functions c_a, c_b, $c_{c'}$ and $c_{a'}$ at the sections indicated in Fig. 8.4(a). These are defined by c restricted to the indicated lines orthogonal to the unperturbed separatrix. In our use of these functions we shall adopt the boundary-layer approximation that field is there approximately parallel to the unperturbed separatrix, and so the functions are determined by a distribution of flux sheets. Then, if c satisfies (8.2.21), we have (suppressing the spatial dependence)

$$c_a(t) = \rho R^3 c_a(t) = \rho c_b(t + \pi/\omega), \tag{8.2.22a}$$

$$c_{a'}(t) = \rho^{-1} R c_{a'}(t) = \rho^{-1} c_{c'}(t + \pi/\omega). \tag{8.2.22b}$$

For example in (8.2.22a), measuring the field c at section b, time $t + \pi/\omega$, is the same as measuring the field $R^3 c$ at section a, time t. Solution of the eigenvalue problem then requires that we construct c_b and $c_{c'}$ from c_a and $c_{a'}$, one example being the approximate procedure of §8.3.

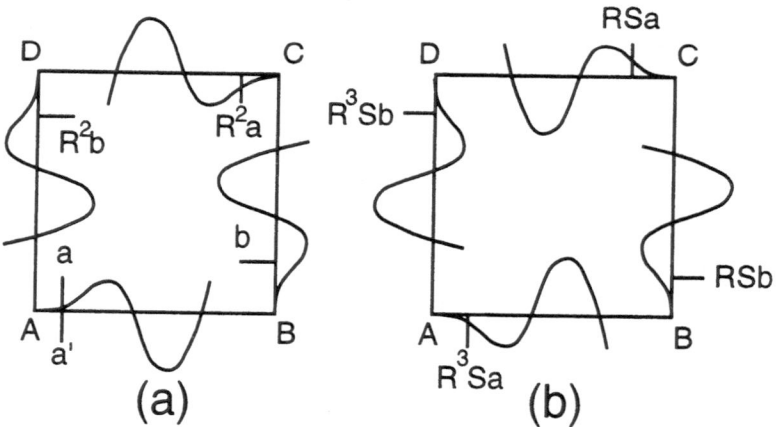

Fig. 8.5 Manifold structure for MW$-\varepsilon$ at times (a) t, and (b) $t + \pi/\omega$. We indicate the mapping of half-sections a and b of the separatrix.

For MW$-\varepsilon$ the construction is similar but somewhat more complicated because of the change of sign of w at adjacent hyperbolic points. The symmetry of the manifolds is shown in Fig. 8.5. It is convenient in this case to define $P = R^2$ and $Q = RS$. The fields corresponding to (8.2.20) are now

---

[65] The boundary-layer theory of §5.5.1 constructs an eigenfunction with the symmetry of c_2 for a steady flow.

$$\mathbf{c}_1 = (I + P + Q + PQ)\mathbf{c}, \tag{8.2.23a}$$
$$\mathbf{c}_2 = (I + P - Q - PQ)\mathbf{c}, \tag{8.2.23b}$$
$$\mathbf{c}_3 = (I - P + Q - PQ)\mathbf{c}, \tag{8.2.23c}$$
$$\mathbf{c}_4 = (I - P - Q + PQ)\mathbf{c}. \tag{8.2.23d}$$

Thus $P\mathbf{c}_k = \alpha_k \mathbf{c}_k$ and $Q\mathbf{c}_k = \beta_k \mathbf{c}_k$ where $(\alpha_1, \alpha_2, \alpha_3, \alpha_4) = (1, 1, -1, -1)$ and $(\beta_1, \beta_2, \beta_3, \beta_4) = (1, -1, 1, -1)$. The connection (8.2.21) is thus replaced by

$$P\mathbf{c} = \alpha\mathbf{c}, \quad Q\mathbf{c} = \beta\mathbf{c}, \quad \alpha^2 = \beta^2 = 1. \tag{8.2.24}$$

To obtain expressions analogous to (8.2.22) it is necessary to consider four critical points rather than two. The more involved expressions are postponed until they are used in §8.3.3.

For the circularly polarized flows, the preceding discussion may be repeated but there are some essential differences. Since from (8.2.9a) the streamfunction now has a secular spatial growth, the symmetry refers only to the velocity field. Also the symmetry transformation must preserve orientation, since the temporal perturbation is circularly polarized, and this eliminates the last four rows of table 8.1. Finally, the rotation of the mean flow is such that R must now be defined with a time shift of $\pi/2\omega$ instead of π/ω. The surviving entries of table 8.1 are shown in table 8.2.

Table 8.2 Symmetries of CP$\pm\varepsilon$.

Operator	Image of (x, y, t)	w $(+)$	w $(-)$
I	(x, y, t)	$+1$	$+1$
R	$(\pi - y, x, t + \pi/2\omega)$	$+1$	-1
R^2	$(-x, -y, t + \pi/\omega)$	$+1$	$+1$
R^3	$(y, \pi - x, t + 3\pi/2\omega)$	$+1$	-1

The connection equations for CP$+\varepsilon$ are then again (8.2.21) with the new R, and (8.2.22) with the time shift of $\pi/2\omega$ instead of π/ω. For CP$-\varepsilon$ we may define $\mathbf{c}_1 = (I + R^2)\mathbf{c}$, $\mathbf{c}_2 = (I - R^2)\mathbf{c}$, so that the connection equation becomes $R^2\mathbf{c} = \rho\mathbf{c}$ for $\rho^2 = 1$. The resulting eigenvalue problem has the same symmetries as the slow dynamo problem of §5.5.2, and CP$-\varepsilon$ will not be considered further.

8.2.4 Numerical Studies

We are aware of only one published study of fast dynamo action near integrability, by Ponty *et al.* (1993, 1995). The flow is CP$+\varepsilon$ with $\varepsilon = 0.1$. The frequency ω is varied from 0 to 3. The calculated growth rates are shown in Fig. 8.6, where we also include the results of calculations to be described in §8.4. Additional data for $R = 5000$ and 10000 may be found in Ponty *et al.* (1995).

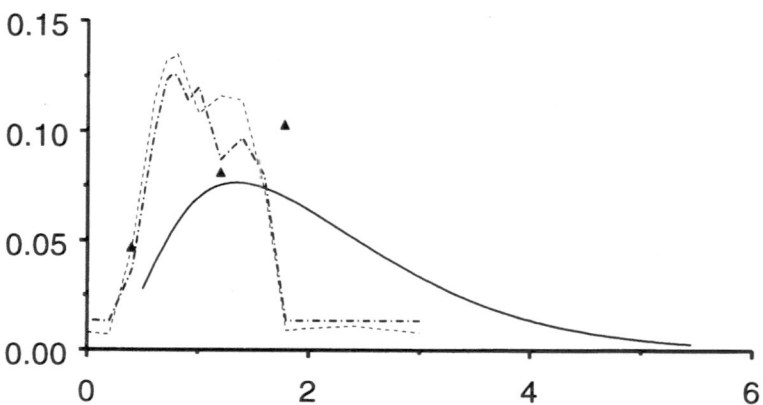

Fig. 8.6 Growth rates for the wave form of CP+ε as a function of ω with $\varepsilon = 0.1$, $k = 0.57$ and $R = 1000$ (dot–dash), and 2000 (dash), from Ponty *et al.* (1993). The triangles are points obtained in the limit of large period, in the perfectly conducting model of §8.4.2. The solid curve is the small-period result given in §8.4.3.

Of special interest here is the clear indication of dynamo activity relatively insensitive to R in the interval between $\omega = 0.2$ and $\omega = 1.8$, approximately. The dynamo action outside this interval is viewed as 'slow' by Ponty *et al.*, while the band of activity within it is termed 'fast'. At these magnetic Reynolds numbers, the slow dynamo action swamps the fast component outside of this relatively narrow range of frequencies. Note that the growth rate decreases with increasing R in the slow regime. The structure of the magnetic field is observed to be quite different in the two regimes. In the case of slow dynamo action, the magnetic field concentrates in large eddies within the non-chaotic regions and surrounding the elliptic stagnation points within the cells; see Fig. 8.7. As R is increased, these eddies become more elongated and tend to circumscribe the elliptic points. This picture is moreover consistent with the known structure of slow helical dynamos, such as those of Ponomarenko type; see §5.4.1. On the other hand, when $\omega = 0.8$ and the dynamo appears to be fast, the magnetic field is confined to the chaotic separatrix region. Calculations of the Liapunov exponent as a function of ω indicate that in the range $1 < \omega < 3$ the exponent fluctuates with an average of about 0.15. There is thus little correlation with the Melnikov function $G(\omega)$ given in (8.2.10), or the fast dynamo growth rate. The conclusion is that the fast dynamo mechanism involves cancellation that is strongly frequency dependent.

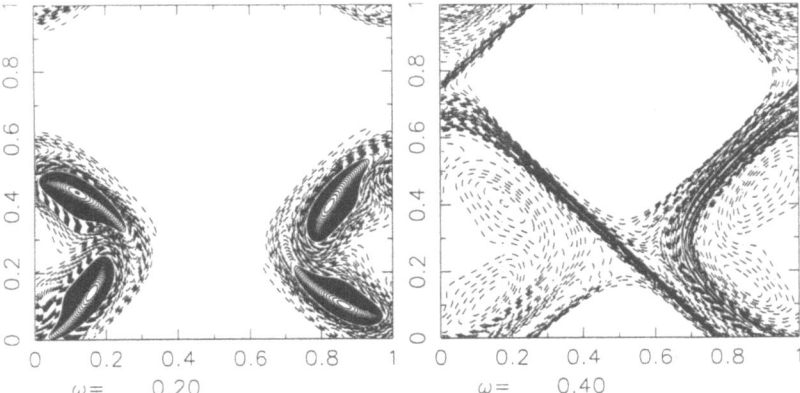

Fig. 8.7 Magnetic structure in CP+ε with $\varepsilon = 0.1$, $R = 2000$, and $\omega = 0.2$ (left), 0.4 (right). (From Ponty *et al.* (1993), reproduced with permission of Cambridge University Press.)

8.3 A Model for Perfect Dynamo Action

We describe a perfectly conducting model of fast dynamo action for these nearly-integrable flows, based upon the flux sheet approximation suggested by the analysis of §8.1.1 (Childress 1993a). The idea is to assign to the unstable manifold $W^u(A)$, emerging from a given hyperbolic stagnation point A, a *total flux* of the field distributed within it. The flux, which for the field **B** we denote by ϕ, will be a function of time t, as well as the position on the manifold, determined by the parameter t_0 introduced above. For large time the manifold will have many lobes stretched out along the unstable manifold of a neighboring hyperbolic point, B, as shown in Fig. 8.8. In the nearly-integrable case, the multiply-folded structure occurs very close to B, and we may regard this layered, folded structure as being stretched out into the upper branch of $W^u(B)$. Since we wish to replace the magnetic structure by a flux sheet, we may identify the flux computed at B with a sum of oriented fluxes over all the lobes, taken at some convenient section of $W^u(B)$. We suggest this construction in Fig. 8.8.

Another way of viewing this construction is as a 'boundary layer' model of the separatrix region, one which does not attempt to account for the detailed magnetic structure within the region. In fact, in the chaotic zone, the tangle is extremely complex, any given manifold having portions near every other critical point within the web. The model simplifies this by accounting only for the development of lobes in one step (and neglecting 'lobes consisting of layered lobes', etc.).

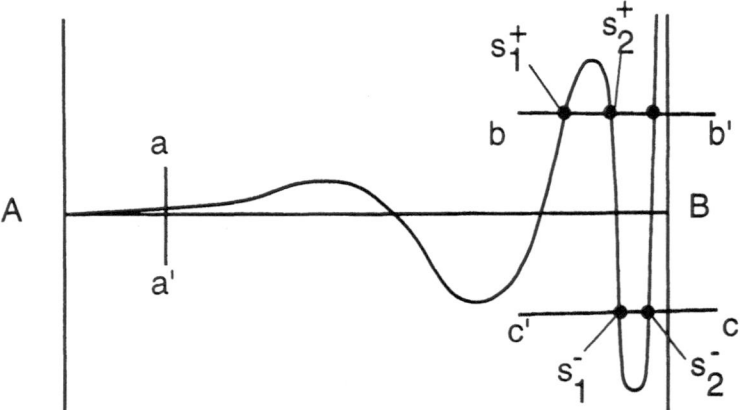

Fig. 8.8 Approximate geometry of the flux in the separatrix region.

8.3.1 The Mapping of Flux

We shall illustrate the model with the flow MW+ε. The manifold $W^u(A)$ given by (8.1.17, 2.12) then has the approximate form when $t - t_0 \gg 1$,

$$y \approx \varepsilon \, \frac{\pi\omega}{\sinh \pi\omega/2} \, \cosh(t - t_0) \cos \omega t_0. \tag{8.3.1}$$

Now suppose that we fix t once and for all, and consider any point on the manifold as determined by t_0. Suppose, in particular, that such a point lies on the section b of Fig. 8.8, i.e., the line segment $x \leq \pi$, $y = \delta \ll \pi$. We shall require also that $\delta \gg \varepsilon$. Although the exact dependence on ε is otherwise arbitrary, a suitable choice is $\delta = \sqrt{\varepsilon}$. In fact a stronger condition on δ may be necessary, as discussed in §8.3.2 below. Suppose that the point lies on b when $t_0 = s_k^+$, where $k = 1, 2, \ldots$ labels the intersection points of $W^u(A)$ with the section b. Now this point coincided with the section $a'a$ near the critical point A, i.e., the line $x = \delta$, at a time t_k where

$$\sin \delta \approx \delta = 1/\cosh (t_k - t_0) \approx 2 \exp(t_k - t_0), \tag{8.3.2}$$

or

$$t_k = s_k^+ + \log \delta/2. \tag{8.3.3}$$

We may thus relate the total flux $\phi_{aa'}$ through the union of a and a' at previous times t_k to the flux of **B** at b at time t. Taking flux as positive in the direction of the unperturbed flow, we see that then

$$\phi_b(t) = \sum_{k=1}^{\infty} (-1)^{k-1} e^{ik\Delta z_k^+} \, \phi_{aa'}(s_k^+ + \log \delta/2). \tag{8.3.4}$$

Note that we include here the crucial phase factors, determined by the vertical drift Δz_k^+ between times t_k and t of the particle given by $t_0 = s_k^+$.

Equation (8.3.4) is the key to our model. Given a similar expression for $\phi_{c'}$ in terms of $\phi_{aa'}$ (see Fig. 8.9 and (8.3.6b) below) we can use the connection formulas (8.2.22) to relate these fluxes to $\phi_{aa'}$, and obtain a single equation for the latter quantity.

We shall now make the assumption that the Floquet scaling for an eigen-function applies to our flux functions, i.e., flux is a periodic function of time with an exponential prefactor; then

$$\phi(t) = e^{pt}\Phi(t), \qquad \Phi(t + 2\pi/\omega) = \Phi(t), \tag{8.3.5}$$

and (8.3.4) becomes

$$e^{pt}\,\Phi_b(t) = \sum_{k=1}^{\infty}(-1)^{k-1}\,e^{ik\Delta z_k^+}\,e^{p(s_k^+ + \log \delta/2)}\,\Phi_{aa'}(s_k^+ + \log \delta/2). \tag{8.3.6a}$$

Similarly, from the intersection of lobes with the section c' in Fig. 8.8 we have

$$e^{pt}\,\Phi_{c'}(t) = \sum_{k=1}^{\infty}(-1)^{k-1}\,e^{ik\Delta z_k^-}\,e^{p(s_k^- + \log \delta/2)}\,\Phi_{aa'}(s_k^- + \log \delta/2), \tag{8.3.6b}$$

where s_k^- and Δz_k^- are calculated below. Using now (8.2.22) applied to the flux functions, we have the connection formulas

$$\Phi_a(t) = \rho\,\Phi_b(t + \pi/\omega), \qquad \Phi_{a'}(t) = \rho^{-1}\,\Phi_{c'}(t + \pi/\omega), \tag{8.3.6c}$$

and using $\Phi_{aa'} = \Phi_a + \Phi_{a'}$ we arrive at an equation for $\Phi_{aa'}(t)$:

$$e^{pt}\,\Phi_{aa'}(t - \pi/\omega) = \sum_{k=1}^{\infty}(-1)^{k-1}\Big[\rho e^{ik\Delta z_k^+}\,e^{p(s_k^+ + \log \delta/2)}\Phi_{aa'}(s_k^+ + \log \delta/2)$$
$$+ \rho^{-1}e^{ik\Delta z_k^-}\,e^{p(s_k^- + \log \delta/2)}\Phi_{aa'}(s_k^- + \log \delta/2)\Big]. \tag{8.3.7}$$

This equation expresses the basic premise of the model, that a sum of fluxes extruded into the manifold at earlier times defines the flux extruded at the current time.

To complete the formulation of the flux equation we must however supply the $\Delta z_k^\pm$ and the intersection parameters $s_k^\pm$. For the latter, we use the defining relation (cf. 8.3.1)

$$\pm\delta = \varepsilon\,\frac{\pi\omega}{\sinh \omega\pi/2}\,\cos \omega s_k^\pm\,\cosh (t - s_k^\pm). \tag{8.3.8}$$

To simplify this, we introduce a reference time $T \gg 1$, satisfying

$$\frac{2\delta}{\varepsilon\omega\pi}\,e^{-T}\,\sinh \omega\pi/2 = 1. \tag{8.3.9}$$

Introducing new variables τ, σ, f,

$$t = -\frac{2\pi}{\omega}\tau + T, \quad s = -\frac{2\pi}{\omega}\sigma, \quad \Phi_{aa'}(t) = f\Big(\frac{\omega}{2\pi}\big(-t + \log\delta/2\big)\Big), \quad (8.3.10)$$

we may rewrite (8.3.7) in the form

$$e^{2\pi p(\tau_0-\tau)/\omega}f(\tau - \tau_0 + 1/2) = \qquad\qquad\qquad (8.3.11)$$

$$\sum_{k=1}^{\infty}(-1)^{k-1}\Big[\rho e^{ik\Delta z_k^+}e^{-2\pi p\sigma_k^+/\omega}f(\sigma_k^+) + \rho^{-1}e^{ik\Delta z_k^-}e^{-2\pi p\sigma_k^-/\omega}f(\sigma_k^-)\Big].$$

Here $\sigma_k^{\pm}$ satisfies

$$\pm e^{2\pi\tau/\omega} = e^{2\pi\sigma_k^{\pm}/\omega}\cos(2\pi\sigma_k^{\pm}), \qquad\qquad\qquad (8.3.12)$$

from (8.3.8, 9), with $t - s_k^{\pm} \gg 1$, and

$$\tau_0 = \frac{\omega}{2\pi}(T - \log\delta/2) = \frac{\omega}{2\pi}\log\Big(\frac{4\sinh\omega\pi/2}{\varepsilon\omega\pi}\Big) \qquad\qquad (8.3.13)$$

from (8.3.9). Note that (8.3.11) does not involve the matching parameter δ, although generally we expect δ to occur in the drifts $\Delta z_k^{\pm}$. The function $f(\tau)$ is periodic with period one.

8.3.2 Conditions for Validity

Equations (8.3.11–13) define the flux equation for MW+ε. We must now ask under what conditions we expect the approximate calculation of the mapping of flux to be valid. There is some ambiguity in the form of the conditions since it is not clear how sensitive the growth rate is to the geometry of the hetero-clinic tangle. The approximation in question is the replacement of a layer of a large number of lobes by an equivalent sheet. The most stringent condition would hold that this identification can be reasonable only if the thickness of the layer, comprising a large number of lobes, is small compared to a typical lateral excursion of the unstable manifold at a distance of order δ from its point of emergence. In other words, we require that the layered structure appear thin relative to the overall manifold geometry, i.e., that it is precisely resolved even in the early phase of growth of the lobes.

If (8.2.11b) is integrated by parts and evaluated on the section $x = \delta$ we obtain a lateral excursion of $W^u(A)$ of order $\varepsilon/\omega\cosh(t-t_0) \approx \varepsilon\delta/\omega$. However, from (8.3.8) we see that the first intersection point with the section b of Fig. 8.8 will occur when

$$\delta = O\left[\varepsilon\pi\omega\cosh(t - t_0)/\sinh\omega\pi/2\right], \qquad\qquad (8.3.14)$$

and this occurs when $\tau - x \approx 1/\cosh(t - t_0)$. Thus this strong form of the condition we seek may be expressed

$$\varepsilon\delta/\omega \gg \varepsilon\pi\omega/\delta \sinh \omega\pi/2, \tag{8.3.15}$$

or

$$\pi\omega^2 / \sinh \omega\pi/2 \ll \delta^2. \tag{8.3.16}$$

Since $\delta \ll \pi$, (8.3.16) requires that ω be either small or large. Given the form of the Melnikov function (8.1.17), the *low frequency limit* of MW$+\varepsilon$ is an attractive one, and we shall take it up in §8.4.2.

An alternative, weaker condition is obtained if it is required only that the folded layer be thin compared to the largest length scale, i.e., the size of the cell. In this case the condition becomes $\varepsilon\omega \ll \delta \sinh \pi\omega/2$, which places no restriction on the frequency. Moreover it is consistent with the idea that δ be small compared to cell size, but not too small. This weak condition might be sufficient simply because, regardless of how the manifold emerges from the originating critical point, the growth properties of the dynamo are dominated by the later stretching out of the lobes. Numerical results seem to indicate that the weak form of the condition is sufficient (although the strong form was invoked by Childress 1993a). We shall explore both large-ω and small-ω approximations in §8.4, but when obtaining numbers will require only that the parameters satisfy the weak form of the condition.

8.3.3 Remarks on MW$-\varepsilon$

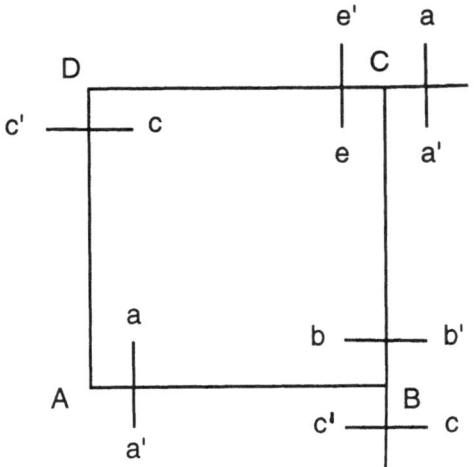

Fig. 8.9 Sections for the flow MW$-\varepsilon$.

We insert a few remarks here concerning the flux equation for MW$-\varepsilon$. We now need to refer to the expanded set of sections shown in Fig. 8.9. From the definition of the operators P and Q and the discussion following (8.2.23), we obtain the following connections between **c** defined at the corresponding sections shown in Fig. 8.9 (see also Fig. 8.5):

$$\mathbf{c}_a(t) = \beta \mathbf{c}_{e'}(t + \pi/\omega) = \alpha \mathbf{c}_e(t) = \alpha\beta \mathbf{c}_{a'}(t + \pi/\omega), \qquad (8.3.17a)$$

$$\mathbf{c}_b(t) = \beta \mathbf{c}_{b'}(t + \tau/\omega) = \alpha \mathbf{c}_c(t) = \alpha\beta \mathbf{c}_{c'}(t + \pi/\omega). \qquad (8.3.17b)$$

The general scheme involves two steps rather than the single one of MW+ε. The magnetic fluxes at a and a' determine those at b and c', while those at b and b' determine those at e and a'. Using (8.3.17) we can then derive a system of two equations for flux functions. The first step gives, in terms of flux functions ϕ,

$$\phi_b(t) = \sum_{k=1}^{\infty} (-1)^{k-1} e^{ik\Delta z_k^+} \phi_{aa'}(s_k^+ + \log \delta/2), \qquad (8.3.18a)$$

$$\phi_{c'}(t) = \sum_{k=1}^{\infty} (-1)^{k-1} e^{ik\Delta z_k^-} \phi_{aa'}(s_k^- + \log \delta/2), \qquad (8.3.18b)$$

$$\phi_{a'}(t) = \sum_{k=1}^{\infty} (-1)^{k-1} e^{-ik\Delta z_k^+} \phi_{bb'}(s_k^+ + \log \delta/2), \qquad (8.3.18c)$$

$$\phi_e(t) = \sum_{k=1}^{\infty} (-1)^{k-1} e^{-ik\Delta z_k^-} \phi_{bb'}(s_k^- + \log \delta/2), \qquad (8.3.18d)$$

(see Figs. 8.5, 9). Note the conjugation of the vertical phase shift occurring in the last two equations, due to the change of sign of w at C relative to B. Using (8.3.17) to derive a system for Φ_b and $\Phi_{a'}$ (for example) given by (8.3.5), we use only (8.3.18a, c) to obtain

$$e^{pt}\Phi_b(t) = \sum_{k=1}^{\infty} (-1)^{k-1} e^{ik\Delta z_k^+} e^{p(s_k^+ + \log \delta/2)}$$
$$\times \left[\Phi_{a'}(s_k^+ + \log \delta/2) + \alpha\beta \Phi_{a'}(\pi/\omega + s_k^+ + \log \delta/2) \right], \qquad (8.3.19a)$$

$$e^{pt}\Phi_{a'}(t) = \sum_{k=1}^{\infty} (-1)^{k-1} e^{-ik\Delta z_k^+} e^{p(s_k^+ + \log \delta/2)}$$
$$\times \left[\Phi_b(s_k^+ + \log \delta/2) + \beta\Phi_b(\pi/\omega + s_k^+ + \log \delta/2) \right]. \qquad (8.3.19b)$$

This system now replaces (8.3.7).

8.4 Analysis

The model defined by (8.3.11–13) is a complicated linear delay equation for a periodic function $f(\tau)$ of unit period. Little is known of a general nature about its solutions. In the next section we give one anti-dynamo result, which indicates the source of possible dynamo action. We shall then discuss various limiting cases where some analysis can be done.

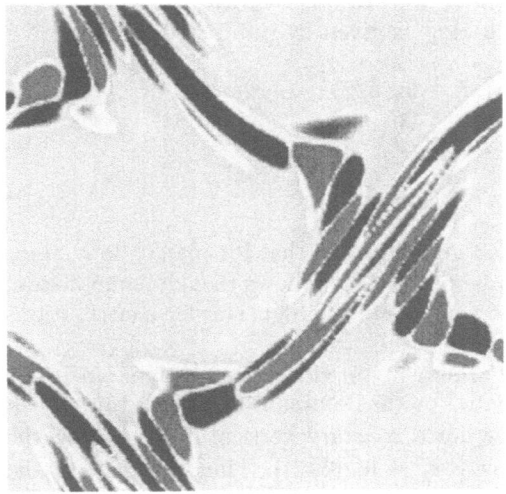

Fig. 8.10 The dominant eigenfunction for MW$-\varepsilon$ with $\varepsilon = 1$ and $R = 1024$. The region near the critical point in the middle right shows alternating bands of positive and negative field (black and dark grey); field near zero is light. Along the unstable manifold sloping downward to the left the bands move apart and tend to alternate *along* the separatrix rather than across it. This is the effect of phase shifting by the vertical motion. The structure indicates an eigenfunction with $\mathrm{Im}\, p \neq 0$.

We shall call attention to the similarity between the model, which requires the evaluation of net flux from an infinite sum of contributions from an oriented manifold, and the methods for estimating growth rates based upon Fredholm determinants (§6.3) and the asymptotic Lefschetz number (§6.2). In fact the intersection points determined by (8.3.12) are periodic points of a time-one Poincaré map provided the delay τ_0 in (8.3.11) takes on suitable values. This is a point we shall exploit in §8.4.2 in the limit of zero frequency. We remark that the underlying premise of the model, that magnetic flux tends to be confined to thin layers, can be tested numerically. Even at modest $R \sim 500$–1000, these manifolds are found to carry the largest field. We show in Fig. 8.10 an example for the MW$-\varepsilon$ flow.

8.4.1 A Non-dynamo Result for MW+ε

We shall argue, on the basis of the model described in §8.3, that the contribution Δz_1 to the vertical drift of the flow MW+ε, which is the dominant contribution for large ω (8.2.16a), cannot produce fast dynamo action in the separatrix region. We observe first that the model can be applied to the nearly-integrable *non-dynamo* obtained by taking the horizontal flow to be identical to that of MW+ε, but the vertical velocity to be zero. From the discussion of §4.2.2 this is a non-dynamo as a planar flow. In this case $\Delta z_k^\pm = 0$ in (8.3.11). We now consider MW+ε for which $\Delta z_k^\pm$ is given by (cf. 8.2.16a)

$$\Delta z_k^\pm = -\varepsilon\sqrt{2}/\omega \left[\sin\omega t + \sin\omega(s_k^\pm + \log\delta/2) \right], \qquad (\omega \gg 1). \qquad (8.4.1)$$

It can then be verified that the substitution

$$f(\tau) = \exp\left[-i(\varepsilon k\sqrt{2}/\omega)\sin(2\pi\tau - \omega\log\delta/2)\right]g(\tau) \qquad (8.4.2)$$

in (8.3.11) produces an equation for g identical to that for planar flow. Since $|f| = |g|$, the real part of p cannot be positive. It follows that dynamo action, should it occur in this model, must come from the higher-order contributions to vertical drift, exponentially small in ω for large ω.

For the flow MW−ε a related argument provides an anti-dynamo result when the vertical drift is approximated by the dominant term (8.2.16b). Then $\Delta z_k^\pm \approx \sqrt{2}(t - s_k^\pm + \log(\delta/2))$. Now for a *constant* vertical flow $w = w_0$, the vertical drift is $w_0(t_2 - t_1) = w_0(t - s_k^\pm - \log(\delta/2))$. Thus the drift in the case of MW−ε is approximately that of a uniform vertical flow, up to a fixed additive constant. Since the origin of z can be of no consequence, we may use the fact that dynamo action is not affected by the addition of a uniform vertical velocity to conclude that (8.2.16b) also fails to produce dynamo action

Note that these arguments require that the model be consistent in allowing no dynamo action in the planar case, a property we have not proved for (8.3.11). This can, however, be established in limiting cases, as we show in the next sections.

8.4.2 Approximations for Large Period

To understand how dynamo action in (8.3.11) is achieved it is useful to consider limiting cases. When the temporal period $2\pi/\omega$ is large, i.e., for sufficiently small frequency ω, the form of the equation simplifies considerably. The actual lateral excursions of $W^u(A)$ remain of order ε in this approximation, from (8.2.11b). This can be understood by considering a slowly varying cats-eye flow having small channel width; see §5.3.3. Particles are effectively confined to the channel region, which remains within an $O(\varepsilon)$ neighborhood of the separatrix of the unperturbed cells. The condition (8.3.16) is difficult to meet for small δ, although as we noted in §8.3.2 it is possible that a less stringent condition $\varepsilon \ll \delta \ll 1$ is sufficient. In any case we shall want to apply this approximation up to $\omega \approx 1$.

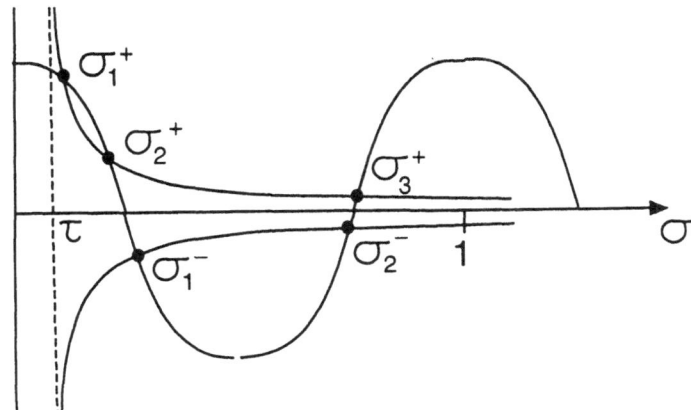

Fig. 8.11 Location of $\sigma_k^{\pm}$ in the low-frequency limit, for $0 \leq \tau \leq 1/4$. The oscillatory curve shows $\cos 2\pi\sigma$, while the exponential curves indicate $\pm \exp(2\pi(\tau - \sigma)/\omega)$.

We consider first MW+ε. When $\omega \ll 2\pi$ the approximate values of the quantities $\sigma_k^{\pm}$ from (8.3.12) are indicated in Fig. 8.11. The first root is essentially at the current τ value, while remaining roots are paired and located at $1/4$ and $3/4$ modulo 1 approximately. We thus extract the following approximate roots valid for $\omega \ll 2\pi$:

$$0 \leq \tau < 1/4 : \quad \begin{cases} \sigma_1^+ = \tau, & 4\sigma_n^+ = 2n - 3, \quad (n \geq 2), \\ & 4\sigma_n^- = 2n - 1, \quad (n \geq 1); \end{cases} \tag{8.4.3a}$$

$$1/4 \leq \tau < 3/4 : \quad \begin{cases} \sigma_1^- = \tau, & 4\sigma_n^- = 2n - 1, \quad (n \geq 2), \\ & 4\sigma_n^+ = 2n + 1, \quad (n \geq 1); \end{cases} \tag{8.4.3b}$$

$$3/4 \leq \tau < 1 : \quad \begin{cases} \sigma_1^+ = \tau, & 4\sigma_n^+ = 2n + 1, \quad (n \geq 2), \\ & 4\sigma_n^- = 2n + 3, \quad (n \geq 1). \end{cases} \tag{8.4.3c}$$

Using these expressions in the flux equation (8.3.11), we obtain, with $G(\tau) = e^{2\pi p(\tau_0 - \tau)/\omega} f(\tau - \tau_0 - 1/2)$,

$$G(\tau) = \begin{cases} \rho F(\tau, \tau) - (\rho - \rho^{-1}) \sum_{n=0}^{\infty} (-1)^n F((2n+1)/4, \tau), \\ \rho^{-1} F(\tau, \tau) - (\rho - \rho^{-1}) \sum_{n=1}^{\infty} (-1)^n F((2n+1)/4, \tau), \\ \rho F(\tau, \tau) - (\rho - \rho^{-1}) \sum_{n=2}^{\infty} (-1)^n F((2n+1)/4, \tau). \end{cases} \tag{8.4.4a}$$

Here the three lines refer to the intervals $0 \leq \tau < 1/4$, $1/4 \leq \tau < 3/4$, $3/4 \leq \tau < 1$, and

$$F(\sigma, \tau) = e^{ik\Delta z(\sigma, \tau)} e^{-2\pi p\sigma/\omega} f(\sigma). \tag{8.4.4b}$$

Since the contribution Δz_1 can be removed as discussed in §8.4.1 we may compute $\Delta z(\sigma, \tau)$ from Δz_2. We then have, from (8.2.15c),

$$
\begin{aligned}
\Delta z(\sigma, \tau) \approx &- \left(\varepsilon w \pi^2 / \sqrt{2} \right) \cosh \left(w \pi / 2 \right) \operatorname{cosech}^2 \left(w \pi / 2 \right) \sin 2\pi\sigma \\
&+ \varepsilon w \pi \sqrt{2} \left(T - \frac{2\pi}{w} (\tau - \sigma) \right) \operatorname{cosech}(w \pi / 2) \cos 2\pi\sigma.
\end{aligned}
\tag{8.4.5}
$$

Using this we may rewrite (8.4.4) in the form

$$
G(\tau) = \begin{cases}
\rho e^{ik \Delta z(\tau, \tau)} e^{-2\pi p\tau / w} f(\tau) \\
\quad -(\rho - \rho^{-1}) e^{ik \Delta z(1/4, \tau)} e^{-\pi p / 2w} f(1/4) K \\
\quad +(\rho - \rho^{-1}) e^{ik \Delta z(3/4, \tau)} e^{-3\pi p / 2w} f(3/4) K, \\
\rho^{-1} e^{ik \Delta z(\tau, \tau)} e^{-2\pi p\tau / w} f(\tau) \\
\quad +(\rho - \rho^{-1}) e^{ik \Delta z(3/4, \tau)} e^{-3\pi p / 2w} f(3/4) K \\
\quad -(\rho - \rho^{-1}) e^{ik \Delta z(5/4, \tau)} e^{-5\pi p / 2w} f(1/4) K, \\
\rho e^{ik \Delta z(-, \tau)} e^{-2\pi p\tau / w} f(\tau) \\
\quad -(\rho - \rho^{-1}) e^{ik \Delta z(5/4, \tau)} e^{-5\pi p / 2w} f(1/4) K \\
\quad +(\rho - \rho^{-1}) e^{ik \Delta z(7/4, \tau)} e^{-7\pi p / 2w} f(3/4) K.
\end{cases}
\tag{8.4.6a}
$$

Here

$$
K = \left[1 - \exp(-2\tau p / w) \right]^{-1},
\tag{8.4.6b}
$$

and we have assumed in summing the series in (8.4.4a) that $\operatorname{Re} p > 0$. It is not difficult to show that (8.4.6a) can be written

$$
e^{2\pi p\tau_0 / w} f(\tau - \tau_0 - 1/2) = f(\tau) g_1(\tau) + c_{1/4} g_2(\tau) + c_{3/4} g_3(\tau),
\tag{8.4.7}
$$

where $c_{1/4} = f(1/4)$ and $c_{3/4} = f(3/4)$. The g_i are periodic but discontinuous functions, with period 1, and $|g_1| = 1$.

Now the only way dynamo action can occur is to balance the decay at fixed time along the flux sheet when $\operatorname{Re} p > 0$, with the constructive reinforcement of folding. Thus a mode exhibiting dynamo action is not possible in the low frequency limit if we choose $\rho = \pm 1$, this being clear from the elimination of the summed terms in (8.4.6) which then occurs. Accordingly, we shall always take $\rho = i$ in the large period limit.

The equation (8.4.7) determining p is an unusual eigenvalue problem, involving a delay and pointwise evaluation of f. Nevertheless there is a straightforward series solution whenever $\operatorname{Re} p > 0$. Note we may express (8.4.7) in the form

$$
\lambda f(\tau - \tau_1) = e^{i\phi(\tau)} f(\tau) + g(\tau, c_{1/4}, c_{3/4}), \qquad (|\lambda| > 1),
\tag{8.4.8}
$$

where $\tau_1 = \tau_0 + 1/2$, and g is periodic in τ and linear in the other two arguments. The formal solution is given by the series

$$f(\tau) = \lambda^{-1} g(\tau + \tau_1, c_{1/4}, c_{3/4}) + \lambda^{-2} e^{i\phi(\tau + \tau_1)} g(\tau + 2\tau_1, c_{1/4}, c_{3/4}) + \ldots . (8.4.9)$$

Since g is bounded, this series converges for any choice of $c_{1/4}$ and $c_{3/4}$. Evaluation at $\tau = 1/4$ and $3/4$ then gives an equation for λ, and hence p, as a solvability condition.

When $\tau_0 = j/2$ for some positive integer j, a condition which fixes ε given ω, the solvability condition may be obtained directly by two evaluations of (8.4.6a). The result then yields $Z = e^{-\pi p/2\omega}$ as a root of a polynomial. For example, if $j = 1$ ($\tau_1 = 1$ in (8.4.8)), we evaluate (8.4.6a) at $\tau = 1/4$, $3/4$ to obtain two equations for $f(1/4)$ and $f(3/4)$. A non-trivial solution yields a 12th order polynomial when $K^{-1} = 1 - Z^4$ is substituted. After the inessential factor of $1 - Z^8$ is discarded, it reduces to

$$1 - 2\sin\theta Z^2 - Z^4 = 0. \tag{8.4.10}$$

Here $\theta = k\Delta z(1/4, \tau) = -k\Delta z(3/4, \tau)$ is obtained from (8.4.5). Note that when $\theta = 0$, (8.4.10) does not given dynamo action. This confirms the consistency of the model, as discussed in §8.4.1, in the limit of large period.

To take a numerical example, if $\omega = 0.942$ we obtain $\varepsilon \approx 0.1$ from (8.3.13) and, with $k = \sqrt{2}$, (8.4.10) has a dominant perfect fast growth rate of $\Gamma = 0.138$ obtained from roots $Z = \pm 0.7946i$.

To compare with the numerical results of Ponty *et al.* we now apply this method to CP+ε. The following modifications of the MW+ε equations are needed. First the asymptotic description of the manifold is different, given by (8.2.10). It is convenient to bring this into the form of the MW+ε expression by a time shift of $+\pi/4\omega$, so that (8.2.11b) becomes

$$y(t, s) \approx \varepsilon G(\omega) \cosh(t - s) \cos(\omega s), \tag{8.4.11}$$

with $G = 2^{-1/2}\omega\pi \operatorname{sech}(\omega\pi/2)$. This changes the definition of T and of τ_0, the latter now being $(\omega/2\pi) \log(4\sqrt{2} \cosh(\omega\pi/2)/\varepsilon\omega\pi)$. Also, with the time shift we now have from (8.2.17)

$$\Delta z(\sigma, \tau) = \varepsilon\omega\pi[T - (2\pi/\omega)(\tau - \sigma)] \operatorname{sech}(\omega\pi/2) \cos 2\pi\sigma$$
$$- (\varepsilon\omega\pi^2/2) \sinh(\omega\pi/2) \operatorname{sech}^2(\omega\pi/2) \sin 2\pi\sigma. \tag{8.4.12}$$

Finally, the phase shift on the left of (8.4.7) is $1/4$ rather than $1/2$. With these changes we may obtain direct values of growth rate for $\tau_0 = (2j - 1)/4$ with $j = 1, 2, \ldots$. The corresponding polynomials in Z may be obtained from (8.4.6), equal to $e^{2\pi p(\tau_0 - \tau)/\omega} f(\tau - \tau_0 - 1/4)$, for the above values of τ_0, by substituting $\tau = 1/4$, $3/4$. The conditions for the existence of a solution for $f(1/4)$, $f(3/4)$ reduce to polynomials of degree $4j + 6$ (j odd) and $4j + 8$ (j even). After elimination of a factor $1 - Z^4$ in the former and $Z^2(1 - Z^4)$ in the latter, the equations for Z become

$$Z^{4j+2} - Z^4 - 4\sin\theta \, Z^{2j+1} - Z^{4j-2} + 1 = 0, \quad (j \text{ odd}), \tag{8.4.13a}$$

$$Z^{4j+2} - Z^{4j-2} + 2\sin\theta\, Z^{2j-1}(1 + Z^4) + Z^4 - 1 = 0, \quad (j \text{ even}), \quad (8.4.13b)$$

where again $\theta = k\Delta z(1/4, \tau)$. Ponty $et\ al.$ adopt $k = 0.57$ as in Galloway & Proctor (1992). In the cell form we then take $k = 0.57\sqrt{2}$. We compute from (8.4.13) the following growth rates for $\varepsilon = 0.1$: $(j, \omega, \Gamma) = (1, 0.393, 0.0468)$, $(2, 1.201, 0.0812)$ and $(3, 1.78, 0.103)$. We compare these values with the data of Ponty $et\ al.$ at the same ε in Fig. 8.6. The agreement is reasonable for the first two points, but small-ω theory does not appear to capture the abrupt fall-off of the growth rate as ω exceeds 1.5. At the same time ω is then very likely outside the range where the theory could be expected to apply. Note that these results are independent of the arbitrary parameter δ. This is because θ is independent of T. For general phase shifts τ_1 in (8.4.8) this dependence emerges and Γ will depend upon δ.

We remarked above that the present model resembles the growth rate estimation from Fredholm determinants and related methods as discussed in §6.3, a point now explicit in the polynomials in Z. It would appear that a very promising line of research would use directly the periodic points in the Melnikov setting to estimate growth rates, without necessarily attempting a description in terms of the folding of flux sheets, or any use of the artificial parameter δ. Recent work by Melnikov[66] concerning periodic points of near-integrable Hamiltonian systems suggests the possibility of such a theory.

8.4.3 Small Period

We have seen that the computations for CP+ε indicate that fast dynamo action turns off rapidly as ω is increased toward values around 2 in the near-integrable limit; see Fig. 8.6. This extinction is understandably not seen in the large-period approximation, but can be studied if we assume the opposite case of small period.

We may use (8.3.12), with the appropriate vertical drifts, for either MW+ε or CP+ε. The roots for $\omega \gg 1$ are given by

$$\sigma^+_{2n-1} \approx n - \gamma_n, \qquad \sigma^+_{2n} \approx n + \gamma_n, \qquad (8.4.14a)$$

$$\sigma^-_{2n-1} \approx n - 1/2 - \gamma_{n-1/2}, \qquad \sigma^-_{2n} \approx n - 1/2 + \gamma_{n-1/2}, \qquad (8.4.14b)$$

where

$$\gamma_n = (1/2\pi)\cos^{-1}\exp((2\pi/\omega)(\tau - n)), \qquad (8.4.14c)$$

and $n = 1, 2, \ldots$. The roots σ^-_1 and σ^-_2 do not appear when $1/2 < \tau < 1$, as can be seen from the graph of (8.3.12) (cf. Fig. 8.11). In any case the variation with τ is slight at large ω. Note that on $W^u(A)$ near the section b the height of the lobes grows slightly with each cycle of the flow. The roots are thus

---

[66]Notes on a lecture given at the Courant Institute in 1994.

initially paired near the extrema of $e^{(2\pi\sigma/\omega)}\cos 2\pi\sigma$, but as n/ω grows they split apart and tend to the zeros of $\cos 2\pi\sigma$.

We seek then to convert the summation in (8.3.11) into an integral with respect to $u \approx n/\omega$, and first apply it to the non-dynamo $w = 0$ to test the consistency assumption of the preceding section. Setting

$$\gamma(u) = \frac{1}{2\pi}\cos^{-1}e^{-2\pi u}, \qquad (0 \le \gamma \le 1/4), \qquad (8.4.15)$$

we obtain from (8.3.11) with $\Delta z_k = 0$,

$$e^{2\pi p(\tau_0-\tau)/\omega}f(\tau - \tau_0 + \Delta\tau) = \omega \int_0^{1/4}(\cos 2\pi\gamma)^{p-1}\sin 2\pi\gamma$$

$$\times \left(\rho[f(-\gamma)e^{2\pi p\gamma/\omega} - f(\gamma)e^{-2\pi p\gamma/\omega}\right] \qquad (8.4.16)$$

$$+ \rho^{-1}\left[f(1/2 - \gamma)e^{2\pi p(\gamma-1/2)/\omega} - f(1/2 + \gamma)e^{-2\pi p(\gamma+1/2)/\omega}\right]\right)d\gamma.$$

Here $\Delta\tau = 1/2$ for MW+ε and $\Delta\tau = 1/4$ for CP+ε. Now for large ω we treat τ/ω as very small, but retain τ_0/ω as possibly $O(1)$. Since for $\mathrm{Re}\,p > 0$ the integral on the right should exist, it follows that $f \approx$ const. This then yields

$$e^{2\pi p\tau_0} \approx \int_0^{1/4}\gamma\,4\pi p(\cos 2\pi\gamma)^{p-1}(\sin 2\pi\gamma)(\rho + \rho^{-1})\,d\gamma. \qquad (8.4.17)$$

Taking $\rho = 1$ and integrating by parts we obtain $e^{2\pi p\tau_0} \approx 1$ and thus zero growth rate.

To see how growth arises from the vertical phase shifting, we consider CP+ε in the small period limit. In terms of the γ variable the flux equation takes the following form:

$$e^{2\pi p(\tau_0-\tau)/\omega}f(\tau - \tau_0 + 1/4) = \omega \int_0^{1/4}(\cos 2\pi\gamma)^{p-1}\sin 2\pi\gamma$$

$$\times \left(\rho[F(-\gamma)e^{2\pi p\gamma/\omega} - F(\gamma)e^{-2\pi p\gamma/\omega}\right] \qquad (8.4.18a)$$

$$+ \rho^{-1}\left[F(1/2 - \gamma)e^{2\pi p(\gamma-1/2)/\omega} - F(1/2 + \gamma)e^{-2\pi p(\gamma+1/2)/\omega}\right]\right)d\gamma.$$

Here

$$F(\gamma) = f(\gamma)\,e^{ik\Delta z(\gamma)}, \qquad (8.4.18b)$$

where Δz is the vertical drift in the CP+ε flow. Using (8.4.12), we find

$$\Delta z(\gamma) = \varepsilon\omega\pi[T - (2\pi/\omega)(\tau - \log|\cos 2\pi\gamma|)]\,\mathrm{sech}(\omega\pi/2)\cos 2\pi\gamma$$
$$- (\varepsilon\omega\pi^2/2)\sinh(\omega\pi/2)\,\mathrm{sech}^2(\omega\pi/2)\sin 2\pi\gamma. \qquad (8.4.19)$$

Here we suppress the dependence on τ. We now use this expression in (8.4.18a) while taking ω to be large and ε to be small. This allows us to expand the exponential under the integral, $e^{2\pi p\gamma/\omega} \approx 1 + 2\pi p\gamma/\omega$ and

$e^{ik\Delta z(\gamma)} \approx 1 + ik\Delta z(\gamma)$. We also choose $\rho = i$ to obtain growth. We may again take $f \approx$ const. The powers of $e^{2\pi p/\omega}$ under the integral in (8.4.18a), with $\rho = i$, do not contribute at leading order. The resulting equation for p may be written, after an integration by parts,

$$ Pe^P = C \int_0^{\pi/2} \cos^{p+1} u\, du \approx C, \tag{8.4.20} $$

where $P = 2\pi\tau_0 p/\omega$ and $C = 2k\varepsilon\omega\pi^2\tau_0 \sinh(\omega\pi/2)\operatorname{sech}^2(\omega\pi/2)$. This equation admits real roots $p > 0$. Taking $k = 0.57\sqrt{2}$ and $\varepsilon = 0.1$ we compare in Fig. 8.6 the $\Gamma(\omega)$ so obtained with the numerical results of Ponty *et al.* (obtained using the wave form at $k = 0.57$ and $\varepsilon = 0.1$ with $R = 1000$ and 2000) and with the large period computations. The main conclusion, given the assumptions made, is that the agreement for the magnitude of the growth rate is reasonable, even though the abrupt fall-off with frequencies above 1.5 is not reproduced. Note that we have used large ω to simplify the integral, mainly in the neglect of the effect of the factors which account for exponential decay of the flux from lobe to lobe as σ increases, even though we account for the overall decay of flux along the full length of the separatrix. It is again a feature of these results that they do not depend upon δ.[67]

---

[67]We take this opportunity to correct a large ω calculation for MW+ε given in Childress (1993a); see equation (51) in that paper. Dynamo action was incorrectly attributed to the choice $\rho = 1$. In fact, as we have just seen, reinforcement rather than cancellation occurs when $\rho = i$. The version of (8.4.20) appropriate to MW+ε applies.

9. Spectra and Eigenfunctions

The kinematic dynamo problem is an eigenvalue problem. The fast dynamo problem thus encompasses the study of the spectrum of the induction operator in the perfectly conducting limit. The more abstract literature on spectral theory has had relatively little impact on dynamo theory, however, in part because of the relatively simple discrete spectrum at finite magnetic Reynolds numbers. The discrete eigenvalues largely disappear in the perfectly conducting limit, and concepts of continuous and residual spectra are needed. There is a considerable literature on the spectrum of linearized MHD equations, recently surveyed in the book by Lifschitz (1989). While this theory must in some sense 'contain' a restriction to kinematic dynamo theory, the emphasis is sufficiently distinct to make a separate treatment desirable. There are also similar issues in the theory of hydrodynamical stability in the inviscid limit. However it is only recently that inviscid stability theory has been extended beyond the classical examples, for example parallel flow or flow with circular streamlines, to more complicated two- and three-dimensional flows, including stretching flows. Recent discussions of the stability of ABC and other Beltrami flows (Arnold 1972, Galloway & Frisch 1987, Moffatt 1986, Friedlander & Vishik 1992, Friedlander, Gilbert & Vishik 1993) have in fact been closely associated with developments in MHD and dynamo theory. Finally, recent concerns about the role of eigenvalue problems in hydrodynamic stability theory, where significant transient growth can occur when nominally stable flows are perturbed, apply here with special force.

It is thus useful to examine just those aspects of spectral theory which pertain to fast dynamo theory. In the present chapter we examine our problem from the point of view of classical operator theory, using very simple examples and in the light of the methods developed in Chaps. 6–8. The models considered here lead up to but do not include pulsed Beltrami waves. The eigenvalue problem arising there is discussed in the anti-integrable limit in the following chapter. Unfortunately the spectral problem of most interest in fast dynamo theory is that of time-periodic flows. In comparison with the extensive results available for the time-separable case of a steady velocity field, the spectral theory of time-periodic flows is relatively undeveloped; for example, apparently lacking is a completeness theorem for the eigenfunctions in the diffusive problem (see §9.2.2).

We begin with a brief review of some of the general ideas which will be useful to us. We assume throughout that the fluid is incompressible, and that the magnetic field is 2π-periodic in all spatial coordinates and has zero mean. The results will hold for other boundary conditions of interest, for example, a bounded spherical conductor surrounded by vacuum (see §1.2.5).

9.1 Preliminaries

We first summarize a number of basic definitions and results from the classical theory of linear operators. Only selected proofs are given. The interested reader can find details in, for example, Kato (1966); see also Friedman (1956), Taylor (1958), and Bollobás (1990).

9.1.1 The Spectrum of an Operator

A *Banach space* is a normed space which is *complete*, i.e., Cauchy sequences have limits in the space. We denote the norm by $\|\cdot\|$. The natural Banach space for dynamo theory is the space $\mathcal{B}(\mathcal{D}) = L^2(\mathcal{D})$ of complex-valued fields $\mathbf{B}$, Lebesgue-integrable over $\mathcal{D}$, with the property that

$$\|\mathbf{B}\|^2 \equiv \int_{\mathcal{D}} |\mathbf{B}|^2 \, dV < \infty. \tag{9.1.1a}$$

The fields $\mathbf{B}$ of interest to us will also be functions of time t for $t \geq 0$. The associated inner product is

$$(\mathbf{B}_1, \mathbf{B}_2) = \int_{\mathcal{D}} \mathbf{B}_1^* \mathbf{B}_2 \, dV. \tag{9.1.1b}$$

The vectors $\mathbf{B}$ will be divergence-free. In $L^2(\mathcal{D})$ this property is defined in a weak sense (cf. (9.2.6) below). A Banach space with norm derived from an inner product is called a *Hilbert space*. In a Banach or Hilbert space, a sequence of functions $\mathbf{B}_k(\mathbf{x})$ for $k = 1, 2, \ldots$ is *complete* in the space if the linear combinations of these functions are dense in the space, i.e., any $\mathbf{B}$ is a limit of a linear combination in the norm of the space.

We are interested in the action of operators $\mathbf{L}$ on the space $\mathcal{B}(\mathcal{D})$, in particular the spatial differential operator $\mathbf{L}_\epsilon$ in the induction equation (4.1.1), which we may write in the form

$$\partial_t \mathbf{B} + \mathbf{L}_\epsilon \mathbf{B} = 0, \quad \mathbf{L}_\epsilon = \mathbf{u} \cdot \nabla \mathbf{B} - \mathbf{B} \cdot \nabla \mathbf{u} - \epsilon \nabla^2 \mathbf{B}. \tag{9.1.2}$$

We shall often drop the subscript and write $\mathbf{L}$ for an operator on $\mathcal{B}(\mathcal{D})$ including, but not necessarily, the induction operator. For a *steady flow* $\mathbf{u}(\mathbf{x})$, $\mathbf{L}_\epsilon$ is independent of time and the *induction operator* $\mathbf{T}_\epsilon(t)$, with the property that $\mathbf{T}_\epsilon(t)\mathbf{B}(\mathbf{x}, 0) = \mathbf{B}(\mathbf{x}, t)$, may be written as $\mathbf{T}_\epsilon(t) = \exp(-\mathbf{L}_\epsilon t)$, where the exponential is, like its matrix counterpart, defined by an infinite series. In this case time may be separated out by setting $\mathbf{B}(\mathbf{x}, t) = e^{-\lambda t}\mathbf{b}(\mathbf{x})$, and λ is obtained as an eigenvalue of $\mathbf{L}_\epsilon$. The magnetic mode $\mathbf{b}(\mathbf{x})$ is the associated eigenfunction of $\mathbf{L}_\epsilon$. Note that for any fixed t, we may also view $\mathbf{b}(\mathbf{x})$ as an eigenfunction for the induction operator $\mathbf{T}_\epsilon(t)$, corresponding to the eigenvalue $e^{-\lambda t}$. We may refer to the latter eigenvalue as the associated *multiplier* (Yudovich 1989).

For the case where $\mathbf{L}_\epsilon$ depends upon time, of particular interest are the periodic flows, $\mathbf{L}_\epsilon(t + T) = \mathbf{L}_\epsilon(t)$. In this case, time is not separable as it is for steady flow. The relevant eigenvalue problem is for the induction operator $\mathbf{T}_\epsilon(T)$, now defined by the effect of the motion from time t to time $t + T$ on an initial field $\mathbf{b}(\mathbf{x})$. The corresponding eigenvalue problem is the *Floquet problem* $\mathbf{T}_\epsilon(T)\mathbf{b}(\mathbf{x}) = e^{pT}\mathbf{b}(\mathbf{x})$, involving the multiplier e^{pT}. The periodic case is discussed separately in §9.2.2. The operator $\mathbf{T}_\epsilon(T)$, carrying the solution forward through one period T, is sometimes called, borrowing a term from the theory of solutions of ordinary differential equations in the complex domain, the *monodromy operator* (Yudovich 1989).

Spectral theory is concerned primarily with the action of operators such as $\mathbf{L}_\epsilon$ on $\mathcal{B}$, but also with how the spectral properties of $\mathbf{L}_\epsilon$ determine the evolution of magnetic field under $\mathbf{T}_\epsilon$. The operator $\mathbf{T}$ or $\mathbf{T}_0$ will denote the corresponding induction operator in a perfect conductor.

An important measure of the 'size' of an operator is its *bound* in $\mathcal{B}$. An operator is bounded if there is a positive number c such that $\|\mathbf{LB}\| \leq c\|\mathbf{B}\|$ for all $\mathbf{B}$ in $\mathcal{B}$. We write $\|\mathbf{L}\|$ for the infimum of all such c. For induction by (9.1.2) an extension of the argument of §1.3 to complex fields shows that $\mathbf{T}_\epsilon(t)$ is bounded with bound $\exp(\Lambda t)$ and that $\|\mathbf{L}_\epsilon\| \leq \Lambda$ where Λ is defined by (1.3.9).

The *resolvent set* of $\mathbf{L}$ is the set of complex numbers μ with the property that $\mu\mathbf{I} - \mathbf{L}$ has a bounded inverse on $\mathcal{B}(\mathcal{D})$. We denote this set by $\rho(\mathbf{L})$. Whenever μ belongs to $\rho(\mathbf{L})$ we call $(\mu\mathbf{I} - \mathbf{L})^{-1}$ the *resolvent* of $\mathbf{L}$, denoted by $\mathbf{R}(\mu)$. For convenience we will drop the $\mathbf{I}$ and write this as $(\mu - \mathbf{L})^{-1}$. All points of the complex plane not in the resolvent set will belong to the *spectrum* of $\mathbf{L}$, denoted by $\sigma(\mathbf{L})$.

Our purpose now is to divide up $\sigma(\mathbf{L})$ in various ways. The *point spectrum* or *discrete spectrum*, $\sigma_d(\mathbf{L})$, consists of all μ such that $\mathbf{Lb} = \mu\mathbf{b}$ has a non-trivial solution in $\mathcal{B}$. We then say that any such μ is an *eigenvalue* of $\mathbf{L}$, and that $\mathbf{b}$ is an associated *eigenfunction*. For an eigenvalue of multiplicity greater than one, the corresponding eigenspace is multi-dimensional. These discrete eigenvalues exhaust the spectrum of an operator in finite dimensional space, but in infinite dimensional space the situation is more complicated since, without μ being an eigenvalue, it is possible that $(\mu - \mathbf{L})\mathbf{b}$ takes on arbitrarily small values as $\mathbf{b}$ ranges over $\mathcal{B}$. Furthermore, even if this does not happen the range of $(\mu - \mathbf{L})\mathbf{b}$ need not, as in the finite-dimensional case, be the whole space. In any event, at such a point the resolvent cannot be bounded and so we have a point distinct from the resolvent set and from the discrete spectrum.

It is helpful to consider first the *approximate spectrum* σ_{ap}. We say that μ belongs to the *approximate spectrum* if there is a sequence $\{b_n\}$ of functions of norm unity such that $\|(\mu - L)b_n\| \rightarrow 0$. Note that the approximate spectrum contains the discrete spectrum, but does not intersect the resolvent set. Generally the approximate spectrum does not fill the non-discrete part of the spectrum, however, and it is customary to divide $\sigma \backslash \sigma_d$ into *continuous* and *residual* spectra. Suppose that the closure of the range of $\mu - L$ is all of $\mathcal{B}$, but its inverse is unbounded. We then say μ belongs to the *continuous spectrum*, $\sigma_c(L)$. A second possibility is that the closure of the range of $\mu - L$ is a proper subset of $\mathcal{B}$, but μ is not in the discrete spectrum. We then say μ is in the *residual spectrum* of L, $\sigma_r(L)$. The approximate spectrum contains the continuous spectrum, and may also contain some of the residual spectrum, but not necessarily all of it. It is clear from these definitions that the spectrum is given by the pairwise disjoint union, $\sigma = \sigma_d \cup \sigma_c \cup \sigma_r$. It is also sometimes useful to introduce the complement of σ_{ap} in σ. This is the *compression spectrum*, σ_{com}, with $\sigma = \sigma_{ap} \cup \sigma_{com}$. These divisions of the spectrum are suggested in Fig. 9.1. In the following sections we give examples from our dynamo models of the various kinds of spectra.

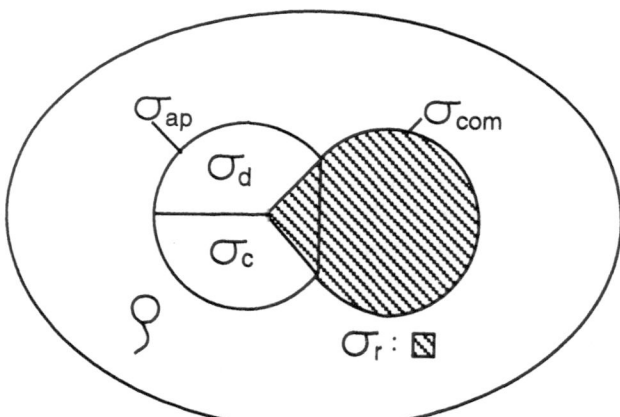

Fig. 9.1 The spectrum of an operator L represented as a connected set. The two flattened discs represent the approximate spectrum σ_{ap} and the compression spectrum σ_{com}. The disc σ_{ap} contains the discrete spectrum σ_d and the continuous spectrum σ_c. The (shaded) residual spectrum σ_r may contain points in σ_{ap} and contains the whole of σ_{com}.

9.1.2 Compact and Closed Operators

An operator is said to be *completely continuous* or *compact* if, for any bounded sequence $\mathbf{b}_n$ in $\mathcal{B}$, $\mathbf{L}\mathbf{b}_n$ contains a convergent subsequence (converging to an element of $\mathcal{B}$ since the space is complete). Compact operators inherit many of the properties of operators on a finite-dimensional space. Central to our discussion is the following unfortunate proposition, which applies to the induction operator $\mathbf{T}$ for the perfect conductor defined in §9.1.1. Perfect induction on $L^2(\mathcal{D})$ for time t_0, say, does not appear to define a compact operator $\mathbf{T}$ for any flow of interest to us. To see this intuitively, consider some initial field $\mathbf{B}_0$ and define the bounded sequence $\mathbf{B}_n = \mathbf{T}(nt_0)\mathbf{B}_0/\|\mathbf{T}(nt_0)\mathbf{B}_0\|$ for $n \geq 1$. Thus $\mathbf{T}(t_0)\mathbf{B}_n$, when normalized to have norm one, is just $\mathbf{B}_{n+1}$, and the stirring action of the flow will not lead to any convergence. (One has to have convergence for every choice of initial field.) Note that in an infinite-dimensional space even the identity operator fails to be compact! The action of diffusion is critical in changing the non-compact operator $\mathbf{T}$ into the compact operator $\mathbf{T}_\epsilon$, as we discuss in §9.2. The difficulties of the fast dynamo problem are associated with this limiting loss of compactness.

We do not know of a general proof of the non-compactness of $\mathbf{T}$, but it can be demonstrated explicitly in certain simple cases. Consider for example a two-dimensional flow containing a circular eddy $\mathbf{u} = (u_r, u_\theta) = (0, r\omega(r))$, where $\omega(r)$ is a continuously differentiable function on some disc $0 \leq r \leq r_0$ with a range of finite extent. The sequence $\mathbf{B}_n$ may then be defined as $\mathbf{a}_n/\|\mathbf{a}_n\|$ where

$$\mathbf{a}_n(r,\theta) = (r^{-1}\partial_\theta A_n, -\partial_r A_n), \tag{9.1.3}$$

and (see §5.2.2)

$$A_n(r,\theta) = A(r,\theta,nt_0) = A_0(r,\theta - n\omega(r)t_0) \tag{9.1.4}$$

is the sequence of magnetic potentials in terms of the initial potential A_0. Suppose now that we had a convergent subsequence for any A_0. Then with $A_0 = r\sin\theta$ and for large n we would have

$$\mathbf{B}_n \approx C(0, r\omega'(r)\cos(\theta - n\omega(r)t_0)), \quad C^{-2} = \pi\int_0^{r_0} r^2\omega'(r)^2 r\,dr. \tag{9.1.5}$$

If $\mathbf{f}(r,\theta)$ is the limit in L^2 of the subsequence, then clearly there is an increasing sequence of positive integers n_k with

$$\int_{\mathcal{D}}\left[f_\theta - Cr\omega'(r)\cos(\theta - n_k\omega(r)t_0)\right]^2 r\,dr\,d\theta \to 0. \tag{9.1.6}$$

It is sufficient to consider f_θ of the form $f_c(r)\cos\theta + f_s(r)\sin\theta$. Using this in (9.1.6) we obtain a positive term and a cross-term involving n_k. Using the Riemann–Lebesgue lemma, the latter tends to zero as n_k tends to infinity since ω' is continuous. Hence there is no such limit f_θ, and there is no subsequence which converges in the L^2 norm.

The important property of a compact operator is that the spectrum is entirely discrete, consisting of a countable number of eigenvalues with no accumulation point different from zero. Each μ in σ is an eigenvalue of $\mathbf{L}$ with finite multiplicity. The same properties are shared by operators with compact resolvents, provided that in addition $\mathbf{L}$ is *closed*. An operator on a Banach space is *closed* if, whenever $\mathbf{b}_n$ and $\mathbf{Lb}_n$ are Cauchy sequences, with $\mathbf{b}_n \to \mathbf{b}$ in $\mathcal{B}$, then $\mathbf{Lb}_n \to \mathbf{Lb}$. Any bounded operator on $\mathcal{B}$ is also closed. We may summarize the result as follows:

Lemma 9.1 *If $\mathbf{L}$ is a closed operator on $\mathcal{B}$ such that the resolvent $\mathbf{R}(\mu)$ exists and is compact for some μ, then $\sigma(\mathbf{L})$ consists of discrete eigenvalues with finite multiplicities, and $\mathbf{R}(\mu)$ is compact for all μ in $\rho(\mathbf{L})$.*

For a discussion and proof see Kato (1966), p. 187. Compactness at only one point need be assumed because of the relation $\mathbf{R}(\mu) - \mathbf{R}(\mu_0) = (\mu_0 - \mu)\mathbf{R}(\mu)\mathbf{R}(\mu_0)$.

9.2 The Spectrum of $\mathbf{L}_\epsilon$

9.2.1 Steady Flow

In this section we outline the proof of the discreteness of the spectrum of $\mathbf{L}_\epsilon$ for a steady incompressible flow $\mathbf{u}(\mathbf{x})$ defined in a bounded domain $\mathcal{D}$. Our aim is to indicate, without going through all the details of the proof, some of the ideas involved, and particularly the effects of diffusion. The method we discuss is derived from analogous results for the hydrodynamic stability problem, and closely follows Sattinger (1970). The reader not interested in these details may want simply to note the content of theorems 9.1 and 9.2 below, and skip to §9.3.

Associated with $\mathbf{L}_\epsilon$ we introduce the linear functional

$$L_\epsilon(\mathbf{a}, \mathbf{b}) = (\mathbf{a}, \mathbf{L}_\epsilon \mathbf{b}) = \int_\mathcal{D} \mathbf{a}^* \cdot \mathbf{L}_\epsilon \mathbf{b} \, dV. \tag{9.2.1}$$

In forming this inner product, we may use boundary conditions and the divergence theorem to express the diffusion in terms of first derivatives,

$$-\epsilon \int_\mathcal{D} a_i^* \nabla^2 b_i dV = \epsilon \int_\mathcal{D} \partial_j a_i^* \partial_j b_i \, dV. \tag{9.2.2}$$

Since these first derivatives will appear in our estimates, we shall then need the norm on first derivatives

$$\|\mathbf{b}\|_1^2 \equiv \int_{\mathcal{D}} \partial_j b_i^* \partial_j b_i \, dV. \tag{9.2.3}$$

We shall assume that $\|\mathbf{b}\|_1 \geq c_1\|\mathbf{b}\|$ for some positive constant c_1, for any $\mathbf{b}$ in $\mathcal{B}$. This, in particular, is a consequence of periodic boundary conditions, given that $\mathbf{b}$ has zero mean. If $\mathbf{u}$ has bounded derivatives, then we have the estimate

$$|L_\epsilon(\mathbf{a}, \mathbf{b})| \leq c_2\|\mathbf{a}\|_1\|\mathbf{b}\|_1, \tag{9.2.4}$$

where c_2 is another positive constant. We also have

$$\operatorname{Re} L_\epsilon(\mathbf{b}, \mathbf{b}) \geq \epsilon\|\mathbf{b}\|_1^2 - \Lambda\|\mathbf{b}\|^2 \tag{9.2.5}$$

from (1.3.10) where Λ is the maximum eigenvalue of the velocity derivative matrix as defined in (1.3.9).

These estimates are sufficient to prove that $\mathbf{L}_\epsilon$ is a closed operator on $\mathcal{B}$. This result, and indeed the solution of the eigenvalue problem $\mathbf{L}_\epsilon\mathbf{b} = \mu\mathbf{b}$, must be understood in the weak sense using (9.2.1). For example, the solenoidal property of the magnetic field is imposed as follows: $\mathcal{B}$ contains complex vector fields from $L^2(\mathcal{D})$ satisfying

$$\int_{\mathcal{D}} \mathbf{b} \cdot \nabla\phi \, dV = 0 \tag{9.2.6}$$

for every continuously differentiable function ϕ on $\mathcal{D}$. To establish, then, the weak closed property of $\mathbf{L}_\epsilon$, assume that that $\mathbf{L}_\epsilon\mathbf{b}_n$ and $\mathbf{b}_n \to \mathbf{b}$ are Cauchy sequences. We want to show that $L_\epsilon(\mathbf{a}, \mathbf{b}_n) \to L_\epsilon(\mathbf{a}, \mathbf{b})$ for every $\mathbf{a}$ in $\mathcal{B}$. It is sufficient to prove this property for $\mathbf{L}_r \equiv \mathbf{L}_\epsilon + r\mathbf{I}$, where r is any real number with $r \geq \Lambda$, and from (9.2.5) we have

$$\epsilon\|\mathbf{b}\|_1^2 \leq \operatorname{Re}(\mathbf{b}, \mathbf{L}_r\mathbf{b}) \leq \|\mathbf{L}_r\mathbf{b}\| \, \|\mathbf{b}\| \leq \frac{1}{c_1}\|\mathbf{L}_r\mathbf{b}\| \, \|\mathbf{b}\|_1, \tag{9.2.7}$$

and therefore $\|\mathbf{L}_r\mathbf{b}\| \geq c_1\epsilon\|\mathbf{b}\|_1$. This last inequality can be used to advantage to establish that the assumed Cauchy convergence $\|\mathbf{L}_r(\mathbf{b}_m - \mathbf{b}_n)\| \to 0$ forces convergence in the $\|\cdot\|_1$ norm. There easily follows from (9.2.4),

$$|L_r(\mathbf{a}, \mathbf{b}_m - \mathbf{b}_n)| \leq c_3\|\mathbf{a}\|_1 \, \|(\mathbf{b}_m - \mathbf{b}_n)\|_1, \tag{9.2.8}$$

for some positive constant c_3. Letting $n \to \infty$ we obtain

$$|L_r(\mathbf{a}, \mathbf{b}_m - \mathbf{b})| \leq c_3\|\mathbf{a}\|_1 \, \|(\mathbf{b}_m - \mathbf{b})\|_1. \tag{9.2.9}$$

From the limit $m \to \infty$ we establish that $\mathbf{L}_r$, and therefore $\mathbf{L}_\epsilon$, is closed.

We now turn to the second important property of $\mathbf{L}_\epsilon$, namely the fact that $\mathbf{L}_\epsilon$ has compact resolvent. Sattinger (1970) establishes this by invoking a basic result of functional analysis, the Lax–Milgram theorem (see Kato 1966), which guarantees in our problem that there is a weak solution of $\mathbf{L}_r\mathbf{b} = \mathbf{f}$ for any $\mathbf{f}$ belonging to $\mathcal{B}$. This shows that $\mathbf{L}_r$ has a bounded inverse on $\mathcal{B}$. Compactness follows from this result. We omit details given in Sattinger (1970) and references therein. Thus we may establish the compactness of $(\mathbf{L}_\epsilon - \mu)^{-1}$ for one value of μ. From lemma 9.1 we may then deduce the discreteness of the spectrum. The main point is that $\mathbf{L}_\epsilon$ is an operator which is equivalent in many respects to $-\epsilon\nabla^2$. This technical property is intuitively at odds with the dominant role of advection at small ϵ, but it controls the discreteness of the spectrum for any positive ϵ.

Restricting attention to the periodic case $\mathbf{b}(\mathbf{x} + L\mathbf{n}) = \mathbf{b}(\mathbf{x})$ for integer $\mathbf{n}$ with the condition $\nabla \cdot \mathbf{b} = 0$, we may solve the eigenvalue problem, $(\mu + \epsilon\nabla^2)\mathbf{b} = 0$ in terms of a potential, writing $\mathbf{b} = \nabla \times \mathbf{a}$, where $\mathbf{a}$ may be taken to have zero mean. We then have the Fourier expansion

$$\mathbf{a} = \sum_{\mathbf{n}} a_{\mathbf{n}}\phi_{\mathbf{n}}(\mathbf{x}), \qquad a_0 = 0, \tag{9.2.10}$$

where the sum is over all integer vectors $\mathbf{n}$. The orthogonal eigenfunctions and associated real eigenvalues are

$$\phi_{\mathbf{n}}^q = \mathbf{e}_q \exp(2\pi i\mathbf{n} \cdot \mathbf{x}/L), \qquad \mu_{\mathbf{n}} = 4\pi^2 n^2 \epsilon/L^2, \tag{9.2.11}$$

where $\mathbf{e}_q$, $q = 1, 2, 3$ are unit vectors parallel to the coordinate axes. The associated *orthonormal* eigenfunctions and the associated eigenvalues may be enumerated as $\mathbf{b}_k$, μ_k for $k = 1, 2, \ldots$ with $\mu_k \to \infty$ as $k \to \infty$. These eigenfunctions are complete in $\mathcal{B}$, and we have for the resolvent,

$$(\mu + \epsilon\nabla^2)^{-1}\mathbf{b} = \sum_{k=1}^{\infty} \frac{(\mathbf{b}_k, \mathbf{b})\mathbf{b}_k}{(\mu - \mu_k)}. \tag{9.2.12}$$

Since

$$\|\mathbf{b}\|_1^2 = \epsilon^{-1} \sum_{k=1}^{\infty} \mu_k (\mathbf{b}_k, \mathbf{b})|^2, \tag{9.2.13}$$

we see that the resolvent in (9.2.12) is a bounded operator in the norm $\|\cdot\|_1$ for any μ with negative real part.

These properties of induction with $\mathbf{u} = 0$ carry over in part to the operator $\mathbf{L}_r$ (with $r \geq \Lambda$ still). In particular the eigenfunctions of $\mathbf{L}_r$ are complete in $\mathcal{B}$ (but not necessarily orthogonal, a point we take up in §9.3). Completeness of the eigenfunctions is a consequence of $\mathbf{L}_r$ possessing an inverse that is 'Hilbert–Schmidt'. An operator $\mathbf{T}$ on $\mathcal{B}$ is said to be *Hilbert–Schmidt* (HS) if $\sum_k \|\mathbf{T}\mathbf{b}_k\|^2 < \infty$ where $\mathbf{b}_k$ is an orthonormal set of functions, for example (9.2.11). The proof of the completeness of eigenfunctions of $\mathbf{L}_r$ depends upon, first, $\mathbf{A} \equiv (r - \epsilon\nabla^2)^{-1}$ being HS, and, second, on a representation of $\mathbf{L}_r$ as a product of a bounded operator and an HS operator. The first statement follows from the estimate (9.2.12). For the second, let $\mathbf{L}_\epsilon = -\epsilon\nabla^2 + \mathbf{N}$; then, with $\mathbf{A}$ as defined above we have

$$\mathbf{L}_r^{-1} = ((\mathbf{I} + \mathbf{NA})\mathbf{A}^{-1})^{-1} = \mathbf{A}(\mathbf{I} + \mathbf{NA})^{-1}. \tag{9.2.14}$$

This expresses $\mathbf{L}_r^{-1}$ as the product of an HS operator, namely $\mathbf{A}$, and one that can be shown to be bounded, provided r is taken to be sufficiently large. Indeed, $\mathbf{NA}$ is an operator on $\mathcal{B}$ with the property that $\|\mathbf{NAb}\| \leq c_4\|\mathbf{Ab}\|_1$ in terms of a positive constant c_4. From (9.2.12) we then have

$$\|\mathbf{Ab}\|_1^2 \leq \sum_{k=1}^\infty \frac{\mu_k|(\mathbf{b}_k, \mathbf{b})|^2}{\epsilon(r + \mu_k)^2} \leq \frac{1}{4r\epsilon}\|\mathbf{b}\|^2, \tag{9.2.15}$$

after maximizing over μ. Thus $\|\mathbf{NAb}\|_1 \leq (c_4/2r^{1/2}\epsilon^{1/2})\|\mathbf{b}\|$. For r sufficiently large $\|\mathbf{NA}\| < 1$, in which case $(\mathbf{I} + \mathbf{NA})^{-1}$ is bounded.

We may summarize these results as follows:

Theorem 9.1 *When $\epsilon > 0$ the spectrum of $\mathbf{L}_\epsilon$ is discrete and consists of a countable number of eigenvalues μ_k, $k = 1, 2, \ldots$, each of finite multiplicity. The corresponding eigenfunctions are complete in $\mathcal{B}$.*

Another result of interest concerns the location of the eigenvalues. Following Sattinger (1970), we may obtain the following estimate:

Theorem 9.2 *The eigenvalues μ_k of theorem 9.1 lie within the parabola*

$$\operatorname{Re}\mu \geq -\Lambda - 1/2 + (\epsilon/2)\,\Gamma^{-2}(\operatorname{Im}\mu)^2, \tag{9.2.16}$$

where Γ is any real number for which $\operatorname{Im} L_\epsilon(\mathbf{b}, \mathbf{b}) \leq \Gamma\|\mathbf{b}\|\,\|\mathbf{b}\|_1$ holds. In particular we may take, from an estimation of the imaginary part of (9.2.1) with $\mathbf{a} = \mathbf{b}$, $\Gamma = \max_{\mathcal{D}}|u| + \Lambda/c_1$.

To prove this we suppose μ is in the resolvent set of $\mathbf{L}_\epsilon$ and consider solving $(\mathbf{L}_\epsilon - \mu)\mathbf{b} = \mathbf{f}$. From this, form $L_\epsilon(\mathbf{b}, \mathbf{b}) - \mu(\mathbf{b}, \mathbf{b}) = (\mathbf{b}, \mathbf{f})$. Separating into real and imaginary parts we have for the real part, using (9.2.5),

$$\epsilon\|\mathbf{b}\|_1^2 - \|\mathbf{b}\|^2 \operatorname{Re}\mu \leq \|\mathbf{f}\|\,\|\mathbf{b}\| + \Lambda\|\mathbf{b}\|^2$$
$$\leq (1/2)\|\mathbf{f}\|^2 + (\Lambda + 1/2)\|\mathbf{b}\|^2. \tag{9.2.17}$$

The last step follows from completing a square. From the imaginary part we obtain similarly

$$\|\mathbf{b}\|^2 \operatorname{Im}\mu \leq \Gamma\|\mathbf{b}\|\,\|\mathbf{b}\|_1 + \|\mathbf{b}\|\,\|\mathbf{f}\|, \tag{9.2.18}$$

and therefore

$$\epsilon(\operatorname{Im}\mu)^2\|\mathbf{b}\|^2/2\Gamma^2 \leq \epsilon\|\mathbf{f}\|^2/\Gamma^2 + \epsilon\|\mathbf{b}\|_1^2. \tag{9.2.19}$$

Adding (9.2.17) and (9.2.19) we have

$$\left[\frac{\epsilon(\text{Im}\,\mu)^2}{2\Gamma^2} - \left(\text{Re}\,\mu + \Lambda + \frac{1}{2}\right)\right]\|\mathbf{b}\|^2 \leq \left(\frac{1}{2} + \frac{\epsilon}{\Gamma^2}\right)\|\mathbf{f}\|^2. \tag{9.2.20}$$

If

$$\frac{\epsilon(\text{Im}\,\mu)^2}{2\Gamma^2} - \left(\text{Re}\,\mu + \Lambda + \frac{1}{2}\right) > 0 \tag{9.2.21}$$

then (9.2.20) yields a bound on $(\mathbf{L}_\epsilon - \mu)^{-1}$. Thus $\mathbf{L}_\epsilon - \mu\mathbf{I}$ has a bounded inverse and so μ is in the resolvent set of $\mathbf{L}_\epsilon$. Consequently any eigenvalue must be in the parabola given by (9.2.16). Note that the effect of high frequency $\text{Im}\,\mu$ in forcing large positive $\text{Re}\,\mu$ (recall $\text{Re}\,\mu < 0$ for dynamo action) disappears in the perfectly conducting limit. For *large* ϵ the eigenvalues tend to lie near the real axis, since the operator $\mathbf{L}_\epsilon$ is then close to self-adjoint (§9.3). It is perhaps worth remarking here that, since our operators are always real-valued, μ^* will belong to the spectrum whenever μ belongs to the spectrum.

We consider the implications of theorem 9.1 for the operator $\mathbf{T}_\epsilon$ on $\mathcal{B}$. The spectrum $\sigma(\mathbf{L}_\epsilon)$ gives us the information we need, since an initial condition $\mathbf{B}_0$ in $\mathcal{B}$ can be obtained as a limit of a linear combination of the eigenfunctions of $\mathbf{L}_\epsilon$. Since eigenvalues have finite multiplicity, $\mathbf{T}_\epsilon\mathbf{B}_0$ will be a sum of terms of the form

$$\mathbf{B}_k(\mathbf{x}, t) = e^{-\mu_k t} \sum_{j=1}^{m_k} t^{j-1} \mathbf{B}_{jk}(\mathbf{x}). \tag{9.2.22}$$

Dynamo action occurs when $\text{Re}\,\mu_k < 0$ for some k. Here we note that the compactness of $\mathbf{T}_\epsilon$ is ensured as a special case of the result of Yudovich (1989) for time-periodic flow, mentioned in the next section.

9.2.2 Time-periodic flow

Time-periodic flows offer a much richer family of stretching flows suitable for dynamo action, and therefore it is essential to enlarge the theory to encompass them. However, like the theory of hydrodynamic stability of time-periodic flows (Drazin & Reid 1981), the spectral theory of the time-periodic induction operator is far less developed than for steady flows, and we know of few general results dealing with the issues of interest in fast dynamo theory. Markus & Matsaev (1984) review comparison theorems for the spectra of a family of non-self-adjoint operators, with references to earlier work of Keldysh (1951). We rely primarily on a chapter from the monograph of Yudovich (1989). We suppose that the period of the flow is $T = 2\pi/\omega$. Given a field $\mathbf{b}(\mathbf{x}, t)$ with this period, we write

$$\mathbf{b}(\mathbf{x}, t) = \sum_{k=-\infty}^{\infty} \mathbf{b}_k e^{ik\omega t}, \quad \mathbf{b}_{-k} = \mathbf{b}_k^*, \quad \sum_{k=-\infty}^{\infty} \int_{\mathcal{D}} |\mathbf{b}_k|^2 \, dV < \infty. \tag{9.2.23}$$

The underlying space can then be redefined as $\mathcal{B}_\omega$, by which we mean a countable set of Fourier coefficient vectors $\mathbf{b}_k(\mathbf{x})$ each belonging to $\mathcal{B}$. The associated norms must then be redefined by summing over the Fourier coefficients,

$$\|\mathbf{b}\|_\omega^2 \equiv \sum_{k=-\infty}^{\infty} \int_{\mathcal{D}} |\mathbf{b}_k|^2 \, dV. \qquad (9.2.24)$$

In this way, $\mathbf{L}_\epsilon$ may be interpreted as an operator on $\mathcal{B}_\omega$.[68]

We can now look for eigenfunctions of the monodromy operator of the form $\mathbf{B}(\mathbf{x},t) = e^{pt}\mathbf{b}(\mathbf{x},t)$ where $\mathbf{b}(\mathbf{x},t)$ belongs to $\mathcal{B}_\omega$. In this case the time derivative acts on the periodic part as well as the exponential. The induction equation gives $\mathbf{L}_\epsilon \mathbf{b} + \mathbf{Db} = -p\mathbf{b}$ where $\mathbf{Db} = \partial_t \mathbf{b}$. This defines the *Floquet problem* and the Floquet exponent p in terms of operators on $\mathcal{B}_\omega$. Note that $\mathbf{D}$ is an unbounded operator on $\mathcal{B}_\omega$. The alternative statement using the monodromy operator on $\mathcal{B}$ is $\mathbf{T}_\epsilon(T)\mathbf{B} = e^{pT}\mathbf{B}$, involving the multiplier e^{pT}.

Once time-dependence of $\mathbf{L}_\epsilon$ is allowed, the result for the steady case that there exists a complete set of eigenfunctions need no longer prevail. In §9.5.3 below we shall show for the folded baker's map that the diffusive monodromy operator has no period-one eigenfunctions at all! This extreme disruption of the spectrum appears to be directly associated with the formation of small-scale structure as a cascade process. For a steady flow, even one exhibiting Lagrangian chaos, the small-scale structure is smoothed by diffusion and eigenfunctions are possible. For time-dependent flows the processes generating small scales need not be arrested in this way, as the folded baker's map illustrates. The abrupt, super-exponential decay exhibited for example in (3.1.8) is associated with the stretch–fold process without any compensating shear. It seems to be this aspect of $\mathbf{T}_\epsilon$ in the time-dependent case which can lead to loss of spectrum, and complicates the finding of eigenfunctions even in the presence of dynamo action.

To be sure, one expects the diffusive monodromy operator to be well-behaved for many periodic flows. Yudovich (1989) establishes for the periodic case the compactness of $\mathbf{T}_\epsilon$ relative to the first-derivative norm (9.2.3). From this follows the discreteness of the spectrum, and the existence of at most a finite number of dynamo modes with $\gamma(\epsilon) > 0$.

To summarize, the situation is that, while discreteness of the spectrum of the diffusive monodromy operator $\mathbf{T}_\epsilon$ is again ensured for the periodic case, the existence of a complete set of eigenfunctions is not. Our attitude

---

[68]In kinematic slow dynamo theory for space-time periodic flows, the magnetic field is allowed a secondary time variation with frequency Ω distinct from any multiple of ω, with $\mathbf{B}$ written as $\exp(i\Omega t)$ times a function of period $2\pi/\omega$ (see Roberts 1972, Ghil & Childress 1987). This is analogous to the secondary spatial variation discussed in §1.2.5. Although we have excluded the secondary spatial variation, the secondary temporal variation is automatically retained, the frequency being given by $|\,\text{Im}\,p|$.

in the following sections is to examine examples for dynamo action and to exhibit eigenfunctions explicitly when they exist. In approaching fast dynamo action with arbitrary time dependence, this gap in the theory is an important issue which probably has physical significance and perhaps deep mathematical implications.

9.3 Adjoints and Normality

9.3.1 Adjoint Operators

We define the *adjoint* $\mathbf{L}_\epsilon^*$ of $\mathbf{L}_\epsilon$ on $\mathcal{B}$ by

$$(\mathbf{a}, \mathbf{L}_\epsilon \mathbf{b}) = (\mathbf{L}_\epsilon^* \mathbf{a}, \mathbf{b}), \qquad (\mathbf{a}, \mathbf{b} \in \mathcal{B}). \tag{9.3.1}$$

Here time is simply a parameter, and nothing is said about the temporal evolution. However we may similarly define an adjoint of the induction operator $\mathbf{T}_\epsilon$ for induction over a given time. Both of these adjoints exist. For $\mathbf{T}_\epsilon$ this is ensured by the fact that any bounded operator has an adjoint (see, e.g., Friedman 1956 for theorems concerning the adjoint operator). For $\mathbf{L}_\epsilon$ we cannot use this theorem, but we can construct $\mathbf{L}_\epsilon^*$ explicitly using the divergence theorem, and the vector identity $\nabla \cdot (\mathbf{A} \times \mathbf{B}) = \nabla \times \mathbf{A} \cdot \mathbf{B} - \nabla \times \mathbf{B} \cdot \mathbf{A}$. We see from (9.3.1) that

$$\mathbf{L}_\epsilon^* \mathbf{a} = \mathbf{u} \times (\nabla \times \mathbf{a}) - \epsilon \nabla^2 \mathbf{a} + \nabla \phi = 0, \tag{9.3.2}$$

where ϕ must be determined by the condition that $\nabla \cdot \mathbf{a} = 0$, the vector fields we consider being in $\mathcal{B}$. Here we have used the boundary or periodicity conditions to eliminate the term which integrates out. We note that

$$\nabla \times \mathbf{L}_\epsilon^* \mathbf{a} = -\mathbf{u} \cdot \nabla \mathbf{b} + \mathbf{b} \cdot \nabla \mathbf{u} - \epsilon \nabla^2 \mathbf{b}, \qquad (\mathbf{b} = \nabla \times \mathbf{a}). \tag{9.3.3}$$

When set equal to zero, this is almost the steady induction equation, the direction of the flow velocity being reversed. The equation $-\mathbf{u} \times (\nabla \times \mathbf{a}) - \epsilon \nabla^2 \mathbf{a} + \nabla \phi = 0$ is the steady form of the equation for the vector potential, discussed in §4.2.1.

For the diffusive problem and a steady velocity field the adjoint eigenvalue problem has all the properties of the *direct* problem $\mathbf{L}_\epsilon \mathbf{b} = \mu \mathbf{b}$. The term $\mathbf{u} \times (\nabla \times \mathbf{a})$ in $\mathbf{L}_\epsilon^*$ is a first derivative term analogous to the $\mathbf{u} \cdot \nabla \mathbf{b}$ term in $\mathbf{L}_\epsilon$. The term in $\nabla \mathbf{u}$ which is present in $\mathbf{L}_\epsilon$ but absent in $\mathbf{L}_\epsilon^*$ does not affect the properties. That is, $\mathbf{L}_\epsilon^*$ has a discrete spectrum and a complete family of eigenfunctions. It can further be shown that if μ is an eigenvalue for $\mathbf{L}_\epsilon$, then μ^* is an eigenvalue of $\mathbf{L}_\epsilon^*$. We omit these proofs. It follows that for a steady flow, the growth rate $\gamma(\epsilon)$ is the same for the forward and reversed flow (Roberts 1960).

Of particular interest is the following result, which applies to any operator **L** on $\mathcal{B}$ possessing an adjoint:

Lemma 9.2 *If μ belongs to the residual spectrum of* **L**, *then μ^* belongs the the discrete spectrum of* **L***.

Again we only outline the proof. The null vectors of **L*** $- \mu^*$ are orthogonal to every vector in the range of **L** $- \mu$. By assumption μ belongs to the residual spectrum, so this range is a proper subset and hence its orthogonal complement is not empty. Hence μ is in the point spectrum of **L***.

We are also interested in the adjoints of **T**$_\epsilon$ and its perfectly-conducting counterpart **T**. If **B**$_0$ in $\mathcal{B}$ is an initial field, then **B** $=$ **T**$_\epsilon(t)$**B**$_0$ is the field evolved to time t. The possibilities for defining an adjoint operator for an evolving field can be illustrated by solutions of the heat equation $B_t - B_{xx} = 0$. On the one hand we can consider functions $B(x, t)$ on the set $[-\infty, +\infty] \times [0, T]$ in the (x, t)-plane, introduce an inner product $\int_0^T \int_{-\infty}^{\infty} A^* B \, dx \, dt$, and define a formal adjoint accordingly. The adjective 'formal' indicates that upon integration by parts in the computation of the adjoint equation, all boundary terms are discarded. In this case the adjoint heat equation is $A_t + A_{xx} = 0$, which is the ill-posed backwards heat equation. Proceeding similarly for the induction equation, we obtain a reversal of sign of the time derivative to match that of the advection term, yielding the equation for the magnetic potential but with the sign of the diffusion terms reversed.

The heat equation analog of our operator **T**$_\epsilon$ is however different, in that it and its adjoint involve physically reasonable *forward* diffusion. We define the operator T in terms of the Poisson kernel,

$$B(x, t) = TB(x, 0) = \int_{-\infty}^{+\infty} \frac{1}{\sqrt{4\pi t}} e^{-(x-y)^2/4t} B(y, 0) \, dy. \tag{9.3.4}$$

For any fixed time, T is in fact self-adjoint on $L^2(-\infty, +\infty)$. So, in terms of the operator T, the adjoint and direct problems coincide.

We now show similarly that the adjoint of **T**$_\epsilon$ on $\mathcal{B}$ with inner product (9.1.1b) corresponds to the following evolution problem: $\mathbf{A}(\mathbf{x}, t) =$ **T**$_\epsilon^*$$\mathbf{A}(\mathbf{x}, 0)$ is defined by

$$\partial_t \mathbf{A} + \mathbf{L}_\epsilon^* \mathbf{A} = 0, \qquad (\mathbf{A}(\mathbf{x}, 0) \text{ given}). \tag{9.3.5}$$

This means that the adjoint is equivalent to evolution of the magnetic potential in the reversed flow $-\mathbf{u}(\mathbf{x}, t)$, i.e., the movement of the fluid is that obtained by reversal of the velocity field. Note this is *not* the 'time-reversed' flow $-\mathbf{u}(\mathbf{x}, -t)$ defined in §6.3.4. The two are generally distinct for unsteady flows. For steady flows they are the same, and in this case we recover the result that $\gamma(\epsilon)$ is invariant under time-reversal of the flow field.

To establish (9.3.5) we let $\mathbf{A}(\mathbf{x}, 0) = \mathbf{A}_0$ and $\mathbf{B}(\mathbf{x}, 0) = \mathbf{B}_0$ where $\mathbf{A}_0$ and $\mathbf{B}_0$ belong to $\mathcal{B}$ and can be differentiated twice. Observe that

$$(\mathbf{A}_0, \mathbf{T}_\epsilon \mathbf{B}_0) = (\mathbf{T}_\epsilon^* \mathbf{A}_0, \mathbf{B}_0) = (\mathbf{A}(\mathbf{x}, t), \mathbf{B}_0) = (\mathbf{A}_0, \mathbf{B}(\mathbf{x}, t))$$
$$= \int_0^t (\mathbf{A}_0, \partial_t \mathbf{B}) \, dt + (\mathbf{A}_0, \mathbf{B}_0). \qquad (9.3.6)$$

Since $\mathbf{B}$ evolves according to the induction equation,

$$(\mathbf{A}(\mathbf{x}, t), \mathbf{B}_0) - (\mathbf{A}_0, \mathbf{B}_0) = - \int_0^t (\mathbf{A}_0, \mathbf{L}_\epsilon \mathbf{B}(\mathbf{x}, t)) \, dt$$
$$= - \int_0^t (\mathbf{L}_\epsilon^* \mathbf{A}_0, \mathbf{B}). \qquad (9.3.7)$$

Dividing both sides by t and letting $t \to 0$ we obtain

$$(\partial_t \mathbf{A}(\mathbf{x}, 0) + \mathbf{L}_\epsilon^* \mathbf{A}(\mathbf{x}, 0), \mathbf{B}_0) = 0. \qquad (9.3.8)$$

Since $\mathbf{B}_0$ is arbitrary, this establishes the result at time zero and hence for all time.

A different argument can be employed for a pulsed flow, defined in §3.1.2, where $\mathbf{T}_\epsilon$ is a product of an $\epsilon = 0$ Lagrangian map $\mathbf{T}$ followed by diffusion in a stationary conductor. We must first consider the adjoint operator $\mathbf{T}_\epsilon^*$ to be defined on vectors unconstrained to be divergence-free. This allows free choice of the scalar field ϕ in the equation for the magnetic potential,

$$\partial_t \mathbf{A} - \mathbf{u} \times (\nabla \times \mathbf{A}) + \nabla \phi - \epsilon \nabla^2 \mathbf{A} = 0. \qquad (9.3.9)$$

One common choice is $\phi = \mathbf{u} \cdot \mathbf{A}$, in which case (9.3.9) becomes

$$\frac{\partial A_i}{\partial t} + u_j \frac{\partial A_i}{\partial x_j} + A_j \frac{\partial u_j}{\partial x_i} - \epsilon \nabla^2 A_i = 0. \qquad (9.3.10)$$

The latter, with $\epsilon = 0$, is the equation for a *material surface* with $\mathbf{A}$ its local normal vector, or equivalently the cross product of two fields of material lines; see, e.g., Roberts (1967). In certain of our examples, for example the SFS map, this interpretation of the adjoint in terms of area was indicated previously; see Fig. 3.9.

From the Cauchy solutions (1.2.17, 4.2.6), it may be checked that the adjoint of the Lagrangian map of magnetic field corresponds to evolution of magnetic potential (in this gauge) in the inverse, i.e., 'time-reversed', map. The diffusive operator is just multiplication by a Poisson kernel and integration over the domain, and is self-adjoint. We denote this operator by the 'heat' operator H_ϵ. Writing $\mathbf{T}_\epsilon = H_\epsilon \mathbf{T}$, we see that in this case $\mathbf{T}_\epsilon^* = \mathbf{T}^* H_\epsilon$. Iteration of these operators produces alternating diffusion and so the position of H_ϵ is unimportant in determining $\gamma(\epsilon)$. It follows that $\gamma(\epsilon)$ is the same for the mapping and for the inverse mapping, provided diffusion is pulsed.

For general time-dependent flows with continuous diffusion, there appears to be no immediate relation between $\gamma(\epsilon)$ for a flow and for its time-reversed counterpart. However it seems likely that for smooth flows the fast dynamo growth rate γ_0 should be invariant; assuming that γ_0 is insensitive to the type of diffusion, we can replace continuous diffusion for pulsed diffusion when $\epsilon \ll 1$ and then use the argument in the previous paragraph. However a proof of this result is lacking.

9.3.2 Normal and Non-normal Operators

An operator $\mathbf{L}$ having an adjoint is said to be *normal* if $\mathbf{LL}^* = \mathbf{L}^*\mathbf{L}$. This is equivalent to $\|\mathbf{Lb}\| = \|\mathbf{L}^*\mathbf{b}\|$ for all $\mathbf{b} \in \mathcal{B}$. The implications of normality are of interest to us because neither $\mathbf{L}_\epsilon$ nor $\mathbf{T}_\epsilon$ are normal operators. This is a consequence of the different forms taken by the advection terms in the operator and its adjoint.

To get a feel for the property of normality it is helpful to consider first the finite-dimensional case (Trefethen 1992). Here it is known that there exist orthonormal eigenfunctions $\mathbf{v}_k$ associated with (possibly repeated) eigenvalues μ_k such that $(\mathbf{v}_j, \mathbf{Lv}_k) = \mu_k \delta_{jk}$; so $\mathbf{L}$ can be diagonalized in this basis (Kato 1966, p. 59). In the infinite-dimensional setting similar conclusions apply. Indeed from $\|\mathbf{Lb}\| = \|\mathbf{L}^*\mathbf{b}\|$ we see that $\mathbf{L} - \mu$ and $\mathbf{L}^* - \mu^*$ have the same null vectors. Moreover if $\mathbf{b}_j$ and $\mathbf{b}_k$ are two eigenvectors of $\mathbf{L}$ corresponding to distinct eigenvalues, then

$$0 = ((\mathbf{L}^* - \mu_j^*)\mathbf{b}_j, \mathbf{b}_k) = (\mu_k - \mu_j)(\mathbf{b}_j, \mathbf{b}_k) \tag{9.3.11}$$

and so these eigenvectors must be orthogonal. This property can be extended to the eigenspace for each eigenvalue.

For a normal operator with a spectrum which is discrete and with orthonormal eigenfunctions which are complete, the solution of the initial value problem is

$$\mathbf{B}(\mathbf{x}, t) = \sum_k (\mathbf{B}(\mathbf{x}, 0), \mathbf{b}_k)\,\mathbf{b}_k(\mathbf{x})\,e^{-\mu_k t}, \tag{9.3.12}$$

and therefore

$$\|\mathbf{B}\|^2(t) = \sum_k |(\mathbf{B}(\mathbf{x}, 0), \mathbf{b}_k)|^2 \|\mathbf{b}_k\|^2\, e^{-2(\mathrm{Re}\,\mu_k)t}. \tag{9.3.13}$$

Thus the growth or decay of magnetic energy may be decided from the eigenvalues. For a non-normal operator, the eigenfunctions are not in general orthogonal and there is no easy characterization of the evolution of energy analogous to (9.3.13). Physically, we obtain here the now familiar result that magnetic energy can be increased significantly by induction, over some finite time interval, even though (say), all eigenvalues have negative real parts and dynamo action does not occur. This point has been emphasized, in connection with hydrodynamical stability, in the recent work of Reddy & Henningson (1993), Trefethen *et al.* (1993), and Henningson & Reddy (1994).

A related property of the resolvent is that, for a normal operator

$$\|(\mu - \mathbf{L})^{-1}\| = (\mathrm{dist}(\mu, \sigma))^{-1}, \tag{9.3.14}$$

where $\mathrm{dist}(\mu, \sigma)$ is the distance from μ to the spectrum of $\mathbf{L}$ (see Kato 1966, p. 277). For a non-normal operator the equality in (9.3.14) is replaced by $\geq$, and the norm of the resolvent can be much larger than indicated by the distance to the spectrum This means in practice that extremely large amplitudes can occur in the initial value problem (Henningson & Reddy 1994). In the next section we examine the planar non-dynamo from this viewpoint to see explicitly how this can happen.

9.4 A Model Problem

We shall first introduce a simple problem of advection–diffusion of magnetic potential by a circular eddy, in order to illustrate some of the spectral theory outlined above. We consider

$$\partial_t A + L_\epsilon A = 0, \qquad L_\epsilon A = w(r)\partial_\theta A - \epsilon\nabla^2 A, \tag{9.4.1}$$

in the disc $r \leq 1$, with $A = 0$ on $r = 1$. The corresponding eigenvalue problem can be written, after separating out θ by setting $A = a(r)e^{im\theta}$, as

$$(imw(r) + \epsilon m^2/r^2)a - \epsilon(a_{rr} + a_r/r) = \mu a, \quad a(1) = 0. \tag{9.4.2}$$

This is still a difficult eigenvalue problem, and since we seek explicit results we replace (9.4.2) by an analogous Cartesian layer problem. We study $a(y)$ on $-1 \leq y \leq 1$, satisfying

$$imw(y)a - \epsilon a_{yy} = \mu a, \quad w(-y) = -w(y), \quad a(\pm 1) = 0. \tag{9.4.3}$$

Here the vector potential A is periodic in x, with $A(x, y, t) = a(y)e^{imx}e^{-\mu t}$. This model can also be viewed as an asymptotic approximation to (9.4.2) in a thin annular region with zero conditions on both boundaries. Note that we have neglected diffusion in the x-direction (although this was present in (9.4.2) as the m^2 term on the left).

The spectrum consists of two parts, one obtained when $m = 0$, giving the decaying structures which are functions of y alone (and corresponding to the rotationally invariant eigenfunctions for the circular eddy (9.4.2)), the second associated with rapidly decaying structure having $m \neq 0$. For the former, we have eigenfunctions and eigenvalues

$$a_k^0(y) = \sin k\pi y, \quad \mu_k^0 = 4\pi^2 k^2 \epsilon, \quad (k = 1, 2, \ldots). \tag{9.4.4}$$

For the latter, we have a two-point boundary value problem for a second-order equation with complex-valued variable coefficients. The choice $w = y$ can be solved in terms of Airy functions, and is the basic choice illustrating the effect of shear.

Indeed, with $\omega = y$ we have the eigenvalue problem

$$(\mu - imy)a + \epsilon a_{yy} = 0, \qquad a(\pm 1) = 0. \tag{9.4.5}$$

Setting

$$y \frac{m^{1/3} e^{\pi i/6}}{\epsilon^{1/3}} + \frac{\mu e^{2\pi i/3}}{\epsilon^{1/3} m^{2/3}} = z, \tag{9.4.6}$$

we obtain Airy's equation $a_{zz} - za = 0$. Two linearly independent solutions are denoted $\mathrm{Ai}(z)$ and $\mathrm{Bi}(z)$ (Abramowitz & Stegun 1972). Forming a linear combination of these two solutions we must have $\alpha\,\mathrm{Ai}(z_\pm) + \beta\,\mathrm{Bi}(z_\pm) = 0$ where $z_\pm$ are the values of z corresponding to $y = \pm 1$. The solvability condition $\mathrm{Ai}(z_+)\,\mathrm{Bi}(z_-) - \mathrm{Ai}(z_-)\,\mathrm{Bi}(z_+) = 0$ provides the eigenvalue equation. To solve this eigenvalue problem for small ϵ, tentatively set

$$\frac{\mu e^{2\pi i/3}}{\epsilon^{1/3} m^{2/3}} = \frac{m^{1/3} e^{\pi i/6}}{\epsilon^{1/3}} - c_k, \tag{9.4.7}$$

so that

$$z_- = -c_k, \qquad z_+ = 2(m/\epsilon)^{1/3} e^{\pi i/6} - c_k. \tag{9.4.8}$$

If c_k is real, positive, and $O(1)$ in ϵ, then z_- can be located near a zero of Ai to make $\mathrm{Ai}(z_-)\,\mathrm{Bi}(z_+)$ as small as we like. At the same time, $\mathrm{Ai}(z_+)\,\mathrm{Bi}(z_-)$ has an exponentially small factor which is changed negligibly by $O(1)$ variations in c_k. Thus for small ϵ eigenvalues are given by (9.4.7) where c_k is close to a zero of Ai. The associated eigenfunction is essentially $\mathrm{Ai}(z)$. Note that if μ is an eigenvalue with eigenfunction $a(y)$, then μ^* is an eigenvalue with eigenfunction $a^*(-y)$. We thus have, for $m = 1, 2, \ldots$, the asymptotic spectrum,

$$\mu \sim -im + \epsilon^{1/3} m^{2/3} e^{\pi i/3} c_k, \qquad (k \to \infty), \tag{9.4.9}$$

and their complex conjugates.

Consider now the perfectly conducting problem corresponding to (9.4.5), that is $(\mu - imy)a(y) = 0$. Here there is only *continuous* spectrum $\sigma_c = \{\mu : |\mathrm{Im}\,\mu| \leq m, \mathrm{Re}\,\mu = 0\}$, with corresponding solutions $a = \delta(y - \mu/im)$. These solutions are not square-integrable, and so are not eigenfunctions corresponding to discrete spectrum. That σ_c is indeed continuous spectrum can be proved by introducing a sequence of smooth L^2 functions approximating the delta function, and thus verifying that we have approximate spectrum. This cannot be residual spectrum since the adjoint operator (here simply the complex conjugate) does not have any discrete spectrum. Hence $\sigma = \sigma_c$ is entirely continuous spectrum. This situation seems to be typical of many fast dynamos: points of the continuous spectrum in L^2 correspond to eigenfunctions in a larger space of distributions; see de la Llave (1993) and Núñez (1993, 1994).

What is the role of small diffusion in this model? We see that the effect of $\epsilon > 0$ is to remove completely the continuous spectrum and replace it by point eigenvalues clustered near $\mathrm{Im}\,\mu = \pm m$. Physically, the continuous spectrum corresponds to concentrated layers of A advected by the flow. The magnetic field is a bilayer of intense field in opposing directions (see de la Llave 1993). We can think of these bilayers as removed by the slightest diffusion, essentially by cancellation of flux. The endpoints $y = \pm 1$ have the special significance of allowing 'half' of a delta-function, thus in principle removing the canceling flux, but the structure of the diffusive eigenfunctions is actually much more complicated, involving an alternating pattern (in y) of what are actually long (in x) thin flux loops. The small y-field gets stretched by the flow, and thus maintains the form of the structure, but there is overall decay. This general picture will reappear below in our discussion of baker's maps; see §9.5.2.

It is worth recalling here that the asymptotic solution of the initial-value problem for

$$a_t + im\omega(y)a - \epsilon a_{yy} = 0, \quad \omega(-y) = -\omega(y), \quad a(\pm 1) = 0, \qquad (9.4.10)$$

can be obtained by the method described in §5.2.2. For large t, and $m \neq 0$,

$$A(x, y, t) \sim \sum_m a_m(y)e^{im(x-\omega(y)t)-\epsilon(\omega'(y))^2 m^2 t^3/3}. \qquad (9.4.11)$$

The functions $a_m(y)$ are determined from initial conditions. Putting $\epsilon = 0$ in (9.4.11) we can see the result of the continuous spectrum. The $\epsilon^{1/3}$ dependence in (9.4.9) is consistent with the decay rate of (9.4.11).

Lastly, we observe that the differentiation of (9.4.11) with respect to y introduces the amplitude factor $-im\omega'(y)t$. These factors can lead to substantial growth of magnetic energy before the time of order $\epsilon^{-1/3}$ when decay ensues. This is, in this example, an explicit verification of the role of non-normality in producing significant transient growth; see §9.3 and the remarks following (9.3.14).

9.5 Baker's Maps Revisited

A related investigation of the spectral representation of a non-dynamo can be carried out for the folded baker's map. In a sense this is a simpler example than the continuous advection–diffusion problem just considered, but the spectrum of the induction operator in the perfectly conducting case turns out to be completely residual instead of continuous; so it has some unusual features which are of interest to us. We shall consider also the diffusive spectrum, and both the direct and adjoint maps.

We first recall that the folded baker's map was defined in Chap. 3 by

$$M : (x, y) \rightarrow \begin{cases} (2x, y/2), & (0 \leq x < 1/2), \\ (2 - 2x, 1 - y/2), & (1/2 \leq x < 1); \end{cases} \qquad (9.5.1)$$

see Fig. 3.1. The perfectly conducting induction operator $\mathbf{T}$ is defined here on fields $(B_1(x,y), B_2(x,y), 0)$ periodic with period one in (x,y). We consider first the proposition discussed in Chap. 4, that eigenfunctions cannot be smooth in the perfectly conducting limit except in certain simple cases of perfect alignment, such as the stacked baker's map. We argue now that the folded baker's map has no simple perfect growing eigenfunction, that is, a function of bounded variation or a function differentiable on an interval of positive length. Indeed, we recall that

$$\mathbf{TB}(x,y) = \mathbf{J}(M^{-1}(x,y)) \cdot \mathbf{B}(M^{-1}(x,y)), \tag{9.5.2}$$

where

$$M^{-1} : (x,y) \to \begin{cases} (x/2, 2y), & (0 \le y < 1/2), \\ (1 - x/2, 2(1 - y)), & (1/2 \le y < 1). \end{cases} \tag{9.5.3}$$

Since

$$\mathbf{J} = \operatorname{sign}(1/2 - y) \begin{pmatrix} 2 & 0 \\ 0 & 1/2 \end{pmatrix}, \tag{9.5.4}$$

we see that y-components are reduced by half while x-components are doubled. Iterating many times, it follows that the y-components must vanish in any growing eigenfunction. But, if the eigenfunction is the divergence-free function $(f(y), 0)$ and the eigenvalue is λ, then clearly $\lambda = 2$ since $\mathbf{T}$ quadruples the magnetic energy. Thus we have $f(y) = f(2y)$ for $0 \le y < 1/2$ and $f(y) = -f(2 - 2y)$ for $1/2 \le y < 1$. Iterating each of these expressions, we obtain four relations, one being $f(y) = f(4y)$ for $0 \le y < 1/4$, and this can be continued. We thus see that any eigenfunction must have the property that all of its values, and their negatives, are contained in any arbitrarily small finite subinterval. Such a function, if non-constant, oscillates on an arbitrarily fine length scale, and is differentiable on no interval of finite length.

Intuitively, the difficulty here is indeed one of scales of spatial variation. Whatever scales exist in a trial eigenfunction, the folded baker's map reduces the maximum scale by $1/2$, the fold introducing in addition a change of sign and a reflection of half of the image. The only way to construct an eigenfunction is to have no finite scale at all! When one enlarges the class of permissible functions, and in particular allows distributions, it becomes possible to again assign a meaning to the notion of an eigenfunction of a baker's map. Examples will be given in the following sections.

9.5.1 The Parallel Maps T and S

In §3.1 we introduced the baker's maps and discussed examples of their action on fields of the form $(b(y), 0)$. For the folded baker's map, we may for these fields replace $\mathbf{T}$ as defined by (9.5.2) by the map T:

$$Tb(y) = \begin{cases} 2b(2y), & (0 \le y < 1/2), \\ -2b(2-2y)), & (1/2 \le y < 1). \end{cases} \tag{9.5.5a}$$

We shall refer to T as the *parallel map* corresponding to **T**. The expression (9.5.5a) is sometimes abbreviated, using the tent map $\tau(y) = \min(2y, 2-2y)$, as

$$Tb(y) = 2\,\text{sign}(1/2 - y)\,b(\tau(y)). \tag{9.5.5b}$$

For T we then define the corresponding adjoint map S, using the inner product (3.2.6), by the condition (Friedman 1956, Bollabás 1990)

$$(b_1, Tb_2) = (Sb_1, b_2) \tag{9.5.6}$$

for all $b_1(y)$, $b_2(y)$. Using (9.5.5) in the last equation, a short calculation gives

$$Sb(y) = b(y/2) - b(1 - y/2). \tag{9.5.7}$$

We see immediately from (9.5.7) that S has the eigenvalue zero with associated eigenfunction $b = 1$. This shows that the adjoint operator has a smooth eigenfunction, and that zero is a point eigenvalue.

This property depends upon our restricting the spaces to parallel fields. The adjoint to **T**, defined as in §9.3.1 using an inner product integral over the unit square, takes the form

$$\mathbf{SB}(\mathbf{x}, t) = \mathbf{J}^{\mathsf{T}}(M(x,y), t) \cdot \mathbf{B}(M(x,y), 0). \tag{9.5.8}$$

This operator behaves differently in $0 \le x < 1/2$ and $1/2 \le x < 1$. The 'smoothing' effect of S comes from the averaging over x and therefore the summing of these two parts of the image. The map M in (9.5.8) has the effect of 'unfolding and compressing' an initial field $(b(y), 0)$. Even though there are still the factors ± 2 affecting the amplitude, it is the averaging in x which produces the well-behaved operator S. We suggest this geometry of the adjoint in Fig. 9.2.

We note that

$$TSb(y) = 2[b(y) - b(1 - y)], \quad STb(y) = 4b(y); \tag{9.5.9}$$

so $TS - ST = -2[b(y) + b(1 - y)]$. Thus T is not a normal operator. It is true however that $TS - ST$ annihilates any $b(y)$ which is odd with respect to $y = 1/2$. Since T maps odd functions to odd functions, T is normal there. On the other hand S does not map odd functions into odd functions; so we can not restrict the operators to this subspace.

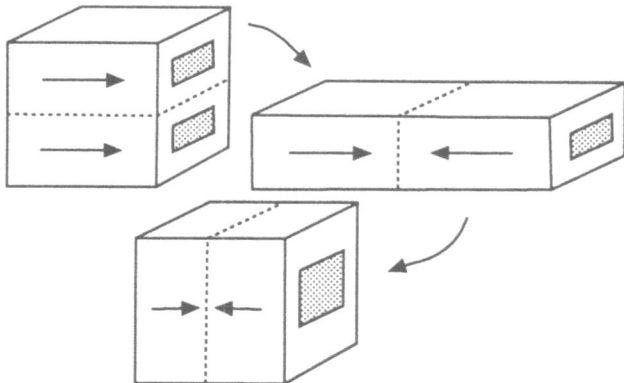

Fig. 9.2 The adjoint map for the folded baker's map applied to parallel fields **B** = $(b(y), 0)$. An initial field is unfolded and compressed, then increased in amplitude by 4 to yield **SB**. Averaging in the x-direction defines Sb. The factor of 4 is equivalent to operating on area elements with normal parallel to the field, as indicated by the shaded squares.

9.5.2 The Spectra of S and T

It will be instructive to consider first the $\epsilon = 0$ spectra of the parallel maps, since this will give us a clue as to the simplest construction of diffusive spectra. Our objective remains the analysis of growth or decay of iterates of a direct operator (here T) making use of a weak formulation. For the folded baker's map the issues can be highlighted by noting that whereas $(1, T^n \cdot 1) = 0$ for $n = 1, 2, 3, \ldots$ as one expects from the folding, it is less obvious but equally true that $(1 - y, T^n \cdot 1) = 1/2$ and that $(y^3 - 3y^2 + 2y, T^n \cdot 1) = 1/4^{n+1}$ for $n = 0, 1, 2, \ldots$. These measures of, in this case, decay, disagree, yet all are reasonable functionals which respond to different aspects of the large oscillations of the field. Analogous results for growth of magnetic field under the slide map were noted in §3.4. These results and their implications for the structure of $T^n b$ can be explored in a rather straightforward way from the eigenfunctions of S

Considering first the adjoint operator (9.5.7), we have indicated the eigenfunction $b = 1$ corresponding to the eigenvalue 0. In fact this eigenvalue is infinitely degenerate, since $Sb = 0$ implies $b(y/2) = b(1 - y/2)$, which is satisfied by any function even with respect to $y = 1/2$. The direct operator T converts any function into an odd one, hence orthogonal to any even function.

It is also easy to guess an eigenfunction of S which is a linear function, namely $b_1 = y - 1$, with eigenvalue 1. Since the functions even in $y - 1/2$ are already accounted for, this suggests looking next at a polynomial of degree 3. We see from Sy^3 that the corresponding eigenvalue must be $1/4$. The remaining coefficients are readily found, yielding $b_3 = y^3 - 3y^2 + 2y = (y-1)^3 - (y-1)$. In this way, we obtain an infinite sequence of eigensolutions of the form

$$b_{2n+1} = \sum_{m=0}^{n} c_m^{(n)}(y-1)^{2m+1}, \quad \lambda_{2n+1} = 2^{-2n}, \quad (n = 0, 1, 2, \ldots), \quad (9.5.10)$$

where we may set $c_n^{(n)} = 1$. The coefficients satisfy the triangular system

$$c_m^{(n)} = \sum_{k=m}^{n} 4^{n-k} \binom{2k+1}{2m+1} c_k^{(n)}, \quad (9.5.11)$$

and may be found by elimination; see the more general results for the slide map in §9.6. By setting $y = 0, 1$ in $Sb = \lambda b$ it is seen that all of these eigenfunctions except b_1 must vanish at 0 and 1.

Now it is not difficult to see that any smooth function $b(y)$ defined on $[0, 1]$ may be written as the sum of an even function (with respect to $y = 1/2$) and a linear combination of eigenfunctions. Thus

$$b(y) = b_{\text{even}}(y) + \sum_{n=1}^{\infty} P_{2n+1}(b)\, b_{2n+1}(y), \quad (9.5.12)$$

where the P_{2k+1} are the (non-orthogonal) projections onto the eigenfunctions. That is, we have a complete set of eigenfunctions of S, with point eigenvalues at $1, 1/4, 1/16, \ldots$, and an eigenvalue 0 of infinite degeneracy. Any other value of λ is in the resolvent set of S, with

$$(\lambda - S)^{-1}b = \lambda^{-1}b_{\text{even}}(y) + \sum_{n=1}^{\infty} P_{2n+1}(b)(\lambda - \lambda_{2n+1})^{-1}b_{2n+1}(y). \quad (9.5.13)$$

We thus have a relatively simple example of a direct construction of eigenfunctions of an adjoint induction operator. Certain of their properties can be established without direct construction, such as the fact that they must be C^∞ (see §9.6.3) but this is one of the few cases where explicit results are known.

We now turn to the spectrum of the direct operator T for the folded baker's map. We remark first that T is not a compact operator on $L^2[0, 1]$ Since Tb is orthogonal to 1 for any b it is also clear that the range of T is a proper subset of $L^2[0, 1]$ and so $\lambda = 0$ is in the residual spectrum of T; see §9.1.1. Similarly, $(T - \lambda_{2k+1})b$ as defined above is orthogonal to the eigenfunction b_{2k+1} of S; so λ_{2k+1} for $k = 0, 1, 2, \ldots$ are also in the residual spectrum of T.

We can also easily identify the region $|\lambda| > 2$ as a subset of the resolvent set, because $\|T\| = 2$ (recall T quadruples magnetic energy). It is more difficult to decide the status of the remaining points of the λ-plane. Note first that if we seek to solve $(T - \lambda)b = f$ with $\lambda \neq 0$ we might as well take f to be odd about $y = 1/2$, since otherwise we can write $(T - \lambda)b = f_{\text{odd}} + f_{\text{even}}$ and instead solve $(T - \lambda)(b + f_{\text{even}}/\lambda) = f_{\text{odd}} + Tf_{\text{even}}/\lambda$. Thus we can work in the subspace of odd fields. But on this subspace we have seen that T is normal, and so we find $\|(T - \lambda)b\| = \|(S - \lambda^*)b\|$. It follows that if λ is in the approximate spectrum of T, λ^* is in the approximate spectrum of S. Since the spectrum of S is discrete and identified above, we have also the approximate spectrum of T. We know the latter contains no eigenvalues but all of the continuous spectrum. It follows that the spectrum of T is the same as that of S and is entirely *residual*.

To take a specific example of inversion, consider $(T - 4)b = 1$. From the Neumann series we see that we get considerable structure in b, with oscillations on all scale. For $\lambda < 2$ we do not have a general formula for the inverse of the resolvent. One special case for $\lambda = 0$ is $Tb = 2y - 1$. Here the right-hand side is chosen to be orthogonal to the eigenfunction 1 of S. We must have $2b(2y) = 2y - 1$ for $0 \leq y < 1/2$ and so $b = (y - 1)/2$. Checking this for $1/2 \leq y < 1$, we verify $-2b(2 - 2y) = 2y - 1$. The choice $\lambda = 0$ is probably the only one not giving small-scale oscillations.

What can then be said about 'eigenfunctions' of the direct operator T, if they are not required to be L^2 functions? This is an interesting question because the physics suggests that smooth eigenfunctions should develop concentrations of flux in the perfectly conducting limit which might be constructed as distributions or generalized functions (Friedman 1956, Lighthill 1962). Because of the dominant eigenvalue of unity, it is tempting to introduce a sequence $\beta_1^{(n)} = T^n b_0$, $b_0 = 2$, and define our eigenfunction $\beta_1(y)$ as the limit (in the sense of distributions) of this sequence. From the definition of the adjoint, however, for any smooth test function $f(y)$ defined on $[0,1]$,

$$(f, T^n b_0) = (S^n f, b_0) \rightarrow f(0) - f(1), \tag{9.5.14}$$

since $[f(1) - f(0)](y - 1)$ is the projection of f onto $y - 1$.[69] We obtain this since a high power of S acts like this projection. We can also obtain the result directly by a tangent line approximation to f on each subinterval of length 2^{1-n}. After n applications of T, evaluate the derivative of f at the midpoints of n intervals of this length and compute the integral of the resulting piecewise linear approximation, giving

$$(f, T^n b_0) = 2^n \sum_{j=1}^{2^n-1} \frac{df}{dy}((2j - 1)/2^n)\left(-4 \int_0^{2^{-n}} y\, dy\right)$$

$$= -2^{-(n-1)} \sum_{j=1}^{2^n-1} \frac{df}{dy}((2j - 1)/2^n) \approx -\int_0^1 \frac{df}{dy}\, dy, \tag{9.5.15}$$

in agreement with (9.5.14).

---

[69]Recall that $y - 1$ is the only eigenfunction with non-zero eigenvalue not vanishing at both 0 and 1. Thus $f(0) = -P_1(f) + b_{\text{even}}(0)$ and $f(1) = b_{\text{even}}(1) = b_{\text{even}}(0)$, and so $P_1(f) = f(1) - f(0)$.

To obtain a second, distinct 'eigendistribution' corresponding to the eigenvalue $1/4$, we should choose b_0 to be orthogonal to b_1 but not to b_3, and consider the sequence $(f, (4T)^n b_0)$. A suitable choice (anticipating a convenient normalization) is $b_0 = 120(y^2 - y + 1/6)$. To obtain the effect of the distribution upon f we must now take a three-term Taylor's series about the points $(2j - 1)/2^n$. Denoting the distribution by β_3, we obtain

$$(f, \beta_3) = -\frac{1}{6} \int_0^1 \frac{d^3 f}{dy^3} \, dy = \frac{1}{6} \left[\frac{d^2 f}{dy^2}(0) - \frac{d^2 f}{dy^2}(1) \right]. \tag{9.5.16}$$

This checks with a direct calculation of the projection P_3. In (9.5.12) we can take the second derivative of both sides to obtain on the right the term $6P_3(y - 1)$, and by our previous argument we recover (9.5.16). By this construction, moreover, $(f, T\beta_3) = (1/4)(f, \beta_3)$ from a calculation involving (9.5.16) and the adjoint S.

This process can be continued and used to construct the eigendistributions

$$\beta_{2n+1}: \quad (f, \beta_{2n+1}) = \frac{1}{(2n+1)!} \left[\frac{d^{2n} f}{dy^{2n}}(0) - \frac{d^{2n} f}{dy^{2n}}(1) \right]. \tag{9.5.17}$$

Note from (9.5.5b) that if $g = Tf$ for any test function f, then $g(0+) = -g(1-)$ and similarly for all even derivatives, and $g'(0+) = g'(1-)$ and similarly for all odd derivatives. We can then see how the degenerate zero eigenvalue of S reflects the vanishing under T of those distributions which produce odd derivatives at the endpoints.

Rado (1993) has represented these distributions in a slightly different way, which we describe in the case of β_1. Define the 'half' delta functions $\delta_\pm$ by

$$\int f(y) \delta_\pm(y) \, dy = \frac{1}{2} f(0\pm). \tag{9.5.18a}$$

We are allowing here test functions which are only piecewise smooth. For smooth test functions, $\delta_-(y) + \delta_+(y) = \delta(y)$. It is also easy to show that

$$\int_0^c f(y) \delta_-(y) \, dy = \int_{-c}^0 f(y) \delta_+(y) \, dy = 0, \tag{9.5.18b}$$

for any $c > 0$, and that

$$\delta_+(-y) = \delta_-(y). \tag{9.5.18c}$$

We now set

$$\beta_1 = 2(\delta_+(y) - \delta_-(y - 1)), \tag{9.5.19}$$

and restrict test functions to $[0,1]$ as smooth functions with $f(0)$ and $f(1)$ arbitrary. Then

$$T\beta_1 = \begin{cases} 4\delta_+(2y) - 4\delta_-(2y-1), & (0 \leq y < 1/2), \\ -4\delta_-(2-2y) + 4\delta_+(1-2y)), & (1/2 \leq y < 1). \end{cases} \qquad (9.5.20)$$

Using (9.5.20) and the scaling property $\delta_\pm(y) = 2\delta_\pm(2y)$, we see that $T\beta_1 = \beta_1$ plus a multiple of the difference of two half δ functions. The latter, when applied to a test function, gives $f(1/2+) - f(1/2-) = 0$. Note that in our inner product this distributional eigenfunction is orthogonal to *adjoint* eigenfunctions for eigenvalues other than 1, as it must be on general grounds.

We thus have a fairly complete picture of the folded baker's map. When T is applied to a smooth function defined on $[0,1]$, we obtain a function whose behavior near $0-$ and $1+$ are connected by the symmetry of the map. The only non-trivial eigensolutions are distributions allowed by this symmetry. If we extend the functions outside $[0,1]$ as smooth and periodic with period one, and impose this periodicity on the test functions, then the distributions vanish, that is, T does not even have eigenfunctions in the form of non-zero distributions. In a sense the folded map is the extreme form of a non-fast dynamo, in that any smooth periodic field is quickly mixed to zero, as measured by averaging over essentially any small interval. The residual distributions at the endpoints cancel adjacent distributions under a periodic extension. What then remains is fine-scale structure which is a zero distribution. We shall return to this description of decay when we consider the diffusive spectra in §9.5.3. Similar considerations arise in the analysis of baker's maps applied to the transport of a scalar field (Childress & Klapper 1991).

We conclude by again emphasizing the role of these spectra on the measurement of growth and decay. We see that it is a non-trivial matter to determine an appropriate measure of the growth of the 'average field' arising under the parallel map T. A reasonable choice is clearly obtained from the eigenvalue of maximum modulus of the adjoint operator. In that case we compute $(y-1, T^n b)$. As $n \to \infty$ this leads to the formal result that field in the baker's map neither grows nor decays in an average sense. This is intuitively satisfying if only because of the difficulties we have noted elsewhere connected with the flux conjecture when dynamo action does not occur (see §§3.1, 4.1.4). On the other hand net flux through the unit interval is zero after the first application of T. The function $y-1$ produces an $O(1)$ integral because it selectively favors one sign of flux in each bilayer of canceling flux. Clearly diffusion will control the ultimate emergence of a flux average.

This situation is quite different in maps which preserve flux. The three-piece baker's map with two folds stretches by a factor 3, the corresponding parallel map arranging the layers in the order $+-+$. The corresponding adjoint map S_3 is given by

$$S_3 b(y) = b(y/3) - b(2/3 - y/3) + b(2/3 + y/3). \qquad (9.5.21)$$

Here 1 and y are eigenfunctions associated with the dominant eigenvalue unity, the next smallest being the eigenvalue $1/9$ with eigenfunction $y^3 - y^2$. In this case a simple flux average always gives the correct result.

9.5.3 The Spectra of S_ϵ and T_ϵ

We turn now to the diffusive spectra, obtained by allowing diffusion for time one after each application of the maps **T** or **S**. We have noted previously the effect of repeated application of this sequence on a class of periodic fields (see 3.1.8), leading to decay after $O(\log(1/\epsilon))$ steps. Out aim now is to study the diffusive spectrum of the combined induction–diffusion operator $\mathbf{T}_\epsilon$. Although the general results for smooth flows do not apply to discontinuous maps, we shall try to construct the spectrum directly and verify that it remains discrete. In view of the effect of the map in producing parallel magnetic field, we again restrict $\mathcal{B}$ to complex square-integrable fields of the form $\mathbf{B} = (b(y), 0)$, periodic in y with period unity.[70]

The diffusive parallel operators are T_ϵ and S_ϵ, defined as $H_\epsilon T$ and SH_ϵ respectively, where if $b = \sum_k e^{2\pi iky} b_k$ then the diffusion operator H_ϵ is defined by (cf. 3.2.4)

$$H_\epsilon b = \sum_k e^{2\pi iky} b_k \exp(-4\pi^2 \epsilon k^2). \tag{9.5.22}$$

Because diffusion in baker's maps is conveniently represented by its effect on the Fourier series, it is of interest also to represent T and S as operators in Fourier space. The operator $T : b \to b'$ is determined by the matrix $\mathbf{A}$, $b'_j = A_{jk} b_k$ where

$$A_{jk} = \begin{cases} (4ij/\pi)(4k^2 - j^2)^{-1}, & (j \text{ odd}), \\ \pm 1, & (j = \pm 2k), \\ 0, & (j = k = 0), \\ 0, & (j \text{ even and } j \neq \pm 2k). \end{cases} \tag{9.5.23}$$

This shows that the the the basic halving of scales is accompanied by the formation of other structure with a $1/k$ decay of Fourier coefficients. These structures are of course the folds, which introduce discontinuities of slope in the field. These are absent in the Fourier form of the corresponding operator for the stacked map (see 3.1.8); in that case A_{mn} is empty except along the line $m = 2n$ where it is 2. The adjoint map can be similarly represented, and its matrix is just the complex conjugate of the transpose of $\mathbf{A}$. From these definitions we have the following action on Fourier coefficients:

$$S_\epsilon T_\epsilon : b_n \to \sum_n D_{mn} b_n, \quad D_{mn} = \sum_k A^*_{km} A_{kn} e^{-8\pi^2 \epsilon k^2}, \tag{9.5.24a}$$

$$T_\epsilon S_\epsilon : b_n \to \sum_n E_{mn} b_n, \quad E_{mn} = \sum_k A_{mk} A^*_{nk} e^{-4\pi^2 \epsilon (m^2 + n^2)}. \tag{9.5.24b}$$

---

[70] Alternatively, we can define $T = \mathbf{TPQ}$ where $\mathbf{P}$ projects onto the x-component and $\mathbf{Q}$ averages over lines $y = \text{const}$. Both $\mathbf{P}$ and $\mathbf{Q}$ are self-adjoint on the complex space of fields $\mathbf{b} = (b_1(x,y), b_2(x,y))$ with the L^2 inner product over the unit square.

We see explicitly from (9.5.23) and (9.5.24) that T_ϵ is non-normal.

Consider first the spectrum of S_ϵ. The property of being even with respect to $y = 1/2$ is invariant under diffusion; so zero remains an eigenvalue of S_ϵ with the same set of eigenfunctions, even in $y - 1/2$. From the periodic extension of $b(y)$ we then see that, since any eigenfunction has $b(1) = 0$, any diffusive eigenfunction, which must be continuous as a periodic function, will have $b(0) = 0$. It follows that we need only consider the Fourier modes $\sin \pi n y$ for integer n. Since S in effect divides the wavenumber in a given mode by two, we can generate a possible eigenfunction $\phi_q(y)$ from any odd integer q of the form $\phi_q(y) = \sum_{k=0}^{\infty} b_k \sin(\pi q 2^k y)$ with $b_0 = 1$. The eigenvalue equation is then

$$\lambda \phi_q = S_\epsilon \phi_q = 2 \sum_{k=0}^{\infty} b_{k+1} \exp(-\pi^2 q^2 4^{k+1} \epsilon) \sin(\pi q 2^k y), \qquad (9.5.25)$$

which requires that $2 \exp(-\pi^2 q^2 4^{k+1} \epsilon) b_{k+1} = \lambda b_k$ for $k = 1, 2, \ldots$. There is no non-zero solution that produces a convergent Fourier series. Thus for $\epsilon > 0$, S_ϵ has no eigenfunctions except for those even in $y - 1/2$.[71] The reason for this abrupt loss of spectrum can be understood from the way the operator S_ϵ acts on functions which vanish at both end points. We see that two halves of the graph are stretched, and one is then reflected about $y = 1/2$, and subtracted from the other. This stretching works in Fourier space to shift magnetic energy from high to low modes. In the absence of diffusion a balance can be effected by taking advantage of the algebraic decay of Fourier coefficients. With diffusion and extension to period 1, this is no longer possible and the shifts drive the energy to zero.

We now see that the resolvent set of S_ϵ indeed contains all points except the origin. For, the function $\lambda^{-1} \sin \pi n y$ with n odd maps into $\sin \pi n y$ under $\lambda - S_\epsilon$. If n is even, a pre-image is $\lambda^{-1} \sin \pi n y + \lambda^{-2} S_\epsilon \sin \pi n y + \ldots$, the series necessarily terminating. Hence $(\lambda - S_\epsilon)^{-1}$ exists on the relevant Fourier modes (a set dense in $L^2[0, 1]$), provided only that $\lambda \neq 0$.

These results, although for a simple model, carry some interesting physical implications. For the direct operator the magnetic energy is clearly carried to small scales where it is dissipated, but this obvious property of the diffusive system gives us no direct way of dealing with the perfect conductor, except to say that 'energy is lost to small scales'. For the adjoint operators we have a quite different representation of the same phenomenon in the case of the folded baker's map. Although $(y - 1, T^n b)$ is in general non-vanishing for all n, indicating the large oscillations in the field, for arbitrary positive ϵ the energy in the sine modes begins to be swept toward wavenumber one where

---

[71]By the same token if the diffusion is backwards in time, so the sign in the exponent of the exponential factors is changed, an eigensolution exists for each q. These are relevant to an initial-value problem where a time-integration is involved in the definition of the adjoint of the heat operator; see §9.3.1.

it is removed. Considered from the standpoint of the images of a single such Fourier mode, the removal of energy has the form

$$\text{even} \rightarrow \text{even} \rightarrow \ldots \rightarrow \text{even} \rightarrow \text{odd} \rightarrow 0,$$

where we again refer to terms even and odd relative to $y = 1/2$. This simple point is of considerable significance in understanding the generalizations of the folded baker's map considered in §9.6 and will be taken up again there.

A final point concerning S_ϵ is the possibility (mentioned in §3.1.1) of extending the functions to have period two, by making $b(y)$ even with respect to $y = 1$. This does not change our conclusion regarding the resolvent of S_ϵ, however, since any dense set of Fourier modes can be used.

For the direct operator T_ϵ the effect of the stretching of field tends to oppose the diffusion of flux away from concentrations and it is of interest to find diffusive eigenfunctions which represent a broadening of the distributions obtained at $\epsilon = 0$. It turns out that these eigenvalues *will* depend upon the nature of the periodic extension. For a period 1 extension they do not exist. In this case zero is in the residual spectrum of T_ϵ, since T maps L^2 into functions odd with respect to $y = 1/2$, and all other values of λ are in the resolvent set. To show this we again need consider only the odd sine functions and the argument is essentially the same as for S_ϵ, except that the series $\lambda^{-1} \sin \pi n y + \lambda^{-2} T_\epsilon \sin \pi n y + \cdots$ is now infinite but strongly convergent. Thus the destruction of flux is total after $O(\log(1/\epsilon))$ iterations of T_ϵ. On the other hand the absence of eigenfunctions does not reveal the increase of magnetic energy to a value of order $4^{\log(1/\epsilon)}$ which occurs prior to this decay.

If the periodic extension is with period 2 rather than period 1, Bayly & Childress (1989) suggested that T_ϵ might have a non-trivial eigenfunction for an eigenvalue near 1 (if $\epsilon \ll 1$), representing persistent flux concentrations at the boundary points. The diffusive eigenfunction is then a broadening of delta functions of opposite signs at $y = 0$ and 1, plus additional structure obtained from the mapping of the half-delta functions as defined by (9.5.18a).

To construct this eigenfunction for $\epsilon \ll 0$, we try, near $y = 0$, a function of the Gaussian form e^{-ay^2}. Under T we then obtain $2e^{-4ay^2}$, and, diffusing for time 1,

$$T_\epsilon e^{-ay^2} = 2(1 + 16\epsilon a)^{-1/2} e^{-4ay^2/(1+16\epsilon a)} = \lambda e^{-ay^2}, \qquad (9.5.26)$$

provided that $a = 3/16\epsilon$ and $\lambda = 1$. To accommodate the periodicity condition we must sum over an infinite array of such concentrations; so we expect the actual eigenfunction to be obtained in the form

$$b(y) = \sum_{n=-\infty}^{+\infty} (-1)^n e^{-3(y-n)^2/16\epsilon} + b'(y), \qquad (9.5.27)$$

where b' is a correction which accounts for the interaction between the exponentially decaying concentrations, together with the smooth, largely canceling structures near the points $j/2^k$ modulo 1 for arbitrary integers j, k. The latter represent the effect of diffusion on the canceling half-delta functions discussed in §9.5.2. When $\epsilon \ll 1$ (9.5.27) can be written approximately

$$b \approx f_0 + f_1 + T_\epsilon f_1 + T_\epsilon^2 f_1 + \cdots,$$
$$f_0 = e^{-3y^2/16\epsilon} - e^{-3(y-1)^2/16\epsilon},$$

(9.5.28)

where f_1 is a structure (two canceling half-Gaussians) localized near $y = 1/2$.

If ϵ is *large* compared to 1, the existence of this eigenfunction can be checked asymptotically. The eigenfunction must then be the odd mode $\cos \pi y$ approximately, and the mapping T_ϵ then gives the eigenvalue as approximately $(4/3\pi) \exp(-\pi^2 \epsilon)$. We conjecture that there is a family depending on ϵ, with $\lambda(\epsilon)$ extending from 1 at $\epsilon = 0$ down to zero at $\epsilon = \infty$. While the existence of this eigenfunction is something of a peripheral issue since no growth of field is involved, it is of interest in interpreting diffusive numerical computations where a growth rate of near zero, rather than near $-\infty$, results.

9.6 The SFSl and SFS Maps

We now take up the important examples of spectra of elementary fast dynamo maps. The examples will serve to illustrate the general principle that the less cancellation there is in folded and stretched field, the easier it is to measure the average flux. In the models to be examined, the degree of cancellation is variable and we can see what happens to the resulting spectra.

9.6.1 The Slide Map in a Perfect Conductor

The slide map, introduced by in one form by Finn *et al.* (1989, 1991) and studied by Klapper (1991) as a variant of the SFS map, will be our first example. We have noted in §3.4 some properties of the spectrum of of the adjoint map S^{SFSl}, which we shall now examine in more detail. Recall that the adjoint is defined by

$$S^{\text{SFSl}} b(y) = e^{-\pi i \alpha} b(y/2) - e^{\pi i \alpha} b(1 - y/2).$$

(9.6.1)

This is not quite an inverse of the direct map T^{SFSl} of (3.4.2), since $S^{\text{SFSl}} T^{\text{SFSl}} b(y) = 4b(y)$. Much of our discussion will follow closely that for the folded baker's map, to which (9.6.1) reduces when $\alpha = 0$. We may compute a complete family of eigenfunctions for S^{SFSl} in the form $b_n(y) = \sum_{m=0}^{n} c_m^{(n)} y^m$ for $n = 0, 1, \dots$. The coefficients satisfy the triangular system

$$\lambda_n^* c_m^{(n)} = \frac{e^{-\pi i \alpha}}{2^m} c_m^{(n)} - \sum_{k=m}^{n} \frac{(-1)^m e^{\pi i \alpha}}{2^m} \binom{k}{m} c_k^{(n)}.$$

(9.6.2)

Setting $c_n^{(n)} = 1$ we recover the eigenvalues (3.4.4),

$$\lambda_n^* = 2^{-n}[e^{-\pi i \alpha} - (-1)^n e^{\pi i \alpha}]. \tag{9.6.3}$$

The infinitely degenerate zero eigenvalue of S also survives, but only if non-smooth functions are allowed. The associated eigenspace for S^{SFSl} consists of all functions of the form

$$b_{\text{null}}(y) = \exp(-\pi i \alpha \, \text{sign}(y - 1/2)) \, f(\tau(y)). \tag{9.6.4}$$

Thus smoothness is lost at $y = 1/2$ if $\alpha \neq 0$ mod 2. We then no longer have eigenfunctions necessarily vanishing at $y = 1$; so the periodic extension of period 1 is discontinuous at 0 as well. The orthogonal complement in $L^2[0,1]$ of this null space contains the range of T^{SFSl}, as can be quickly verified, and we thus have the generalizations of the even and odd functions with respect to $y = 1/2$ for the folded baker's map. Since the eigenfunctions just described consist of polynomials of distinct integer degree, we can approximate *continuous* functions on $[0,1]$, and thus the eigenfunctions are dense in $L^2[0,1]$. We may divide up any $b(y)$ into these two orthogonal components, and represent the element in the range of T^{SFSl} by its eigenfunction expansion,

$$b(y) = b_{\text{null}}(y) + \sum_{n=0}^{\infty} a_n b_n(y). \tag{9.6.5}$$

If $\lambda^* \neq \lambda_k^*$, 0, we then have

$$(\lambda^* - S^{\text{SFSl}})^{-1} b = \frac{1}{\lambda^*} b_{\text{null}}(y) + \sum_{n=0}^{\infty} (\lambda^* - \lambda_n^*)^{-1} a_n b_n(y). \tag{9.6.6}$$

This establishes the resolvent set of S^{SFSl} as all complex numbers excluding zero and the λ_k^*.

For the direct slide operator, defined by (3.4.2), we remark that it, like the folded baker's map, is not compact, even in the special case where α is 1/2 mod 1 and T^{SFSl} becomes similar to the direct operator for the *stacked* baker's map. We next show that T^{SFSl} has no eigenvalues of modulus less than $\sqrt{2}$. For, if $2e^{\pi i \alpha} b(2y) = \lambda b(y)$ then

$$2 \int_0^1 |b(y)|^2 \, dy = |\lambda|^2 \int_0^{1/2} |b(y)|^2 \, dy \leq |\lambda|^2 \int_0^1 |b(y)|^2 \, dy \tag{9.6.7}$$

and this proves the result. It is also clear from the Neumann expansion of $(\lambda - T)^{-1}$ that $|\lambda| > 2$ is in the resolvent set (as $\|T\| = 2$). Actually there is a stronger statement: if λ^* is in the resolvent set of S^{SFSl}, then λ is in the resolvent set of T^{SFSl}. This follows from a basic theorem concerning closed operators, and the fact that any bounded operator on a complete Hilbert space is closed (Kato 1966, pp. 163, 183).

The final possibility is that $2i \sin \pi\alpha$ is an eigenvalue of T^{SFSl} provided its modulus exceeds $\sqrt{2}$. It then follows from (3.4.2) that any associated eigenfunction $b(y)$ must satisfy $b(y) = (b(y/2) + b(1 - y/2))/2$. Iterating the operator here, the smoothing effect produces a limit and the only possible eigenfunction as a constant. This works only if $2e^{\pi i\alpha} = 2i \sin \pi\alpha$, or $\alpha = 1/2 \mod 1$, where in fact the slide map resembles the stacked baker's map and has this point eigenvalue. Since it is sufficient to consider $0 \leq \alpha \leq 1/2$, we may summarize our results for these slide operators as follows:

Theorem 9.3 *If $0 < \alpha < 1/2$ the spectrum of the slide map consists of residual spectrum at the points $\lambda_n = 2^{-n}[e^{\pi i\alpha} + (-1)^{1+n}e^{-\pi i\alpha}]$ for $n = 0, 1, \ldots$, as well as at the point $\lambda = 0$. The λ_n^* are point eigenvalues of the adjoint of the slide map, with associated eigenfunctions complete in $L^2[0, 1]$. The origin is a point eigenvalue of S^{SFSl} of infinite degeneracy.*

From the point of view of energy in Fourier space the adjoint slide map differs from the adjoint baker's map S in the existence of a growing mode when $2 \sin \pi\alpha > 1$. However it is instructive to first try to follow the argument for the energy for the folded map, by expanding $b(y)$ in sine functions (an expansion which does not take into account the new features of the slide map). We then see that $S^{\text{SFSl}} \sin 2n\pi y = 2 \cos \pi\alpha \sin n\pi y$ so there is the same cascade to low wavenumber. The difference comes when we reach a mode $\sin (2m + 1)\pi y$ by this cascade. For the folded baker's map, such a mode is in the null space of S. Now however, so long as $\alpha \neq 0 \mod 1$, such a mode maps under S^{SFSl} into a function having a cosine series with coefficients

$$b_n = -4i \sin \alpha\pi \int_0^1 \cos \pi n y \sin[(2m + 1)\pi y/2] \, dy. \tag{9.6.8}$$

This contributes a reverse cascade to high wavenumbers, which feeds into the growing eigenfunction but in a way that is difficult to assess. It is probably simplest to return to Fourier series in $e^{2\pi i n y}$ so that the growing eigenfunction is identified with the $n = 0$ mode. The transfer of energy from mode to mode in this case is suggested in Fig. 9.3.

The action of the direct operator T^{SFSl}, given by (3.4.2), in Fourier space is defined by an extension of (9.5.23) for the folded baker's map. Using the same notation, the matrix $\mathbf{A}^{\text{SFSl}}$ becomes

$$A_{jk}^{\text{SFSl}} = \begin{cases} (4ij \cos \pi\alpha - 8k \sin \pi\alpha)/\pi(4k^2 - j^2), & (j \text{ odd}), \\ \pm e^{\pm i\alpha}, & (j = \pm 2k \neq 0), \\ 2i \sin \pi\alpha, & (j = k = 0) , \\ 0, & (j \text{ even and } j \neq \pm 2k). \end{cases} \tag{9.6.9}$$

Dynamo action comes from direct mapping of zero modes, and such contributions eventually result from iterations on any sine mode.

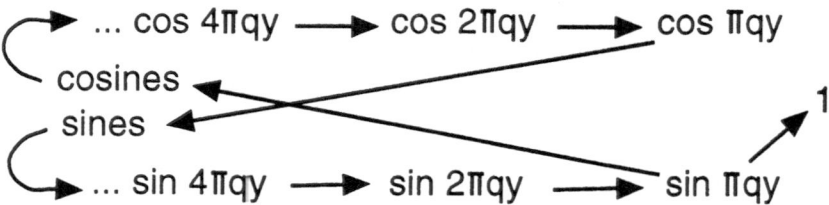

Fig. 9.3 The mapping of Fourier modes by the slide map S^{SFSl}. Here q is any odd integer.

9.6.2 The Slide Map with Diffusion

We have already noted (in §3.4) that the slide map is unusual in the fact that the eigenfunction $b = 1$ of S^{SFSl} is also an eigenfunction of the diffusion operator H_ϵ (9.5.22). For the slide map we use the separation in z, $\mathbf{B} = e^{2\pi i z}(b(y), 0)$. Since the vertical wavenumber is unity we see that 1 is also a diffusive eigenfunction of S_ϵ^{SFSl} with the eigenvalue $\lambda_\epsilon^* = \lambda_0^* e^{-4\pi^2 \epsilon}$. Physically, diffusion does not affect total flux significantly and therefore the slide map is a non-diffusive fast dynamo, with $\Gamma(0) = \gamma_0$ provided $2|\sin \pi\alpha| > 1$. An initial field b_0 with non-zero flux satisfies $\lambda_\epsilon^N(1, b_0) = (1, (T_\epsilon^{\text{SFSl}})^N b_0)$.

But we see that $\cos \pi\alpha$ is an eigenvalue of S^{SFSl} which can dominate for some α. Although dynamo action is not involved, it is still of interest to establish whether or not $\gamma_0 = |\cos \pi\alpha|$ in its range of dominance. In this case T_ϵ^{SFSl} should have an eigenfunction $b_\epsilon(y)$ of maximal growth rate, carrying now zero flux, with an associated eigenvalue λ_ϵ different from that associated with λ_0. Even for this simple model no explicit results are available, and we must resort to numerical computations using the power method. The adjoint property implies that,

$$\gamma(\epsilon) = \sup_{f,g} \lim_{N \to \infty} \frac{1}{N} \log |((S_\epsilon^{\text{SFSl}})^N f, g)|. \tag{9.6.10}$$

It is important that we take the limit here before letting $\epsilon \to 0$! The eigenfunction can be sought by taking an appropriate f and renormalizing after each step. We show some results in Figs. 9.4 and 9.5.

For α above about 0.18 the iterations converge quickly to the constant eigenfunction of the fast dynamo. The direct eigenfunction is of course complicated; see Fig. 9.4(a). In the range $0.15 < \alpha < 0.18$ approximately, iterations do not settle down, or else appear to converge to a periodic cycle. We interpret this as a region of *mode crossing* where there are nearby eigenvalues or an eigenspace of dimension exceeding one (see Rado 1993).

Once $\alpha < 0.15$ we again obtain convergence and good agreement between adjoint and direct calculations when $\epsilon = 10^{-4}$; see Fig. 9.5. The direct

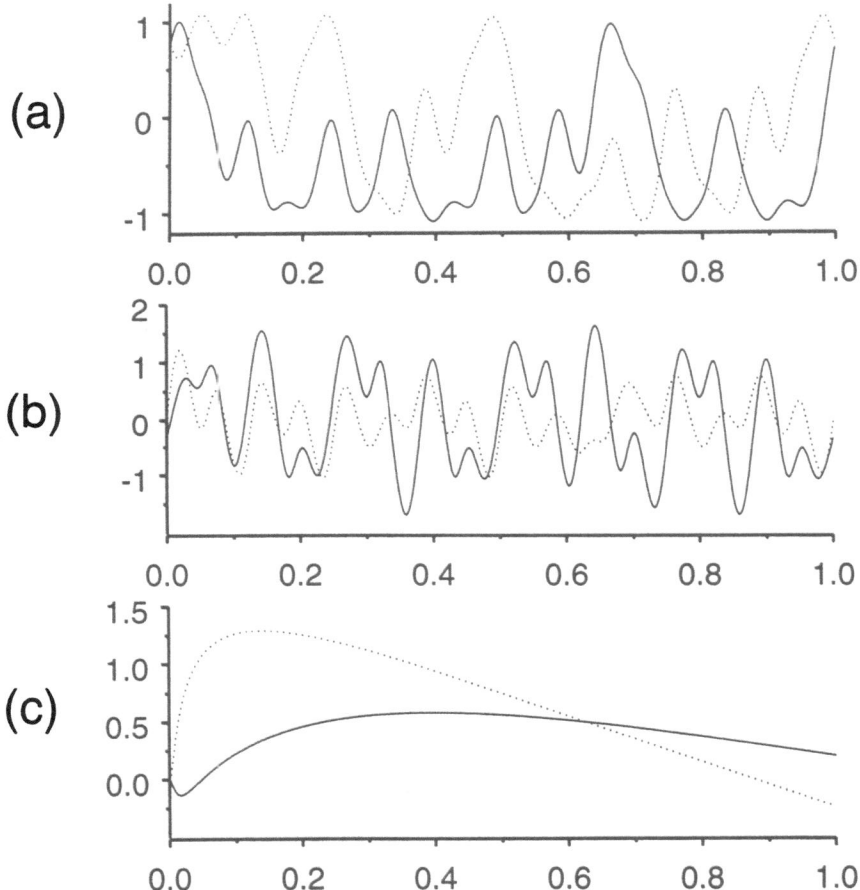

Fig. 9.4 Eigenfunctions for the slide map on 512 points. (a) Direct map with $\epsilon = 10^{-4}$ and $\alpha = 0.3$. (b) Direct map for $\epsilon = 10^{-4}$ and $\alpha = 0.05$. (c) Adjoint map, with $\epsilon = 10^{-5}$ and $\alpha = 0.05$.

and adjoint eigenfunctions for $\alpha = 0.05$ are shown in Figs. 9.4(b, c) respectively. Note that the adjoint eigenfunction resembles to some extent the perfect eigenfunction b_1 linear in y, but there is no indication of convergence to the corresponding eigenvalue $\cos \pi \alpha$.[72] In fact there is a region of fast dynamo action in this range, where none is predicted by the perfect theory. This is

---

[72]There is a discontinuity at 0 in the eigenfunction shown in Fig. 9.4(c) because of the nature of the diffusive adjoint. The diffusion operator H_ϵ is applied *first*. This smooths the discontinuity. Applying the adjoint map we then see that $b(0+)$ need not equal $b(1-)$ after one cycle.

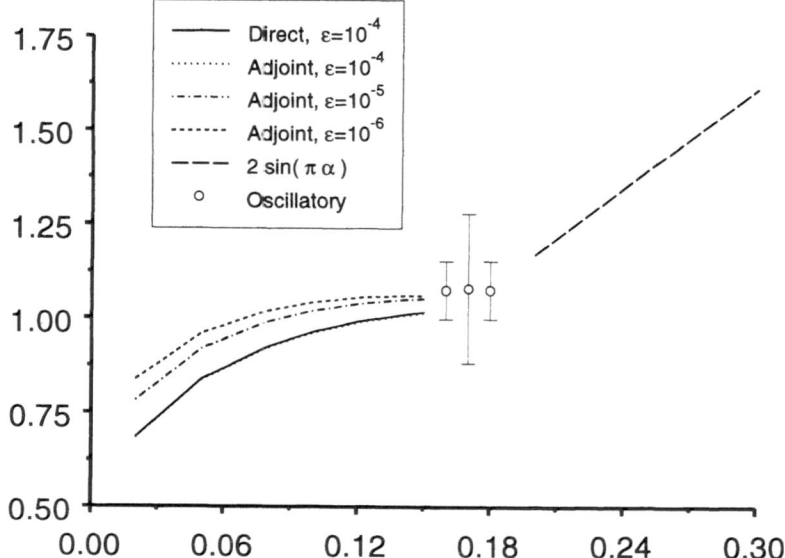

Fig. 9.5 Growth factor e^γ versus α for the slide map, for various ϵ. Computations are on 512 points, using both direct and adjoint operators.

an unexpected feature of the slide map which illustrates the subtlety of the perfectly conducting limit. We suggest that it provides an example of how diffusive and non-diffusive fast dynamo action can occur for different parameter values within the same flow.

We thus conjecture that the curves in Fig. 9.5 to the left of $\alpha = 0.16$ represent diffusive dynamo action analogous to such diffusive fast dynamos as the Ponomarenko dynamo (§5.4.1). (Here of course the vertical wavenumber is unity, not large as it is in the Ponomarenko dynamo for fast dynamo action at small ϵ.) This conclusion is supported by the form of the direct eigenfunctions, which tend to have oscillations down to the diffusive scale $\sqrt{\epsilon}$, but no net flux, when α is in this range. It is also worth recalling that for $\alpha = 0$ the adjoint eigenfunction $b_1 = y - 1$ for $\epsilon = 0$ disappears completely for positive ϵ. For small positive α there is apparently a comparable disruption of the adjoint spectrum, with diffusive decay or slight growth determined by a discontinuous eigenfunction of $S_\epsilon^{\mathrm{SFSI}}$. The *discontinuity* in the eigenfunction b_1 for S^{SFSI} is presumably similarly disrupted for positive ϵ, however small, but now there is a slight reverse cascade in adjoint space from small to large wavenumbers, evidenced by the discontinuity which remains in the diffusive eigenfunction.

One test for this idea is again a *large* ϵ result. If we consider an eigenfunction of $S_\epsilon^{\mathrm{SFSI}}$ projected onto functions of the form $A\cos 2\pi y + B\sin 2\pi y$, we obtain an eigenvalue λ satisfying $\lambda^2 = (64i/9\pi^2)\sin 2\pi\alpha\, e^{-16\pi^2\epsilon}$. We must here take ϵ sufficiently large to make the constant eigenfunction, with eigenvalue $2i \sin \pi\alpha\, e^{-4\pi^2\epsilon}$, subdominant. For $\alpha = 0.05$, for example, it is sufficient that $\epsilon = 0.005$, which is not large enough to justify neglecting higher harmonics. The approximation gives a growth factor of 0.317 in this case, while numerical calculations give 0.398.

More interesting is the asymptotic limit $\epsilon \to 0$. The numerical results (in particular, Fig. 9.4c) suggest that we are dealing with a eigenfunction with a boundary layer and an interior (or 'outer') limit which is linear in y. We accordingly look for a solution of the form (see Fig. 9.6)

$$b(y) = f(y/\sqrt{\epsilon}) - \beta_m y - \beta_1(1 - y) + f((y - 1)/\sqrt{\epsilon}). \qquad (9.6.11)$$

Here $f(\bar{y})$ with $\bar{y} = y/\sqrt{\epsilon}$ is a complex-valued function defined on the real line, and β_1, β_m are complex constants. As shown in Fig. 9.6, just prior to diffusion $f(\bar{y}) = \beta_1$ for $\bar{y} < 0$, $f(0+) = \beta_0$, and $f(\infty) = \beta_m$. Following diffusion $f(-\infty) = \beta_1$, $f(0) = \beta_s$, and $f(\infty) = \beta_m$.

When the function shown in Fig. 9.6, including the effect of diffusion shown by the dotted lines, is mapped under S^{SFSl}, the midpoints are mapped into the right endpoints, which accounts for the absence of a right boundary layer in the solid curve. The mapping of this function into the right endpoint implies that

$$\lambda\beta_1 = -2\beta_{1/2}\, i \sin \pi\alpha, \qquad (9.6.12a)$$

where $\beta_{1/2} = b(1/2)$ and λ is the associated eigenvalue. The smoothing by diffusion produces the values β_s at $y = 0, 1$ and so we have

$$\lambda\beta_0 = -2\beta_s\, i \sin \pi\alpha. \qquad (9.6.12b)$$

Also, because of the linearity and the thinness of the boundary layers,

$$\beta_{1/2} = (\beta_m + \beta_1)/2. \qquad (9.6.12c)$$

In the boundary-layer approximation, we can neglect the interaction of boundary layers, relying on the expected exponential decay at infinity.

By applying diffusion and S^{SFSl} locally to the boundary layer, we obtain the following eigenvalue problem for the function $g(\bar{y}) = df/d\bar{y}$ evaluated just after the map:

$$\int_0^\infty \frac{1}{2}\left[e^{-\pi i\alpha}G(\bar{y}/2 - \eta) + e^{\pi i\alpha}G(\bar{y}/2 + \eta)\right] g(\eta)\, d\eta$$
$$+ \cos\pi\alpha\,(\beta_0 - \beta_1)e^{-\bar{y}^2/16}/\sqrt{4\pi} = \lambda g(\bar{y}). \qquad (9.6.13)$$

Here $G(y) = (4\pi)^{-1/2}\exp(-y^2/4)$. Once we solve (9.6.13), β_s and β_m are determined by

$$\beta_s = \frac{1}{2}(\beta_0 + \beta_1) + \frac{1}{\sqrt{\pi}}\int_0^\infty \mathrm{Erfc}(\eta/2)g(\eta)\, d\eta, \qquad (9.6.14a)$$

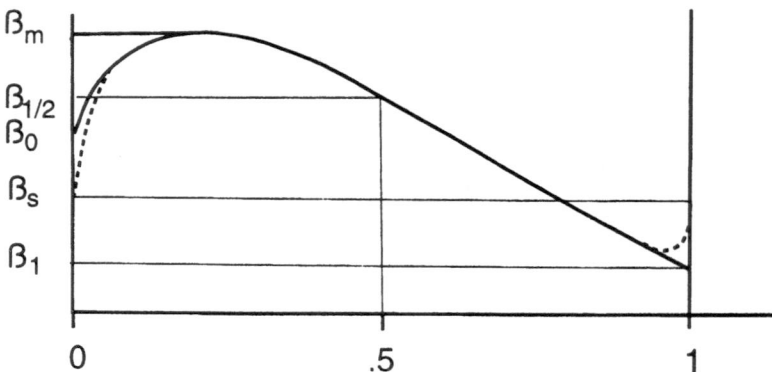

Fig. 9.6 Boundary-layer structure of the limiting eigenfunction for the slide map. The function $f(y)$ just prior to diffusion is given by the solid line. The dotted lines show the effect of diffusion, prior to the map by S^{SFSl}.

$$\mathrm{Erfc}(y) = \int_y^\infty e^{-t^2}\, dt, \qquad\qquad (9.6.14b)$$

$$\beta_m = \beta_0 + \int_0^\infty g(\eta)\, d\eta. \qquad\qquad (9.6.14c)$$

To get an idea of how a solution might look, consider the crude approximation, essentially one term of a Neumann expansion, given by

$$g(y) \approx \frac{(\beta_0 - \beta_1)}{\sqrt{4\pi\,\lambda}}\, e^{-y^2/16} \cos \pi\alpha. \qquad\qquad (9.6.15)$$

Then, (9.6.12) and (9.6.14) yield the two equations

$$\lambda\beta_0 = -2i \sin \pi\alpha \left[\frac{\beta_0 + \beta_1}{2} + \frac{(\beta_0 - \beta_1)K \cos \pi\alpha}{\lambda}\right], \qquad\qquad (9.6.16a)$$

$$\lambda\beta_1 = -2i \sin \pi\alpha \left[\frac{\beta_0 + \beta_1}{2} + \frac{(\beta_0 - \beta_1) \cos \pi\alpha}{2\lambda}\right], \qquad\qquad (9.6.16b)$$

where

$$K = \frac{1}{\pi}\int_0^\infty e^{-y^2/4}\, \mathrm{Erfc}(y)\, dy \approx 0.148. \qquad\qquad (9.6.16c)$$

We see there are two solutions, one with $\beta_0 = \beta_1$ and $\lambda = -2i \sin \pi\alpha$, the other (seen from the difference of (9.6.16a, b)) having

$$\lambda^2 = (1/2 - K)i \sin 2\pi\alpha. \qquad\qquad (9.6.17)$$

The growth factor for the latter solution dominates for sufficiently small α, below about 0.055 in this approximation. At this truncation we cannot explain the numerical results of Fig. 9.5, but there is clear indication of a distinct root qualitatively similar to that revealed by the numerical calculation.

To summarize, the slide map appears to exhibit a disruption of the spectrum, similar to that for the folded baker's map, when diffusion is added. The calculations indicate that diffusive dynamo action can occur. Non-diffusive fast dynamo action, consistent with the flux conjecture, is obtained whenever $2|\sin \pi \alpha| > 1$.

9.6.3 The SFS Map

The analogous spectral results for the SFS map have largely emerged from numerical experiments and there are few explicit calculations one can make. The direct perfect operator T^{SFS} is given by (3.2.2) and the perfect adjoint S^{SFS} by (3.2.8). B.J. Bayly (personal communication) observed that $\sin \pi y$ is an eigenfunction of S^{SFS} for the eigenvalue $-i$. The numerical calculations indicate that this marks the onset of perfect dynamo action as α is increased from zero. Eigenfunctions for the direct and adjoint maps for the case $\alpha = 1$ are given in Fig. 9.7.

We have shown in Fig. 3.5 diffusive calculations of the growth rate obtained with $T_{\epsilon}^{\mathrm{SFS}}$, compared with perfect results using the adjoint map S^{SFS}. The main points of interest here are (1) the evidence for diffusive dynamo action in the range $\alpha = 0.5$–0.7, and (2) the indication of mode crossing near $\alpha = 2.1$. We have discussed above the extensive mode crossing apparent in calculations at higher α, but little is known concerning the mechanisms of dynamo action at work in these crossing regions, or exactly what the spectral properties of the operators are there. There is no reason to believe that the diffusive instability of the spectrum we have found in the slide map is avoided by the constant shear in SFS, since it has not been linked directly to the discontinuities of the vertical motion in the slide map. We suspect that generalized SFS maps with arbitrary shear as in (3.1.6) all have similar properties in this regard.

We shall consider here only one simple general property of the SFS maps, namely the smoothness of the adjoint eigenfunctions for the perfect conductor (Bayly & Childress 1989).

Theorem 9.4 *Let b be an eigenfunction of the adjoint SFS map for general C^{∞} shear $f(y)$, obtained by the power method starting from a C^{∞} function, and corresponding to an eigenvalue with modulus greater than unity. Then b possesses derivatives of all orders.*

We shall prove this property for

$$S^{\mathrm{SFS}}b(y) = e^{2\pi i \alpha f(y/2)} b(y/2) - e^{2\pi i \alpha f(1-y/2)} b(1-y/2), \qquad (9.6.18)$$

with $f(y)$ the shear function, equal to $y - 1/2$ in the constant shear case. Let $\|\cdot\|$ denote in the present discussion the supremum norm on $[0,1]$, and consider

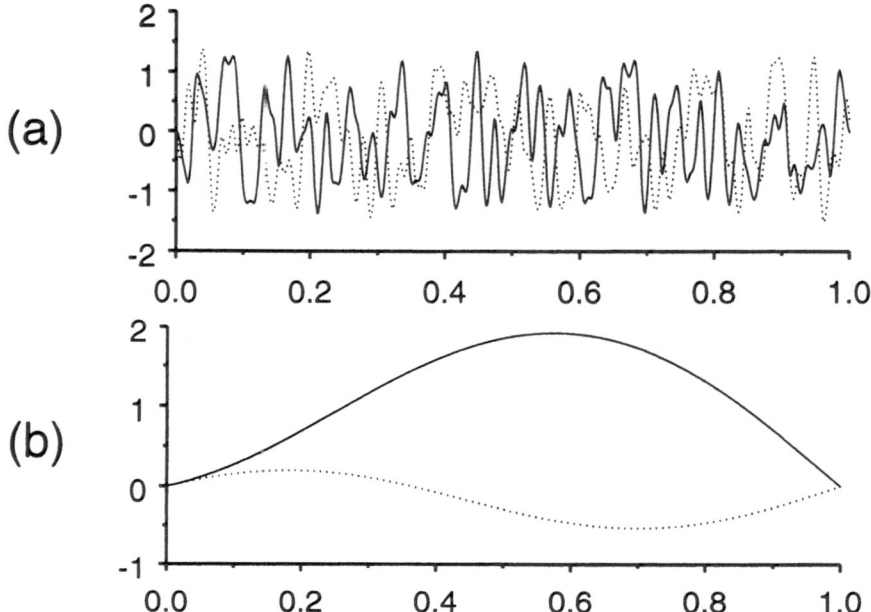

Fig. 9.7 (a) Eigenfunction for T_ϵ^{SFS} for $\alpha = 1$ and $\epsilon = 10^{-5}$. (b) Eigenfunction for S^{SFS} and $\alpha = 1$.

a sequence of iterates b_n with $b_{n+1} = S^{\text{SFS}} b_n$, such that $\|b_{n+1}\| \geq a\|b_n\|$ for $n = 1, 2, \ldots$, where $a > 1$ and b_1 is a C^∞ function. Let $R_n^{(m)} = \|b_n^{(m)}\|/\|b_n\|$, $b_n^{(m)}$ being the mth derivative of b_n. We shall show that, for any m, $R_n^{(m)}$ has an upper bound independent of n. Indeed, this result is obvious for $m = 0$. For $m = 1$ we have, from (9.6.18),

$$\|[S^{\text{SFS}} b]'\| \leq 2\pi\alpha\|f'\| \, \|b\| + \|b'\|. \tag{9.6.19}$$

If we use (9.6.19) for the sequence b_n then we obtain

$$aR_{n+1}^{(1)} \leq 2\pi\alpha\|f'\| + R_n^{(1)}, \tag{9.6.20}$$

and from this it follows easily that

$$R_n^{(1)} < 2\pi\alpha\|f'\|/(a-1) + R_1^{(1)}. \tag{9.6.21}$$

Now assume that $R_n^{(k)} \leq C^{(k)}$ for $k \leq m-1$. Differentiating (9.6.18) m times, there is a constant B such that

$$aR_{n+1}^{(m)} \leq B + R_n^{(m)}, \tag{9.6.22}$$

from which we obtain the bound $C^{(m)} = B/(a-1) + R_1^{(m)}$ and we are done.

9.6.4 A Horseshoe Map

The well-known horseshoe map (Smale 1967) is an area-preserving map of the plane which which takes a subdomain of the unit square, here $-1/2 \le x \le 1/2$, $0 \le y \le 1$, into a proper subset. In the perfectly-conducting limit, this means that the support of the magnetic field decreases geometrically under iteration, although flux can grow and fast dynamo action is possible. In particular magnetic field is discarded in the process. The spectrum of a model of this kind, incorporating the piecewise constant vertical motion of a slide map, is discussed by Finn *et al.* (1991). The map is defined by

$$M(x,y,z) = \begin{cases} ((x-1)/3, 3y, z - \alpha/2), & (0 \le y \le 1/3), \\ ((1-x)/3, 3(1-y), z + \alpha/2), & (2/3 \le y \le 1). \end{cases} \tag{9.6.23}$$

The horizontal part of essentially the same map, shifted by $1/2$ in x, was described in Chap. 7; see Fig. 7.7. With $\mathbf{B} = e^{ikz}(0, b(x), 0)$ the horseshoe operator T^H is then defined by

$$b' = T^H b = \begin{cases} 3e^{ik\alpha/2} b(3x+1), & (-1/2 \le x < -1/6), \\ -3e^{-ik\alpha/2} b(1-3x), & (1/6 \le x < 1/2). \end{cases} \tag{9.6.24}$$

The field is by definition zero for all other values of x. With

$$\hat{b}(\kappa) = \int_{-1/2}^{+1/2} e^{-i\kappa x} b(x)\, dx, \tag{9.6.25}$$

the equation for the magnetic field in the Fourier domain is

$$\hat{b}'(\kappa) = e^{i(k\alpha/2 + \kappa/3)} \hat{b}(\kappa/3) - e^{-i(k\alpha/2 + \kappa/3)} \hat{b}(-\kappa/3). \tag{9.6.26}$$

We immediately have the flux map

$$\hat{b}'(0) = 2i \sin(k\alpha/2)\, \hat{b}(0) \equiv \lambda \hat{b}(0), \tag{9.6.27}$$

determining the eigenvalue λ for perfect fast dynamo action.

The Fourier transform for the corresponding eigenfunction, $\hat{\beta}(\kappa)$ say, can be derived as a convergent infinite product. Writing (9.6.26) in the form

$$\begin{pmatrix} \hat{b}'(\kappa) \\ \hat{b}'(-\kappa) \end{pmatrix} = \begin{pmatrix} e^{ik\alpha/2 + i\kappa/3} & -e^{-ik\alpha/2 - i\kappa/3} \\ -e^{-ik\alpha/2 + i\kappa/3} & e^{ik\alpha/2 - i\kappa/3} \end{pmatrix} \begin{pmatrix} \hat{b}(\kappa/3) \\ \hat{b}(-\kappa/3) \end{pmatrix}, \tag{9.6.28}$$

and denoting the 2×2 matrix on the right of (9.6.28) by $L(\kappa/3)$, the eigenfunction is seen to have the representation

$$\begin{pmatrix} \hat{\beta}(\kappa) \\ \hat{\beta}(-\kappa) \end{pmatrix} = \left(\prod_{n=1}^{\infty} \lambda^{-1} L(\kappa/3^n) \right) \begin{pmatrix} 1 \\ 1 \end{pmatrix}, \tag{9.6.29}$$

with the normalisation $\hat{\beta}(0) = 1$. This eigenfunction is a distribution, since by the construction of the map the field vanishes except on a Cantor set.

The diffusive map T_ϵ^H may be analyzed similarly and the eigenfunction is found to have the Fourier transform

$$\hat{\beta}_\epsilon(\kappa) = e^{-9\kappa^2\epsilon/8}\hat{\beta}(\kappa); \tag{9.6.30}$$

the factor comes from summation of the diffusive prefactors associated with all scales. The associated eigenvalue is $\lambda_\epsilon = 2i \sin(k\alpha/2)e^{-k^2\epsilon}$.

The horseshoe map is thus a perfect and non-diffusive fast dynamo as defined in Chap. 4. It has the advantage of a relatively simple and explicit eigenfunction, with nice convergence properties. Indeed, if $f(x)$ is a test function, we obtain weak convergence of $\beta_\epsilon(x)$ to $\beta(x)$ in the L^2 sense. For, using the Parseval relation,

$$\int_{-\infty}^{+\infty} f^*(x)(\beta_\epsilon(x) - \beta(x))\, dx = \int_{-\infty}^{+\infty} \hat{f}^*(\kappa)(e^{-9\kappa^2\epsilon/8} - 1)\hat{\beta}(\kappa)\, \frac{d\kappa}{2\pi}. \tag{9.6.31}$$

When $\epsilon \ll 1$ the middle factor on the right removes the middle portion of the integral, corresponding to $o(\epsilon^{-1/2})$ values of κ. What remains is small provided that the integral $\int_{-\infty}^{\infty} \hat{f}^*(\kappa)\hat{\beta}(\kappa)\, d\kappa$ converges, as it must if f is a suitable test function. The limit of the right-hand side tends to zero as $\epsilon \to 0$, and the eigenfunction with diffusion converges weakly to the perfect eigenfunction.

The main difficulty with the horseshoe model is that a portion of the conductor is discarded at each step, and the embedding of the map in a more realistic bounded flow would have to follow these discards (cf. §1.4). This point and the application of the model to steady flow were discussed in §7.3.1.

9.7 Pulsed Waves

While the discontinuous baker's map provides a building block for a variety of fast dynamo models, the significant advantages offered by the adjoint formulation need not carry over to smooth flows. This problem is revealed by the simplest examples of pulsed Beltrami waves of the kind introduced in Chap. 2. We have seen in §9.3.1 that the adjoint to $\mathbf{T}_\epsilon$ in a pulsed flow involves advection of vector potential by the time-reversed flow, i.e., by a flow obtained by changing $\mathbf{u}(\mathbf{x}, t)$ to $-\mathbf{u}(\mathbf{x}, -t)$. Now, while for the folded baker's map forward and reversed operations are clearly quite different, for pulsed flows involving two Beltrami waves they need not be. Forward and reversed pulsing can be indistinguishable, for example if the wave is simply turned on and off.[73] The fact that stretching and folding then occurs in a 'reversible'

---

[73]If two distinct waves of amplitudes $\alpha(t)$ and $\beta(t)$ are involved, a reversible flow will be one in which the forward and backward motions of the point $(\alpha(t), \beta(t))$ are

way means that a direct attack on the eigenvalue problem, which can deal with the complexity of the field, is needed.

A related point is the prominence of *continuous spectrum* for pulsed waves in a perfect conductor. In cases with no simplifying adjoint formulation, the 'eigenfunctions' necessarily possess structure on all length scales. This fractal or multi-fractal structure, which was discussed in Chap. 7, can at best be approximated by sequences of smooth functions, and eigenfunctions defined as the limits of such sequences (see, e.g., Lighthill 1962). In cases where the direct and adjoint operators are essentially equivalent, the associated spectrum must be continuous, the residual property being no longer a possible alternative.

Because of this difficulty, the spectral theory of pulsed flows is at this writing relatively undeveloped. The main advances, to be described in Chap. 10, depend upon asymptotic methods. To this end we summarize next a procedure arising in the perturbation of spectra, which will be used to advantage in Chap. 10. We then discuss briefly a smooth map using a trigonometric function and contrast it with the folded baker's map.

9.7.1 Perturbation of Spectra

We shall in Chap. 10 take up the problem of solving an eigenvalue problem by comparison with a 'nearby' problem whose solution is known. The general theory of perturbation of spectra is a rich subject, with quite distinct methods for discrete and continuous spectra (see, e.g., Friedrichs 1965, Kato 1966). In the present section we summarize a general procedure based on an *associated problem*, which encompasses the perturbation of isolated point eigenvalues. In another guise it is applicable to techniques of smoothing and of homogenization (Ghil & Childress 1987, Bensoussan, Lions, & Papanicolaou 1978). It has been applied to dynamo theory in the context of spatially periodic flows (Childress 1967, 1970, Roberts 1972).

We do not single out any one of the operators discussed above and so introduce some operator L on a Banach space V of vectors v, on which a projection P is defined. Recall that a projection is any operator with the property that $P^2 = P$. Our object is to solve the fixed-point problem $Lv = v$ in V. Consider the space $F = PV$ of projections f of vectors in V. Our object is to formulate an associated problem in F, whose solution will generate the desired fixed point. The method is contained in the following result.

Lemma 9.2 *Let* $L' = (I - P)L$, *and assume that* $\|L'\| < 1$. *Then the vector* $v = (I - L')^{-1}f$ *is the desired fixed point, provided that* f *is a vector of* F *satisfying the associated problem* $f = PL(I - L')^{-1}f$.

---

indistinguishable. Motion in one direction around a circle *would* be distinguishable, for example; so for such cases the adjoint formulation might offer advantages. As far as we are aware no examples of this have been studied.

To prove this, we compute

$$Lv = L(I - L')^{-1}\tilde{f} = PL(I - L')^{-1}f + L'(I - L')^{-1}f$$
$$= f + L'(I + L' + L'^2 + \cdots)f = (I - L')^{-1}f = v. \tag{9.7.1}$$

Here we used the Neumann series expansion of $(I - L')^{-1}$ twice.

One application of interest, to be used in Chap. 10, involves a projection which is a flux average, i.e., onto a scalar. We introduce the eigenparameter by setting $L = \lambda^{-1}L_1$. Then $f = 1$ and the equation for f becomes an equation for λ. The series for $(I - L')^{-1}f$ is then a convergent representation of the eigenfunction. The main problem is then to localize the eigenvalue sufficiently to be able to obtain the necessary estimate on $\|L'\|$. In general the method is useful when the projected problem is a simplification. In examples where L involves rapidly oscillating or small-scale components, the projection is usually on smooth or large-scale fields.

In the canonical problem of perturbation of eigenvalues of a self-adjoint operator T_0 (Friedrichs 1965), we assume that v belongs to a Hilbert space H and that P is defined by an inner product on H. Suppose that T_0 is known to have an isolated point eigenvalue λ_0 with associated eigenspace spanned by the unit vector f_0. The natural projection is then onto this space; so $Pv = (f_0, v)f_0$. For λ any point in the resolvent set of T_0 we set $T = T_0 + U$ and define the operator L by $L = (\lambda - T_0)^{-1}U$. Here the associated problem reduces to the scalar equation

$$\lambda = \lambda_0 + (f_0, U[I - (\lambda - T_0)^{-1}(I - P)U]^{-1}f_0). \tag{9.7.2}$$

This is the equation for the perturbed eigenvalue. Notice that in (9.7.2) we have interchanged $(\lambda - T_0)^{-1}$ and $I - P$, a step that is allowed by the self-adjoint property of T_0 and the definition of P. This puts the equation in a form where $(\lambda - T_0)^{-1}$ acts on the image of $I - P$, and so it is clear that the equation now applies to a small neighborhood of λ_0 where $\|(\lambda - T_0)^{-1}(I - P)U\| < 1$. The associated perturbed eigenvector is then given by

$$v = [I - (\lambda - T_0)^{-1}(I - P)U]^{-1}f_0. \tag{9.7.3}$$

9.7.2 The Cosine Map

We seek a smooth counterpart to the folded baker's map that introduces some of the properties of a Beltrami wave. The idea is to turn on the flow $\mathbf{u} = (0, \cos x)$ in two dimensions for time τ, then rotate to align the stretching direction with the x-axis. In the present discussion we shall use the rotation $(x, y) \to (y, -x)$ for this purpose.[74] The cosine map M is thus given by

---

[74]Bayly & Childress (1988) use the reflection $(x, y) \to (y, x)$, which introduces an awkward change of orientation of the plane coordinates.

$$M: \quad (x, y) \rightarrow (y + \tau \cos x, -x), \tag{9.7.4a}$$

with inverse

$$M^{-1}: \quad (x, y) \rightarrow (-y, x - \tau \cos y). \tag{9.7.4b}$$

A scalar magnetic potential $A(x, y)$ in a perfect conductor thus has the image $A'(x, y) = A(M^{-1}(x, y)) = A(-y, x - \tau \cos y)$. A magnetic field $\mathbf{b}(x, y) = (b_x, b_y)$ transforms according to the induction map $T^{\cos}$ defined by

$$T^{\cos}\mathbf{b}(x, y) = (\tau \sin y \, b_x + b_y, -b_x)(M^{-1}(x, y)). \tag{9.7.5}$$

The last expression can also be obtained by differentiating $A(-y, x - \tau \cos y)$, and we shall also use $T^{\cos}$ to indicate induction of a potential, $T^{\cos}A(x, y) = A(M^{-1}(x, y))$. The corresponding adjoint operator $S^{\cos}$ is defined using the complex inner product on $[0, \pi] \times [0, \pi]$. This means the field $A(x, y)$ is mapped by $S^{\cos}$ into the field $A(M(x, y))$.

These last expressions imply a simple result for any incompressible flow in two dimensions generating a map M, and the corresponding operators T and S acting on vector potentials. If f is an eigenfunction of T, so that $Tf = \lambda f$ or here $f(-y, x - \tau \cos y) = \lambda f(x, y)$, then f is also an eigenfunction of S for the eigenvalue λ^{-1}. Since the operators share the same eigenfunction, we may restrict the problem to that eigenspace and conclude that $\lambda^* = \lambda^{-1}$ and therefore λ has modulus unity. This implies that T and S are *isometries* on L^2. It also means that there is no difference between the eigenfunctions of the direct and adjoint induction problems, in contrast to the situation for the folded baker's map. Of course we may anticipate that the discrete spectrum for T is here empty, but the same remark also applies to the approximate spectrum (see §9.1.1), and this will be relevant to pulsed flows. That is, given an approximating sequence f_n associated with the eigenvalue λ of T, the same sequence is approximating for an eigenvalue $\lambda^{-1} = \lambda^*$ of S.

On the other hand we see that the cosine map has an effect on $\mathbf{b}$ which is very similar to the stretch and fold of the baker's map. Consider for example the effect of $T^{\cos}$ on the field $\mathbf{b} = (\text{sign}(\sin y), 0)$. For large τ, $T^{\cos}\mathbf{b}$ is dominated by $\tau \sin y \, \text{sign}(\sin(x - \tau \cos y))\mathbf{e}_x$, which oscillates on a scale $O(1/\tau)$ with amplitude of order τ. The variation is fast in y but slow in x. The field is smoothly folded back on itself and is two-dimensional, although elongated in the x-direction in the sense that $|b_x| \gg |b_y|$ except near the folds. Despite these similarities with a baker's map, the continuous folding present here lies at the core of the difficulties presented by Beltrami waves. The continuous folds contaminate the structure everywhere under repeated iteration.

We can expel this difficulty by 'parallelizing' the map as follows. Define $\tau_c(y) = \tau \cos y$, and set $T_\parallel a(y) = a(\tau_c(y))$. The adjoint $S_\parallel$ can then be computed on $L^2[-\pi, \pi]$ for any τ. For the special case $\tau = \pi$, we obtain

$$S_\parallel b(y) = (\pi^2 - y^2)^{-1/2} [b(\cos^{-1}(y/\pi)) + b(-\cos^{-1}(y/\pi))] \tag{9.7.6}$$

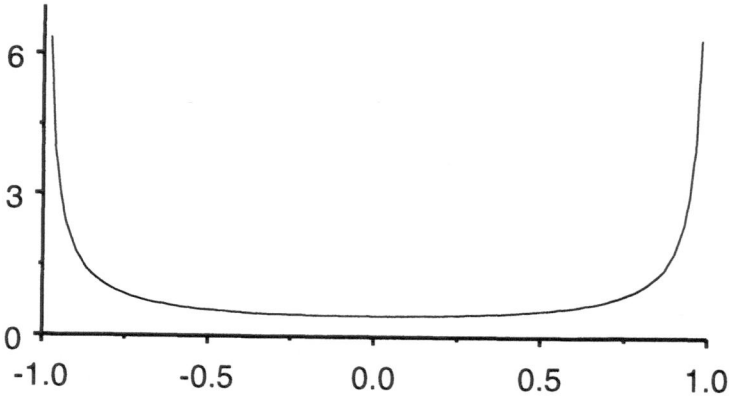

Fig. 9.8 The invariant measure for the cosine map plotted against y/π. This is the eigenfunction for the eigenvalue 1 of the adjoint $S_\parallel$ given by (9.7.6), with $\tau = \pi$.

where the branch of the inverse cosine is in the interval $[0, \pi]$. The equivalent formula for vector potential in the folded baker's map is

$$S_\parallel b(y) = (1/2) \left[b(y/2) + b(1 - y/2) \right]. \tag{9.7.7}$$

The eigenfunction $b = 1$ of (9.7.7) with eigenvalue 1 gives the invariant measure (or invariant probability distribution) of the tent map (Lichtenberg & Lieberman 1983, p. 412), and similarly a solution of $S_\parallel b = b$ gives an invariant measure for the map $y \to \pi \cos y$. This eigenfunction can be computed by the power method and is shown in Fig. 9.8.

Thus the elimination of the folds puts us back into a theoretical framework similar to that for baker's maps. It is evident that the elimination of dependence upon x was the crucial step. The methods to be developed next are designed to deal with the flux dependence upon x and y, while taking some advantage from the elongation of structure which is realized at large τ.

10. Strongly Chaotic Systems

In this chapter we explore fast dynamo action in the limit of *strong chaos*. This limit was first used by Soward (1993a) who takes the pulsed Beltrami waves of Bayly & Childress (1988, 1989),

$$\mathbf{u} = \begin{cases} (0, \sin x, \cos x), & (0 < t \bmod 2\tau < \tau), \\ (-\sin y, 0, \cos y), & (\tau < t \bmod 2\tau < 2\tau), \end{cases} \qquad (10.0.1)$$

and considers the limit of very long pulses, $\tau \gg 1$. In the opposite limit $\tau \to 0$, the rapid pulsing of waves generates the average of the two waves, which is the Roberts' flow of §5.1, an integrable flow. As τ is increased from zero, separatrices split, leaving thin bands of chaos in which the stretching and folding of magnetic field can be understood using the techniques of Chap. 8. As τ is increased further, the phase space becomes more chaotic, until eventually, for large τ, Poincaré sections show chaos everywhere. This is the limit of strong chaos or the *anti-integrable limit*, and is an important limit for proving results about chaotic systems (Aubry & Abramovici 1990). For example, using the standard map with parameter K in this limit $K \to \infty$, Rechester & White (1980) have studied diffusion of particles, and Aubry & Abramovici (1990) have shown how periodic orbits bifurcate from a $K = \infty$ Bernoulli system (see §3.3.2).

We do not have in mind particular physical systems for which the limit of strong chaos is appropriate: for example, the long time-correlations in (10.0.1) when $\tau \to \infty$ are certainly not characteristic of fluid turbulence. Rather we see the limit $\tau \to \infty$ as a mathematical convenience for establishing analytical fast dynamo results for classes of flows, by somehow sidestepping the more natural $\epsilon \to 0$ limit. This is reminiscent of the early work of Herzenberg (1958) and Backus (1958), who used artificial flows and artificial asymptotic limits (of large distance and infrequent pulsing) to establish the principle of a dynamo in a homogeneous fluid. Analogously the existence of smooth flows that are fast dynamos may perhaps be established in the anti-integrable limit; although this has not yet been achieved, there are indications that this limit is a fruitful one.

Soward (1993a) studies the flow above in the limit $\tau \to \infty$ and obtains an asymptotic formula for the fast dynamo growth rate (§10.2). This formula agrees well with growth rates obtained numerically, but has not been justified mathematically. We approach Soward's work by first sketching Oseledets'

(1993) elegant analysis of fast dynamo action for general maps of $\mathbb{T}^2$ (§10.1). This exact analysis has a number of features which motivate our approach to the limit of strong chaos.

In §§10.3, 4 we discuss models which, although much more idealized than Soward's, can be studied rigorously in the limit of strong chaos: we first analyze the SFS map in the limit of a large number of folds, and then introduce fast dynamo models based on cat maps and *pseudo-Anosov (pA) maps* with shear. Those based on pA maps function by the SFS mechanism, but have the advantage over the SFS model that for $\epsilon = 0$ there is no cutting and reconnection of field lines.

10.1 Diffeomorphisms of $\mathbb{T}^2$

Oseledets (1993) considers a general diffeomorphism M of $\mathbb{T}^2$ — a differentiable map with differentiable inverse, typified by the cat map of §3.5. We take $\mathbb{T}^2$ as $\mathbf{x} = (x, y)$ identified modulo one. A diffeomorphism M of $\mathbb{T}^2$ can be written as

$$M(\mathbf{x}) = \mathbf{Mx} + \boldsymbol{\phi}(\mathbf{x}), \tag{10.1.1}$$

where $\mathbf{M}$ is a 2×2 matrix with integer entries, $\det \mathbf{M} = \pm 1$ and $\boldsymbol{\phi}(\mathbf{x}) = (\phi_1(\mathbf{x}), \phi_2(\mathbf{x}))$, with ϕ_1 and ϕ_2 periodic, period one, in x and y.

Formula (10.1.1) essentially says that any diffeomorphism M can be achieved by applying a matrix $\mathbf{M}$, followed by a continuous (and differentiable) distortion under a flow on the torus, represented by the function $\boldsymbol{\phi}(\mathbf{x})$. From §3.5 we know that the action of a matrix $\mathbf{M}$ can amplify mean field provided it is hyperbolic, i.e., has an eigenvalue λ with $|\lambda| > 1$, and in this case there are smooth eigenfunctions. Let us refer to the action of a hyperbolic $\mathbf{M}$ as a 'cat map' for brevity, without implying a particular choice of matrix. We also know that a smooth flow on a torus just generates fluctuations and no large-scale amplification; in fact the mean field is constant (§4.2.1). For the combination $M(\mathbf{x}) = \mathbf{Mx} + \boldsymbol{\phi}(\mathbf{x})$, we shall see that it is the matrix $\mathbf{M}$ that can amplify mean fields, while the action of the distortion $\boldsymbol{\phi}(\mathbf{x})$ is to create fluctuations that are passive. Under the dynamo operator $\mathbf{T}_\epsilon$ corresponding to M, we shall establish that

$$\text{mean field} \xrightarrow{\mathbf{T}_\epsilon} \text{mean field} + \text{fluctuating field}, \tag{10.1.2a}$$

$$\text{fluctuating field} \xrightarrow{\mathbf{T}_\epsilon} \text{fluctuating field}. \tag{10.1.2b}$$

The fluctuating field is *slaved* to the mean field; it cannot affect the mean field, and cannot grow without mean field. The amplification of the mean field gives the fast dynamo growth rate, which is that of the cat map $\mathbf{x} \to \mathbf{Mx}$, while the fluctuations cascade passively to small scales and diffusive annihilation.

This model, more than any other, exemplifies the schematic Fig. 1.5 and allows explicit calculation of growth rates and eigenfunctions, with and without diffusion. Unfortunately the complete slaving of the fluctuations to the mean field is unrealistic for physically reasonable flows, as we shall see when we study pulsed Beltrami waves in §10.2. The fluctuations generally are coupled back to the mean field and there is some element of an *inverse cascade* from the fluctuations to the mean field, affecting the growth rate; we shall seek to quantify and minimize the effect.

10.1.1 Growth Rates

First we show that M takes the form (10.1.1). Note that $M(x + n, y + m) = M(x, y) \bmod 1$ for any integers n, m, and so the Jacobian $\mathbf{J}(\mathbf{x}) = \partial M\mathbf{x}/\partial \mathbf{x}$ is periodic, period one; take $\mathbf{M}$ as the mean part of $\mathbf{J}$,

$$\mathbf{M} = \int_{\mathbb{T}^2} \mathbf{J}(\mathbf{x}) \, d\mathbf{x}, \tag{10.1.3}$$

and the form (10.1.1) follows. The inverse of M takes a similar form,

$$M^{-1}(\mathbf{x}) = \mathbf{M}^{-1}\mathbf{x} + \psi(\mathbf{x}). \tag{10.1.4}$$

Both $\mathbf{M}$ and $\mathbf{M}^{-1}$ must have integer entries to preserve the periodicity and so $\det \mathbf{M} = \det \mathbf{M}^{-1} = \pm 1$.

We restrict our attention to area- and orientation-preserving maps, so that $\det \mathbf{M} = 1$ and $\det \mathbf{J} = 1$ below; however Oseledets (1993) also considers general diffeomorphisms. We use a pulsed model, applying the map and then diffusion for $\epsilon \geq 0$; magnetic field $\mathbf{B}(\mathbf{x})$ evolves according to

$$\mathbf{T}_\epsilon \mathbf{B}(\mathbf{x}) = H_\epsilon \left(\mathbf{J}(M^{-1}\mathbf{x})\mathbf{B}(M^{-1}\mathbf{x}) \right), \quad \mathbf{J}(\mathbf{x}) = \mathbf{M} + \partial\phi/\partial\mathbf{x}. \tag{10.1.5}$$

Here H_ϵ is the heat operator,

$$H_\epsilon \mathbf{B}(\mathbf{x}) = \int_{\mathbb{T}^2} h_\epsilon(\mathbf{x} - \mathbf{y})\mathbf{B}(\mathbf{y}) \, d\mathbf{y}, \quad h_\epsilon(\mathbf{x}) = \sum_{\mathbf{j}} \exp(2\pi i\mathbf{j} \cdot \mathbf{x} - \epsilon\mathbf{j}^2), \tag{10.1.6}$$

with $\mathbf{j}$ ranging over integer vectors. This gives the effect of applying diffusion, diffusivity $\epsilon/(2\pi)^2$, on $\mathbb{T}^2$ for a time unity; this choice of diffusivity avoids numerous factors of 2π in calculations. For $\epsilon = 0$, H_ϵ reduces to the identity. Oseledets takes $\epsilon > 0$ and constructs a growing eigenfunction, $\mathbf{B}$:

$$\mathbf{T}_\epsilon \mathbf{B} = \mu\mathbf{B} \quad \text{with } |\mu| > 1. \tag{10.1.7}$$

We take $\nabla \cdot \mathbf{B} = 0$ and so $\mathbf{B}$ can be written

$$\mathbf{B} = \overline{\mathbf{B}} + (\partial_y a, -\partial_x a) \equiv \overline{\mathbf{B}} + \mathbf{R}\nabla a. \tag{10.1.8}$$

Here $\overline{\mathbf{B}}$ is a constant vector, the *mean field*. The field $(\partial_y a, -\partial_x a)$, with vector potential $a(\mathbf{x})$, is the *fluctuating field*. Note that $a(\mathbf{x})$ is periodic in $\mathbf{x}$. The matrix $\mathbf{R}$ is given by

$$\mathbf{R} = \begin{pmatrix} 0 & 1 \\ -1 & 0 \end{pmatrix}, \tag{10.1.9}$$

and we shall later use the property that for any 2×2 matrix $\mathbf{Q}$ with $\det \mathbf{Q} = 1$,

$$\mathbf{Q}^{-1} = -\mathbf{R}\mathbf{Q}^{\mathrm{T}}\mathbf{R}, \qquad (\det \mathbf{Q} = 1). \tag{10.1.10}$$

Suppose now that $\overline{\mathbf{B}} = 0$ in (10.1.8) and the field is entirely fluctuating field; in this case a is a vector potential for $\mathbf{B}$ which may be taken to be transported as a passive scalar under the advection–diffusion operator:

$$S_\epsilon a(\mathbf{x}) = H_\epsilon(a(M^{-1}\mathbf{x})) \tag{10.1.11}$$

(see §4.2.2). Thus

$$\mathbf{T}_\epsilon \mathbf{R} \nabla a = \mathbf{R} \nabla S_\epsilon a \tag{10.1.12}$$

and from fluctuations $\mathbf{T}_\epsilon$ generates further fluctuations and no mean field, establishing (10.1.2b). Furthermore for an eigenfunction, $S_\epsilon a = \mu a$, which implies $|\mu| \leq 1$, contradicting the assumption of growth, $|\mu| > 1$. Therefore $\mathbf{B}$ must take the form (10.1.8) with non-zero mean field $\overline{\mathbf{B}} \neq 0$.

We now consider the evolution of the mean field $\overline{\mathbf{B}}$; first define $\mathbf{K}$ as the Jacobian of M^{-1},

$$\mathbf{K}(\mathbf{x}) = \partial M^{-1}\mathbf{x}/\partial \mathbf{x} = \mathbf{M}^{-1} + \partial \psi/\partial \mathbf{x}, \tag{10.1.13}$$

and express $\mathbf{J}$ in terms of $\mathbf{K}$ using (10.1.10):

$$\begin{aligned} \mathbf{J}(M^{-1}\mathbf{x}) = \mathbf{K}^{-1}(\mathbf{x}) &= -\mathbf{R}\mathbf{K}^{\mathrm{T}}(\mathbf{x})\mathbf{R} \\ &= -\mathbf{R}(\mathbf{M}^{-1} + \partial\psi/\partial\mathbf{x})^{\mathrm{T}}\mathbf{R} = \mathbf{M} - \mathbf{R}(\partial\psi/\partial\mathbf{x})^{\mathrm{T}}\mathbf{R}. \end{aligned} \tag{10.1.14}$$

Now apply $\mathbf{T}_\epsilon$ to the mean part $\overline{\mathbf{B}}$ of $\mathbf{B}$,

$$\begin{aligned} \mathbf{T}_\epsilon \overline{\mathbf{B}} &\equiv H_\epsilon(\mathbf{J}(M^{-1}\mathbf{x})\overline{\mathbf{B}}) \\ &= H_\epsilon(\mathbf{M}\overline{\mathbf{B}} - \mathbf{R}(\partial\psi/\partial\mathbf{x})^{\mathrm{T}}\mathbf{R}\overline{\mathbf{B}}) \\ &= \mathbf{M}\overline{\mathbf{B}} - \mathbf{R}H_\epsilon(\nabla(\psi^{\mathrm{T}}\mathbf{R}\overline{\mathbf{B}})) = \mathbf{M}\overline{\mathbf{B}} - \mathbf{R}\nabla H_\epsilon(\psi \cdot \mathbf{R}\overline{\mathbf{B}}). \end{aligned} \tag{10.1.15}$$

$\mathbf{T}_\epsilon$ generates from $\overline{\mathbf{B}}$ a mean field $\mathbf{M}\overline{\mathbf{B}}$ and a fluctuating field with potential $-H_\epsilon(\psi(\mathbf{x}) \cdot \mathbf{R}\overline{\mathbf{B}})$, confirming (10.1.2$a$).

Substituting a general magnetic field (10.1.8) into (10.1.7) and using (10.1.12, 15) gives

$$\mathbf{M}\overline{\mathbf{B}} - \mathbf{R}\nabla H_\epsilon(\psi \cdot \mathbf{R}\overline{\mathbf{B}}) + \mathbf{R}\nabla S_\epsilon a = \mu\overline{\mathbf{B}} + \mu\mathbf{R}\nabla a. \tag{10.1.16}$$

If we average this over $\mathbb{T}^2$ all gradients vanish to leave $\mathbf{M}\overline{\mathbf{B}} = \mu\overline{\mathbf{B}}$. This is an exact mean field equation; the fluctuations are entirely decoupled. For growth we require that $\overline{\mathbf{B}} = \mathbf{e}_+$ (up to normalization) and $\mu = \lambda$, where λ is an eigenvalue of $\mathbf{M}$ with $|\lambda| > 1$, and $\mathbf{e}_+$ the corresponding eigenvector. Thus $\mathbf{M}$ must be a *hyperbolic matrix* and λ is real.

Given this we still need to satisfy the fluctuating part of (10.1.16),

$$-\nabla H_\epsilon(\psi \cdot \mathbf{Re}_+) + \nabla S_\epsilon a = \lambda\nabla a \tag{10.1.17}$$

or, integrating and absorbing any constant of integration into a,

$$-H_\epsilon(\psi \cdot \mathbf{Re}_+) + S_\epsilon a = \lambda a. \tag{10.1.18}$$

This can be rewritten

$$\begin{aligned}
a &= -\lambda^{-1}(1 - \lambda^{-1}S_\epsilon)^{-1} H_\epsilon(\psi \cdot \mathbf{Re}_+) \\
&= -\lambda^{-1}\sum_{n=0}^{\infty} \lambda^{-n} S_\epsilon^n H_\epsilon(\psi \cdot \mathbf{Re}_+),
\end{aligned} \tag{10.1.19}$$

where we have expanded $(1 - \lambda^{-1}S_\epsilon)^{-1}$ in a Neumann series. This series for $a(\mathbf{x})$ converges uniformly since $\lambda > 1$ and $\|S_\epsilon\|_\infty \leq 1$ for $\epsilon \geq 0$, where $\|\cdot\|_\infty$ is the supremum norm. We have constructed a growing fast dynamo eigenfunction explicitly, and finally obtain Oseledets' result:

Theorem 10.1 *(Oseledets 1993) Let M be a diffeomorphism on the two-torus; then the fast dynamo exponent γ_0 is positive if and only if the matrix*

$$\mathbf{M} = \int_{\mathbb{T}^2} \mathbf{J}(\mathbf{x})\,d\mathbf{x}, \qquad (\mathbf{J}(\mathbf{x}) = \partial M\mathbf{x}/\partial\mathbf{x}) \tag{10.1.20}$$

has an eigenvalue λ with $|\lambda| > 1$. In this case $\gamma_0 = \log|\lambda|$.

This result is true for any diffeomorphism; the general proof proceeds by constructing an eigenfunction for the adjoint to $\mathbf{T}_\epsilon$. It is easily seen that the topological entropy (here the rate of line stretching) is an upper bound on the growth rate: follow the evolution of a curve C from $(0,0)$ to $(1,1)$ in the covering space $\mathbb{R}^2$ and note that its length satisfies

$$\begin{aligned}
|M^n C| &\geq |M^n(0,0) - M^n(1,1)| \\
&= |\mathbf{M}^n(0,0) - \mathbf{M}^n(1,1)| \sim \text{const.} \times \lambda^n,
\end{aligned} \tag{10.1.21}$$

and so $h_{\text{top}} \equiv h_{\text{line}} \geq \log|\lambda|$. Equality holds when M is an Anosov diffeomorphism (Oseledets 1993).

In the case when the whole of the phase space $\mathbb{T}^2$ is chaotic under the action of M^n, Oseledets' conjecture (conjecture 5 of §6.2) can be checked (Oseledets 1993). We give only a brief sketch. The map M^n has a trivial action on constant functions and on two-forms; on the first cohomology space of one-forms, the action of M^n is given by the matrix $(\mathbf{M}^n)^{\mathrm{T}}$ and formula (6.2.7) becomes

$$L(M^n) = 1 - \mathrm{trace}(\mathbf{M}^n)^{\mathrm{T}} + 1 = 2 - \lambda^n - \lambda^{-n}, \qquad (10.1.22)$$

from which $h_{\mathrm{tw}} = \log|\lambda| = \gamma_0$, as required.

10.1.2 Eigenfunctions

For $\epsilon > 0$ the eigenfunction is smooth. However the arguments of §10.1.1 go through for $\epsilon = 0$ and allow us to define a continuous function $a(\mathbf{x})$, since the series (10.1.19) remains uniformly convergent. With $\overline{\mathbf{B}} = \mathbf{e}_+$ an eigenfunction can thus be constructed for $\epsilon = 0$, but in this case $a(\mathbf{x})$ is not generally differentiable, and so the field $\mathbf{B}$ is generally a distribution. In the limit $\epsilon \to 0$ the function $a|_\epsilon$ tends to $a|_{\epsilon=0}$ and in this sense the diffusionless eigenfunction is the limit of eigenfunctions as $\epsilon \to 0$ (Oseledets 1993).

More formally a weak eigenfunction $\mathbf{d} = \mathbf{e}_+ + (\partial_y a, -\partial_x a)$ can be defined as a distribution acting on any C^1 test function $\mathbf{f}(\mathbf{x})$ using integration by parts:

$$\begin{aligned}
(\mathbf{f}, \mathbf{d}) &= \int_{\mathbf{T}^2} \mathbf{f}^* \cdot [\mathbf{e}_+ + (\partial_y a, -\partial_x a)] \, dx \\
&= \int_{\mathbf{T}^2} [\mathbf{f}^* \cdot \mathbf{e}_+ + (-\partial_y f_1^* + \partial_x f_2^*)a] \, dx.
\end{aligned} \qquad (10.1.23)$$

The weak eigenfunction $\mathbf{d}$ obeys equation (10.1.7) in a weak sense, i.e., for any C^1 function $\mathbf{f}$,

$$\int_{\mathbf{T}^2} \mathbf{f}^* \cdot (\lambda \mathbf{d} - \mathbf{T}\mathbf{d}) \, dx = 0, \qquad (10.1.24)$$

where $\mathbf{T} \equiv \mathbf{T}_0$. Furthermore since $a|_\epsilon \to a|_{\epsilon=0}$ as $\epsilon \to 0$ then in this limit the eigenfunction with diffusion tends weakly to $\mathbf{d}$.

Since we know the eigenfunctions for $\epsilon \geq 0$ more or less explicitly we can check that (A) evolution from general initial conditions approaches $\mathbf{d}(\mathbf{x})$ in a weak sense, and (B) eigenfunctions $\mathbf{B}_\epsilon$ for $\epsilon > 0$ may be obtained from the $\epsilon = 0$ weak eigenfunction $\mathbf{d}$ by a smoothing process.

To sketch a proof of (A), write any initial condition in the form $\mathbf{B} = B_+ \mathbf{e}_+ + B_- \mathbf{e}_- + \mathbf{R}\nabla\alpha(\mathbf{x})$ with $B_\pm$ constants; then

$$\lambda^{-n}(\mathbf{f}, \mathbf{T}^n \mathbf{B}) \to B_-(\mathbf{f}, \mathbf{d}), \qquad (n \to \infty). \qquad (10.1.25)$$

This is a direct consequence of the three limits:

$$\begin{aligned}
&\lambda^{-n}(\mathbf{f}, \mathbf{T}^n \mathbf{e}_+) \to (\mathbf{f}, \mathbf{d}), \quad \lambda^{-n}(\mathbf{f}, \mathbf{T}^n \mathbf{e}_-) \to 0, \\
&\lambda^{-n}(\mathbf{f}, \mathbf{T}^n \mathbf{R}\nabla\alpha) \to 0
\end{aligned} \qquad (10.1.26a, b, c).$$

as $n \to \infty$. To prove the last of these, note that from (10.1.12) with $S \equiv S_0$:

$$\lambda^{-n}(\mathbf{f}, \mathbf{T}^n \mathbf{R}\nabla\alpha) = \lambda^{-n}(\mathbf{f}, \mathbf{R}\nabla S^n \alpha) = \lambda^{-n}(\nabla \cdot (\mathbf{R}^{\mathrm{T}}\mathbf{f}), S^n \alpha)$$

$$\leq \lambda^{-n}(\max|\alpha|) \int_{\mathbb{T}^2} |\nabla \cdot (\mathbf{R}^{\mathrm{T}}\mathbf{f})| \, d\mathbf{x} \to 0. \tag{10.1.27}$$

For the first, (10.1.26a), we use (10.1.12, 15) repeatedly, then (10.1.19),

$$\lambda^{-n}(\mathbf{f}, \mathbf{T}^n \mathbf{e}_+) = \left(\mathbf{f}, \mathbf{e}_+ - \mathbf{R}\nabla\lambda^{-1}\sum_{m=0}^{n-1}\lambda^{-m}S^m(\psi \cdot \mathbf{Re}_+)\right)$$

$$= (\mathbf{f}, \mathbf{d}) + \left(\mathbf{f}, \mathbf{R}\nabla\lambda^{-1}\sum_{m=n}^{\infty}\lambda^{-m}S^m(\psi \cdot \mathbf{Re}_+)\right) \tag{10.1.28}$$

$$= (\mathbf{f}, \mathbf{d}) + O(\lambda^{-n-1}),$$

and finally argue as in (10.1.27) to obtain the $O(\lambda^{-n-1})$ estimate: the proof of (10.1.26b) is similar.

To establish (B) note that $a|_\epsilon = (1 - \lambda^{-1}S_\epsilon)^{-1}H_\epsilon(1 - \lambda^{-1}S_0)a|_{\epsilon=0}$ from (10.1.19). Thus if we set $PB = \int_{\mathbb{T}^2}\mathbf{B}\,d\mathbf{x}$, then we can write

$$\mathbf{B}_\epsilon = \left[P + \mathbf{R}\nabla(1 - \lambda^{-1}S_\epsilon)^{-1}H_\epsilon(1 - \lambda^{-1}S_0)(\mathbf{R}\nabla)^{-1}(1 - P)\right]\mathbf{d}. \tag{10.1.29}$$

The presence of the heat kernel H_ϵ in the middle of this expression indicates that $\mathbf{B}_\epsilon$ is a smoothed version of the $\epsilon = 0$ eigenfunction $\mathbf{d}$. Problems with similar features are studied by Childress & Klapper (1991) and Finn *et al.* (1991) (see §9.6.4). For further discussion on the structure of eigenfunctions, see Oseledets (1993).

10.2 Pulsed Beltrami Waves

In terms of finding astrophysically reasonable flows that are fast dynamos, diffeomorphisms of $\mathbb{T}^2$ suffer the same limitation as the cat maps in §3.5: they cannot be realized by a fluid flow on $\mathbb{T}^2$ but only on somewhat abstract Riemannian manifolds. At the heart of the mapping (10.1.1) is a cat map with matrix $\mathbf{M}$, and it is this that amplifies the mean field. By contrast, for a flow $\mathbf{u}$ in periodic three-dimensional space the mean magnetic field cannot grow. Nevertheless there are features of Oseledets' study that one might hope to exploit for more realistic fast dynamos, in the setting of an approximate, rather than exact, analysis. Indeed these examples of fast dynamos have all the good properties with regard to growth rates and eigenfunctions that we should like a general fast dynamo to possess.

Let us therefore try to apply these ideas to the flow (10.0.1); take a growing magnetic field $\mathbf{B}(\mathbf{x}, t) = \exp(ikz)\mathbf{b}(x, y, t)$, for which necessarily $k \neq 0$. We would like to establish something similar to the slaving of fluctuating to mean field shown in (10.1.2); however the mean field $<\mathbf{B}>$ is zero, the average being taken over all space. Since there is no mean field to use, the next best quantity is the Fourier mode of largest scale, $\mathbf{B}_{\mathrm{LS}} = <\mathbf{b}> \exp(ikz)$. We can then put all the remaining modes into the fluctuating field, $\mathbf{B}_{\mathrm{fluc}} = \mathbf{B} - \mathbf{B}_{\mathrm{LS}}$. The problem is that under the action of a Beltrami wave, the fields $\mathbf{B}_{\mathrm{LS}}$ and $\mathbf{B}_{\mathrm{fluc}}$ are inextricably coupled: schematically,

$$\mathbf{B}_{\mathrm{LS}} \xrightarrow{\mathbf{T}_\epsilon} \mathbf{B}_{\mathrm{LS}} + \mathbf{B}_{\mathrm{fluc}}, \qquad \mathbf{B}_{\mathrm{fluc}} \xrightarrow{\mathbf{T}_\epsilon} \mathbf{B}_{\mathrm{LS}} + \mathbf{B}_{\mathrm{fluc}}, \qquad (10.2.1a, b)$$

and so we do not achieve (10.1.2). As well as the cascade from large-scale field to fluctuations, there is an inverse cascade. We cannot avoid this, but might hope to minimize its effect in some limit.

The appropriate limit turns out to be that of large τ (Soward 1993a). Consider the action of two waves on an initial constant b-field for $\epsilon = 0$,

$$\mathbf{B}_0(\mathbf{x}) = \mathbf{B}_{\mathrm{LS}} = \overline{\mathbf{b}}\,e^{ikz}, \qquad \overline{\mathbf{b}} = (\overline{b}_x, \overline{b}_y) = \text{const.}; \qquad (10.2.2)$$

we drop z-components of field for brevity. After the action of the first Beltrami wave in (10.0.1), we have

$$\mathbf{B}(\mathbf{x}, \tau) = \begin{pmatrix} 1 & 0 \\ \tau \cos x & 1 \end{pmatrix} \overline{\mathbf{b}}\,e^{ik(z - \tau \cos x)}; \qquad (10.2.3)$$

strong y-directed field is generated in the long pulse. After the action of the second wave,

$$\mathbf{B}(\mathbf{x}, 2\tau) = \begin{pmatrix} 1 & -\tau \cos y \\ 0 & 1 \end{pmatrix} \begin{pmatrix} 1 & 0 \\ \tau \cos(x + \tau \sin y) & 1 \end{pmatrix} \overline{\mathbf{b}} \qquad (10.2.4)$$
$$\times \exp\left[ik(z - \tau \cos(x + \tau \sin y) - \tau \cos y)\right].$$

This field contains a large-scale component, which can be found by averaging over x and y, together with fluctuations which vary rapidly with y because of the $-ik\tau \cos(x + \tau \sin y)$ term in the exponential. (We have in mind $k\tau = O(1)$ while $\tau \to \infty$.) There is *scale separation* between the mean field and the fluctuating field, which has a length scale of $O(1/\tau)$. Note, however, that this scale separation is non-uniform near to zeros of $\sin y$, i.e., close to the folds in the field, and this is a major obstacle to analysis.

Given that the fluctuating field is small-scale, the mean field generated *from the fluctuations* during the next wave ($2\tau < t < 3\tau$) will be very weak, because the effect of the wave in disentangling fine structure to give large-scale structure is insignificant; the principal transfer is to send the fluctuations to yet smaller scales. In the limit $\tau \to \infty$ the direct cascade is very strong and the inverse cascade is weak. We appeal to intuition rather than writing this down explicitly for the flow (10.0.1), since the weak transfer $\mathbf{B}_{\mathrm{fluc}} \to \mathbf{B}_{\mathrm{LS}}$ occurs for $t > 2\tau$, by which time the field is quite complicated.

We can summarize the situation schematically:

$$\mathbf{B}_{\mathrm{LS}} \xrightarrow{\mathbf{T}_\epsilon} \mathbf{B}_{\mathrm{LS}} + \mathbf{B}_{\mathrm{fluc}}, \qquad \mathbf{B}_{\mathrm{fluc}} \xrightarrow{\mathbf{T}_\epsilon} \delta\mathbf{B}_{\mathrm{LS}} + \mathbf{B}_{\mathrm{fluc}}, \qquad (\delta \ll 1). \qquad (10.2.5a, b)$$

Here δ is a small parameter related to the limit of strong chaos. We stress that the small parameter is there not because the fluctuating fields are weak (they are generally strong compared with the mean field), but because they are small-scale. Now if δ were zero, we would be back in the situation where $\mathbf{B}_{\text{fluc}}$ is slaved to $\mathbf{B}_{\text{LS}}$, and the fast dynamo problem would become easy. When δ is small the perturbation theory of §9.7.1 can be used to obtain approximate growth rates and eigenfunctions. This is the key to Soward's (1993a) study, on which §10.2 is based.

10.2.1 Formulation and Symmetries

We generalize a little by considering flows of the form

$$\mathbf{u} = \begin{cases} (0, f(x), g(x)), & (0 < t \bmod 2\tau < \tau), \\ (-f(y), 0, g(y)), & (\tau < t \bmod 2\tau < 2\tau), \end{cases} \tag{10.2.6}$$

where f and g satisfy

$$f(x) = -f(-x) = f(x + 2\pi), \qquad g(x) = g(-x) = g(x + 2\pi). \tag{10.2.7}$$

For the Beltrami wave flow (10.0.1), $f(x) = \sin x$ and $g(x) = \cos x$. This is the flow considered by Soward, up to an inessential translation; we prefer the form (10.2.6) for discussing symmetries. The magnetic field is taken to be

$$\mathbf{B}(\mathbf{x}, z, t) = (\mathbf{b}(\mathbf{x}, t), b_3(\mathbf{x}, t)) \exp(ikz), \tag{10.2.8}$$

with $\mathbf{x} = (x, y)$. The flow is periodic in x and y and the magnetic field is taken to have the same spatial periodicity.

Since the flow has time period 2τ, we can seek eigenfunctions for $\epsilon > 0$ having a Floquet form (1.3.5),

$$\mathbf{b}(\mathbf{x}, t) = \exp(pt)\mathbf{c}(\mathbf{x}, t), \tag{10.2.9}$$

where $\mathbf{c}$ is periodic in time, period 2τ, and $p(\epsilon)$ is the complex growth rate. In order to simplify the problem, we consider symmetries of the flow and magnetic field.

Define a rotation R through $\pi/2$ and a time shift S:

$$R(x, y, z) = (-y, x, z), \qquad S(t) = t - \tau. \tag{10.2.10}$$

R maps each helical wave into the other, and so $\mathbf{u}$ is invariant under a rotation and time shift, $RS\mathbf{u} = \mathbf{u}$. Given a Floquet solution (10.2.9) for a given growth rate p, other solutions for the same p are $(RS)^n\mathbf{c}$ for $n = 0, 1, 2, 3, \ldots$. Since $R^4 = 1$ and $S^2\mathbf{c} = \mathbf{c}$, just four symmetries are relevant here, forming the cyclic group of order four. This is analogous to the situation for the flow MW+, discussed in §8.2.3, and as there we can, without loss of generality, take $\mathbf{c}(\mathbf{x}, t)$ to satisfy

$$RS\mathbf{c} = \rho\mathbf{c}, \qquad (\rho^4 = 1). \tag{10.2.11}$$

Note that $\rho^2\mathbf{c} = (RS)^2\mathbf{c} = R^2\mathbf{c}$. It follows that for $\rho^2 = +1$, the field $\mathbf{c}$ must have zero mean. Since the dynamo mechanisms we seek involve primarily amplification of mean $\mathbf{c}$-field, we only consider the case $\rho^2 = -1$, or $\rho = \pm i$, for which the mean $\mathbf{c}$ is generally non-zero.[75]

Let $\mathbf{T}_\epsilon$ be the induction operator for the evolution of the field $\mathbf{b}(\mathbf{x}, t)$ in the first wave of (10.2.6) over a time τ:

$$(x, y, z) \to (x, y + \tau f(x), z + \tau g(x)), \quad \mathbf{b}(\mathbf{x}, \tau) = \mathbf{T}_\epsilon \mathbf{b}(\mathbf{x}, 0). \tag{10.2.12}$$

In terms of $\mathbf{c}$, from (10.2.9),

$$e^{p\tau}\mathbf{c}(\mathbf{x}, \tau) = \mathbf{T}_\epsilon \mathbf{c}(\mathbf{x}, 0). \tag{10.2.13}$$

Given that $\mathbf{c}$ belongs to one of the four symmetry classes, we can write

$$\mathbf{c}(\mathbf{x}, \tau) = S\mathbf{c}(\mathbf{x}, 0) = R^{-1}(RS)\mathbf{c}(\mathbf{x}, 0) = \rho R^{-1}\mathbf{c}(\mathbf{x}, 0). \tag{10.2.14}$$

We can eliminate $\mathbf{c}(\mathbf{x}, \tau)$ between (10.2.13, 14) to obtain the simplified eigenvalue problem, $\rho \exp(p\tau)\mathbf{c}(\mathbf{x}, 0) = R\mathbf{T}_\epsilon \mathbf{c}(\mathbf{x}, 0)$. Writing $\mathbf{c}(\mathbf{x}) \equiv \mathbf{c}(\mathbf{x}, 0)$ the eigenvalue problem and corresponding growth rate are

$$\mu\mathbf{c}(\mathbf{x}) = R\mathbf{T}_\epsilon \mathbf{c}(\mathbf{x}), \quad \gamma(\epsilon) = \tau^{-1}\log|\mu|. \tag{10.2.15}$$

10.2.2 Mean-field Problem

We now set $\epsilon = 0$; since we have in mind the construction of generalized eigenfunctions for zero diffusion, the above symmetry arguments still apply. From the Cauchy solution, for $\epsilon = 0$ we have

$$\mathbf{T}_0\mathbf{c}(x, y) = \begin{pmatrix} 1 & 0 \\ \tau f'(x) & 1 \end{pmatrix} \mathbf{c}(x, y - \tau f(x)) e^{-ik\tau g(x)} \tag{10.2.16}$$

and, writing $\mathbf{T} \equiv R\mathbf{T}_0$ for brevity, from (10.2.10),

$$\mathbf{T}\mathbf{c}(x, y) = \begin{pmatrix} -\tau f'(y) & -1 \\ 1 & 0 \end{pmatrix} \mathbf{c}(y, -x - \tau f(y)) e^{-ik\tau g(y)}. \tag{10.2.17}$$

If fluctuating fields can be neglected in the large-τ limit, we can find approximate eigenvalues by projecting them out. Let P be a projection of $\mathbf{c}$ onto its (x, y) average,

$$P\mathbf{c}(x, y) = <\mathbf{c}> \equiv (2\pi)^{-2} \int_0^{2\pi} \int_0^{2\pi} \mathbf{c}(x, y)\, dx\, dy, \tag{10.2.18}$$

---

[75]The situation $\rho = \pm i$ is reminiscent of helical waves $(\pm i, 1)\exp(ikz)$ with parity ∓ 1, since under R such waves are multiplied by $\pm i$. The situation for $\mathbf{c}$ is not entirely analogous, as the symmetry is RS, rather than R, and so relates fields at different times.

and set $\overline{\mathbf{T}} = P\mathbf{T}$; this operator applies a wave, a rotation and then projects away fluctuations.

Eigenfunctions of $\overline{\mathbf{T}}$ with non-zero eigenvalue are constant fields. Starting with constant field $\overline{\mathbf{c}}$,

$$\overline{\mathbf{T}}\overline{\mathbf{c}} \equiv <\mathbf{T}\overline{\mathbf{c}}> = < \begin{pmatrix} -\tau f'(y) & -1 \\ 1 & 0 \end{pmatrix} e^{-ik\tau g(y)} > \overline{\mathbf{c}}. \tag{10.2.19}$$

For Beltrami waves (10.0.1), this amounts to

$$\overline{\mathbf{T}}\overline{\mathbf{c}} = \begin{pmatrix} i\tau J_1(k\tau) & -J_0(k\tau) \\ J_0(k\tau) & 0 \end{pmatrix} \overline{\mathbf{c}}. \tag{10.2.20}$$

The eigenvalues of the operator $\overline{\mathbf{T}}$ are $\mu_{0\pm}$ where

$$\mu_{0+} = i\tau J_1(k\tau) + O(\tau^{-1}), \quad \mu_{0-} = J_0^2(k\tau)/i\tau J_1(k\tau) + O(\tau^{-3}), \tag{10.2.21}$$

corresponding to growth rates $\gamma_{0\pm}$ (Soward 1993a). Here we are thinking of $k\tau$ as fixed while $\tau \to \infty$. This approximation is essentially the same as Otani's (1993) for the MW+ flow; see equation (2.3.11).

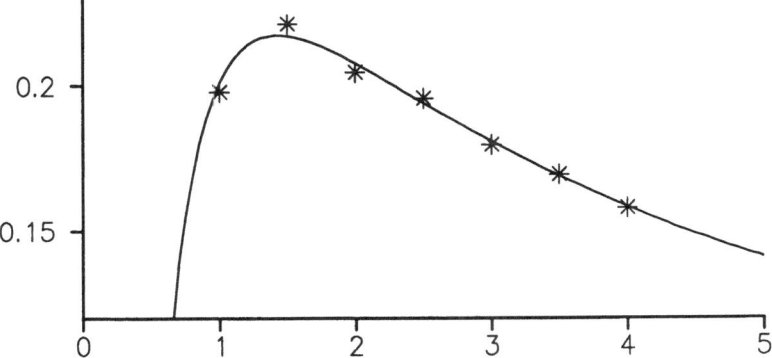

Fig. 10.1 Growth rates $\gamma(\epsilon)$ plotted against τ/π. Shown are numerical growth rates $\gamma(\epsilon)$ (markers) for $\epsilon = 10^{-4}$ and $k\tau = 1.8412$. The solid line shows the leading order theoretical growth rate γ_{0+}.

If we neglect diffusion and fluctuations, $\gamma_{0\pm}$ should be good approximations to fast dynamo growth rates for large τ. To verify this, Fig. 10.1 shows growth rates obtained numerically for the flow (10.0.1)[76] with various values

---

[76]In the code the waves are in fact switched off and on smoothly; the flow field used is not $\mathbf{u}$ from (10.0.1), but $2\sin^2(\pi t/\tau)\mathbf{u}$.

of τ/π, $\epsilon = 10^{-4}$ and $k\tau = 1.8412$, which maximizes $J_1(k\tau) \simeq 0.5819$ in μ_{0+} (10.2.21). The agreement is impressive even for moderate τ.

10.2.3 Fluctuations

Given that the mean-field approximation appears to work, how might it be justified? This appears to be a difficult problem, and all we can do here is to give some suggestions. There appear to be two approaches: the first is to construct approximate eigenfunctions (in L^2 say) and so show that the approximate spectrum of $\mathbf{T}$ contains a point close to μ_{0+} (Soward 1993a, b, 1994a). The second approach is to work in a space large enough to contain the $\epsilon = 0$ eigenfunction, which we guess will be a distribution of some kind; in this case the eigenvalue close to μ_{0+} will lie in the point spectrum of $\mathbf{T}$ in this big space. The problem here is that the appropriate space is not known. This second approach has however been carried through for the models of §§10.3, 4.

We explore the second approach first. We observe that diffusion can have little effect on the evolution of large-scale fields under $\overline{\mathbf{T}}_\epsilon$. The modification of the growth rates $\gamma_{0\pm}$ will be of order ϵk^2, which is uniformly small for $\epsilon \to 0$ with $k \leq O(1)$. The principal effect of diffusion on fluctuations is damping; thus if fluctuations may be neglected then probably so can diffusion. Furthermore if the fluctuations can be neglected with $\epsilon = 0$, it is likely that they can be neglected for $\epsilon > 0$. We therefore set $\epsilon = 0$ and discuss the effect of fluctuations by developing a perturbation series expansion; see §9.7.1. Set $\mathbf{T} = \overline{\mathbf{T}} + \mathbf{T}'$ and seek an eigenfunction $\mathbf{c}(\mathbf{x})$ of $\mathbf{T}$, eigenvalue μ, as in (10.2.15). By formal manipulation

$$\mu\mathbf{c} = (\overline{\mathbf{T}} + \mathbf{T}')\mathbf{c}, \tag{10.2.22a}$$

$$\mu(1 - \mu^{-1}\mathbf{T}')\mathbf{c} = \overline{\mathbf{T}}\mathbf{c}, \tag{10.2.22b}$$

$$\mu\mathbf{c} = (1 - \mu^{-1}\mathbf{T}')^{-1}\overline{\mathbf{T}}\mathbf{c}. \tag{10.2.22c}$$

Applying $\overline{\mathbf{T}}$ to both sides and setting $\overline{\mathbf{c}} = P\mathbf{c} \equiv \mu^{-1}\overline{\mathbf{T}}\mathbf{c}$, a constant vector, we obtain

$$\mu\overline{\mathbf{c}} = \overline{\mathbf{T}}(1 - \mu^{-1}\mathbf{T}')^{-1}\overline{\mathbf{c}}. \tag{10.2.23}$$

This is a nonlinear equation for the eigenvalue μ, which we expand in a series:

$$\mu\overline{\mathbf{c}} = \overline{\mathbf{T}}\overline{\mathbf{c}} + \mu^{-1}\overline{\mathbf{T}}\mathbf{T}'\overline{\mathbf{c}} + \mu^{-2}\overline{\mathbf{T}}\mathbf{T}'^2\overline{\mathbf{c}} + \mu^{-3}\overline{\mathbf{T}}\mathbf{T}'^3\overline{\mathbf{c}} + \cdots. \tag{10.2.24}$$

Each term $\overline{\mathbf{T}}\mathbf{T}'^n\overline{\mathbf{c}}$ of the series is a 2×2 matrix acting on $\overline{\mathbf{c}}$, and can be written in terms of sums of Bessel functions (Soward 1993a).

Truncating the series at leading order yields $\mu \simeq \mu_0$, where μ_0 satisfies $\mu_0\overline{\mathbf{c}} = \overline{\mathbf{T}}\overline{\mathbf{c}}$ and so recovers the mean-field problem (10.2.20, 21). Successive

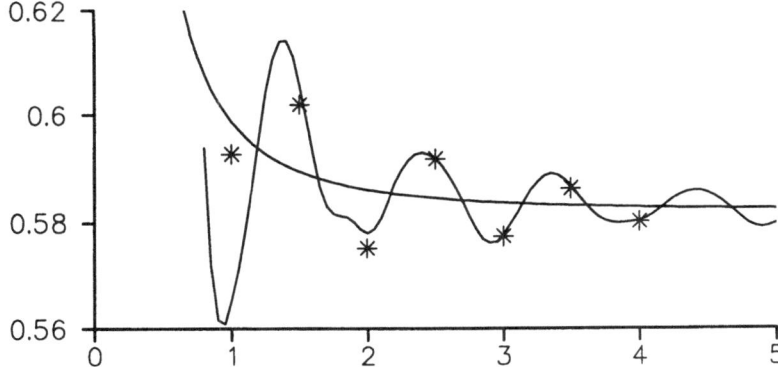

Fig. 10.2 Comparison of numerical growth rates and asymptotic results. Numerical values of $|\mu_{+}|/\tau$ (markers) are plotted against τ/π for $\epsilon = 10^{-4}$ and $k\tau = 1.8412$. The curving line shows the leading order theoretical value $|\mu_{0+}|/\tau$, while the oscillatory line shows the high-order approximation $|\mu_{4+}|/\tau$. (After Soward 1993a; theoretical values courtesy A.M. Soward.)

approximations may be found by truncating the series at higher orders.[77] Soward (1993a) has calculated growth rates taking into account terms up to $\overline{\mathbf{T}}\mathbf{T}'^{4}\overline{\mathbf{c}}$. Comparison with numerical growth rates is given in Fig. 10.2, which shows the leading approximation $|\mu_{0+}|/\tau$ (curving line) and the higher order approximation $|\mu_{4+}|/\tau$ (oscillatory line), together with numerical results (markers). The agreement is excellent in view of the scale of the ordinate.

The approximations certainly appear to work. Mathematically, the problem is to justify the manipulations in (10.2.22) and show that the expansion (10.2.24) converges. This remains to be proved, although numerical and other evidence is strong; see Soward (1993a, b) for further discussion. As discussed in §9.7.1, it is sufficient to find a norm $\|\cdot\|$ on magnetic fields for which $\mu^{-1}\|\mathbf{T}'\| \ll 1$ in the limit of strong stretching. This would justify the inversion in (10.2.22c), and ensure that the series (10.2.24) converges rapidly. Since the fluctuating fields are unimportant because they are small-scale (and not weak), such a norm must give a low weighting to the high wave-number fluctuations generated by $\mathbf{T}$.

Such a norm has not yet been constructed for the Beltrami waves, but can be done for related problems in the next sections. During a long pulse, the cascade of field to small scales is strong over the domain, except near the folds of the field, as alluded to after (10.2.4) and discussed in Soward (1993a).

---

[77]Note that Oseledets' growth rate for maps of $\mathbb{T}^2$ (§10.1) can be recovered here. For such maps $\overline{\mathbf{T}}\mathbf{T}'^{n}\overline{\mathbf{c}} = \mathbf{0}$ for $n \geq 1$, since mean field cannot be generated from fluctuations.

This weak stretching near the folds and the corresponding lack of uniformity is a serious obstacle to analysis. This is perhaps surprising, since the field generated near the folds is relatively weak, and one might hope it could be ignored; how to do this here and in other contexts is not known. Note that a stronger cascade to small scales near the folds can be achieved by using piecewise linear functions for f and g.

Assuming a solution μ and $\bar{c}$ of (10.2.23), an eigenfunction is

$$\mathbf{c} = (1 - \mu^{-1}\mathbf{T}')^{-1}\bar{\mathbf{c}} = \bar{\mathbf{c}} + \mu^{-1}\mathbf{T}'\bar{\mathbf{c}} + \mu^{-2}\mathbf{T}'^2\bar{\mathbf{c}} + \cdots, \qquad (10.2.25)$$

from (10.2.22c). Given a Banach space in which $\mu^{-1}\|\mathbf{T}'\|$ is small, this series will converge to an eigenfunction in this space. The eigenfunction, and its properties in real space and in Fourier space, are discussed in Soward (1993b, 1994a). It appears to have a self-similar structure: it seems likely that it has well-defined Fourier coefficients, but that these do not tend to zero fast enough at high wave-numbers to determine a 'reasonable' function in real space.

Since the above perturbation theory, when it can be made to work, gives the construction of an eigenfunction (10.2.25) as a convergent series, this highlights the need to work in a space large enough to contain the eigenfunction. The other approach, mentioned at the start of §10.2.3, is to construct a sequence of candidate approximate eigenfunctions $\mathbf{B}_n$, in say L^2, which would then converge to a distributional eigenfunction in some weak sense. We now outline Soward's (1993a) method of doing this.

We construct $\mathbf{B}_n$ for each n from a sequence of functions $\mathbf{b}_m$ for $m = 0, \ldots, n$ starting with $\mathbf{b}_0 = \bar{c}_0$. Given any complex number μ, first set $\mathbf{b}_m = \mathbf{T}\mathbf{b}_{m-1} + \bar{c}_m$ for $m = 1, \ldots n$; we apply the dynamo operator and add a constant field. We demand that $P\mathbf{b}_m = \mu P\mathbf{b}_{m-1}$, exactly. Given $\bar{c}_0$ and μ, it may be checked that this determines $\bar{c}_m$ by

$$\overline{\mathbf{T}\mathbf{T}'^{m-1}}\bar{c}_0 + \bar{c}_m = \mu\bar{c}_{m-1}, \qquad (m = 1, \ldots n). \qquad (10.2.26)$$

Now we take the nth approximate eigenfunction as $\mathbf{B}_n = \mathbf{b}_n$ and to fix the parameters $\bar{c}_0$ and μ we demand that $\overline{\mathbf{T}}\mathbf{B}_n = \mu\mathbf{B}_n$. This is an eigenvalue problem, and is equivalent to applying the dynamo operator, but not adding a mean field, that is, setting $\bar{c}_{n+1} = 0$ in (10.2.26) with $m = n + 1$ to give $\overline{\mathbf{T}\mathbf{T}'^n}\bar{c}_0 = \mu\bar{c}_n$. Using (10.2.26) repeatedly to eliminate $\bar{c}_m$ we obtain, with $\bar{c} \equiv \bar{c}_0$,

$$\mu\bar{c} = \overline{\mathbf{T}}\bar{c} + \mu^{-1}\overline{\mathbf{T}\mathbf{T}'}\bar{c} + \mu^{-2}\overline{\mathbf{T}\mathbf{T}'^2}\bar{c} + \cdots + \mu^{-n}\overline{\mathbf{T}\mathbf{T}'^n}\bar{c}. \qquad (10.2.27)$$

We therefore recover the eigenvalue problem (10.2.24), truncated to n terms. The corresponding eigenfunction is also familiar; explicitly,

$$\mathbf{B}_n = \mu^n(\bar{\mathbf{c}} + \mu^{-1}\mathbf{T}'\bar{\mathbf{c}} + \mu^{-2}\mathbf{T}'^2\bar{\mathbf{c}} + \cdots + \mu^{-n}\mathbf{T}'^n\bar{\mathbf{c}}), \qquad (10.2.28)$$

which up to normalization is (10.2.25), truncated to n terms.

Since the sequence of functions $\mathbf{B}_n$ satisfies $\overline{\mathbf{T}}\mathbf{B}_n = \mu_n\mathbf{B}_n$ where μ_n are the solutions of the truncated eigenvalue problem (10.2.27), this is clearly a

good candidate for a sequence of approximate eigenfunctions as defined in
§9.1.1; however to establish this property we would need to prove that μ_n
converges to μ, and to consider carefully $\|\mathbf{TB}_n - \mu\mathbf{B}_n\|$.

We will not consider pulsed Beltrami wave models further, but instead
move on to models which can be analyzed by the techniques mentioned. These
are based on baker's maps and cat maps, which are hyperbolic and so have
a more effective cascade to small scales than Beltrami waves. There are close
parallels with Soward's study, and the fact that they can be analyzed suggests
that Soward's work may perhaps soon be put on a rigorous basis, leading to
the first proof of fast dynamo action in a realistic flow.

10.3 SFS in the Limit of Strong Chaos

The SFS map can be extended to a large-amplitude or strong chaos limit by
increasing the stretching factor from two to a large positive even integer $2N$.
Such a model is of interest since we can use the simplifications inherent in
the adjoint formulation. We briefly consider the results obtained by Childress
(1993b) for one model of this kind. We consider a combination of the stacked
and folded baker's maps of §3.1.1. Following the stretch by the factor $2N$, we
divide up into $2N$ slabs of thickness $1/2N$ and length 1, and stack N of them
with positive orientation to fill $0 \leq y \leq 1/2$, and stack the remaining N slabs
with negative orientation to fill $1/2 \leq y \leq 1$. The stacking is in the natural
order, with slabs N and $N + 1$ adjacent to the plane $y = 1/2$. The parallel
operator T^{SFS} introduced in §3.2 is easily computed for this geometry. We do
not give this but instead consider the corresponding parallel adjoint operator
(see §3.2.2), which we now denote (suppressing the superscript SFS) by S:

$$
\begin{aligned}
Sb(y) = &\sum_{j=0}^{N-1} b\left(\frac{j}{2N} + \frac{y}{2N}\right) \exp\left[2\pi i\alpha k \left(\frac{j}{2N} + \frac{y}{2N} - \frac{1}{2}\right)\right] \\
&- \sum_{j=N}^{2N-1} b\left(\frac{j+1}{2N} - \frac{y}{2N}\right) \exp\left[2\pi i\alpha k \left(\frac{j+1}{2N} - \frac{y}{2N} - \frac{1}{2}\right)\right].
\end{aligned}
\tag{10.3.1}
$$

For $N = 1$, (10.3.1) reduces to (3.2.8). Now the main innovation in considering
large N is the observation that then (10.3.1) is an approximating sum to a
Riemann integral:

$$
\begin{aligned}
Sb \approx S_0 b &\equiv 2N \int_0^1 b(y)\,\mathrm{sign}(1/2 - y)\,\exp[2\pi i\alpha k(y - 1/2)]\,dy \\
&\equiv 2N \int_0^1 b(y)g(y)\,dy;
\end{aligned}
\tag{10.3.2}
$$

this defines the limiting operator S_0, which maps $b(y)$ into a constant. Thus
S_0 has an eigenfunction 1 with eigenvalue λ_0^*:

$$\lambda_0^* = \frac{4N}{i\pi\alpha k} \sin^2(\pi\alpha k/2) = 2N \int_0^1 g(y)\, dy. \tag{10.3.3}$$

Note that the proximity to a Riemann integral is set up by the stacking of slabs, with S_0 and $g(y)$ reflecting the local average orientation of slabs.

The idea is now to use the perturbation theory for eigenvalues discussed in §9.7.1 applied to $S' = S - S_0$ as a small perturbation of S_0. The nature of the approximation is that of a Riemann sum to a Riemann integral, and this makes it natural to work in the space of functions of bounded variation. The projection we use is just the flux integral

$$Pb = \int_0^1 b(y)\, dy, \tag{10.3.4}$$

in which case we have from (10.3.1, 2) that $PS = S_0$, and so $PS' = 0$. For a norm we take the function $g(y)$ defined by (10.3.2) and set

$$\|b\| = \left| \int_0^1 b(y)g(y)\, dy \right| + \mathcal{V}(b), \tag{10.3.5}$$

where $\mathcal{V}(b)$ is the total variation of b on $[0,1]$, that is, in terms of partitions $\mathcal{P} = \{0 \le y_0 < y_1 < \ldots < y_n \le 1\}$ of the interval,

$$\mathcal{V}(b) = \sup_{\mathcal{P}} \sum_k |\Delta b_k|, \qquad (\Delta b_k = b(y_k) - b(y_{k-1})). \tag{10.3.6}$$

This defines a norm provided that

$$\left| \int_0^1 g(y)\, dy \right| > 0, \tag{10.3.7}$$

in which case λ_0 is a number of order N for large N.

The perturbation theory of §9.7.1 can thus be used provided we show that $\|S'\|$ is small compared to N. Now the contribution from $\mathcal{V}(S'b)$ to $\|S'b\|$ can be estimated as follows. From the definition (10.3.1) of S we may write $Sb(y) = \sum_{k=0}^{2N-1} b(y_k(y))g(y_k(y))$, where for each y, $\{y_k(y)\}$ is a partition of the interval. Thus $\mathcal{V}(S'b)$ will involve differences taken at points $y_k(y_j)$ where the y_j belong to a partition. However any set of such points constitutes another partition, and so we have

$$\mathcal{V}(S'b) \le M_g \mathcal{V}(b) + M_b \mathcal{V}(g), \tag{10.3.8}$$

where M_g, M_b are maxima of $|g|, |b|$ over $[0,1]$.

Now the contribution from the first term on the right of (10.3.5) can be estimated using results from the theory of integration of functions of bounded variation (see, e.g., Apostol 1957). For any $\delta > 0$, there exists an N_1 such that for $N > N_1$,

$$\left| \frac{1}{2N} \sum_{k=0}^{2N-1} b(y_k) g(y_k) - \int_0^1 b(y) g(y) \, dy \right| \le \delta (M_g \mathcal{V}(b) + M_b \mathcal{V}(g)). \quad (10.3.9)$$

If we can show that

$$M_b \le C \|b\|, \quad (10.3.10)$$

for some positive constant C, then (10.3.8–10) combine to yield

$$\|S'\| \le 2N (M_g \delta + 1/2N)[M_g + C\mathcal{V}(g)]. \quad (10.3.11)$$

Taking δ sufficiently small, and N sufficiently large, we obtain $\|S'\|/N$ as small as desired.

To show that (10.3.10) holds for some C, we need to bring in the role of the function g in more detail. Of course if $g = \text{sign}(1/2 - y)$, without the phase factor, we are dealing with a *folded* map with no dynamo action. Clearly (10.3.7) has to make an explicit appearance in the estimates, and it is now that this happens. The maximum field M_b occurs at some point y_0 in $[0,1]$. Using a one-point partition we then have

$$M_b - \left| \int_0^1 b(y) \, dy \right| \le \left| b(y_0) - \int_0^1 b(y) \, dy \right| \le \mathcal{V}(b), \quad (10.3.12)$$

and so

$$M_b \le \left| \int_0^1 b(y) \, dy \right| + \mathcal{V}(b). \quad (10.3.13)$$

Thus it is sufficient to establish an estimate of the form (10.3.10) for the *total flux*. With this aim we consider a discrete relation and derive from it the needed inequality on Riemann integrable functions. Take a large even integer K and consider the partition $(1/K, 2/K, ..., 1)$. If $g_i = g(i/K)$ we seek to bound $|g_1 b_1 + g_2 b_2 + \cdots + g_K b_K|$ from below. Let $a_i = b_i - b_{i+1}$ for $i = 1, \ldots, K - 1$ and set $a_K = b_1 + b_2 + \cdots + b_K$. Forming the column vectors $\mathbf{a}$ and $\mathbf{b}$ we have $\mathbf{a} = G\mathbf{b}$ where

$$G = \begin{pmatrix} 1 & -1 & 0 & \cdots & 0 \\ 0 & 1 & -1 & \cdots & 0 \\ \vdots & \vdots & \vdots & \ddots & \vdots \\ 1 & 1 & 1 & \cdots & 1 \end{pmatrix}. \quad (10.3.14)$$

The inverse is

$$G^{-1} = K^{-1} \begin{pmatrix} K-1 & K-2 & K-3 & \cdots & 1 & 1 \\ -1 & K-2 & K-3 & \cdots & 1 & 1 \\ -1 & -2 & K-3 & \cdots & 1 & 1 \\ \vdots & \vdots & \vdots & \ddots & \vdots & \vdots \\ -1 & -2 & -3 & \cdots & -(K-1) & 1 \end{pmatrix}. \quad (10.3.15)$$

Computing $\mathbf{g} \cdot \mathbf{b} = \mathbf{g} \cdot G^{-1}\mathbf{a}$ we obtain the estimate

$$|g_1 b_1 + g_s b_2 + \ldots + g_K b_K| \geq |m_g||a_K| - KM_g \sum_{k=1}^{K-1} |a_k|, \qquad (10.3.16)$$

where

$$m_g = K^{-1} \sum_{k=1}^{K} g_k. \qquad (10.3.17)$$

Passing to the limit of infinite K there results

$$\left| \int_0^1 b\,dy \right| \leq \left| \int_0^1 g\,dy \right|^{-1} \left[\int_0^1 gb\,dy + M_g \mathcal{V}(b) \right]$$
$$\leq \left| \int_0^1 g\,dy \right|^{-1} (1 + M_g)\,\|b\|. \qquad (10.3.18)$$

This establishes that the bound (10.3.10) holds with the constant

$$C = \left| \int_0^1 g\,dy \right|^{-1} (1 + M_g) + 1. \qquad (10.3.19)$$

We again stress the main advantage of working with a baker's map in the case of strong chaos: the entire field structure is mapped into numerous small-scale images with a net flux dictated by orientation and shear. By exploiting the adjoint formulation, the collection of small replicas becomes a Riemann sum, closely approximating a simple integral operator. The method also works for continuous maps of the form $Tb(y) = \alpha r(y)b(\alpha s(y))$ where α is a large parameter and $s(y)$ is monotone or a finite union of monotone functions.

10.4 Cat Maps and Pseudo-Anosov Maps with Shear

One obstacle to progress in fast dynamo theory is a lack of volume-preserving flows or maps for which much is known about the chaotic trajectories. There are many significant differences between a baker's map or cat map, and flows such as MW+ or ABC. Baker's maps involve unphysical cutting of field lines, and the folding in cat maps is trivial but unrealizable in ordinary space. We need to draw from dynamical systems theory classes of maps or flows that are more realistic than these, but yet which offer some hope of analysis.

One such class is *pseudo-Anosov* (pA) maps (see, e.g., Boyland & Franks 1989, Poénaru 1979 and other papers in the journal Astérisque, vols. 66–67). These maps are used in extending the Sarkovskii ordering of periodic orbits in maps from an interval to itself (see, e.g., Guckenheimer & Holmes 1983), to maps from the unit disc D^2 to itself. In studying how periodic orbits are ordered it proves useful to define maps with minimal collections of periodic orbits. These are pA maps on D^2, and have the property that except at a number of singular points on D^2 there are well-defined stretching and contracting directions at a point. Except at the singular points these maps resemble Anosov maps, hence the name pseudo-Anosov.

As far as we are concerned the important features of a pA map are that it is continuous, has relatively simple chaotic properties, and yet allows folding of magnetic field. While these maps are still somewhat simplified they represent a step beyond cat and baker's maps in the direction of more realistic flows. We shall discuss only the simplest pA maps, which can be obtained from cat maps by taking a quotient. The models we study comprise a pA map, which gives stretching and folding, and shear to allow an SFS mechanism; the models can be analyzed in the limit of strong chaos (Gilbert 1993).

10.4.1 Dynamo Action in a pA Map with Shear

The properties of pA maps are best seen by an example. Consider a cat map given by the integer matrix $\mathbf{M}$,

$$M : \mathbf{x} \to \mathbf{M}\mathbf{x}, \qquad (\det \mathbf{M} = 1), \qquad (\mathbf{x} = (x_1, x_2)). \qquad (10.4.1)$$

This acts on $\mathbb{T}^2$, which we think of interchangeably either as the square $[0, 1] \times [0, 1]$ with opposite edges identified, or as $\mathbb{R}^2$ with points identified modulo one. The map M has a symmetry $J : \mathbf{x} \to -\mathbf{x}$, which is a π-rotation of the square. We can write any two-dimensional magnetic field on $\mathbb{T}^2$ as the sum of an even field and an odd field,

$$\mathbf{b}(-\mathbf{x}) = \mathbf{b}(\mathbf{x}), \qquad (\text{even}), \qquad \mathbf{b}(-\mathbf{x}) = -\mathbf{b}(\mathbf{x}), \qquad (\text{odd}), \qquad (10.4.2)$$

and the map M preserves these forms. Even fields can have non-zero mean; examples are the fields $\mathbf{b} = \mathbf{e}_\pm$, which are dynamo eigenfunctions for a pulsed cat map, growth rates $\gamma(\epsilon) = \pm \log |\lambda|$. Here we adopt the notation of §3.5 for eigenvectors $\mathbf{e}_\pm$ and eigenvalues $\lambda^{\pm 1}$ of $\mathbf{M}$ with $|\lambda| > 1$.

Odd fields have zero mean, $<\mathbf{b}> = 0$, and any odd field of finite energy decays super-exponentially when the cat map and non-zero diffusion are pulsed (M.M. Vishik, personal communication). To see this consider a Fourier mode, wave-vector $\mathbf{k}$,

$$\mathbf{b}(\mathbf{x}, 0) = \mathbf{b_k} \exp(2\pi i \mathbf{k} \cdot \mathbf{x}). \qquad (10.4.3)$$

After n periods of map–diffuse, with heat kernel (10.1.6), the field is

$$\mathbf{b}(\mathbf{x}, n) = \mathbf{M}^n \mathbf{b_k} \exp(2\pi i \mathbf{k} \cdot \mathbf{N}^n \mathbf{x}) e^{-\epsilon((\mathbf{k} \cdot \mathbf{N})^2 + (\mathbf{k} \cdot \mathbf{N}^2)^2 + \cdots + (\mathbf{k} \cdot \mathbf{N}^n)^2)}, \qquad (10.4.4)$$

where we define $\mathbf{N} \equiv \mathbf{M}^{-1}$. The field decays rapidly as $n \to \infty$ for $\mathbf{k} \neq 0$. We can think of the pA map acting on even fields as analogous to a stacked baker's map, and acting on odd fields as analogous to a folded baker's map.

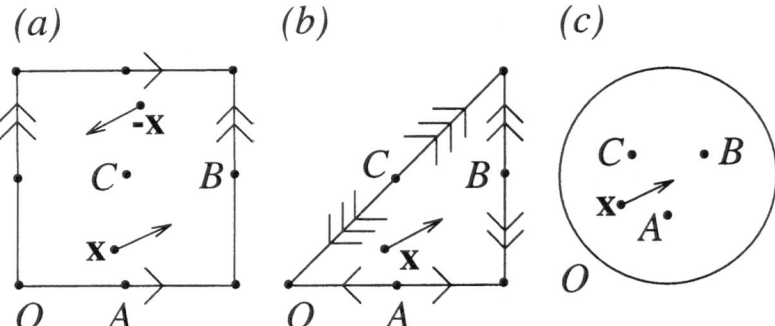

Fig. 10.3 Spaces: (a) the torus $\mathbb{T}^2$, (b) the quotient space $\mathbb{T}^2/J$, which is the sphere S^2, and (c) the space $\mathbb{T}^2/J\backslash\{O\}$, which is the disc D^2. (Fig. 3 of Gilbert (1993), reproduced with permission of the Royal Society.)

There is a close connection between the decay of these odd fields and the anti-dynamo theorems for planar flows (§4.2.2). Since an odd field is invariant under J, we can consider it as lying in the *quotient space* $\mathbb{T}^2/J$ and being acted on by the *quotient map* M/J. The torus $\mathbb{T}^2$ is shown in Fig. 10.3(a) and the quotient space — half a torus — in 10.3(b). The edges are identified according to the arrows and if we imagine sewing them up, we see that $\mathbb{T}^2/J$ is topologically a sphere S^2. Puncturing this sphere at the origin O gives a disc D^2 (Fig. 10.3c). The quotient map M/J can be considered as a map from D^2 to D^2 which preserves the boundary (as O is a fixed point of M).

The map M/J is one of the simplest pseudo-Anosov maps of the disc. It inherits the grid of stretching and contracting directions of the cat map M, except at the four *branch points* O, $A \equiv (1/2,0)$, $B \equiv (0,1/2)$ and $C = (1/2,1/2)$. At these points J is one-to-one rather than two-to-one, the grid has singularities and folding of magnetic field takes place. To illustrate this, and the taking of the quotient, Fig. 10.4 shows a loop of field $b(\mathbf{x}) = \mathbf{e}_x \sin 2\pi y$ in $\mathbb{T}^2$ and the corresponding fields in $\mathbb{T}^2/J$ and D^2, under 0, 1 and 2 iterations of the cat map. We observe that the field lines are stretched and folded on the disc D^2, but not cut or reassembled. The pA map of the disc is continuous (and so can be effected by a flow on the disc), and the stretching of line elements is everywhere finite. We therefore see that the decay of odd fields under the cat map M corresponds to the anti-dynamo theorem for planar flows in §4.2.2. Compared with a realistic flow (say MW+), the folding is still oversimplified, as it occurs only at the branch points rather than folds accumulating everywhere; nevertheless the folding is more realistic than that in baker's maps or cat maps.

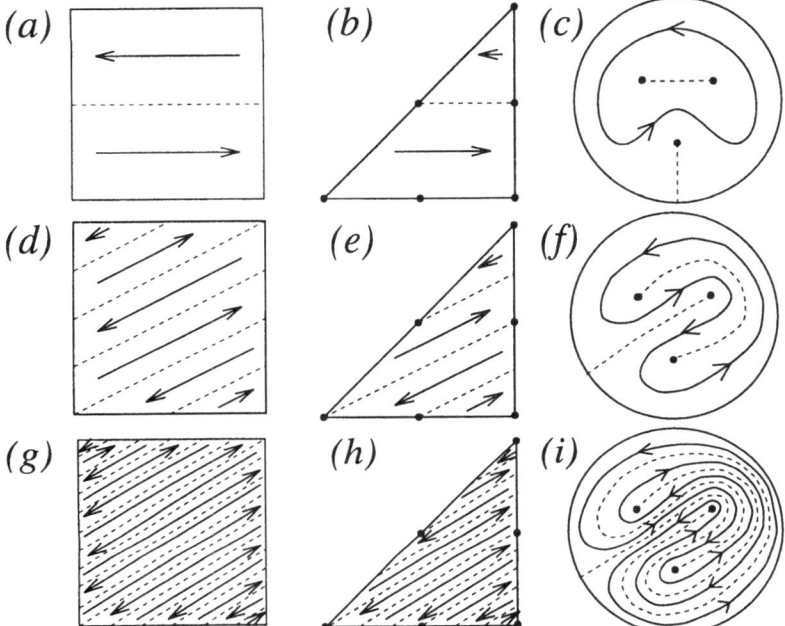

Fig. 10.4 Evolution of an odd magnetic field in $\mathbb{T}^2$ and the quotient spaces, S^2 and D^2, under the cat map (10.4.1) with matrix (3.5.3). $(a,\ b,\ c)$ show the initial field in each space, $(d,\ e,\ f)$ the field after one iteration and $(g,\ h,\ i)$ the field after two iterations. The field is zero on the dashed lines. (Fig. 6 of Gilbert (1993), reproduced with permission of the Royal Society.)

To turn the planar pA map from a non-dynamo to a dynamo, we include shear in a third direction to avoid cancellations and allow growth by the familiar SFS mechanism. We can think of this as the result of a flow in the cylinder $D^2 \times \mathbb{R}$. The flow in D^2 achieves the pA map, and is followed by shear in the $\mathbb{R}$ direction. For analysis, however, it is easiest to work on $\mathbb{T}^2$ and consider models in which the magnetic field is odd or even. When it is even, we have a cat map, which is a dynamo with no shear; shear will modify the growth rate by introducing cancellations. When the field is odd, we could take the quotient and imagine the field lying on the disc and acted on by a pA map; in this case we obtain a dynamo only with shear, as for SFS.

Working on $\mathbb{T}^2$ we apply the cat map and shear, giving an induction operator

$$\mathbf{Tb}(\mathbf{x}) = \phi(\mathbf{x})\,\mathbf{Mb}(\mathbf{Nx}), \qquad (\mathbf{N} \equiv \mathbf{M}^{-1}), \qquad (\mathbf{x} = (x_1, x_2)). \qquad (10.4.5)$$

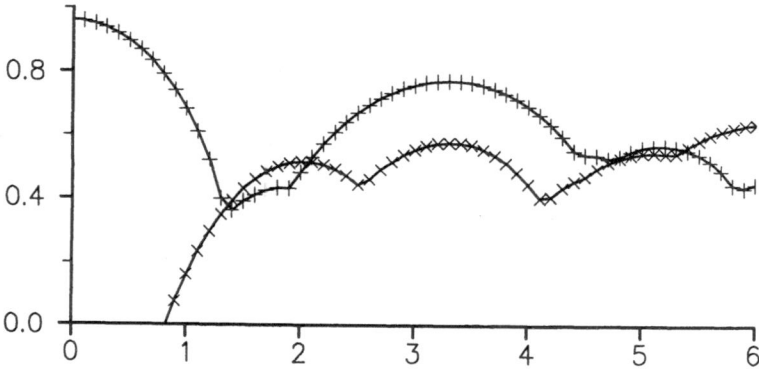

Fig. 10.5 Plot of the growth rate $\gamma(\epsilon)$ as a function of α with $\epsilon = 10^{-3}$ for even fields (+) and odd fields (×). (After Gilbert 1993.)

Here $\phi(\mathbf{x})$ is a complex phase shift (with $|\phi(\mathbf{x})| = 1$) resulting from motion in the x_3 direction. We also require that $\phi(\mathbf{x}) = \phi(-\mathbf{x})$ to ensure that the shear operation preserves the even/odd symmetry, for example

$$\phi(\mathbf{x}) \equiv \phi(x_1, x_2) = \exp(i\alpha \cos 2\pi x_1), \qquad (\alpha = \text{const.}). \qquad (10.4.6)$$

The field $\mathbf{b}$ in (10.4.5) is an element of the true magnetic field which is $\mathbf{B} = (\mathbf{b}(\mathbf{x}), b_3(\mathbf{x})) \exp(ikx_3)$ with $k = 1$; as usual the b_3 component is ignored and only the horizontal field $\mathbf{b}(\mathbf{x})$ considered. As a final simplification set $\mathbf{b}(\mathbf{x}) = b(\mathbf{x})\mathbf{e}_+$; then a dynamo operator for the sequence diffuse–map–shear is

$$T_\epsilon b(\mathbf{x}) = \lambda \phi(\mathbf{x})(H_\epsilon b)(\mathbf{N}\mathbf{x}) \equiv \lambda \phi(\mathbf{x}) \int_{\mathbf{T}^2} h_\epsilon(\mathbf{N}\mathbf{x} - \mathbf{y}) b(\mathbf{y}) \, d\mathbf{y}. \qquad (10.4.7)$$

Figure 10.5 shows growth rates as a function of α obtained numerically using (10.4.6, 7) and the standard cat map (3.5.3), for even (+) and odd (×) fields. At $\alpha = 0$ there is growth for even fields; this corresponds to a cat map with eigenfunction $\mathbf{e}_+$ and growth rate $\gamma(\epsilon) = \log \lambda$. As α is increased, the phase shifts give cancellations which reduce the growth rate of even fields below $\log \lambda$, and there is a series of mode crossings. The mechanism could be described as stretch–shear, rather than SFS: the shear acts to reduce the growth rate below that of the basic cat map.

More interesting is the case of odd fields. These show super-exponential decay for no shear, $\alpha = 0$. As the shear is increased the vertical motion allows some constructive reinforcement and the growth rate eventually becomes positive; the dynamo operates by an SFS mechanism. Numerical results as $\epsilon \to 0$ support a fast dynamo (Gilbert 1993).

10.4.2 Analysis

While for $\alpha \neq 0$ the SFS *mechanism* is operating in these models, the methods used to analyze the SFS *map* in §§3.2, 9.6 do not work here. The problem is that the adjoint operator of (10.4.5) is no simpler than the direct operator. This can be traced to the fact that the scalar field $b(\mathbf{x})$ depends on both stretching and contracting directions. This is unavoidable: the only continuous fields independent of the stretching direction $\mathbf{e}_+$ are constant on $\mathbb{T}^2$ (because a line through, say O, in the direction $\mathbf{e}_+$ winds densely round $\mathbb{T}^2$), and constant fields are too simple to be eigenfunctions.

These models are not straightforward to analyze, despite being generalizations of the cat map and the SFS model. One way of recasting the problem for the case of even field is to start with a field $b(\mathbf{x}) = 1$ and follow the diffusionless evolution of the mean field $<T_0^n 1>$ on the torus. From (10.4.7) with $T_0 \equiv T$,

$$<T^n 1> = \lambda^n <\phi(\mathbf{x})\phi(\mathbf{N}\mathbf{x}) \cdots \phi(\mathbf{N}^{n-1}\mathbf{x})>. \qquad (10.4.8)$$

To simplify this further, note that we can replace the average $<\cdot>$ over $\mathbf{x} \in \mathbb{T}^2$ by setting $\mathbf{x} = x\mathbf{e}_-$ and averaging over all x on the real line, since $x\mathbf{e}_-$ winds densely around the torus as x varies. We therefore obtain

$$<T^n 1> = \lambda^n <\phi(x\mathbf{e}_-)\phi(\lambda x\mathbf{e}_-) \cdots \phi(\lambda^{n-1}x\mathbf{e}_-)>_{x \in \mathbb{R}}, \qquad (10.4.9)$$

or for the phase (10.4.6), with $y = 2\pi x\mathbf{e}_- \cdot \mathbf{e}_1$,

$$<T^n 1> = \lambda^n <\exp(i\alpha[\cos y + \cos \lambda y + \cdots + \cos \lambda^{n-1}y])>_{y \in \mathbb{R}}. \qquad (10.4.10)$$

This is the average of a product of phases of different scales, showing explicitly how the field adopts a self-similar structure. We are interested in how fast the right-hand side increases with increasing n. This is difficult because of the cancellations in the average of the highly oscillatory phase. Related problems are studied in *large deviation theory* (Varadhan 1984, Ellis 1985), but general results appear only to be available for the exponentials of real quantities; see Thompson (1990) and Bayly (1992a) for further discussion.

To obtain information about growth rates and eigenfunctions of T_ϵ, we consider an asymptotic limit of strong stretching, in which the matrix $\mathbf{M} \equiv \mathbf{N}^{-1}$ is varied to make $\lambda \to \infty$. We shall motivate this more below. For here, simply note that in this limit $\lambda \to \infty$, there is an obvious (uniform) scale separation in (10.4.10); each term in the exponential varies rapidly on a distinct scale. If we exploit the scale separation by averaging each exponential term separately, taking a product of terms $<\exp(i\alpha \cos \lambda^m y)>$, we obtain $<T^n 1> \simeq [\lambda J_0(\alpha)]^n$, giving a growth rate

$$\Gamma \simeq \log[\lambda J_0(\alpha)]; \qquad (10.4.11)$$

this is the technique of Rechester & White (1980). We of course recover the growth rate for the cat map with no shear when $\alpha = 0$.

One can pursue this argument to develop a systematic approximation; at the next level one considers also correlations between two adjacent phase terms in (10.4.10) (Rechester & White 1980). Instead of following this route, however, we shall recast the problem in terms of a perturbed operator; the two methods are equivalent, but the latter allows discussion of the structure of eigenfunctions with and without diffusion.

We shall only study even fields in detail, in other words cat maps with shear. A similar analysis can be carried out for the odd case, pA with shear. This is more complicated because the average field is always zero, and so the breaking up of averages in (10.4.10) will not capture a dynamo in this case.

10.4.3 Even Fields

We consider even fields $b(\mathbf{x})$ and for analysis we use a shear–diffuse–map cycle. Set

$$Fb(\mathbf{x}) = [H_\epsilon(\phi b)](\mathbf{Nx}) \equiv \int_{\mathbb{T}^2} h_\epsilon(\mathbf{Nx} - \mathbf{y})\phi(\mathbf{y})b(\mathbf{y})\,d\mathbf{y} \qquad (10.4.12)$$

and rewrite the eigenvalue problem as

$$Fb(\mathbf{x}) = \mu b(\mathbf{x}), \quad \gamma(\epsilon) = \log|\lambda\mu|. \qquad (10.4.13a,b)$$

Write b (and similarly ϕ) as a Fourier series

$$b(\mathbf{x}) = \sum_{\mathbf{j}} b_{\mathbf{j}} \exp(2\pi i \mathbf{j} \cdot \mathbf{x}), \qquad (10.4.14)$$

the sum being taken over all vectors $\mathbf{j} = (j_1, j_2)$ with integer components. The mode of largest scale in the true field $\mathbf{B}$ is $b_0 \mathbf{e}_+ \exp(ikz)$ and its amplitude is b_0; we abuse language somewhat and refer to b_0 as the *mean field*.

We introduce a projection P onto this mean field, with $Pb = \langle b \rangle = b_0$. Define operators $\overline{F} = PF$ and $F' = F - \overline{F}$, which give the result of F projected onto the mean and fluctuating fields, respectively. In terms of the Fourier components of b and ϕ,

$$Fb(\mathbf{x}) = \sum_{\mathbf{j}} e^{2\pi i \mathbf{j} \cdot \mathbf{Nx}} e^{-\epsilon \mathbf{j}^2} \sum_{\mathbf{l}} \phi_{\mathbf{j}-\mathbf{l}} b_{\mathbf{l}}, \qquad (10.4.15)$$

and thus

$$\overline{F}b(\mathbf{x}) = \sum_{\mathbf{l}} \phi_{-\mathbf{l}} b_{\mathbf{l}}, \quad F'b(\mathbf{x}) = \sum_{\mathbf{j} \neq 0} e^{2\pi i \mathbf{j} \cdot \mathbf{Nx}} e^{-\epsilon \mathbf{j}^2} \sum_{\mathbf{l}} \phi_{\mathbf{j}-\mathbf{l}} b_{\mathbf{l}}. \qquad (10.4.16)$$

Take ϕ such that $\phi_0 \neq 0$; then the operator $\overline{F}$ has a single non-zero eigenvalue $\mu_0 = \phi_0$ corresponding to constant field $b(\mathbf{x}) = 1$. This would give the dynamo growth rate if we could justify ignoring the fluctuations.

To mitigate the effect of fluctuations we shall impose various conditions; to motivate these, assume that the fluctuations have a small effect and follow the argument in equation (10.2.22) to obtain from (10.4.13a) equations,

$$\mu \bar{b} = \overline{F}(1 - \mu^{-1} F')^{-1} \bar{b}, \qquad b = (1 - \mu^{-1} F')^{-1} \bar{b}, \qquad (10.4.17a, b)$$

for the eigenfunction b and eigenvalue μ, analogous to (10.2.23, 22c), where $\bar{b}$ is a constant which we can take as unity.

Analogous to (10.2.24) we expand (10.4.17a):

$$\mu = \overline{F}(1 + \mu^{-1} F' + \mu^{-2} F'^2 + \mu^{-3} F'^3 + \cdots)1 \qquad (10.4.18a)$$
$$= \mu_0 + \mu^{-1} \overline{F} F' 1 + \mu^{-2} \overline{F} F'^2 1 + \mu^{-3} \overline{F} F'^3 1 + \cdots. \qquad (10.4.18b)$$

The series gives μ in terms of μ_0 and corrections. We want to make this series converge, by proving that $\overline{F} F'^n 1$ tends to zero rapidly with n. Consider the leading order correction $\mu^{-1} \overline{F} F' 1$:

$$\overline{F} F' 1 = \overline{F} \sum_{\mathbf{j} \neq \mathbf{0}} e^{2\pi i \mathbf{j} \cdot \mathbf{N} \mathbf{x}} e^{-\epsilon \mathbf{j}^2} \phi_{\mathbf{j}} = \sum_{\mathbf{j} \neq \mathbf{0}} e^{-\epsilon \mathbf{j}^2} \phi_{-\mathbf{j} \cdot \mathbf{N}} \phi_{\mathbf{j}}. \qquad (10.4.19)$$

For a 'reasonable' ϕ, such as (10.4.6), the Fourier coefficients $\phi_{\mathbf{j}}$ tend to zero rapidly as, say, $\|\mathbf{j}\|_1 \equiv |j_1| + |j_2|$ increases. We can make each term of the sum small if we ensure that either $\|\mathbf{j}\|_1$ or $\|\mathbf{j} \cdot \mathbf{N}\|_1$ is large for any non-zero wave-vector $\mathbf{j}$. By the triangle inequality, $\|\mathbf{j} \cdot \mathbf{N}\|_1 + \|\mathbf{j}\|_1 \geq \|\mathbf{j} \cdot \mathbf{N} - \mathbf{j}\|_1$, and so we can ensure this if we choose $\mathbf{N}$ so that $\|\mathbf{j} \cdot \mathbf{N} - \mathbf{j}\|_1$ is large for any $\mathbf{j} \neq \mathbf{0}$. These two ingredients — rapid decay of Fourier coefficients of ϕ and the property that $\mathbf{N}$ moves non-zero wavevectors $\mathbf{j}$ a large distance — are crucial, and we give specific conditions:

Conditions on ϕ. We take ϕ to satisfy $|\phi(\mathbf{x})| = 1$ and $\phi_0 \neq 0$. We require that the Fourier coefficients decay according to

$$|\phi_{\mathbf{j}}| \leq \zeta \exp(-\rho \|\mathbf{j}\|_1), \qquad (\|\mathbf{j}\|_1 \equiv |j_1| + |j_2|), \qquad (10.4.20)$$

for some positive constants ζ and ρ. This is equivalent to requiring that ϕ be analytic and bounded in a strip of width ρ in x_1 and x_2, when extended as a function of $\mathbb{C}^2$.

Form of $\mathbf{M}$ and $\mathbf{N}$. We choose a large positive integer K and take

$$\mathbf{M} = \begin{pmatrix} K^2 + 1 & K \\ K & 1 \end{pmatrix}, \qquad \mathbf{N} \equiv \mathbf{M}^{-1} = \begin{pmatrix} 1 & -K \\ -K & K^2 + 1 \end{pmatrix}. \qquad (10.4.21)$$

Our approximations will be valid in the limit $K \to \infty$. Eigenvectors and eigenvalues of $\mathbf{M}$ are given in (10.4.31, 32) below.

The conditions on ϕ are satisfied by (10.4.6) provided $J_0(\alpha) \neq 0$. We have

$$\phi_{(j_1, 0)} = i^{j_1} J_{j_1}(\alpha) \quad \text{and} \quad \phi_{(j_1, j_2)} = 0 \text{ for } j_2 \neq 0, \qquad (10.4.22)$$

and since the phase function is entire (analytic in all of $\mathbb{C}$) the Fourier coefficients fall off faster than exponentially: for any $\rho > 0$ a ζ can be found so that (10.4.20) holds. The matrix $\mathbf{N}$ has the desired property, since

$$\|\mathbf{j} \cdot \mathbf{N} - \mathbf{j}\|_1 = K(|\dot{j}_1 - Kj_2| + |j_2|), \tag{10.4.23}$$

and so is at least K for $\mathbf{j} \neq \mathbf{0}$. In fact a little experimentation shows that $\|\mathbf{j} \cdot \mathbf{N} - \mathbf{j}\|_1$ takes positive values mK with multiplicity $4m$.

We can now bound the size of the first correction; from (10.4.19, 20),

$$|\overline{F}F'1| \leq \zeta^2 e^{-\epsilon} \sum_{\mathbf{j} \neq \mathbf{0}} e^{-\rho(\|\mathbf{j}\cdot\mathbf{N}\|_1 + \|\mathbf{j}\|_1)} \leq \zeta^2 e^{-\epsilon} \sum_{\mathbf{j} \neq \mathbf{0}} e^{-\rho\|\mathbf{j}\cdot\mathbf{N}-\mathbf{j}\|_1} \tag{10.4.24a}$$

$$= \zeta^2 e^{-\epsilon} \sum_{m=1}^{\infty} 4m e^{-\rho m K} = O(e^{-\epsilon - \rho K}). \tag{10.4.24b}$$

Thus $|\overline{F}F'1|$ is of order $\exp(-\epsilon - \rho K)$ and tends to zero for large K. Ignoring higher corrections in (10.4.18b), the eigenvalue $\mu = \mu_0 + O(\exp(-\epsilon - \rho K))$. The corresponding growth rate (10.4.13b) is $\gamma(\epsilon) = \log|\lambda\mu_0| + O(\exp(-\epsilon - \rho K))$. Since $\lambda = O(K^2)$, from (10.4.31) below, the growth rate is positive for K sufficiently large, uniformly in ϵ, and so this model is a fast dynamo, provided we can ignore higher corrections. Note that the presence of diffusion only assists convergence and the calculation goes through for $\epsilon = 0$.

This approach can be used to bound higher order terms $\overline{F}(F')^n 1$ in (10.4.18b). However a more natural way to proceed is to define a norm $\|\cdot\|$ on magnetic fields and to calculate the norms of the operators $\overline{F}$ and F' (see §9.7.1). In particular we need to show that $\|F'\|$ is small for large K to quantify precisely the unimportance of fluctuating fields. These fluctuations are small-scale and may be strong. A suitable norm that suppresses the fluctuations is

$$\|b\| = \sum_{\mathbf{j}} e^{\nu(|j_+| - |j_-|)} |b_{\mathbf{j}}|, \qquad (\mathbf{j} = j_+\mathbf{e}_+ + j_-\mathbf{e}_-), \tag{10.4.25}$$

where ν is a positive constant to be fixed. Each Fourier wave-vector $\mathbf{j}$ is written in terms of its components $j_\pm$ in the stretching (+) and contracting (−) directions of $\mathbf{M}$ (see (10.4.32) below). The norm gives low weight to fields whose wave-vectors $\mathbf{j}$ point principally in the expanding direction $\mathbf{e}_-$ of $\mathbf{N}$, where fluctuations are carried by the cat map. By contrast wave-vectors pointing principally in the contracting direction $\mathbf{e}_+$ of $\mathbf{N}$ are strongly weighted, because these are carried to large scales by the map and so can influence the mean field, before they cascade harmlessly in the $\mathbf{e}_-$ direction.

This norm is a weighted l_1 norm on the Fourier coefficients, and we work in the Banach space $\mathcal{B}$ of magnetic fields having finite norm. At the end of this section we establish that

$$\|\overline{F}\| \leq \zeta, \qquad \|F'\| = O(e^{-\rho K - \epsilon}). \tag{10.4.26a, b}$$

For sufficiently large K, these results allow the inversion of $1 - \mu^{-1} F'$ in (10.4.17a), imply that equation (10.4.17a) has a solution μ close to μ_0, and justify the series expansion (10.4.18). This series converges rapidly since $\|\overline{F} F'^n 1\| \leq \|\overline{F}\| \|F'^n\| \|1\| = [O(e^{-\rho K - \epsilon})]^n$.

Given the eigenvalue μ, the corresponding eigenfunction is given by (10.4.17b), which can be expanded as a convergent series (cf. 10.2.25). Thus the eigenfunction exists in $\mathcal{B}$ and is close to uniform field $b = 1$ in norm. Of course, since this norm suppresses small scales, generally for $\epsilon = 0$ the eigenfunction will have Fourier coefficients that do not tend to zero with large wavenumber. We can think of the eigenfunction as a distribution on test functions $f(\mathbf{x})$ that are analytic and bounded in a strip of width ρ_1. We must take ρ_1 sufficiently large (say $\rho_1 = 2\nu$) so that the exponential decay of the Fourier coefficients $f_{\mathbf{j}}$ dominates the possible growth in $b_{\mathbf{j}}$ allowed by (10.4.25); then by Parseval's theorem,[78]

$$(f, b) \equiv \int_{\mathbb{T}^2} f^*(\mathbf{x}) b(\mathbf{x}) \, d\mathbf{x} = \sum_{\mathbf{k}} f_{\mathbf{k}}^* b_{\mathbf{k}} < \infty \qquad (10.4.27)$$

for any $b \in \mathcal{B}$. The eigenfunction obeys (10.4.13a) in the corresponding weak sense.

Diffusion plays a passive role in the above analysis. It can be shown that $\|F|_\epsilon - F|_{\epsilon=0}\| = O(\epsilon e^{-\rho K})$ and so diffusion is a small perturbation for large K. It follows that as $\epsilon \to 0$ the fast dynamo eigenvalue $\mu|_\epsilon \to \mu|_{\epsilon=0}$, and the eigenfunction $b|_\epsilon \to b|_{\epsilon=0}$ in the norm (10.4.25), or weakly in the sense of (10.4.27). Growth rates with and without diffusion can also be obtained from sums over periodic orbits (Aurell & Gilbert 1993). For $\epsilon > 0$, it can be shown that the eigenfunctions are square integrable; however it is not clear that these can be obtained from the singular $\epsilon = 0$ eigenfunction by a straightforward smoothing process.

We sketch briefly a proof of the flux conjecture in this system. Take $\epsilon = 0$ and define first the quantities $s_n \equiv (1, F^n 1) = PF^n 1$; these obey the relation,

$$s_n = s_{n-1} \overline{F} 1 + s_{n-2} \overline{F} F' 1 + \cdots + s_1 \overline{F} F'^{n-2} 1 + s_0 \overline{F} F'^{n-1} 1. \qquad (10.4.28)$$

On dividing through by s_{n-1} this becomes very similar to (10.4.18a), and from this and (10.4.26) it can be established that $s_n / s_{n-1} \to \mu$ as $n \to \infty$. Thus starting with constant field, the average field grows asymptotically at the correct growth rate γ_0. The average also grows at this rate from general initial conditions g (in $\mathcal{B}$), since we may write

$$(1, F^n g) \equiv PF^n g = \sum_{m=1}^{n} s_{n-m} \overline{F} F'^{m-1} g, \qquad (n \geq 1), \qquad (10.4.29a)$$

---

[78]This requirement on f is much stronger than the condition of C^∞ suggested in §4.1. However the class of admissible functions f can probably be enlarged for the particular phase function (10.4.6), because its Fourier coefficients decay super-exponentially whereas we are only requiring exponential decay.

and this sum is dominated by the leading terms. In fact

$$\lim_{n\to\infty} \mu^{-n}(1, F^n g) = P_1 g \equiv \sum_{m=1}^{\infty} \sigma\mu^{-m}\overline{F}F'^{m-1}g, \tag{10.4.29b}$$

where $\sigma = \lim_{n\to\infty} \mu^{-n}s_n$. Up to normalization, P_1 represents a projection onto the diffusionless eigenfunction. Finally using a test function f, we can express

$$(f, F^n g) = \sum_{m=0}^{n-1}(f, F'^m 1)PF^{n-m}g + (f, F'^n g) \tag{10.4.30a}$$

and so

$$\lim_{n\to\infty} \mu^{-n}(f, F^n g) = (P_2 f)(P_1 g), \quad P_2 f \equiv \sum_{m=0}^{\infty} \mu^{-m}(f, F'^m 1). \tag{10.4.30b}$$

Here P_2 represents, up to normalization, a projection onto an adjoint eigenfunction. Thus a functional f of an initial condition g grows at the rate γ_0 provided $P_2 f$ and $P_1 g$ are non-zero. We conclude that the flux conjecture holds in this system.

The analysis for fields of odd symmetry is similar and is discussed in Gilbert (1993). In this case there is no mean field b_0 and instead of P, a projection is taken onto the four non-zero modes which minimize $\|\mathbf{j}\cdot\mathbf{N} - \mathbf{j}\|_1$ (see 10.4.23); the phase function ϕ has to be modified to allow these modes to grow in the limit of large K.

We conclude this chapter with a proof of (10.4.26). In preparation, note that $\mathbf{M}$ has eigenvalues $\lambda^{\pm 1}$ with

$$2\lambda = (K^2 + 2) + \sqrt{(K^2 + 2)^2 - 4}, \tag{10.4.31}$$

corresponding to orthonormal eigenvectors

$$\mathbf{e}_\pm = \pm(K, \lambda^{\pm 1} - K^2 - 1)[K^2 + (\lambda^{\pm 1} - K^2 - 1)^2]^{-1/2}. \tag{10.4.32}$$

The eigenvectors $\mathbf{e}_\pm$ nearly coincide with the x- and y-axes for large K; the angle between $\mathbf{e}_+$ and the x-axis is

$$\theta = \tan^{-1}\left[K^{-1}(\lambda - K^2 - 1)\right] = K^{-1} + O(K^{-3}). \tag{10.4.33}$$

To establish (10.4.26), we need to move between (j_1, j_2) components and (j_+, j_-) components. For this purpose set $\|\mathbf{j}\|_\pm = |j_+| + |j_-|$ and note that

$$\chi\|\mathbf{j}\|_1 \le \|\mathbf{j}\|_\pm \le \chi^{-1}\|\mathbf{j}\|_1 \tag{10.4.34}$$

with $\chi = \cos\theta - \sin\theta = 1 - K^{-1} + O(K^{-2})$. Then from (10.4.20),

$$|\phi_\mathbf{j}| \le \zeta e^{-\rho\chi\|\mathbf{j}\|_\pm} \tag{10.4.35}$$

and we fix $\nu = \rho\chi$ in (10.4.25).

Equation (10.4.26 a) follows easily from (10.4.16, 35):

$$\|\overline{F}b\| = \left|\sum_1 \phi_{-1}b_1\right| \leq \zeta \sum_1 e^{-\rho\chi(|l_+|+|l_-|)}|b_1|$$

$$\leq \zeta \sum_1 e^{\rho\chi(|l_+|-|l_-|)}|b_1| = \zeta\|b\|. \tag{10.4.36}$$

and for (10.4.26 b):

$$\|F'b\| = \sum_{j\neq 0} e^{\rho\chi(\lambda^{-1}|j_+|-\lambda|j_-|)}e^{-\epsilon j^2}\left|\sum_1 \phi_{j-1}b_1\right|,$$

$$\leq \sum_{j\neq 0} e^{\rho\chi(\lambda^{-1}|j_+|-\lambda|j_-|)}e^{-\epsilon}\sum_1 \zeta e^{-\rho\chi(|j_+-l_+|+|j_--l_-|)}|b_1|$$

$$\leq \sum_{j\neq 0} e^{\rho\chi(\lambda^{-1}|j_+|-\lambda|j_-|)}e^{-\epsilon}\sum_1 \zeta e^{-\rho\chi(|j_+|-|l_+|-|j_-|+|l_-|)}|b_1| \tag{10.4.37}$$

$$= \sum_{j\neq 0} e^{\rho\chi((\lambda^{-1}-1)|j_+|+(1-\lambda)|j_-|)}e^{-\epsilon}\sum_1 \zeta e^{\rho\chi(|l_+|-|l_-|)}|b_1|$$

$$= \sum_{j\neq 0} e^{-\rho\chi\|j\cdot N-j\|\pm}e^{-\epsilon}\zeta\|b\|,$$

from (10.4.16, 25) and the triangle inequality. Using (10.4.34) and arguing as in (10.4.24) we obtain

$$\|F'\| \leq e^{-\epsilon}\zeta \sum_{j\neq 0} e^{-\rho\chi^2\|j\cdot N-j\|_1} \leq e^{-\epsilon}\zeta \sum_{m=1}^{\infty} 4me^{-\rho\chi^2 mK} \tag{10.4.38}$$

$$= O(e^{-\epsilon-\rho\chi^2 K}) = O(e^{-\epsilon-\rho K}),$$

as required.

11. Random Fast Dynamos

One way to attempt to simplify the fast dynamo problem is to study the evolution of magnetic fields in random flows. We say that an ensemble of random flows is a fast dynamo if the ensemble-averaged magnetic field grows at a rate independent of the diffusivity ϵ in the limit $\epsilon \to 0$, that is, if

$$\gamma_0 = \liminf_{\epsilon \to 0} \gamma(\epsilon) > 0, \tag{11.0.1}$$

where the growth rate $\gamma(\epsilon)$ can be defined for $\epsilon > 0$ by, for example, the growth of magnetic energy or linear functionals:

$$\gamma(\epsilon) = \sup_{\mathbf{B}_0} \limsup_{t \to \infty} \frac{1}{2t} \log <E_M(t)> \tag{11.0.2a}$$

$$= \sup_{\mathbf{B}_0} \sup_f \limsup_{t \to \infty} \frac{1}{t} \log <L_f \mathbf{B}(t)>. \tag{11.0.2b}$$

Here $<\cdot>$ is an ensemble average. This definition is the most tractable mathematically; we return to discuss it in §11.5, but here we simply remark that it is not the only definition, and perhaps not the best definition.

The advantage of using random flows is that taking an average over an ensemble of random flows acts rather like strong diffusion, destroying the small-scale fluctuating fields that make the fast dynamo problem difficult. However there are corresponding disadvantages: it becomes difficult to find out how these small scales are in fact evolving! Furthermore information is lost in taking an average over many different flows, and one has to ask whether the field generated in a typical realisation of a random flow bears any resemblance to the ensemble-averaged field, not only in terms of structure, but also in terms of growth rate (Hoyng 1987a, b, 1988).

There are many parallels with deterministic fast dynamo theory. For $\epsilon = 0$ it is relatively easy to study the behavior of magnetic field moments and the growth of small-scale structure and spatial intermittency (§§11.2, 3); a related discussion for baker's maps may be found in §7.1. This issue is somewhat tangential to the fast dynamo problem, however, and for $\epsilon > 0$ we can say far less about the detailed magnetic field structure. It is possible to follow the ensemble-averaged magnetic field at a point, $<\mathbf{B}(\mathbf{x}, t)>$, and to obtain its growth rate (§11.4). This is insensitive to diffusion in the limit $\epsilon \to 0$; in fact one can set $\epsilon = 0$. Since one can use $<\mathbf{B}(\mathbf{x}, t)>$ to construct a flux or linear functional of the ensemble-averaged field by integration, it may then be seen that a flux or linear functional of the ensemble-averaged field has the same growth rate for $\epsilon = 0$ and in the limit $\epsilon \to 0$. We thus recover an analog of the flux conjecture for random flows. Finally in §11.5 we return to the question of whether the ensemble average reflects the behavior of typical realisations, and report numerical simulations.

Our discussion is based largely on examples; we use flows made up of Beltrami waves (§11.1), for which many results can be obtained analytically. We have not attempted to cover the broad literature concerning alpha and other transport effects, linear and nonlinear, in random and turbulent flows (see Moffatt 1978, Krause & Rädler 1980), but instead focus only on aspects directly relevant to the fast dynamo problem.

11.1 Random Renewing Flows

We shall work with ensembles $\mathcal{E}$ of *renewing flows* (or *renovating flows*) $\mathcal{E} = \{\mathbf{u}(\mathbf{x}, t)\}$. These flows were introduced in dynamo studies by Steenbeck & Krause (1969), and are used by Kraichnan (1974, 1976a) and Vainshtein (1982). The reviews of Molchanov, Ruzmaikin & Sokoloff (1985) and Zeldovich *et al.* (1988) discuss more recent work.

First divide time t into intervals $I_n = [n\tau, (n+1)\tau)$ of length τ for $n \in \mathbb{Z}$. We generate a random flow $\mathbf{u} \in \mathcal{E}$ by setting

$$\mathbf{u}(\mathbf{x}, t) = \mathbf{u}_\tau(\mathbf{x}, t - n\tau), \qquad (t \in I_n) \qquad (11.1.1)$$

for each time interval I_n, where the flow $\mathbf{u}_\tau$ is chosen at random from an ensemble $\mathcal{E}_\tau$ of smooth flows of length τ,

$$\mathcal{E}_\tau = \{\mathbf{u}_\tau(\mathbf{x}, t) : 0 \leq t < \tau\}. \qquad (11.1.2)$$

The flows chosen in different intervals are taken to be independent and identically distributed. For the ensembles we use, $\mathcal{E}_\tau$ contains a collection of steady Beltrami waves: the random flows then correspond to switching on a randomly chosen wave in each time interval I_n.

A renewing flow $\mathbf{u} \in \mathcal{E}$ is *piecewise deterministic*; it has a sharp loss of memory at times $t = n\tau$ but is deterministic in between. This property, although obviously artificial, allows exact equations for the evolution of magnetic field moments. These equations involve averages of the Lagrangian properties of the flows $\mathbf{u}_\tau$ over the time interval $0 \leq t < \tau$. The key simplification here is that this is a finite time interval, whereas fast dynamos in deterministic flows are sensitive to the long-time behavior of trajectories. These finite-time averages may be calculated analytically for simple flows.

The ensemble $\mathcal{E}$ is given by specifying the distribution of flows $\mathbf{u}_\tau \in \mathcal{E}_\tau$ for a time interval. We take each flow to be a single steady Beltrami wave:

$$\mathbf{u}_\tau(\mathbf{x}) = \mathbf{v}\sin(\mathbf{q}\cdot\mathbf{x} + \psi) + \mathbf{w}h\cos(\mathbf{q}\cdot\mathbf{x} + \psi). \tag{11.1.3}$$

Here $\mathbf{q}$ is the wave-vector and ψ is a phase. We take $\{\mathbf{q}, \mathbf{v}, \mathbf{w}\}$ to be a right-handed orthogonal set, which ensures incompressibility, with

$$\mathbf{v}\cdot\mathbf{q} = \mathbf{w}\cdot\mathbf{q} = 0, \qquad \mathbf{w} \equiv \mathbf{q}\times\mathbf{v}/q. \tag{11.1.4}$$

The parameter $h = \pm 1$ for Beltrami waves of parity ± 1. The energy density of the wave is $E_V = v^2/2$ and we define a turnover time by $T = 1/vq$.

Over a period of time τ a particle moves from $\mathbf{x}$ to

$$M\mathbf{x} = \mathbf{x} + \mathbf{v}\tau\sin(\mathbf{q}\cdot\mathbf{x} + \psi) + \mathbf{w}h\tau\cos(\mathbf{q}\cdot\mathbf{x} + \psi). \tag{11.1.5}$$

The corresponding Jacobian matrix is

$$\mathbf{J}(\mathbf{x}) = \mathbf{I} + \mathbf{v}\mathbf{q}^T\tau\cos(\mathbf{q}\cdot\mathbf{x} + \psi) - \mathbf{w}\mathbf{q}^Th\tau\sin(\mathbf{q}\cdot\mathbf{x} + \psi), \tag{11.1.6}$$

and for $\epsilon = 0$ the Cauchy solution yields

$$\mathbf{B}(M\mathbf{x}, (n+1)\tau) = \mathbf{J}(\mathbf{x})\mathbf{B}(\mathbf{x}, n\tau), \tag{11.1.7}$$

over the period τ of a single random wave.

We need to specify the probability of choosing a given Beltrami wave for each time interval, in other words a probability distribution for $\mathbf{q}$, ψ, $\mathbf{v}$ and h. We consider ensembles for which the helicity parameter h and the magnitudes $|\mathbf{v}| = |\mathbf{w}| = v$, $|\mathbf{q}| = q$ are given non-random constants. We take the phase ψ to be uniformly distributed in $[0, 2\pi]$, which ensures that these ensembles are *homogeneous*: if $\mathbf{u} \in \mathcal{E}$ then any translation of $\mathbf{u}$ is also in $\mathcal{E}$. We now need to fix the possible directions of $\mathbf{q}$, $\mathbf{v}$ and $\mathbf{w}$. Suppose for the moment that we have chosen $\mathbf{q}$ somehow. We then take the direction of the vector $\mathbf{v}$ to be distributed uniformly in the plane perpendicular to $\mathbf{q}$; possible choices of $\mathbf{v}$ make up a circle of radius v in this plane. Once $\mathbf{q}$ and $\mathbf{v}$ are chosen, $\mathbf{w}$ is given by (11.1.4).

The ensemble is then completely specified by the parameters τ, h, v, q and the distribution of directions of $\mathbf{q}$. Two parameters can be scaled away by a choice of length and time scales. In the fast dynamo context it makes sense to choose scales such that

$$T \equiv 1/vq = 1, \qquad L \equiv 1/q = 1, \qquad v = 1; \tag{11.1.8}$$

so the convective time-scale T and the spatial scale L of the flow are both unity. The only parameters left are $h = \pm 1$ and τ, which is the *correlation time* non-dimensionalised by the turnover time, together with a distribution of directions of the unit vector $\mathbf{q}$; these define the ensemble of renewing flows.

Two examples of ensembles, from Gilbert & Bayly (1992), are:

- the *3-d isotropic renewing ensemble* $\mathcal{E}_{3d}(\tau, h)$. The direction of the vector $\mathbf{q}$ is distributed isotropically in three-dimensional space. The ensemble is *isotropic*: if $\mathbf{u}$ is in the ensemble then so is any rotation of $\mathbf{u}$.

- *the 111 renewing ensemble* $\mathcal{E}_{111}(\tau, h)$. The direction of $\mathbf{q}$ is chosen to be one of the three coordinate axes $\mathbf{e}_x$, $\mathbf{e}_y$ and $\mathbf{e}_z$ with probability $1/3$. This is a random version of the steady ABC flow with parameter values 111.

We now consider the evolution of magnetic field; first without diffusion and then with diffusion.

11.2 Diffusionless Evolution of Field

In this section we consider the growth in magnetic field moments in the case of zero diffusion $\epsilon = 0$; the results of this section do not extend to non-zero diffusion, however weak. For zero diffusion, following a magnetic field amounts to multiplying random Jacobian matrices giving the stretching of field along Lagrangian trajectories (Zeldovich *et al.* 1984). Much of the analysis in this section is taken from Kraichnan (1974), Molchanov, Ruzmaikin & Sokoloff (1984)[79] and Drummond & Münch (1990); mathematical background can be found in Furstenberg (1963).

It is only tractable to follow high-order moments in ensembles that are isotropic and homogeneous, for example $\mathcal{E}_{3d}(\tau, h)$ above. We use $<\cdot>$ or sometimes an overbar to denote an ensemble average. Consider first the second-order single-point moment,

$$\overline{B^2}(t) \equiv <|\mathbf{B}(\mathbf{x}, t)|^2> = <(\mathbf{B}^{\mathsf{T}}\mathbf{B})(\mathbf{x}, t)>, \qquad (11.2.1)$$

which is twice the average magnetic energy density at time t. Over the time interval I_n a single wave (11.1.3) acts and, by (11.1.7),

$$(B_i B_j)(M\mathbf{x}, (n+1)\tau) = (J_{ik} J_{jl})(\mathbf{x})(B_k B_l)(\mathbf{x}, n\tau), \qquad (11.2.2)$$

where the parameters defining M and $\mathbf{J}$ depend on the random wave chosen in interval I_n.

The flows in different time intervals I_m are independent, and so let us average (11.2.2) first over I_m with $m = 0, 1, 2, \ldots n-1$, which takes us up to time $t = n\tau$; since $\mathbf{J}$ and M in (11.2.2) are independent of what happened during these earlier intervals we obtain

$$<(B_i B_j)(M\mathbf{x}, (n+1)\tau)>_{I_1,\ldots,I_{n-1}} = (J_{ik} J_{jl})(\mathbf{x})<(B_k B_l)(\mathbf{x}, n\tau)>. \ (11.2.3)$$

Now suppose the initial condition $\mathbf{B}_0(\mathbf{x})$ for the magnetic field is random, and also homogeneous and isotropic. For example, we could take the initial field to be a constant unit vector isotropically distributed in space. Let us

---

[79]In this paper, when handling diffusion of field, it is assumed after equation (19) that an average of orthogonal matrices is itself orthogonal. This is not true in general, and so the results of §5 for the limit $\epsilon \to 0$ appear limited to flows without significant cancellations, for example Anosov flows.

average over these initial conditions as well as over the ensemble of flows. Since the ensemble of flows is also homogeneous and isotropic any single-point average of the field at a later time must be independent of position and direction. Therefore by (11.2.1),

$$<(B_k B_l)(\mathbf{x}, n\tau)> = \tfrac{1}{3}\overline{B^2}(n\tau)\delta_{kl}, \qquad (11.2.4)$$

where the average includes an average over the initial conditions. Substituting in (11.2.3) gives

$$<(B_i B_j)(M\mathbf{x}, (n+1)\tau)>_{I_1,\dots,I_{n-1}} = (J_{ik}J_{jl})(\mathbf{x})\tfrac{1}{3}\overline{B^2}(n\tau)\delta_{kl}, \qquad (11.2.5)$$

and then averaging over I_n implies that

$$\begin{aligned}
\overline{B^2}((n+1)\tau) &= \tfrac{1}{3} <J_{ik}(\mathbf{x})J_{ik}(\mathbf{x})> \overline{B^2}(n\tau) \\
&= \tfrac{1}{3} <\mathrm{trace}(\mathbf{J}^{\mathrm{T}}\mathbf{J})(\mathbf{x})> \overline{B^2}(n\tau).
\end{aligned} \qquad (11.2.6)$$

To calculate the growth in $\overline{B^2}$ for the ensemble $\mathcal{E}_{3d}(\tau, h)$ we insert the Jacobian (11.1.6) and average over the ensemble, i.e., over ψ, $\mathbf{v}$, $\mathbf{w}$ and $\mathbf{q}$, to obtain

$$\overline{B^2}((n+1)\tau)/\overline{B^2}(n\tau) = \tfrac{1}{3} <\mathrm{trace}(\mathbf{J}^{\mathrm{T}}\mathbf{J})(\mathbf{x})> = 1 + 2\tau^2/3 > 1, \qquad (11.2.7)$$

using (11.1.8) and $h = \pm 1$. The ensemble-averaged energy of the field grows in time by a constant factor per time interval I_n because of stretching in the flow. Note that this conclusion is only correct in the absence of diffusion; one would obtain similar growth of magnetic energy for planar chaotic flows with $\epsilon = 0$, but for $\epsilon > 0$ the ensemble-averaged magnetic energy would ultimately decay by the anti-dynamo theorem for planar flows.

The case of the pth order single-point moment,

$$\overline{B^p}(t) \equiv <|\mathbf{B}(\mathbf{x}, t)|^p>, \qquad (11.2.8)$$

in an isotropic, homogeneous ensemble, is similar and in the following proof of Molchanov et al. (1984) we can relax the assumption about the initial conditions being isotropic, although they must still be homogeneous. For example it applies to the deterministic initial condition of uniform field. From (11.1.7),

$$|\mathbf{B}(M\mathbf{x}, (n+1)\tau)|^p = |\mathbf{J}(\mathbf{x})\mathbf{B}(\mathbf{x}, n\tau)|^p; \qquad (11.2.9)$$

let us write $\mathbf{B}(\mathbf{x}, n\tau)$ in terms of its magnitude $|\mathbf{B}(\mathbf{x}, n\tau)|$ and direction $\mathbf{e}(\mathbf{x}, n\tau)$,

$$|\mathbf{B}(M\mathbf{x}, (n+1)\tau)|^p = |\mathbf{J}(\mathbf{x})\mathbf{e}(\mathbf{x}, n\tau)|^p |\mathbf{B}(\mathbf{x}, n\tau)|^p. \qquad (11.2.10)$$

We fix $\mathbf{x}$ and average first over the random flows in the final interval I_n only:

$$<|\mathbf{B}(M\mathbf{x}, (n+1)\tau)|^p>_{I_n} = <|\mathbf{J}(\mathbf{x})\mathbf{e}(\mathbf{x}, n\tau)|^p>_{I_n} |\mathbf{B}(\mathbf{x}, n\tau)|^p. \qquad (11.2.11)$$

In this equation $\mathbf{e}$ and $|\mathbf{B}|$ are functions of position, given by the evolution in the flow up to time $t = n\tau$, and over which we have yet to average. The random flows used for the final interval I_n comprise an isotropic collection: the random matrices $\mathbf{J}(\mathbf{x})$ are isotropic and so the average $<|\mathbf{J}(\mathbf{x})\mathbf{e}(\mathbf{x}, n\tau)|^p>$ is independent of the direction $\mathbf{e}(\mathbf{x}, n\tau)$. One could equally well use a fixed unit vector $\mathbf{e}$ here and obtain

$$<|\mathbf{B}(M\mathbf{x}, (n+1)\tau)|^p>_{I_n} = <|\mathbf{J}(\mathbf{x})\mathbf{e}|^p>_{I_n} |\mathbf{B}(\mathbf{x}, n\tau)|^p. \qquad (11.2.12)$$

The quantity $<|\mathbf{J}(\mathbf{x})\mathbf{e}|^p>_{I_n}$ is independent of $\mathbf{x}$ by homogeneity and so is simply a number, independent of the structure of $\mathbf{B}(\mathbf{x}, n\tau)$. Finally average over all the earlier time intervals I_m, $m = 0, 1, \ldots, n-1$ to obtain

$$<|\mathbf{B}(M\mathbf{x}, (n+1)\tau)|^p> = <|\mathbf{J}(\mathbf{x})\mathbf{e}|^p>_{I_n} <|\mathbf{B}(\mathbf{x}, n\tau)|^p>. \qquad (11.2.13)$$

Using homogeneity, this can be written

$$\overline{B^p}((n+1)\tau) = <L^p> \overline{B^p}(n\tau), \qquad L^2 = |\mathbf{J}\mathbf{e}|^2 = \mathbf{e}^\mathrm{T}\mathbf{J}^\mathrm{T}\mathbf{J}\mathbf{e}, \qquad (11.2.14)$$

where $\mathbf{e}$ is any unit vector. The average is taken over the random collection of flows for any interval I_n. Note that $\mathbf{J}^\mathrm{T}\mathbf{J}$ is the matrix Λ introduced in §6.1.

For comparison with the results of Drummond & Münch (1990) and Molchanov et al. (1984) we introduce stretching exponents Λ_p defined by

$$\Lambda_p = (p\tau)^{-1} \log <L^p>, \qquad (p > 0),$$
$$\Lambda_0 = \lim_{p \to 0} \Lambda_p = \tau^{-1} <\log L>. \qquad (11.2.15)$$

Similar exponents are defined for deterministic flows in (7.1.8, 9). The asymptotic rate of increase of the pth moment for $p > 0$ is

$$p\Lambda_p = \lim_{t \to \infty} t^{-1} \log \overline{B^p}(t). \qquad (11.2.16)$$

As in §7.1, we can identify Λ_1 with the line-stretching exponent h_{line} and Λ_0 with the Liapunov exponent, Λ_{Liap}.

In the next section we discuss general properties of the Λ_p; here we calculate them for the ensemble $\mathcal{E}_{3\mathrm{d}}$. We use the Jacobian (11.1.6) to obtain

$$L^2 = 1 + 2\tau(\mathbf{e} \cdot \mathbf{q})(\mathbf{e} \cdot \mathbf{v} \cos \psi' - h\mathbf{e} \cdot \mathbf{w} \sin \psi') + (\mathbf{e} \cdot \mathbf{q})^2 \tau^2, \qquad (11.2.17)$$

with $\psi' = \mathbf{q} \cdot \mathbf{x} + \psi$. Now $\mathbf{e}$ is a fixed vector, while in different realisations the frame $\{\mathbf{q}, \mathbf{v}, \mathbf{w}\}$ takes all possible orientations. Instead one can consider the frame to be fixed while $\mathbf{e}$ takes all possible orientations and use polar angles θ and ϕ to give the direction of $\mathbf{e}$:

$$\mathbf{e} \cdot \mathbf{q} = \cos \theta, \qquad \mathbf{e} \cdot \mathbf{v} = \sin \theta \cos \phi, \qquad \mathbf{e} \cdot \mathbf{w} = \sin \theta \sin \phi, \qquad (11.2.18)$$

as shown in Fig. 11.1. Using these angles we obtain

$$L^2 = 1 + \tau \sin 2\theta \cos \psi'' + \tau^2 \cos^2 \theta, \qquad (11.2.19)$$

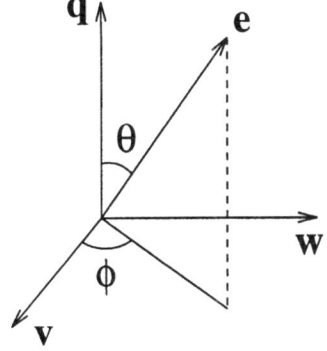

Fig. 11.1 The direction of **e** is given by the angles θ and ϕ.

where $\psi'' = \psi' + h\phi$. In taking the ensemble average $\cos\theta$ is uniformly distributed between -1 and $+1$, while ψ'' is uniformly distributed from 0 to 2π. Note that any explicit dependence on ϕ has disappeared: since $h = \pm 1$, rotating **v** and **w** about **q** corresponds to a change of the phase ψ. More formally the Beltrami wave (11.1.3) is invariant under the transformation

$$\mathbf{v} \leftarrow \mathbf{v}\cos\alpha - \mathbf{w}\sin\alpha, \quad \mathbf{w} \leftarrow \mathbf{v}\sin\alpha + \mathbf{w}\cos\alpha, \quad \psi \leftarrow \psi - h\alpha. \quad (11.2.20)$$

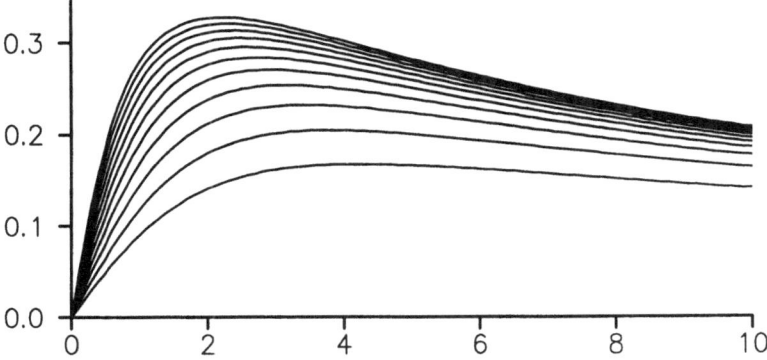

Fig. 11.2 Behavior of the stretching exponents Λ_p as functions of τ for the ensemble $\mathcal{E}_{3d}(\tau, h)$. The curves, reading upwards, correspond to integer values of p in the range 0 to 10. (After Gilbert & Bayly 1992.)

To obtain Λ_p (11.2.15) for $\mathcal{E}_{3d}$ we must take the average, $<L^p>$, using (11.2.19); in the absence of an analytical formula valid for general p, we do this numerically. Figure 11.2 shows Λ_p plotted as a function of τ for this ensemble. For any fixed τ the exponents Λ_p increase with p, indicating the presence of uneven stretching and spatial intermittency, as in the deterministic flows of §7.1. For small τ the growth rates Λ_p tend to zero. For large τ each Λ_p decays slowly, reflecting the fact that each Beltrami wave is integrable and so has relatively poor stretching properties if applied for a long time.

11.3 Stretching in Random Flows

The stretching exponents Λ_p have a number of properties. Firstly $\Lambda_p \geq \Lambda_0$ for $p > 0$ (Orszag 1970). This follows from the convexity of the exponential function: given any average $<\cdot>$, we have the two equivalent inequalities

$$<\exp X> \geq \exp<X>, \qquad <\log X> \leq \log<X> \qquad (11.3.1a,b)$$

(for X real). Substituting $X = L^p$ into the second of these implies that $\Lambda_p \geq \Lambda_0$. Furthermore $\Lambda_p \geq \Lambda_q$ if $p > q$, which follows from the inequality

$$<X^p>^{1/p} \geq <X^q>^{1/q}, \qquad (p > q > 0); \qquad (11.3.2)$$

see, for example, Hardy, Littlewood & Pólya (1967). Thus Λ_p is an increasing function of p; it is bounded above for $\tau > 0$:

$$\Lambda_p \leq (p\tau)^{-1} \log(\sup L)^p = \tau^{-1} \log \sup L. \qquad (11.3.3)$$

The supremum here is taken over all the flows $\mathbf{u}_\tau \in \mathcal{E}_\tau$; provided the rate of strain of these flows is bounded above, the supremum exists. The exponent Λ_p then has a finite limit as $p \to \infty$ (Zeldovich et al. 1988).

The fact that there is stretching at all in incompressible isotropic random fluid flows was established by Cocke (1969), who showed that $\Lambda_0 \geq 0$. To prove this for isotropic renewing flows, consider the average of $\log L = \frac{1}{2} \log(\mathbf{e}^T \mathbf{J}^T \mathbf{J}\mathbf{e})$. Since the ensemble is isotropic it is harmless first to fix $\mathbf{J}$ and to average over $\mathbf{e}$ as $\mathbf{e}$ covers the unit sphere $|\mathbf{e}| = 1$. The matrix $\mathbf{J}^T \mathbf{J}$ is symmetric and has eigenvalues λ_1, λ_2 and λ_3 with respect to some set of principal axes. These eigenvalues are positive and non-zero, as line elements cannot be shrunk to zero or negative lengths. Also $\lambda_1 \lambda_2 \lambda_3 = 1$ from the incompressibility of the fluid flow. Averaging over directions of $\mathbf{e}$:

$$<\log L>_\mathbf{e} = \frac{1}{4\pi} \int_{|\mathbf{e}|=1} \frac{1}{2} \log(\lambda_1 e_1^2 + \lambda_2 e_2^2 + \lambda_3 e_3^2) \, dS \qquad (11.3.4a)$$

$$\geq \frac{1}{4\pi} \int_{|\mathbf{e}|=1} \frac{1}{2} (e_1^2 \log \lambda_1 + e_2^2 \log \lambda_2 + e_3^2 \log \lambda_3) \, dS \qquad (11.3.4b)$$

$$= \frac{1}{6} (\log \lambda_1 + \log \lambda_2 + \log \lambda_3) = \frac{1}{6} \log(\lambda_1 \lambda_2 \lambda_3) = 0. \qquad (11.3.4c)$$

The inequality used in obtaining (11.3.4b) from (11.3.4a) is (11.3.1b): think of X taking each of the three values λ_i with probability e_i^2. To evaluate the integral we used

$$3 \int_{|e|=1} e_1^2 \, dS = \int_{|e|=1} \mathbf{e}^2 \, dS = 4\pi, \tag{11.3.5}$$

and similarly for e_2 and e_3. Under further averaging over different Jacobians, $<\log L>$ will remain non-negative. Thus $\varLambda_0 = \tau^{-1}<\log L> \geq 0$, showing that stretching generally occurs in isotropic renewing flows. Equality holds only when all the λ_i are unity, which is the special case when every $\mathbf{J}$ is a rotation.

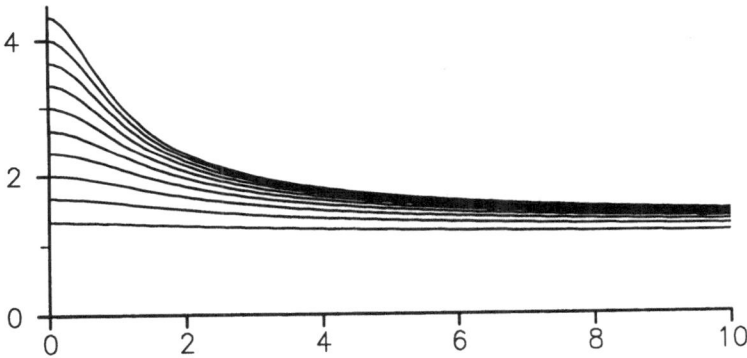

Fig. 11.3 Behavior of the ratios $\varLambda_p/\varLambda_0$ as functions of τ for the ensemble $\mathcal{E}_{3d}(\tau, h)$; the curves are for p from 1 to 10, reading upwards. (After Gilbert & Bayly 1992.)

Approximate results for the growth rates of magnetic moments can be obtained in various limiting cases; we consider here only the popular limit of short correlation time, $\tau \to 0$. Consider again the three-dimensional isotropic ensemble $\mathcal{E}_{3d}$ and using (11.2.19) expand L^p binomially for p fixed and $\tau \ll 1$:

$$L^p = 1 + \tfrac{1}{2}p(\tau \sin 2\theta \cos \psi'' + \tau^2 \cos^2 \theta)$$
$$+ \tfrac{1}{8}p(p-2)\tau^2 \sin^2 2\theta \cos^2 \psi'' + O(\tau^3). \tag{11.3.6}$$

Averaging this over ψ'' and θ yields eventually

$$\varLambda_p = (\tau/30)\,(5 + (p-2)) + O(\tau^2). \tag{11.3.7}$$

This result has been obtained by Kraichnan (1974) and Drummond & Münch (1990) in more general contexts. To confirm it Fig. 11.3 plots the ratios $\varLambda_p/\varLambda_0$ as a function of τ. The values at $\tau = 0$ are given by the approximation (11.3.7).

There is good agreement as $\tau \to 0$; however the approximation is plainly not uniformly good as p increases for fixed small τ.

The limit of zero correlation time allows approximation and analytical results, but all the exponents Λ_p tend to zero. To obtain finite growth rates in this limit, it is necessary to allow the velocity v to increase while the correlation time τ tends to zero. Instead of fixing $v = 1$ (11.1.8), as we have done, *for this paragraph only* we take $v^2\tau = \kappa$ as $\tau \to 0$, where κ is a constant with the dimensions of diffusivity. With this scaling

$$\Lambda_p = (\kappa/30)(5 + (p - 2)) + O(\tau^{1/2}), \tag{11.3.8}$$

and one can take the limit $\tau \to 0$ to obtain finite stretching rates

$$\Lambda_p|_{\tau=0} = (\kappa/30)(5 + (p - 2)). \tag{11.3.9}$$

In this limit the flow is called *delta-correlated* and particle trajectories take the form of Brownian motion. The magnetic field vectors have a log-normal distribution (Kraichnan 1974). While this is an attractive limit for approximations, some caution is required in extrapolating results for $\tau = 0$ to finite correlation times, however small. For example if $\tau = 0$ then the Λ_p increase linearly and unboundedly with p from (11.3.9). However for any fixed $\tau > 0$, Λ_p is bounded above as p increases, by (11.3.3): with the present scalings

$$\sup L \le 1 + \tau v = 1 + \sqrt{\kappa\tau}, \tag{11.3.10}$$

and so from (11.3.3),

$$\Lambda_p \le \tau^{-1}\log(1 + \sqrt{\kappa\tau}) \le \sqrt{\kappa/\tau}. \tag{11.3.11}$$

This odd behavior arises because in the delta-correlated limit, $\tau \to 0$, the kinetic energy and rate of strain of the flow field tend to infinity. For recent results on delta-correlated flows see Baxendale & Rozovskii (1993).

We now return to the usual scalings (11.1.8) with $v = 1$. Instead of using L from (11.2.14) in (11.2.15) for calculating Λ_p we can in fact use L', where

$$L'^2 = |\mathbf{J}^T\mathbf{e}|^2 = \mathbf{e}^T\mathbf{J}\mathbf{J}^T\mathbf{e}, \tag{11.3.12}$$

since

$$<L^p> = <L'^p> \tag{11.3.13}$$

for any p (Kraichnan 1974, Drummond & Münch 1990). We explain this result using a geometrical argument. Since the ensemble is isotropic it is harmless to average over all directions of $\mathbf{e}$. To prove (11.3.13) it therefore suffices to fix $\mathbf{J}$ and show that the two averages, with $\mathbf{e}$ running over the unit sphere,

$$<f(L^2)>_{\mathbf{e}} = \int_{|\mathbf{e}|=1} f(\mathbf{e}^T\mathbf{J}^T\mathbf{J}\mathbf{e})\,dS, \tag{11.3.14a}$$

$$<f(L'^2)>_{\mathbf{e}} = \int_{|\mathbf{e}|=1} f(\mathbf{e}^{\mathrm{T}}\mathbf{J}\mathbf{J}^{\mathrm{T}}\mathbf{e})\,dS \qquad (11.3.14b)$$

coincide for any function f. Now $\mathbf{J}^{\mathrm{T}}\mathbf{J}$ is a symmetric matrix and so can be diagonalised:

$$\mathbf{e}^{\mathrm{T}}\mathbf{J}^{\mathrm{T}}\mathbf{J}\mathbf{e} = \lambda_1 e_1^2 + \lambda_2 e_2^2 + \lambda_3 e_3^2, \qquad (11.3.15)$$

where λ_i are its eigenvalues and e_i are the components of $\mathbf{e}$ with respect to a certain orthonormal basis. Therefore

$$<f(L^2)>_{\mathbf{e}} = \int_{|\mathbf{e}|=1} f(\lambda_1 e_1^2 + \lambda_2 e_2^2 + \lambda_3 e_3^2)\,dS \qquad (11.3.16)$$

and depends on the symmetric matrix $\mathbf{J}^{\mathrm{T}}\mathbf{J}$ only through its eigenvalues. However the symmetric matrix $\mathbf{J}\mathbf{J}^{\mathrm{T}}$ has the same characteristic polynomial and eigenvalues as $\mathbf{J}^{\mathrm{T}}\mathbf{J}$ since

$$\mathrm{trace}\big(\mathbf{J}\mathbf{J}^{\mathrm{T}}\big)^p = \mathrm{trace}\big(\mathbf{J}^{\mathrm{T}}\mathbf{J}\big)^p = \lambda_1^p + \lambda_2^p + \lambda_3^p \qquad (11.3.17)$$

for $p = 1, 2, \ldots$; thus the two averages $(11.3.14a, b)$ are the same.

This surprising result has interesting implications (Kraichnan 1974, Drummond & Münch 1990). An area element $\mathbf{A}$ frozen in the flow evolves by

$$\mathbf{A}(M\mathbf{x}, (n+1)\tau)) = (\mathbf{J}^{-1})^{\mathrm{T}}(\mathbf{x})\,\mathbf{A}(\mathbf{x}, n\tau), \qquad (11.3.18)$$

which is also the equation for the evolution of vector potential (4.2.6) in an appropriate gauge. We can define stretching exponents for areas $\mathbf{A}$ just as for vectors $\mathbf{B}$,

$$\Lambda_p^{\mathrm{area}} = (p\tau)^{-1}\log <L_{\mathrm{area}}^p> = (p\tau)^{-1}\log <L'^p_{\mathrm{area}}>, \quad (p > 0), \qquad (11.3.19)$$

where

$$L_{\mathrm{area}}^2 = |(\mathbf{J}^{-1})^{\mathrm{T}}\mathbf{e}|^2, \qquad L'^2_{\mathrm{area}} = |(\mathbf{J}^{-1})\mathbf{e}|^2, \qquad (11.3.20)$$

given by replacing $\mathbf{J}$ in (11.2.14, 3.12) by $(\mathbf{J}^{-1})^{\mathrm{T}}$. Now suppose the ensemble of flows is *statistically time-reversible*: if $\mathbf{u}(\mathbf{x}, t) \in \mathcal{E}$, then $-\mathbf{u}(\mathbf{x}, -t) \in \mathcal{E}$. For example the renewing ensembles defined in §11.1 have this property. In this case the probability distribution for $\mathbf{J}$ is the same as that for $\mathbf{J}^{-1}$,

$$<L_{\mathrm{area}}^p> = <L'^p_{\mathrm{area}}> = <L^p> = <L'^p>, \qquad (11.3.21)$$

and so $\Lambda_p = \Lambda_p^{\mathrm{area}}$. All moments of $\mathbf{A}$ and $\mathbf{B}$, i.e., of line elements and surface elements, grow at the same rate if the ensemble is time-reversible. Numerical work of Drummond & Münch (1990) bears out this picture; furthermore they find that for ensembles that are not time-reversible, area elements and line elements have different growth rates. This is relevant to Navier–Stokes flows, which are not time-reversible, because of the irreversible dissipation of energy by viscosity (Kraichnan 1974). One can also study the evolution of the curvature of field lines, which is related to the Lorentz force, and their torsion (see Drummond 1992 and references therein).

While renewing fluid flows are somewhat artificial because of the sudden loss of memory at times $n\tau$, simulation of line-stretching in more general random flows (Drummond & Münch 1990) gives qualitatively similar results; these authors consider flows made up of a sum of Beltrami waves,

$$\mathbf{u}(\mathbf{x}, t) = \sum_{n=1}^{N} (\mathbf{v}_n \sin(\mathbf{q}_n \cdot \mathbf{x} + \omega_n t + \psi_n)$$

$$+ \mathbf{w}_n h \cos(\mathbf{q}_n \cdot \mathbf{x} + \omega_n t + \psi_n)),$$

(11.3.22)

using our notation (Kraichnan 1970). Here the quantities ω_n, $\mathbf{q}_n$, $\mathbf{v}_n$ and ψ_n are chosen from Gaussian distributions.

Other unrealistic aspects of the flows used in this chapter are their time-reversibility, and their long spatial correlations. The flows are not good models of turbulence, as they are characterised by a single length-scale, rather than a range of scales. This is not likely to be a serious problem, however, since under the Kolomogorov picture of turbulence the most intense stretching occurs at scales close to the dissipation scale, rather than throughout the whole inertial range (Batchelor 1952). A random model for dynamo action in a flow with many eddies is discussed by Ruzmaikin, Liewer & Feynman (1993).

11.4 Diffusion and Dynamo Action

We now consider the evolution of magnetic fields with non-zero diffusion $\epsilon > 0$. This complicates matters considerably because the Cauchy solution no longer applies, and a Green's function is needed to evolve the field over one renewing time interval: $\mathbf{B}(\mathbf{x}, (n+1)\tau) = \mathbf{T}_\epsilon \mathbf{B}(\mathbf{x}, n\tau)$ with

$$B_i(\mathbf{x}, (n+1)\tau) = \int G_{\epsilon ij}(\mathbf{x}, \mathbf{y}) B_j(\mathbf{y}, n\tau) \, d\mathbf{y}.$$

(11.4.1)

Here the induction operator $\mathbf{T}_\epsilon$ and the corresponding Green's function $\mathbf{G}_\epsilon(\mathbf{x}, \mathbf{y})$ depend on the random choice of wave used in the interval I_n. For $\epsilon = 0$, the kernel contains a delta function

$$\mathbf{G}_\epsilon(\mathbf{x}, \mathbf{y}) = \delta(\mathbf{x} - M\mathbf{y})\mathbf{J}(\mathbf{y}), \qquad (\epsilon = 0),$$

(11.4.2)

giving the Cauchy solution, but for $\epsilon > 0$ the kernel is smooth.

Because the $\epsilon > 0$ Green's function couples field on different Lagrangian trajectories, we can no longer follow high-order single-point moments easily. The only quantity that is easily followed is the first moment, i.e., the ensemble-averaged field $<\mathbf{B}(\mathbf{x}, t)>$ at a point $\mathbf{x}$. Below we shall use this quantity to test for fast dynamo action.

We consider a renewing flow, and require it to be homogeneous. No other restrictions are imposed; the ensembles may be isotropic or anisotropic. The initial conditions are arbitrary, and we shall think of $\mathbf{B}_0(\mathbf{x})$ as some given, non-random field. Average (11.4.1) up to time $t = n\tau$,

$$<\mathbf{B}(\mathbf{x}, (n+1)\tau)>_{I_1,...,I_{n-1}} = \int \mathbf{G}_\epsilon(\mathbf{x}, \mathbf{y}) <\mathbf{B}(\mathbf{y}, n\tau)> d\mathbf{y}, \tag{11.4.3}$$

and then over the interval I_n:

$$<\mathbf{B}(\mathbf{x}, (n+1)\tau)> = \int <\mathbf{G}_\epsilon(\mathbf{x}, \mathbf{y})><\mathbf{B}(\mathbf{y}, n\tau)> d\mathbf{y}. \tag{11.4.4}$$

The average splits up because the flows in different time intervals are independent, and so the random choice of Green's function in the interval I_n is independent of the evolution up to time $n\tau$.

As the ensemble is homogeneous, $<\mathbf{G}_\epsilon(\mathbf{x}, \mathbf{y})>$ depends only on $\mathbf{x} - \mathbf{y}$ and (11.4.4) is a convolution integral, which can be simplified by taking Fourier transforms. Define the Fourier transform pair

$$\begin{aligned} \mathbf{B}(\mathbf{x}, t) &= \int \widetilde{\mathbf{B}}(\mathbf{k}, t) \exp(i\mathbf{k} \cdot \mathbf{x}) \, d\mathbf{k}, \\ (2\pi)^3 \widetilde{\mathbf{B}}(\mathbf{k}, t) &= \int \mathbf{B}(\mathbf{x}, t) \exp(-i\mathbf{k} \cdot \mathbf{x}) \, d\mathbf{x} \end{aligned} \tag{11.4.5}$$

and note that since the magnetic field is solenoidal,

$$\mathbf{k} \cdot \widetilde{\mathbf{B}}(\mathbf{k}, t) = 0. \tag{11.4.6}$$

In Fourier space the convolution (11.4.4) becomes

$$<\widetilde{\mathbf{B}}(\mathbf{k}, (n+1)\tau)> = (2\pi)^3 \widetilde{\mathbf{G}}_\epsilon(\mathbf{k}) <\widetilde{\mathbf{B}}(\mathbf{k}, n\tau)>, \tag{11.4.7}$$

with

$$(2\pi)^3 \widetilde{\mathbf{G}}_\epsilon(\mathbf{k}) = \int <\mathbf{G}_\epsilon(\mathbf{x}, \mathbf{y})> \exp[-i\mathbf{k} \cdot (\mathbf{x} - \mathbf{y})] \, d\mathbf{x} \tag{11.4.8}$$

(Kraichnan 1974). This is independent of $\mathbf{y}$ by homogeneity. The quantity $(2\pi)^3 \widetilde{\mathbf{G}}_\epsilon(\mathbf{k})$ is the *response tensor* or *transfer function*. Because of the homogeneity of the ensemble, the Fourier modes $<\widetilde{\mathbf{B}}(\mathbf{k}, n\tau)>$ of the average field evolve independently by the simple equation (11.4.7). The small-scale fluctuations generated by each random wave are eradicated on averaging over the ensemble; in this sense ensemble averaging is like strong diffusion.

The response tensor gives the growth of Fourier modes of the ensemble-averaged field, and so determines the growth of linear functionals, since

$$L_{\mathbf{f}}<\mathbf{B}> = \int \mathbf{f}(\mathbf{x})^* \cdot <\mathbf{B}(\mathbf{x}, t)> d\mathbf{x} = (2\pi)^3 \int \widetilde{\mathbf{f}}^*(\mathbf{k}) \cdot <\widetilde{\mathbf{B}}(\mathbf{k}, t)> d\mathbf{k}, \tag{11.4.9}$$

by Parseval's theorem. From (11.0.2b), the dynamo growth rate is given in terms of the response tensor by

$$\gamma(\epsilon) = \tau^{-1} \sup_{\mathbf{k}} \max_{i} \log |\mu_i(\epsilon, \mathbf{k})|, \qquad (11.4.10)$$

where $\mu_i(\epsilon, \mathbf{k})$ for $i = 1, 2$ are the eigenvalues of $(2\pi)^3 \widetilde{\mathbf{G}}_\epsilon(\mathbf{k})$ in the space of fields satisfying the solenoidal condition (11.4.6).

As an example, consider one of the renewing ensembles introduced in §11.1. We need the Green's function for the evolution of field in Beltrami waves with diffusion $\epsilon > 0$. This is messy to calculate; it could be obtained from Vishik's (1989) asymptotic expansion, or by using Weiner noise techniques (Dittrich et $al.$ 1984). For the moment we take the simplest option, which is to pulse diffusion. This, although artificial, gives us an exact Green's function and, as we shall see shortly, little is lost by doing this.

Therefore suppose that in each interval I_n we apply a wave instantaneously using the Cauchy solution, and then apply the heat kernel

$$H_\epsilon(\mathbf{x}) = (4\pi\epsilon\tau)^{-3/2} \exp(-\mathbf{x}^2/4\epsilon\tau) \qquad (11.4.11)$$

to all components of the field, this being equivalent to allowing diffusion ϵ to act for a time τ. The Green's function for this pulsed model is

$$\mathbf{G}_\epsilon(\mathbf{x}, \mathbf{y}) = (4\pi\epsilon\tau)^{-3/2} \exp[-(\mathbf{x} - M\mathbf{y})^2/4\epsilon\tau] \, \mathbf{J}(\mathbf{y}), \quad (\epsilon > 0), \qquad (11.4.12)$$

where the map M and Jacobian $\mathbf{J}$ depend on the choice of random flow. The corresponding response tensor is

$$(2\pi)^3 \widetilde{\mathbf{G}}_\epsilon(\mathbf{k}) = (4\pi\epsilon\tau)^{-3/2} \int <e^{-(\mathbf{x}-M\mathbf{y})^2/4\epsilon\tau} \, \mathbf{J}(\mathbf{y})> e^{-i\mathbf{k}\cdot(\mathbf{x}-\mathbf{y})} \, d\mathbf{x}. \quad (11.4.13)$$

Integrating over $\mathbf{x}$ yields

$$(2\pi)^3 \widetilde{\mathbf{G}}_\epsilon(\mathbf{k}) = <\mathbf{J}(\mathbf{y}) \exp(-i\mathbf{k} \cdot (M\mathbf{y} - \mathbf{y}))> \exp(-\epsilon\tau k^2). \qquad (11.4.14)$$

The effect of pulsed diffusion is to contribute a factor of $\exp(-\epsilon\tau k^2)$, which is just the factor by which a wave with wavevector $\mathbf{k}$ would be suppressed by diffusion ϵ over a time τ. For fixed $\mathbf{k}$, in the fast dynamo limit $\epsilon \to 0$ the effect of diffusion on the response tensor and growth rate becomes negligible. This is a result of the ensemble averaging, which has eliminated any small-scale fluctuations, leaving behind large-scale modes for which diffusion is irrelevant in the fast dynamo limit.

This means that we can take the limit $\epsilon \to 0$ in (11.4.10) to obtain

$$\gamma_0 = \tau^{-1} \sup_{\mathbf{k}} \max_{i} \log |\mu_i(0, \mathbf{k})|, \qquad (11.4.15)$$

yielding the fast dynamo growth rate in terms of the eigenvalues $\mu_i(0, \mathbf{k})$ of the diffusionless response tensor $(2\pi)^3 \widetilde{\mathbf{G}}_0$. It is sufficient to obtain growth of a $\mathbf{k}$-mode for $\epsilon = 0$ to obtain fast dynamo action.[80]

The property that diffusion has a uniformly small effect on the growth of the ensemble average is not simply a consequence of artificially pulsing diffusion, but persists if we use the true Green's function or an approximation (Dittrich *et al.* 1984). This is because each wave only acts for a finite time and has a finite rate of strain, and so starting with a given Fourier mode, the effect of diffusion must be uniformly small in the limit $\epsilon \to 0$. This may also be seen using Vishik's (1989) approximate Green's function for continuous diffusion from §6.6: if we substitute the leading order approximation (6.6.9, 12) into (11.4.8), a short calculation (integrating over $\mathbf{y}$ instead of $\mathbf{x}$ and setting $\mathbf{z} = M^{-1}\mathbf{x}$) yields

$$(2\pi)^3 \widetilde{\mathbf{G}}_\epsilon(\mathbf{k}) \simeq \, <\mathbf{J}(\mathbf{z}) \exp(-i\mathbf{k} \cdot (M\mathbf{z} - \mathbf{z})) \exp(-\epsilon \mathbf{k} \cdot \mathbf{Q}(\mathbf{z}, \tau)\mathbf{k})>. \quad (11.4.16)$$

The matrix $\mathbf{Q}$ defined in equation (6.6.10) contains information about the stretching along the trajectory through $\mathbf{z}$, and gives the effect of diffusion in dissipating field, an effect that is small in the limit $\epsilon \to 0$ (for fixed $\mathbf{k}$ and τ). Thus (11.4.15), giving the fast dynamo growth rate, holds in general homogeneous renewing ensembles.

Equation (11.4.15) means that the flux conjecture holds for homogeneous renewing random flows: Γ_{sup}, defined from the growth of linear functionals (11.4.9) with $\epsilon = 0$, is equal to γ_0. Thus conjectures 1–4 of Chap. 4 hold when translated into statements about the ensemble-averaged field $<\mathbf{B}>$ in these flows.

We now return to the simpler case of pulsed diffusion; for the isotropic renewing flow $\mathcal{E}_{3\mathrm{d}}(\tau, h)$ a calculation of the average (11.4.8) yields

$$(2\pi)^3 \widetilde{\mathbf{G}}_{\epsilon ij}(\mathbf{k}) = [j_0(k\tau)\delta_{ij} + j_1(k\tau)(ih\tau/2k)\epsilon_{ijl}k_l] \exp(-\epsilon\tau k^2), \quad (11.4.17)$$

where the spherical Bessel functions j_0 and j_1 are given by

$$j_0(s) = (\sin s)/s, \qquad j_1(s) = (\sin s - s \cos s)/s^2. \quad (11.4.18)$$

This is derived in the next two paragraphs, which the reader may omit.

Taking $\epsilon = 0$, from (11.1.5, 6, 11.4.14) the response tensor for $\mathcal{E}_{3\mathrm{d}}(\tau, h)$ is given by

$$(2\pi)^3 \widetilde{\mathbf{G}}_0(\mathbf{k}) = <(\mathbf{I} + \mathbf{v}\mathbf{q}^{\mathrm{T}}\tau \cos \psi' - \mathbf{w}\mathbf{q}^{\mathrm{T}}h\tau \sin \psi')$$
$$\times \exp(-i\mathbf{k} \cdot (\mathbf{v}\tau \sin \psi' + \mathbf{w}h\tau \cos \psi'))>, \quad (11.4.19)$$

---

[80]One might consider replacing the '=' sign in (11.4.15) by '$\geq$': there is the possibility of non-uniform behavior of the growth rates for large $|\mathbf{k}|$ in the limit $\epsilon \to 0$ that could allow γ_0 to be larger than the right-hand side. In terms of Chap. 4, this would be a fast dynamo that is not a perfect dynamo. However (11.4.16) shows that diffusion has only a damping effect in homogeneous renewing flows and so equality is correct in (11.4.15).

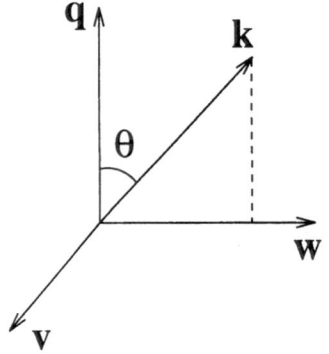

Fig. 11.4 The vector **k** relative to **v**, **w** and **q**.

with $\psi' = \mathbf{q} \cdot \mathbf{y} + \psi$. First recall that $h = \pm 1$ and so a phase shift in ψ corresponds to a rotation of **v** and **w** about **q** (11.2.20). We can therefore choose the direction of **v** so that it is perpendicular to **k**. We let θ be the angle between **k** and **q** as shown in Fig. 11.4, with $0 \le \theta \le \pi$ and $\mathbf{w} \cdot \mathbf{k} = k \sin \theta$. We then obtain

$$(2\pi)^3 \widetilde{\mathbf{G}}_0(\mathbf{k}) = \; <(\mathbf{I} + \mathbf{v}\mathbf{q}^{\mathrm{T}} \tau \cos \psi') \exp(-ikh\tau \sin \theta \cos \psi')>. \qquad (11.4.20)$$

Now we can average over ψ':

$$(2\pi)^3 \widetilde{\mathbf{G}}_0(\mathbf{k}) = \mathbf{I} <J_0(k\tau \sin \theta)> \; - \; <\mathbf{v}\mathbf{q}^{\mathrm{T}} ih\tau J_1(k\tau \sin \theta)>, \qquad (11.4.21)$$

where J_0 and J_1 are Bessel functions (see Abramowitz & Stegun 1972).

The tensor $\widetilde{\mathbf{G}}_0(\mathbf{k})$ must take the form

$$\widetilde{G}_{0ij}(\mathbf{k}) = F_1(k)\delta_{ij} + F_2(k)k_i k_j / k^2 + F_3(k)\epsilon_{ijl}k_l / k, \qquad (11.4.22)$$

because the ensemble is isotropic, with

$$2F_1 = \widetilde{G}_{0ii} - \widetilde{G}_{0ij}k_i k_j / k^2, \qquad 2F_2 = -\widetilde{G}_{0ii} + 3\widetilde{G}_{0ij}k_i k_j / k^2,$$
$$2F_3 = \widetilde{G}_{0ij}\epsilon_{ijl}k_l / k. \qquad (11.4.23)$$

We contract the indices of $\widetilde{\mathbf{G}}_0$ and average over θ to obtain

$$(2\pi)^3 \widetilde{G}_{0ii} = 3 <J_0(k\tau \sin \theta)> \; = 3j_0(k\tau) \qquad (11.4.24)$$

(see §§11.4.10, 10.1.1 of Abramowitz & Stegun 1972). Note that in this average $\cos \theta$ is distributed uniformly in the range $[-1, 1]$. Also

$$(2\pi)^3 \widetilde{G}_{0ij}k_i k_j = k^2 <J_0(k\tau \sin \theta)> \; = k^2 j_0(k\tau), \qquad (11.4.25)$$

using $\mathbf{v} \cdot \mathbf{k} = 0$. Finally

$$
\begin{aligned}
(2\pi)^3 \widetilde{G}_{0ij}\epsilon_{ijl}k_l &= <-ih\tau\, \mathbf{v} \cdot \mathbf{q} \times \mathbf{k}\, J_1(k\tau \sin\theta)> \\
&= ihk\tau<\sin\theta\, J_1(k\tau \sin\theta)> \qquad (11.4.26) \\
&= ihk\tau\, j_1(k\tau).
\end{aligned}
$$

Piecing together these results gives the response tensor in equation (11.4.17) above.

Taking $\mathbf{k} = k\mathbf{e}_z$ for definiteness, eigenvectors of the matrix (11.4.17) are $(0,0,1)$, which is not allowed by (11.4.6), and $(\mp i, 1, 0)$ with eigenvalues

$$
\mu_\pm(\epsilon, k) = [j_0(k\tau) \mp (h\tau/2)j_1(k\tau)]\exp(-\epsilon\tau k^2). \qquad (11.4.27)
$$

The corresponding asymptotic growth rates are

$$
\gamma_\pm(\epsilon, k) = \tau^{-1}\log|\mu_\pm(\epsilon, k)|. \qquad (11.4.28)
$$

The eigenvectors $\frac{1}{2}(\mp i, 1, 0)$ correspond to Beltrami waves $(\pm \sin kz, \cos kz, 0)$ of the ensemble-averaged field $<\mathbf{B}(\mathbf{x})>$ with positive and negative parity. The rejected $(0,0,1)$ eigenvector gives the evolution of a scalar gradient in the ensemble.

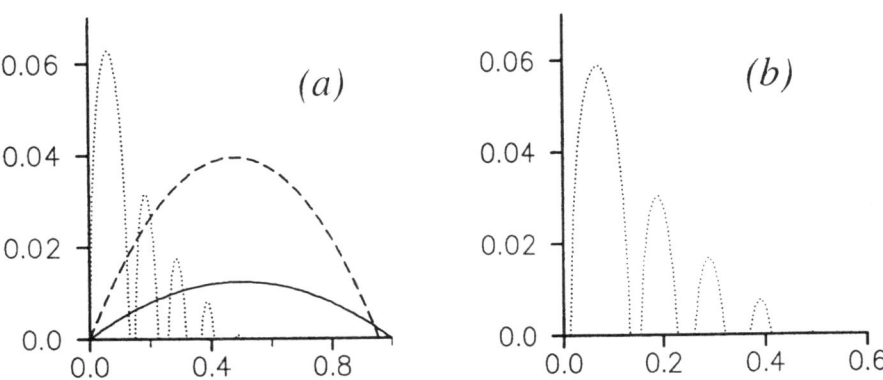

Fig. 11.5 Growth rates for magnetic fields in the ensemble $\mathcal{E}_{3d}(\tau, h)$ with $h = 1$ and $\epsilon = 0$. In (a) $\gamma_-(0, k)$ and in (b) $\gamma_+(0, k)$ are plotted as a function of k for $\tau = 0.3$ (solid), 1.0 (dashed) and 32 (dotted). (After Gilbert & Bayly 1992.)

Figure 11.5 shows growth rates obtained by setting $\epsilon = 0$ in (11.4.27, 28) for $h = 1$, and $\tau = 0.3$ (solid), 1.0 (dashed) and 32 (dotted). In (a) and (b) the growth rates $\gamma_-(0, k)$ and $\gamma_+(0, k)$ of Beltrami waves with negative and positive helicity are shown. The growth rates are quite low compared with similar deterministic flows such as MW+; see Fig. 2.1. This is unsurprising, since the ensemble contains many flows that are unfavorable to dynamo action, as well as effective dynamos, such as alternating x and y Beltrami waves.

To understand the results, consider first the limit of short correlation times with $h = 1$, when (11.4.27) can be approximated to give

$$\gamma_\pm(\epsilon, k) = \mp k\tau/6 - k^2(\epsilon + \tau/6) + O(\tau^3) \tag{11.4.29}$$

as $\tau \to 0$ with k bounded above (Molchanov, Ruzmaikin & Sokoloff 1983). If k is sufficiently small, $\gamma_-(\epsilon, k)$ is positive and there is dynamo action, with growth of large-scale magnetic fields having negative helicity or parity, opposite to the parity $h = 1$ of the fluid flow. This can be seen for the $\tau = 0.3$ curve (solid) in Fig. 11.5(a). Fields with positive helicity decay for small τ.

In this $\tau \to 0$ limit we recover the famous result of Steenbeck, Krause & Rädler (1966). Under certain approximations a large-scale magnetic field in a turbulent flow obeys

$$\partial_t \mathbf{B} = \alpha \nabla \times \mathbf{B} + \beta \nabla^2 \mathbf{B}, \tag{11.4.30}$$

(see, e.g., Moffatt 1978, Krause & Rädler 1980) where α is the alpha effect and β the eddy diffusivity, given in the limit of short correlation time, $\tau_{\mathrm{corr}} \to 0$, by

$$\alpha = -(\tau_{\mathrm{corr}}/3) <\mathbf{u} \cdot \nabla \times \mathbf{u}>, \qquad \beta = \epsilon + (\tau_{\mathrm{corr}}/3) <\mathbf{u}^2>. \tag{11.4.31}$$

A Beltrami wave of magnetic field with wavenumber k grows at a rate $\gamma_\pm(\epsilon, k) = \pm \alpha k - \beta k^2$. Since $\mathbf{u}^2 = 1$ and $\mathbf{u} \cdot \nabla \times \mathbf{u} = h = 1$ in our flows, this agrees with (11.4.29) provided we take $\tau_{\mathrm{corr}} = \tau/2$ (Dittrich et $al.$ 1984).

Another approximation can be obtained in the limit of very long renewing time $\tau \to \infty$. In this limit

$$\gamma_\pm(\epsilon, k) = \tau^{-1} \log |(h/2k)\cos(k\tau)| - \epsilon k^2 + O(\tau^{-2}) \tag{11.4.32}$$

for k bounded below. For a large, fixed time τ and small ϵ, there are a number of windows as k varies where the growth rates $\gamma_\pm(\epsilon, k)$ are positive: growth is obtained for magnetic modes of both helicities. These phenomena are clear in Fig. 11.5(a, b) for the case $\tau = 32$ (dotted).

We have discussed the behavior, with diffusion, of the first moment, or ensemble average, of the magnetic field. This gives little information about the magnetic energy, and its distribution over different length scales, for which we need knowledge of the second moment. It is possible to make some analytical progress with the second moment in isotropic homogeneous ensembles, but there are complications. Diffusion cannot be ignored since it plays a significant role in dissipating energy, and one has to follow the full two-point second moment, $<B_i(\mathbf{x})B_j(\mathbf{x} + \mathbf{r})>$, which, as an isotropic function of $\mathbf{r}$, is equivalent to three scalar functions of $|\mathbf{r}|$. One can in principle write down equations giving the evolution of these scalar functions, but these are complicated unless one takes the limit of delta-correlated flows. We do not pursue this further here but refer the reader to papers by Kazantsev (1967), Kraich-nan & Nagarajan (1967), Vainshtein & Zeldovich (1972), Vainshtein (1982), Novikov, Ruzmaikin & Sokoloff (1983) and Vainshtein & Kichatinov (1986). A recent discussion of the kinematic fast dynamo problem using probabilistic methods may be found in Thompson (1990).

One can question the use of renewing ensembles, rather than more realis-tic random ensembles. There has been little work on calculating the response tensor directly for more general random flows (but see Kraichnan 1976b). Attention has instead focussed on the alpha effect and eddy diffusivity, and it appears that these coefficients do have finite values for random flows of the form (11.3.22), provided the correlation time is finite (Kraichnan 1976b, Drummond & Horgan 1986). This is also confirmed by theoretical studies (Vainshtein 1970, Moffatt 1974, Knobloch 1977) which use various approxi-mations. In conclusion there appears to be no evidence that renewing flows are qualitatively different from more general classes of random flows with finite correlation times. The situation is markedly different for the extreme case of steady random flows, for which the alpha coefficient and eddy diffusiv-ity appear to vanish in the absence of molecular diffusion (Kraichnan 1976b, Drummond & Horgan 1986).

11.5 Numerical Simulations

We have seen that when the definitions (11.0.1, 2) are used, fast dynamo theory for homogeneous ensembles of random flows becomes straightforward; testing for growth of the ensemble-averaged field involves only finite-time averages, and the limit $\epsilon \to 0$ is trivial. The problem comes in interpreting what these results mean: they refer to the ensemble average, but do not tell us what might happen in 'typical' flows in the ensemble, i.e., for typical sequences of random Beltrami waves. The bad news is that they tell us surprisingly little.

This is best illustrated by the following 'toy model' of Zeldovich *et al.* (1987). Suppose we have 'magnetic field' $\tilde{B}(t)$; this might, for example, be the amplitude of one of the Fourier modes $\tilde{\mathbf{B}}(\mathbf{k}, t)$ used in §11.4. Suppose that in each renewing interval I_m it is multiplied by a random number v_m,

$$B(n\tau) = v_{n-1} \ldots v_2 v_1 v_0 B_0, \tag{11.5.1}$$

and take the v_m to be independent and identically distributed. Now consider the case when $v_m = 3$ or $v_m = 0$ with probability $1/2$. Then

$$<B(n\tau)> = <v_{n-1} \ldots v_2 v_1 v_0> B_0 = (3/2)^n B_0. \tag{11.5.2}$$

The ensemble-averaged field grows by a factor of $<v_m> = 3/2$ per interval; from (11.0.2) this corresponds to a growth rate $\gamma = \tau^{-1} \log 3/2 > 0$. Yet in almost all realisations the field is eventually zero. In only one realisation does the field grow indefinitely. As time goes on the ensemble average is dominated by fewer and fewer atypical realisations.

This disturbing behavior can persist even if v_m takes non-zero values, for example 1/3 and 2; the growth in the average is now by $<v_m> = 7/6 > 1$ per interval, giving $\gamma_0 = \tau^{-1} \log 7/6 > 0$. However in typical realisations v_m takes the value 1/3 half of the time and 2 half of the time, giving growth by an average factor of $\sqrt{2/3} < 1$ per iteration, and a growth rate $\gamma_{typ} = \tau^{-1} \log \sqrt{2/3} < 0$. Again the ensemble average is dominated by unlikely realisations while the field in a typical realisation decays.

Returning to dynamos, the problem is that at the outset we defined the dynamo growth rate, by say (11.0.2a), as the growth rate of the ensemble-averaged field; however the growth rate in a typical realisation is not this, but

$$\gamma_{typ}(\epsilon) = \limsup_{t \to \infty} \frac{1}{2t} <\log E_M(t)>. \tag{11.5.3}$$

The two are generally different. For the toy model above, with real amplification factors v_m, $\gamma_{typ} = \tau^{-1} <\log v_m> \leq \gamma_0 = \tau^{-1} \log <v_m>$ by inequality (11.3.1b). Thus the ensemble average can overestimate the growth rate in typical realisations because of *variable amplification*. This is analogous to the inequality $\Lambda_{Liap} \leq h_{top}$ for deterministic flows with uneven stretching.

For the toy model, if $B(t)$ is allowed to be complex (as is a Fourier mode $\tilde{B}(\mathbf{k}, t)$ in a true dynamo), the growth in the ensemble average can underestimate the growth rate in realisations because of *phase mixing*; this point is stressed by Hoyng (1987b). Again use the toy model (11.5.1) and consider the case $v_m = 2$ or -2 with equal probability. Plainly $<B(n\tau)>$ is zero for $n \geq 1$, although in each realisation the field doubles in magnitude over each interval τ. This is a case where there is cancellation between different members of the ensemble, leading to an underestimate of the growth rate.

A less extreme example, typical of what may happen in practice, is to take v_m to be $2 + i$ or $2 - i$ with equal probability. Then the ensemble-averaged field doubles over each period τ, while the field in each realisation grows in magnitude by $\sqrt{5}$. The ensemble average grows relatively slowly because different realisations are getting out of phase and partially cancelling.

So the analytical results in §11.4 about the fast growth of magnetic field modes in the ensemble average tell us virtually nothing about typical realisations: they may overestimate the growth rate because of variable amplification, or underestimate it because of phase mixing. Because of this perhaps the fast dynamo problem might be better formulated using $\gamma_{typ}(\epsilon)$ in (11.5.3), rather than $\gamma(\epsilon)$, to define γ_0 and fast dynamo action. The problem is that $\gamma_{typ}(\epsilon)$ does not appear to be available analytically for dynamos, although it is for the toy model. All we can calculate is $\gamma(\epsilon)$, which can be misleading, but does allow some analytical progress.

We turn to numerical simulations to see what happens for dynamos in particular cases. The ensembles $\mathcal{E}_{3d}$ and $\mathcal{E}_{111}$ are not ideal for numerical studies, since typical flows in these ensembles are three-dimensional and so computationally expensive. Instead we introduce a renewing ensemble containing flows that are two-dimensional and periodic in space. This is defined analogously to $\mathcal{E}_{3d}$ and $\mathcal{E}_{111}$ in §11.1, but with another choice of directions $\mathbf{q}$:

- 110 *renewing ensemble* $\mathcal{E}_{110}(\tau, h)$. The direction of the vector $\mathbf{q}$ is chosen to be one of the two coordinate axes $\mathbf{e}_x$ and $\mathbf{e}_y$ with probability $1/2$.

This example is analogous to flows such as MW+, or pulsed Beltrami waves (10.0.1). We take $h = 1$; the prescription in §11.1 amounts to choosing either an x-wave or a y-wave,

$$\mathbf{u}_\tau(\mathbf{x}, t) = 2\sin^2(\pi t/\tau)(0, \sin(x + \psi), \cos(x + \psi)), \tag{11.5.4a}$$
$$\mathbf{u}_\tau(\mathbf{x}, t) = 2\sin^2(\pi t/\tau)(\cos(y + \psi), 0, \sin(y + \psi)), \tag{11.5.4b}$$

in each renewing interval I_n, probability $1/2$. The phase ψ is also chosen randomly for each wave. Note that for numerical reasons the flow in each renewing interval is taken to be time-dependent and is switched off and on fairly smoothly. In terms of (11.1.4), our random choice of waves corresponds to choosing one of two orthogonal sets, $\{\mathbf{q}, \mathbf{v}, \mathbf{w}\} = \{\mathbf{e}_x, \mathbf{e}_y, \mathbf{e}_z\}$ or $\{\mathbf{e}_y, \mathbf{e}_z, \mathbf{e}_x\}$.

Since flows in the ensemble $\mathcal{E}_{110}(\tau, h)$ are independent of z, the magnetic field can be written in the separable form $\mathbf{B}(x, y, z, t) = \mathbf{b}(x, y, t)\exp(ikz)$ (with $\mathbf{b}$ periodic in x, y, period 2π) and followed numerically. Suppose that initially $\mathbf{b}$ is a constant and so $\mathbf{B}$ comprises a pure $\exp(ikz)$ wave. Then the evolution of the ensemble-averaged field is given by the response tensor $(2\pi)^3 \widetilde{\mathbf{G}}_\epsilon(k\mathbf{e}_z)$ with eigenvalues $\mu_\pm(\epsilon, k)$. A calculation similar to that in §11.4 and following on from equation (11.4.21) with $h = 1$ yields

$$(2\pi)^3 \widetilde{G}_{0ij}(k\mathbf{e}_z) = J_0(k\tau)\delta_{ij} + (i\tau/2)J_1(k\tau)\epsilon_{ij3}, \tag{11.5.5}$$

for $\epsilon = 0$. The corresponding growth rates with diffusion are

$$\gamma_\pm(\epsilon, k) = \tau^{-1}\log|J_0(k\tau) \mp (\tau/2)J_1(k\tau)| - O(\epsilon k^2) \tag{11.5.6}$$

for Beltrami-wave magnetic fields of parity ± 1.

In our simulations we use $\tau = \pi$, $h = 1$, $\epsilon = 10^{-3}$ and choose $k = 0.35$ to maximise the growth rate (11.5.5), with

$$\gamma_-(\epsilon, k) = \tau^{-1}\log|\mu_-(\epsilon, k)| \simeq 0.120, \qquad \mu_-(\epsilon, k) \simeq 1.46. \tag{11.5.7}$$

Let us start with an initial field $\mathbf{b}_0 = \frac{1}{2}(i, 1, 0)$, corresponding to $\mathbf{B}_0 = (-\sin kz, \cos kz, 0)$. In the ensemble average this grows by a factor of $\mu_-(\epsilon, k)$, which is real and positive, over each period I_n. In a simulation other modes will be excited; so we define a projection onto this wave by

$$P(t) = (2\pi)^{-2} \int (-ib_x(x, y, t) + b_y(x, y, t))\, dx\, dy. \tag{11.5.8}$$

For an ensemble average we then have

$$<P(n\tau)> = [\mu_-(\epsilon, k)]^n \qquad\qquad (11.5.9)$$

since $P(0) = 1$, but in a given realisation $P(n\tau)$ will generally become complex and may grow at a different rate.

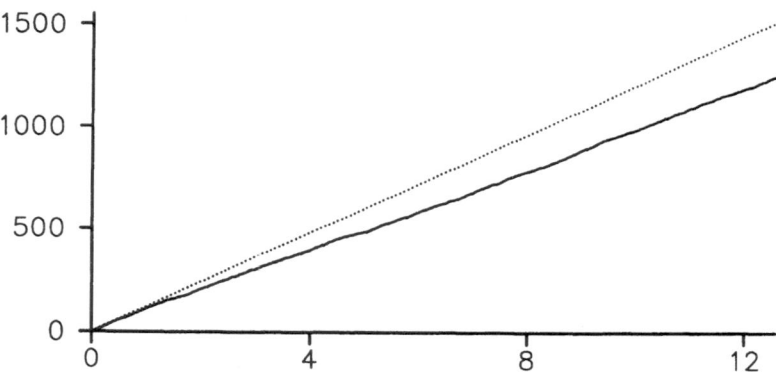

Fig. 11.6 Growth of field in a realisation from the ensemble $\mathcal{E}_{110}(\tau, 1)$. Shown are $\log|P(t)|$ (solid) and $\log <P(t)>$ (dotted), given by (11.5.7, 9), plotted against scaled time $t/1000$. Parameter values are $\tau = \pi$, $h = 1$, $\epsilon = 10^{-3}$ and $k = 0.35$.

In our simulation 4000 random waves were used, from $t = 0$ to $t = 4000\pi$. The code was tested by verifying (11.5.9) for small n and 200 realisations, giving μ_- in good agreement with (11.5.7). Figure 11.6 shows the growth in $|P(t)|$ (solid): there is clear exponential growth at a rate $\gamma_{\text{typ}} \simeq 0.099$, which is shared by other measures of the field strength, for example $E_M^{1/2}$ (not shown). However this growth is considerably slower than that for the ensemble average $<P(t)>$, which occurs at a rate $\gamma_-(\epsilon, k)$ in (11.5.7), shown by the dotted line. From the toy model it appears that variable amplification is acting to reduce the growth rate in a typical realisation below that of the ensemble average.

Nevertheless there is also phase mixing taking place. Figure 11.7 shows the behavior of $\arg P(t)$ as a function of time in our simulation; the argument is measured in degrees, and care is taken to avoid discontinuous behavior in the graph when the phase changes by 360°. We observe very irregular behavior; the phase tends to stay approximately constant for a period, then changes very rapidly, usually by about 180°. However from (11.5.9) the ensemble-averaged phase $\arg <P(t)>$ is zero for all t. Thus we have evidence of phase mixing, which in isolation would tend to increase the growth rate in a typical realisation above the theoretical value γ_-.

The conclusion is that the phenomena of variable amplification and phase mixing are both present in this simulation; these pull the growth rate in

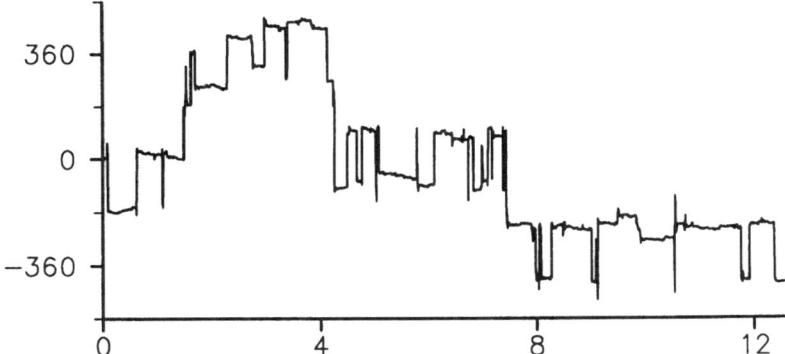

Fig. 11.7 Phase mixing in a realisation from the ensemble $\mathcal{E}_{110}(\tau, 1)$: $\arg P(t)$ is shown plotted in degrees against scaled time $t/1000$. Parameter values are as in Fig. 11.6.

different directions, but the result is a growth rate reduced below that of the ensemble-averaged field. These phenomena will occur generally, and lead to problems in interpreting a statistical theory for physical systems (Hoyng 1987a, b, 1988). Numerical simulations that we shall not report here reveal, however, that in the short correlation time limit, $\tau \to 0$ with $\tau^2 v = \text{const.}$, growth rates for typical realisations do agree well with those based on the ensemble-averaged field.

12. Dynamics

Kinematic dynamo theory is a successful theory in the sense that it is now established that dynamo action occurs commonly in three-dimensional flows of an electrically conducting fluid. This fact bodes well for MHD dynamo theories, as discussed in §1.2.4, but does not carry with it any information about the fields that are ultimately excited in a dynamical model: the structure of the magnetic field must be determined by the coupled system of field and flow, and in particular the feedback of the Lorentz force.

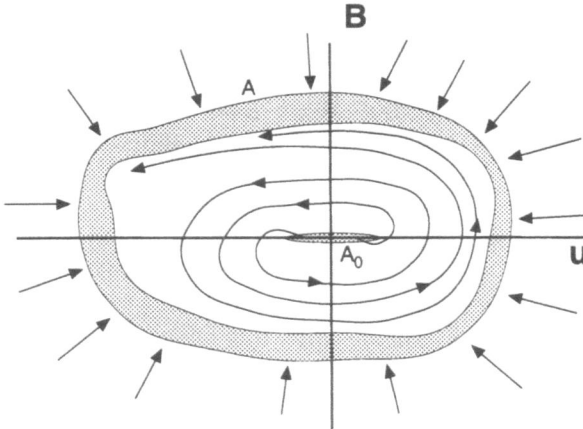

Fig. 12.1 Equilibration of a fast dynamo to an attractor in phase space $(\mathbf{u}, \mathbf{B})$. Kinematic fast dynamo theory studies the neighborhood of the subspace $\mathbf{B} = 0$, represented here by a single sub-attractor A_0. The MHD attractor A is bounded, in terms of total energy, when the power input to the driving is bounded.

The general problem of equilibration can be viewed as evolution in a phase space of $\mathbf{u}$ and $\mathbf{B}$ as indicated in Fig. 12.1. In the absence of magnetic field, i.e., in the subspace $\mathbf{B} = 0$, a given dynamical fluid model with prescribed driving might reach equilibrium at any one of a number of possible flows, steady, periodic, quasiperiodic or chaotic, and these could be fluid dynamically stable or unstable. If every one of the *stable* equilibrated flows is a kinematic fast dynamo then we refer to the coupled system of fluid and magnetic field as an *MHD fast dynamo*.[81] In such a flow a seed field will rapidly adopt a growing kinematic eigenfunction possessing weak large-scale field amidst a sea of strong small-scale fluctuations. Because the Lorentz stresses are quadratic in the magnetic field, these are determined largely by the contribution from the small-scale component. Thus, whereas in the kinematic fast dynamo problem we have sought to neglect the fluctuations, in the dynamical problem they must be reinstated; they will modify the flow significantly and, since the small-scale magnetic field is formed from the stretching, twisting, and folding of the large-scale field, the dynamics of large- and small-scale magnetic fields are inexorably intertwined.

This is a crucial feature of the MHD fast dynamo and is the main focus of the present chapter. The predominance of small-scale structure requires approaches different from the simplest models of dynamical equilibration of slow dynamos, which rely on a small magnetic Reynolds number or a dominant large-scale magnetic field; see, for example, Moffatt (1978) and Krause & Rädler (1980). Numerical simulations are also difficult, because of the need to resolve two vector fields in three-dimensional geometry. Finally, because of the weak dissipative effects, an MHD dynamo usually involves sustaining a magnetic field within a *turbulent* MHD system. In this case dynamo theory becomes part of the larger problem of understanding MHD turbulence. The limitations of the existing theories of turbulent flow, driven far from equilibrium and excited on a great range of scales, do not need to be elaborated here. Suffice it to say that the question of the dynamical equilibration of an MHD fast dynamo is a problem quite distinct from the excitation process that is the focus of the kinematic theory.

The dynamics can be affected by diffusive processes, however large R may be, in particular by reconnection and the associated changes in topology of the magnetic field, even though these occur only on small, diffusive scales. To take a simple example, consider two distinct flux tori lying in a viscous, perfectly conducting, unforced fluid (Fig. 12.2a). Starting with the fluid at rest the tori, frozen in the fluid, will relax, that is, will deform continuously while decreasing magnetic energy, into pencil-shaped toroidal objects. The length of each magnetic field line tends to zero, as does the magnetic energy (see Arnold 1974, who acknowledges Ya.B. Zeldovich). On the other hand if the two tori are linked, the relaxation is towards a tight bundle of linked flux tubes; see Fig. 12.2(b). Topological constraints prevent a field line from shrinking to zero length and the magnetic energy is bounded above zero. Consequently measures of linking of flux tubes, such as the helicity density

---

[81]We emphasize that the distinction between slow and fast kinematic growth does not imply any corresponding classification of saturated states. What we have given here is therefore one of several possible definitions of an MHD fast dynamo. For example, it might be required that the system be periodic on the advective time-scale for arbitrarily large R. Which definition will turn out to be most useful and appropriate remains open.

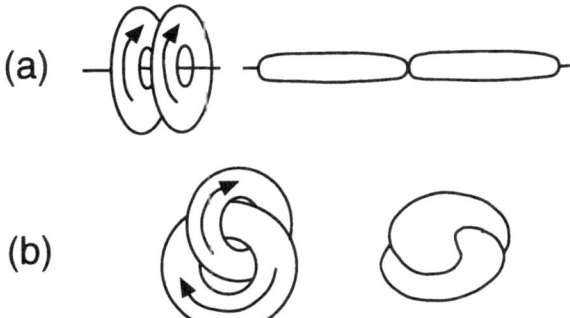

(a)

(b)

Fig. 12.2 The role of topology on relaxation to minimum magnetic energy in a perfect conductor. (*a*) Unlinked tori relax toward their common axis and form thin, long toroidal structures. (*b*) Linked tori relax toward a compact bundle. (After Moffatt 1985.)

and total helicity, have dynamical implications; see §12.4 below (Arnold 1974, Moffatt 1985, 1986, Freedman 1988).

We begin this chapter by formulating the dynamical problem in §12.1, and discussing the evolution of energy and momentum. In §12.2 we consider kinematic eigenfunctions: their structure and distribution of energy give hints as to how saturation will commence, as well as a reference point from which to measure the change of structure of the equilibrated magnetic field. We discuss numerical studies in §12.3, and the modelling of turbulent MHD cascades in §12.4. Finally in §§12.5, 6 we discuss implications for astrophysical systems.

12.1 Formulation

12.1.1 Momentum and Energy

We first summarize the equations of MHD for a geometry of interest in some dynamo models, and one which is in any case sufficient for the discussion of this chapter. Derivation of these equations may be found in Moffatt (1978), Priest (1984), and Biskamp (1993). Consider a periodicity cube $\mathcal{D}$ of side L and introduce a uniform gravitational field $\mathbf{g}$. We allow for the compressibility of the fluid, and write the momentum equation in the form

$$\rho \frac{d\mathbf{u}}{dt} = \mathbf{J} \times \mathbf{B} + \rho\mathbf{g} - \nabla p + \nabla \cdot \boldsymbol{\tau} + \mathbf{F}, \tag{12.1.1}$$

where $\boldsymbol{\tau}$ is the viscous stress tensor and $\mathbf{F}$ is a body force. The energy equation takes the form

$$\rho T \frac{ds}{dt} = \rho \frac{dh}{dt} - \frac{dp}{dt} = \nabla \cdot (K \nabla T) + Q_v + Q_J + Q, \qquad (12.1.2)$$

where s is the specific entropy, and h is the specific enthalpy, a function of temperature T in an ideal gas, and equal to $c_p T$ for a polytropic gas with constant specific heat c_p. Here Q_v, Q_J and Q are entropy sources, derived from viscous dissipation, Joule heating by electric currents, and internal sources or radiative emission; K is a thermal conductivity. The equation of mass conservation (1.2.8), Ampère's law (1.2.1a), the induction equation (1.2.2), and an equation of state complete a system of nine equations for $\mathbf{u}$, $\mathbf{B}$, T, ρ and p. In such a model dynamo action can be sustained by supplying a body force $\mathbf{F}$, that is, a stirring force for the fluid, or by providing a suitable heat source Q, or by invoking both mechanisms simultaneously. Braginsky (1963) has suggested that the Earth's magnetic field is sustained by convection that is non-thermal in origin, arising from buoyancy of light material released during solidification at the boundary of the solid inner core. The idea has been elaborated in some detail; see in particular Loper & Roberts (1983) and Braginsky & Roberts (1995). In galaxies other processes, such as random explosions of supernovae, may be relevant. We may nevertheless adopt thermal convection as a typical form of driving which can supply energy to the magnetic field.

The first step in modelling the magnetic fields of planets and stars, for example generation in the convection zone of the Sun, is to consider the equilibrium stratification of the fluid. To simplify the discussion we shall deal here only with a layer model in which p, ρ and T depend upon a single Cartesian coordinate z, and the velocity and magnetic fields (together with the corresponding entropy sources) vanish. Assuming $\mathbf{g} = (0, 0, -g)$ and an ideal gas law $p = \rho RT$, and setting $\mathbf{F} = 0$, we have the barotropic relation $dp/dz = -\rho g$ and $-d(K dT/dz)/dz = Q$, from which equilibrium distributions $\rho_e(z)$, $p_e(z)$ and $T_e(z)$ can be found, given suitable boundary conditions.

Two characteristics of such an equilibrium are significant in dynamo modelling. First, if it is unstable there is the possibility of convective motions suitable for dynamo action, and no other driving of the system need be invoked. Consider a small parcel of fluid in pressure equilibrium with its surroundings and having density $\rho = \rho_e(z)$. Convective instability occurs if, when moved upward a distance Δz, the parcel acquires adiabatically a new density $\rho + \delta \rho_{ad}$ smaller than the new ambient density $\rho + \delta z (d\rho_e/dz)$, that is when

$$\left(\frac{d\rho}{dz} \right)_{ad} < \frac{d\rho_e}{dz}. \qquad (12.1.3)$$

Using the barotropic relation this condition reads, for a polytropic gas,

$$\frac{dp_e}{dz} \frac{\rho_e}{\gamma p_e} = -\frac{\rho_e^2 g}{\gamma p_e} < \frac{d\rho_e}{dz}, \qquad (12.1.4a)$$

where $\gamma = c_p/c_v$ is the ratio of specific heats. The condition may equally be written in terms of the specific entropy,

$$\frac{ds_e}{dz} = \frac{c_p}{T_e}\frac{dT_e}{dz} - \frac{1}{\rho T_e}\frac{dp_e}{dz} = \frac{c_p}{T_e}\frac{dT_e}{dz} + \frac{g}{T_e} < 0. \qquad (12.1.4b)$$

In the solar convection zone condition (12.1.4b), known as the *Schwarzschild condition*, appears to be fulfilled by a factor of at least 10. The resulting violent instability is the main reason for the complex pattern of magnetic activity observed there.

Secondly, it is necessary to assess the role of compressibility in the stratified equilibrium. Clearly it is essential to include the buoyancy force when this is a primary driving mechanism of the motions. However the variable inertia of the fluid is not necessarily as important. It is therefore sometimes useful to adopt the *Boussinesq approximation*, which neglects departures of density from a constant except in the calculation of buoyancy forces. For this approximation to be valid, the characteristic velocities of the system must be small compared to the equilibrium speed of sound c_e. The Boussinesq approximation requires that the scale L of the motions satisfies $L \ll L_p$ where $L_p = (-d\log p_e/dz)^{-1}$ is the pressure scale height. From the barotropic equilibrium we obtain $gL \ll c_e^2$, a condition that the speed of gravity waves be very subsonic. This restricts the model to sufficiently thin layers.

A second assumption in the Boussinesq approximation, called in isolation the *anelastic approximation*, similarly eliminates sound waves by requiring that $|\mathbf{u}|$ be small compared to c_e. Taken together, these approximations reduce the equation for conservation of mass to $\nabla \cdot \mathbf{u} = 0$, and allow ρ to be taken as constant except in the term $\rho\mathbf{g}$ in (12.1.1). This term can be rewritten as $\rho'\mathbf{g}$, where ρ' is the fluctuating density, after absorbing a constant into the pressure gradient.

For the Earth the Boussinesq theory provides a good first approximation, and more detailed models retain some of its features (Braginsky & Roberts 1995). The Boussinesq approximation is a poor one in the solar convection zone, which has a depth of about 10 scale heights, and the current generation of numerical codes generally allow a compressible gas and place no condition on layer depth.

By itself the anelastic approximation provides a model allowing compressibility but filtering out sound waves. The theoretical basis has been discussed by Gough (1969), and Latour *et al.* (1976). Brandenburg (1988) finds that layer depth should not exceed 10 pressure scale heights. It is thus of marginal value in the solar convection zone, which has a depth of about 10 pressure scale heights. When the anelastic approximation is applied in this way, it reduces the equation for conservation of mass to the approximate form $\nabla \cdot (\rho\mathbf{u}) = 0$. The equation for the pressure is then obtained by taking the divergence of (12.1.1), and does not involve a time derivative of $\mathbf{u}$.

When the Boussinesq approximation is used in an ordinary liquid, the energy equation can expressed as an equation for the temperature,

$$\frac{dT}{dt} - \frac{1}{\rho c_p}\frac{dp}{dt} - \nabla \cdot (\kappa\nabla T) = \text{heat sources}, \qquad (\kappa = K/\rho c_p). \qquad (12.1.5)$$

Given the Boussinesq form of (12.1.1), the system must then be closed by expressing the density fluctuation ρ' in terms of the remaining variables. In general we would have

$$\rho' \approx \left(\frac{\partial \rho}{\partial T}\right)_0 (T - T_0) + \left(\frac{\partial \rho}{\partial p}\right)_0 (p - p_0), \qquad (12.1.6)$$

where the subscript zero now refers to the local value of the equilibrium. In non-magnetic systems the second term in (12.1.6) is usually neglected, but as we indicate below it is the source in MHD systems of *magnetic buoyancy* (Parker 1979, Spiegel & Weiss 1980). It is in (12.1.6) that other sources of buoyancy, for example from phase changes in the material, can be introduced.

Large-scale rotation is an important feature in stellar and planetary cores, and is essential for achieving dynamo action in simple convective flows. The system can be recast in a coordinate frame rotating with constant angular speed Ω about the z-axis. With the velocity $\mathbf{u}$ understood to be relative to the rotating frame, the Coriolis force $2\Omega\mathbf{e}_z \times \mathbf{u}$ must be added to the left-hand side of (12.1 1), and the pressure includes a centrifugal correction. The Boussinesq approximation neglects the centrifugal torques created by the (slight) z-dependence of the centrifugal forces exerted on the equilibrium stratification. In general centrifugal torques will create meridional motions in rotating fluid bodies whenever non-uniform distributions of temperature or entropy are present.

12.1.2 The Lorentz Force

The Lorentz force term in (12.1.1) has the form

$$\frac{1}{\mu}(\nabla \times \mathbf{B}) \times \mathbf{B} = \frac{1}{\mu}\left(\mathbf{B} \cdot \nabla\mathbf{B} - \nabla\frac{B^2}{2}\right), \qquad (12.1.7)$$

expressed as the divergence of a magnetic stress tensor $\mu^{-1}(B_i B_j - \delta_{ij}B^2/2)$. The isotropic part defines the *magnetic pressure* $B^2/2\mu$. The remaining term can be understood as a tension of strength B^2 in the lines of force, whose divergence yields components parallel and perpendicular to the field. The tangential components from pressure and tension cancel, as is obvious from the original form of the Lorentz force. The combination of pressure plus tension is useful for assessing the forces developed in certain magnetic structures. For example, the tension causes hoop stresses in the configurations of Fig. 12.2, which account for the relaxation indicated there. Also, eddies of vortical motion will be affected by the field lines that thread them, as the tension in the lines resists the twisting and stretching of advection.

In a kinematic fast dynamo we have seen that magnetic fields are excited over a range of scales, and we need to quantify their contributions to the Lorentz force. It is first useful to define the spectrum, $E_M(k)$, of magnetic energy (per unit volume), such that $E_M(k)\,dk$ is the energy in the wavenumber shell bounded by k and $k + dk$,[82]

$$E_M = \frac{1}{2\mu} <\mathbf{B}^2> = \int_0^\infty E_M(k)\,dk; \qquad (12.1.8)$$

see, e.g., Batchelor (1953) and Tennekes & Lumley (1989).

In a fast dynamo eigenfunction this integral is typically dominated by the small-scale fields with $Lk = O(\epsilon^{-1/2})$. Quantities such as the Lorentz force, helicity and dissipation are also given by integrals over the range of wave-numbers $O(1) \leq kL \leq O(\epsilon^{-1/2})$. As a convenient shorthand we speak of the field as a sum of large- and small-scale components $\mathbf{B} = \mathbf{B}_{LS} + \mathbf{B}_{SS}$, even though there is no corresponding 'spectral gap' in Fourier space. To say that the energy, or any other integral measure of the field, is dominated by B_{SS} means that the integral is dominated by fields at the smallest scales, $Lk = O(\epsilon^{-1/2})$, and similarly for B_{LS}. In practice 'large scale' will encompass modes out to some moderately large wavenumber, a cut-off which is fixed in the perfectly conducting limit. In a kinematic eigenfunction, $B_{SS} \gg B_{LS}$; the Lorentz force term in (12.1.7) is then given approximately by $\mu^{-1}(\nabla \times \mathbf{B}_{SS}) \times \mathbf{B}_{SS}$ and will in general have a non-trivial average over the large scales. The importance of estimating this feedback has been emphasized in recent years by S.I. Vainshtein and his colleagues, a discussion we take up in §12.2.

A secondary effect of B_{SS} comes from the magnetic buoyancy generated by the pressure field that balances, at least in part, the Lorentz stresses. The magnetic pressure as defined above is immediately identifiable as a significant contributor to (12.1.6), through the term

$$\rho_M' = -\left(\frac{\partial \rho}{\partial p}\right)_0 (B_{SS}^2/2\mu). \qquad (12.1.9)$$

Other contributions to the density perturbation, which in general cannot be neglected relative to (12.1.9), come from the magnetic tension.

If the flow field is sustained by forcing on large scales only, then its small scales will be turbulent, provided that the viscous stresses are sufficiently small. If ν is a characteristic kinematic viscosity appearing in the viscous stress tensor, we can define the (fluid) Reynolds number $Re = UL/\nu$ and the magnetic Prandtl number $Pr_M = \nu/\eta = R/Re$. The value of Pr_M will determine the effects of dissipation on the small scales. In a medium having $Pr_M \gg 1$, the small scales of vorticity will be suppressed more effectively than the small scales of the magnetic field, and the latter will be developed through stretching, twisting, and folding by the vortical eddies over a range of scales (depending on Re) up to L. The large Pr_M case, which applies to MHD models of the solar corona, interstellar gas and galaxy, thus inherits some aspects of

---

[82]When k is discrete, say for periodic boundary conditions, we may nevertheless approximate the spectrum by a continuous function of k.

kinematic theory, since the fluid motions are determined dynamically but the resulting small-scale magnetic structure is not. We describe a model based upon these assumptions in §12.3.2.

In the solar convection zone, however, Pr_M is small compared to unity, estimated values being in the range 10^{-2}–10^{-6}. The same situation probably prevails in the core of the Earth, although there is considerable uncertainty about the value of the viscosity there (Poirier 1988). When $Pr_M \ll 1$ the fluid is essentially inviscid on all scales of the magnetic field, the resistive dissipation occurring on scales large compared to the smallest velocity eddies. The astrophysical applications of fast dynamo theory must recognize these two very different limiting cases. The question of equilibration of fast dynamos is essentially an MHD theory at large R, parameterized by the parameter Pr_M.

12.2 The Post-kinematic Approach

We use this term for the study of dynamical evolution based primarily upon results of kinematic theory. We may imagine an existing flow sustained by a fixed energy input, which at $t = 0$ is seeded with a weak magnetic field having negligible dynamical effect. Initially kinematic theory will apply, and the field takes the form of a kinematic eigenfunction. At some point the Lorentz force will become significant and saturation will commence. A key question, which depends only on the structure of the kinematic eigenfunction, is: what is the magnitude of the large-scale field B_{LS} at the onset of saturation, and how does this compare with the fields observed in astrophysical objects (Moffatt 1961, Vainshtein & Cattaneo 1992, Boozer 1992, Brandenburg et al. 1992, Vainshtein, Parker & Rosner 1993a, Vainshtein et al. 1993b, Ott & Vainshtein 1995)?

Let a length L and velocity U be associated with a fast dynamo $\mathbf{u}(\mathbf{x}, t)$ and consider evolution from a seed magnetic field with $\epsilon \ll 1$. Let us assume that nonlinear effects become important at 'equipartition', that is, when the total kinetic energy is comparable with the total magnetic energy.[83] This suggests that the onset of saturation will occur when $B_{LS}^2 \sim B_{eq}^2 \equiv \mu\rho U^2$; this condition is built into many models which average over the small-scale fields and invoke an alpha effect.

---

[83]This seems appropriate when the flow and field are of the same scale and $Pr_M = O(1)$ (cf. §12.2.3). In situations where there is scale separation between a large-scale field and small-scale flow, super-equipartition magnetic fields can be maintained; this is seen in detailed models and in closure calculations (Pouquet, Frisch & Léorat 1976). For example, in some cases a large-scale field $\mathbf{B}$ will obey an equation of the form $\partial_t \mathbf{B} = \nabla \times [\alpha(|\mathbf{B}|)\mathbf{B}] + \eta\nabla^2\mathbf{B}$, with an alpha effect $\alpha(|\mathbf{B}|)$ modified by the large-scale field and suppressed for large $\mathbf{B}$ (e.g., Moffatt 1978, §10.4). A steady field in the form of a helical wave of scale L, magnitude B will obey $\alpha(B) \sim \eta/L$. The field can be made as strong as desired simply by increasing its length scale, L.

However at large R this procedure is incorrect since E_M is dominated by B_{SS}^2/μ rather than by B_{LS}^2/μ, so that at onset of saturation $E_M \sim B_{SS}^2/\mu \sim E_V \sim \rho U^2$. This could occur when B_{LS} is relatively weak, and to quantify the size of B_{LS}^2 relative to E_M, Vainshtein & Cattaneo (1992) define a scaling exponent ϱ_E (n in their notation) that expresses how the ratio of mean energy to large-scale energy scales with ϵ in a fast dynamo eigenfunction:

$$2E_M \equiv \mu^{-1}<B^2> \sim \mu^{-1}B_{LS}^2 \, \epsilon^{-\varrho_E}. \tag{12.2.1}$$

Here B_{LS} is some measure of the large-scale field, the modulus of any large-scale flux or linear projection (independent of ϵ), for example, the amplitude of the gravest non-zero Fourier mode in a system. The average $<\cdot>$ is a spatial average, and '$\sim$' denotes approximate proportionality. For a fast dynamo we expect that $\varrho_E > 0$.[84]

The equipartition assumption $E_V \sim E_M$ then means that at onset

$$B_{LS} \sim B_{eq}\,\epsilon^{\varrho_E/2}, \qquad (B_{eq} = \sqrt{\mu\rho}\,U). \tag{12.2.2}$$

Since $\varrho_E > 0$ generally, and ϵ can be as small as 10^{-20} in astrophysical objects, the onset of saturation can occur when the large-scale field is weak, far weaker than one might expect based on equipartition estimates involving only B_{LS}. Just how weak we shall investigate by estimating ϱ_E in various models.

Let us for the moment suppose that ϱ_E is not particularly small, and so in astrophysical contexts, the large-scale fields are relatively weak when saturation commences. This has an impact on observations: measurements of the large-scale fields and flow velocities in astrophysical objects might suggest that kinematic dynamo theory is valid, whereas in fact the small-scale fields are dynamically significant. It has been argued further that observations of the Sun and galaxy are incompatible with (12.2.2), the large-scale fields being too strong relative to the observed kinetic energy, and so the question of whether dynamo action is the origin of these fields needs to be reexamined (Vainshtein & Cattaneo 1992, Kulsrud & Anderson 1992, 1993).

On this point, regardless of the magnitude of ϱ_E and astronomical observations, we note that we must be careful to distinguish the structure and magnitude of the magnetic field (a) at the onset of saturation, when kinematic theory ceases to be valid, and (b) when the field finally equilibrates, probably yielding complex spatio-temporal behavior as the field explores the MHD attractor mentioned above. Between onset and equilibration the field may change its structure, length-scale and magnitude considerably, albeit perhaps over quite long time-scales. This is particularly likely in systems such as the MHD fast dynamo, which are highly supercritical and allow a great range of scales to be excited. This change of structure during equilibration can be

---

[84]Note that we have in mind a non-diffusive fast dynamo; for a diffusive fast dynamo (such as the Ponomarenko dynamo of §5.4.1) essentially all the field is at small scales, and either ϱ_E will be large, or the power law (12.2.1) will turn into exponential growth with $\epsilon \to 0$, depending on the means of measuring B_{LS}.

observed in systems discussed below and is seen even in very simple dynamo models (see, e.g., Gilbert & Sulem 1990). Examples of nonlinear modification of unstable modes toward large scales are known in classical convection theory. An example is Bénard convection under the condition of fixed flux; see Chapman & Proctor (1980) and Proctor (1981).

A post-kinematic approach allows one to explore onset (a), but there are clear difficulties in applying criteria such as (12.2.2) to the final equilibrium (b); in fact one can argue that the structure of the field and exponents such as ϱ_E are likely to change during equilibration, because of the modified transport properties of the flow (see §12.2.4 below). It is correct that an equilibrium field violating (a) falls outside the scope of kinematic theory, but such fields may well be possible by nonlinear evolution, even if not readily analyzed or computed. To argue solely from estimates based on kinematic eigenfunctions that the magnetic field in an astrophysical object cannot be generated by dynamo action is to apply kinematic theory beyond its domain of validity. Moreover we shall find that quite small values of ϱ_E are possible in fast dynamos and so the inequality (12.2.2) is not as severe as one might first expect. Further discussion is given in Brandenburg et al. (1992).

A complementary estimate of the effect of small-scale fields is introduced by Boozer (1992), who considers the rate of dissipation of magnetic energy,

$$\Omega_M \equiv \mu\eta <\mathbf{J}^2> \sim \mu^{-1}(U/L)B_{LS}^2\, \epsilon^{-\varrho_\Omega}, \qquad (\nabla \times \mathbf{B} = \mu\mathbf{J}). \qquad (12.2.3)$$

The exponent ϱ_Ω measures how the power required to support dynamo action for a given large-scale field B_{LS} increases as $\epsilon \to 0$; when this becomes comparable with the kinetic energy of the fluid (divided by the advective time), saturation will commence. This criterion is similar to (12.2.2), since during kinematic growth the rate of working of the fluid against the Lorentz force increases the magnetic energy at a rate $dE_M/dt \sim (U/L)E_M$, and a significant fraction of the energy goes into Joule dissipation Ω_M. Thus $\Omega_M \sim (U/L)E_M$ and for compatibility between (12.2.1, 3) we require $\varrho_E = \varrho_\Omega$. With $\varrho_\Omega > 0$ in general, Boozer (1992) points out that, paradoxically, the smaller the resistivity, the more power is required to sustain a given large-scale field, at least kinematically!

The two exponents ϱ_E and ϱ_Ω may be related to the magnetic energy spectrum. Assume this has a scaling behavior over the range of magnetic fluctuations $1 \ll Lk \ll \epsilon^{-1/2}$ with a diffusive cut-off for $Lk \gg \epsilon^{-1/2}$:

$$E_M(k) \sim \mu^{-1}LB_{LS}^2\, F(Lk\epsilon^{1/2})\,(Lk)^{-q}, \qquad (1 \ll Lk), \qquad (12.2.4)$$

where $F(0) = 1$ and $F(s) \to 0$ rapidly as s increases. Since

$$<\mathbf{J}^2> = 2\mu^{-1}\int_0^\infty k^2 E_M(k)\, dk, \qquad (12.2.5)$$

it may be checked that ϱ_E and ϱ_Ω are related to the spectral slope by

$$\varrho_E = \varrho_\Omega = (1 - q)/2, \qquad (q < 1). \tag{12.2.6}$$

In fact any shape of $E_M(k)$ for which the total energy is dominated by the diffusive scales will give equality between ϱ_E and ϱ_Ω.

Our immediate task is to obtain values for ϱ_E and ϱ_Ω from known kinematic dynamos, valid at least in the early phases of saturation. Clearly because of model dependence this is a dangerous game, if a fairly accessible one. We must therefore draw on supporting dynamical calculations whenever possible.

12.2.1 Planar Flow

We review the arguments advanced by Vainshtein & Cattaneo (1992), based upon scaling ideas from kinematic dynamo theory. Consider the kinematic build-up of small-scale magnetic structure in a chaotic planar flow. Let us impose periodic boundary conditions and a magnetic field $\mathbf{B} = \overline{\mathbf{B}} + \mathbf{B}'$, where $\overline{\mathbf{B}}$ is a constant mean field and $<\!\mathbf{B}'\!> = 0$. The velocity field (u_x, u_y) is taken to be divergence-free, and a vector potential $A(x, y)$ for the fluctuating field $\mathbf{B}'$ obeys the advection–diffusion equation,

$$\partial_t A + \mathbf{u} \cdot \nabla A = \epsilon \nabla^2 A + u_x \overline{B}_y - u_y \overline{B}_x, \tag{12.2.7}$$

with a source term from the mean field.

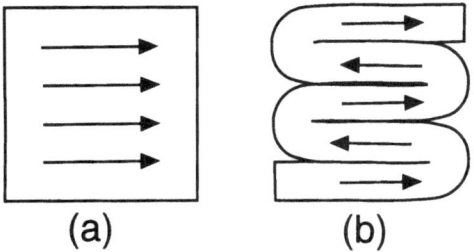

Fig. 12.3 Schematic picture of folding in planar flow: (a) initial mean field, and (b) folded field.

Starting with $A = 0$, the field A is generated from $\overline{\mathbf{B}}$ and evolves in a chaotic flow so that its scale is reduced to $O(L\epsilon^{1/2})$. Its gradient, which is related to $\mathbf{B}'$ by a rotation, is correspondingly increased to $B_{SS} = O(B_{LS}\epsilon^{-1/2})$, where $B_{LS} \equiv |\overline{\mathbf{B}}|$ (Fig. 12.3). Assuming these scales and field strengths are achieved over the whole volume, we obtain

$$\varrho_E = 1, \qquad \text{(2 dimensions, space-filling)} \tag{12.2.8}$$

(Zeldovich 1957) The criterion (12.2.2) then gives a limit on the magnitude of the imposed $\overline{\mathbf{B}}$ for kinematic theory to be valid. Dynamical simulations for planar flow are discussed below in §12.3.3.

This argument involves no dynamics, is independent of the existence of fast dynamo action, and is well-supported by simulations of planar flows. However planar flows are not dynamos, and constructive folding, i.e., the enhancement of B_{LS} relative to B_{SS}. is excluded. We note that the estimate (12.2.8) was earlier derived by Moffatt (1961) for MHD turbulence under the condition that $Pr_M \ll 1$.

12.2.2 Three Dimensions

In three dimensions it is possible to rearrange stretched and folded field in a constructive way, so as to amplify large-scale fluxes and modify the relative orientations of small-scale fields. Starting with a tube of flux, length-scale L and strength B_{LS}, we could draw out a length, stretch it and fold it back and forth leading to space-filling flux tubes with diameters of order $L\epsilon^{1/2}$, and increased field strength of order $B_{LS}\epsilon^{-1}$. Folded inefficiently, that is without changing B_{LS} (cf. Fig. 12.3), this leads to

$$\varrho_E = 2, \qquad \text{(3 dimensions, space-filling tubes, poor folding)} \qquad (12.2.9)$$

(Ott & Vainshtein 1995). Note that the nature of the folding is important. If instead the initial tube of flux were operated on with the idealized STF moves, we would obtain a bundle of flux tubes of the same orientation. The large- and small-scale fields are increased by a factor of 2^n, $B_{LS} \sim B_{SS}$ and

$$\varrho_E \simeq 0, \qquad \text{(3 dimensions, STF)} \qquad (12.2.10)$$

(Vainshtein & Cattaneo 1992). Here the transverse fields generated in the STF picture are bounded in magnitude by the fields in the tube. In practice of course the idealized STF picture is not realized and ϱ_E is small but positive.

Thus ϱ_E measures the efficiency of folding in a chaotic system, as does the cancellation exponent κ of Chap. 7, or $h_{top} - h_{tw}$ in Chap. 6. The exponent ϱ_E is reduced by the increase in B_{LS} relative to B_{SS}. Since we expect folding to be less efficient than STF in natural systems, it appears that ϱ_E could be between zero and two. The larger ϱ_E is, the smaller is the large-scale field, given by (12.2.2), that is compatible with kinematic dynamo theory: stronger fields can only arise through nonlinear processes.

While this discussion gives ranges of possible values of ϱ_E, the values typical of realistic flows must be found by numerical simulations. Flows (or maps) having a single length-scale L and velocity-scale U (derived from the forcing on the system) may be justified by a limit of large Pr_M. The first systems to check are Lagrangian maps. In the SFS map, for example, the intensity and the wavenumber of the small-scale field grow with iteration n at

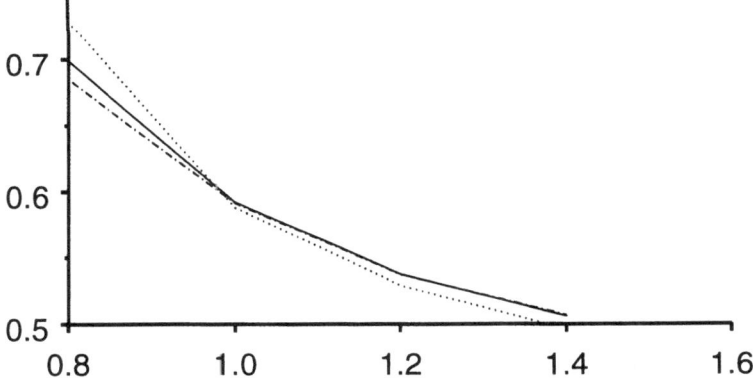

Fig. 12.4 The exponents ϱ_E (solid curve) and ϱ_Ω (dotted curve), versus the shear parameter α for the SFS map. The dot–dash curve is $1 - \Gamma/\log 2$.

the rate $e^{n\log 2}$, while B_{LS} grows at the rate $e^{n\Gamma}$. Setting the former equal to $\epsilon^{-1/2}$ at diffusive saturation, we compute

$$\varrho_E = 1 - \Gamma/\log 2. \tag{12.2.11}$$

We show in Fig. 12.4 a comparison of (12.2.11) with calculations of the diffusive SFS map. The reasonable agreement indicates that ϱ_E can be reduced to the range 0–1 in fast dynamo models at large Pr_M.

Other accessible models are the two-dimensional unsteady flows studied in Chap. 2. Kinematic simulations for given z-wavenumbers k and varying ϵ indicate clear scaling and notably small values of ϱ_E (Fig. 12.5), equal to 0.33 for CP ($k = 0.57$) and 0.34 for LP ($k = 0.62$). For MW+, $\varrho_E = 0.24$ for $k = 0.8$ and 0.27 for $k = 1.0$ (Gilbert, Otani & Childress 1993). These values of ϱ_E are significantly smaller than (12.2.8, 9), and quite close to the ideal value of zero for STF, reflecting the efficiency of these dynamos. It also indicates that the limit (12.2.2) for the validity of kinematic dynamo theory may not be as severe as one might initially think, at least at large Pr_M. Although when more Fourier modes are included in the flow (corresponding to a reduced Pr_M) ϱ_E increases (Cattaneo *et al.* 1995), values obtained so far from fast dynamo simulations have been well in the range $0 < \varrho_E < 1$.

These values of ϱ_E can be compared with the fall-off of the energy spectrum (12.2.4, 6). For the SFS map the data shown in Fig. 12.4 indicate q to be near zero but either positive or negative depending upon α. A computation with $\alpha = 1$ and $\epsilon = 10^{-6}$ for the dissipation spectrum is noisy but yields q in the range -0.1 to -0.2, consistent with Fig. 12.4. Figure 12.6 shows a log–log plot of $E_M(k)$ for MW+ with various values of ϵ. The spectrum has

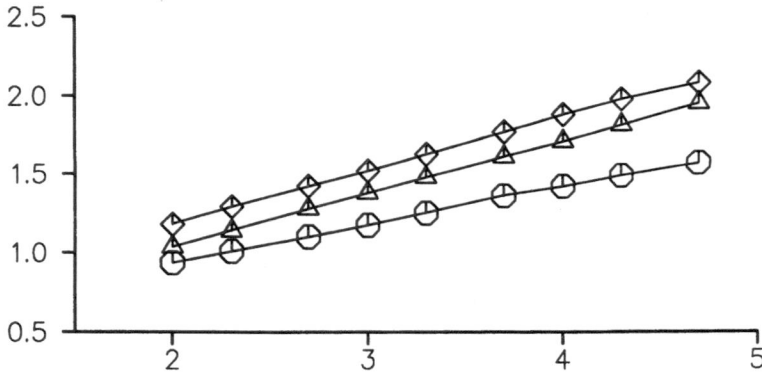

Fig. 12.5 Measurements of $g(\epsilon) = \log_{10}(<\mathbf{B}^2>/B_{LS}^2)$ plotted against $\log_{10} \epsilon^{-1}$. Shown are $g(\epsilon)$ (circles) for MW+ with $k = 0.8$, $g(\epsilon) + 0.2$ (triangles) for CP with $k = 0.57$, and $g(\epsilon) + 0.4$ (diamonds) for LP with $k = 0.62$.

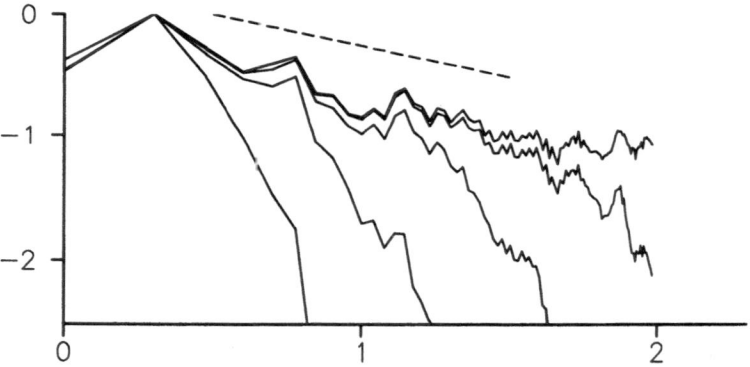

Fig. 12.6 $\log_{10} E_M(k)$ plotted against $\log_{10} k$ for MW+ with vertical wave-number 0.8 and $\epsilon = 10^{-1}$, 10^{-2}, 10^{-3}, 10^{-4} and 10^{-5}, reading curves from left to right. The dashed line has slope $-q = -0.52$.

a very rough power law decay, which is consistent with the slope $-q = -0.52$ predicted from (12.2.6) with $\varrho_E = 0.24$, shown here as a dashed line.

12.2.3 Effects of Intermittency

The exponent ϱ_E can be modified by intermittency, that is, the concentration of magnetic field in small regions of space, a problem considered by Ott & Vainshtein (1995). In planar flow, for example, flux expulsion can concentrate field into isolated sheets of strength $O(B_{LS}\epsilon^{-1/2})$ and width $O(L\epsilon^{1/2})$, yielding $\varrho_E = 1/2$. Similarly in three dimensions concentration of field into a flux rope of radius $O(L\epsilon^{1/2})$ lowers the exponent from $\varrho_E = 2$ to $\varrho_E = 1$. In either case the exponent is reduced to half the 'maximum' value by expulsion of flux and the resulting intermittency.

If intermittency is present one can define a different exponent relating the maximum value B_{max} of the field to B_{LS}, with

$$B_{max}^2 \sim B_{LS}^2 \, \epsilon^{-\varrho_{max}}. \tag{12.2.12}$$

This is the exponent q of Ott & Vainshtein (1995). Here $\varrho_{max} \geq \varrho_E$ with equality if there is no intermittency, that is, $B_{SS} \sim B_{max}$ over a sizable fraction of the volume. This then suggests a limit on the large-scale field in a kinematic dynamo based on equipartition between B_{max}^2 and B_{eq}^2, which would be more stringent than (12.2.2) for intermittent fields; this is proposed by Vainshtein et al. (1993b) and Ott & Vainshtein (1995) for $Pr_M = O(1)$.

The restriction on Pr_M arises because the idea of equipartition of magnetic and kinetic energies becomes less tenable in a viscous fluid when the field is intermittent. In fact peak fields can exceed equipartition strength, as viscosity can effectively smooth out the force distribution from localized field maxima. This is seen in the dynamical study of Galloway, Proctor & Weiss (1978) of an imposed magnetic field B_{LS} in axisymmetric convection (see also Jones & Galloway 1993a, b). The field is concentrated into an axial flux rope. For weak external B_{LS} the flow is unaffected by the field, the flux rope has width $O(L\epsilon^{1/2})$ and strength $B_{max} \sim B_{LS}/\epsilon$. If B_{LS} is increased the flow structure and kinetic energy become affected significantly when

$$(B_{max}/B_{eq})^2 \sim Pr_M/\log R = R/Re \log R, \tag{12.2.13}$$

as shown by Galloway et al. (1978) in an analysis valid for $R \gg 1$ and $Re \ll 1$.

Thus as $R = 1/\epsilon \to \infty$ peak fields much larger than equipartition can exist without dynamical consequences, at least at small Reynolds number (Peckover & Weiss 1978, Galloway & Moore 1979). For this configuration estimates of the dynamical influence of field based on B_{max} would be misleading for $Pr_M \gg 1$. Although the detailed analysis assumes $Re \ll 1$, similar results are likely for larger Re (Galloway et al. 1978); given this, for $Pr_M = O(1)$ dynamical effects become important when $B_{max} \sim B_{eq}$, ignoring the logarithm in (12.2.13); this confirms the importance of the magnetic Prandtl number in discussing saturation when B_{max} is used.

Regardless of the behavior of peak fields, the net flux is small, and the mean field imposed on the system, given by $B_{LS} \sim B_{eq}(R \, Re \log R)^{-1/2}$, is weak. For given Re, B_{LS}/B_{eq} scales as $\epsilon^{1/2}$ (up to a logarithmic term), and so this agrees well with the equipartition estimate (12.2.2) based on total energy (not peak field) for the value $\varrho_E = 1$ obtained above.

These models based on flux ropes are somewhat artificial from the point of view of dynamos. There is no amplification of large-scale field; instead it is imposed and so line-tying may be significant in maintaining strong local fields. Note that in the solar convection zone, sunspot fields are above equipartition values, and suppress convection significantly. This indicates that in a dynamo, it is possible for super-equipartition fields to be maintained locally, even though $Pr_M \ll 1$. Finally note that these organized flux ropes that arise in steady flows only form on extremely long time-scales, $O(L/U\epsilon)$. Their relevance to fast dynamos, evolving on the fast $O(L/U)$ time-scale, is unclear.

12.2.4 Routes to Dynamical Equilibrium

We thus have some indication of how and when saturation will commence in a fast dynamo. The simplest situation in which to explore the subsequent dynamical evolution is the case of small Reynolds number, $Re \leq O(1)$: the fluid is relatively viscous, and supports only a few large-scale modes of motion. Since $Pr_M \equiv \nu/n \equiv R/Re$, this corresponds to large magnetic Prandtl number, $Pr_M \geq O(R)$. In this case, when a fast dynamo saturates, the time-behavior of the large-scale modes must change somehow so as to switch off the constructive folding. This discussion is appropriate to the numerical models in §§12.3.1, 2.

Suppose that a fast dynamo flow field is driven on a scale of order L, velocity U, and that there is initially no magnetic field. Since the flow is a fast dynamo, the magnetic Reynolds number $R = UL/\eta$ (based on the flow velocities in the absence of field, and so on the driving of the system) is large, $R \gg 1$. Furthermore there must be field-line stretching, $h_{line} > 0$; since h_{line} has the dimensions of a reciprocal time, a useful measure of stretching is a stretching magnetic Reynolds number, $R_{line} = h_{line}L^2/\eta$. Finally there must be constructive folding, which is not easily quantified (although perhaps a cancellation exponent could be used here).

Thus in the flow, $R = O(R_{line}) \gg 1$. Now let a seed magnetic field be introduced. This will grow and eventually saturate, leaving a modified flow field that could be complex in space and time. In the saturated state we can define an effective magnetic Reynolds number using the equilibrated velocity U^{eq} by $R^{eq} = U^{eq}L/\eta$. We can also define an effective stretching magnetic Reynolds number R_{line}^{eq} using the value of h_{line} appropriate to the saturated flow field. The following means of saturation suggest themselves.

(i) The first possibility might be termed *flow suppression*, and is typical of slow dynamos operating near the onset of magnetic activity. The action of the Lorentz force is to reduce the speed of the flow so that $R^{eq} = O(1)$. In this case the flow acts as a dynamo close to critical, and so reaches equilibrium.

An important variant of this scenario would allow two or more modes of motion, e.g., poloidal and toroidal motions in a sphere, which can be suppressed independently, but all of which are needed for dynamo action. Typically a product of magnetic Reynolds numbers must then be sufficiently large, and the system can be driven below critical when the kinetic energy is mainly in one mode. Near-critical dynamo action would again result, but in the form of a periodic or chaotic oscillation about equilibrium. Examples of this kind are found in the simplest homopolar dynamos (Cook & Roberts 1970, Robbins 1977). A related equilibration to a limit cycle was observed in a dynamical model of the Roberts cell by McMillan (1988).

(ii) Another possibility is that R^{eq} remains large, but $R^{\mathrm{eq}}_{\mathrm{line}}$ decreases significantly. The flow ceases to be a stretching flow and its line-stretching and Liapunov exponents are strongly suppressed by the presence of magnetic fields which resist extension. We could call this *suppression of stretch*.

One way this could happen is if the flow turned into an integrable flow, without the flow speed decreasing significantly. Perhaps a more likely result would be a flow that is chaotic in the Eulerian sense — showing sensitive dependence to the initial conditions of the fluid motion and having chaotic fluid velocity at a fixed point in space — but does not stretch material lines or vectors exponentially, and so is non-chaotic in a Lagrangian sense. For a simple example, the flow $\mathbf{u} = (f(t)y, 0, 0)$ is a shear flow and, if f is bounded, it stretches vectors linearly in time, with $h_{\mathrm{line}} = 0$ and $\Lambda_{\mathrm{Liap}} = 0$; it is thus non-chaotic in a Lagrangian sense. However it is chaotic in an Eulerian sense if the signal $f(t)$ is chaotic. The independence of the ideas of Eulerian and Lagrangian chaos has been emphasized by Falcioni, Paladin & Vulpiani (1988), and Babiano *et al.* (1994).

(iii) A third mechanism supposes that R^{eq} and $R^{\mathrm{eq}}_{\mathrm{line}}$ remain large; so the flow remains vigorous and stretches fluid elements. For the magnetic field to stop growing, the folding properties of the flow must be modified so as to fold field less constructively, and thus suppress the fast dynamo exponent. We might term this *suppression of twist* to suggest that stretching and folding could remain significant and still fast dynamo action be turned down or off. As an idealized example, the flow could adopt an approximately planar configuration in the saturation regime, while remaining chaotic.

(iv) Finally, a mechanism which is not relevant to our periodic geometry but is of possible importance in astrophysics allows flux to be removed from the system (Parker 1979). For example, in the solar convection zone magnetic buoyancy could lead to the floating up of tubes of flux, which are then transported out into the corona where they are drawn out by the solar wind. Recently Vainshtein & Rosner (1991) have suggested that the magnetic suppression of transport can curtail such loss, essentially as a variant of mechanism (ii). In galaxies field losses can come from the pressure of the cosmic ray gas and from ambipolar diffusion (or ambipolar drift) (Parker 1979).

These routes to equilibration could occur singly or in combination, and in a spatio-temporally complex flow, it could be difficult to isolate any one. For smaller Prandtl numbers, i.e., for $Re \gg 1$, the possibilities for saturation become considerably more complicated, since the small-scale flow field can respond to the small-scale magnetic field, and the fast dynamo can no longer be thought of simply as a large-scale stretching and folding process. We consider this more difficult case in §§12.3.3, 4 and §12.4.

If saturation occurs by mechanism (ii) or (i), the extinguishing of the Lagrangian chaos will lead to a reduction in transport by the flow, as measured by alpha effects and eddy diffusivities (Cattaneo & Vainshtein 1991, Tao, Cattaneo & Vainshtein 1993). This loss of stretching also implies reduced small-scale fields relative to B_{LS}. The ratio $B_{\mathrm{SS}}/B_{\mathrm{LS}}$ will decrease, which then implies that a post-kinematic estimate such as $B_{\mathrm{SS}} \sim B_{\mathrm{eq}}$ will become less tight, allowing B_{LS} to grow. This in turn means further suppression of fluctuations, and so on (Moffatt 1961, Kulsrud & Anderson 1993). The dynamical positive feedback of increasing B_{LS} and decreasing B_{SS} suggests that the process will stop only when $B_{\mathrm{SS}} \sim B_{\mathrm{LS}} \sim B_{\mathrm{eq}}$ (Brandenburg et al. 1992). This will be seen in models in §§12.3.1, 2 below. Thus if mechanisms (i) or (ii) act, there is good reason to believe that the field will evolve nonlinearly from the onset of saturation given by (12.2.2) to a final equilibrium for which the post-kinematic estimate (12.2.2) is irrelevant.

On the other hand if saturation mechanism (iii) occurs, the magnitude of the fluctuations B_{SS} is likely to increase relative to a large-scale field, reflecting the less efficient folding that takes place. The ratio $B_{\mathrm{SS}}/B_{\mathrm{LS}}$ will increase from the kinematic to saturation regimes, as would measures such as the cancellation exponent. This suggests that in the final equilibrium the large-scale field will be reduced relative to an estimate such as (12.2.2).

12.3 Numerical Studies

In this section we discuss some of the numerical results for dynamical fast dynamos. In §12.3.1 we consider a model based on the SFS map, and in §12.3.2 a model based on the two-dimensional flows of Chap. 2. These are essentially low Re (and hence high Pr_{M}) models that only allow a limited range of scales of the velocity field to be excited. For these systems the discussion of §12.2 is relevant and we observe the saturation mechanisms discussed there. In §§12.3.3, 4 we explore planar and three-dimensional flows with $Pr_{\mathrm{M}} = O(1)$.

12.3.1 Energetics of the SFS Map

This simple formulation of a dynamical fast dynamo makes use of a discrete map.[85] Let $M(\alpha)$ be the SFS map of the unit cube with shear parameter α; see §3.1.1. We perform maps only at integer times, but we shall extend the model by introducing an energetic basis for adjusting α and deciding when a map shall be performed. Diffusion occurs between maps. Because of the variation of α, the model thus allows equilibration to occur in various of the ways outlined in §12.2. We use the SFS map as an illustrative example for an approach that could be applied to any map with internal parameters, although the method of handling these will depend upon their meaning in the map.

We distinguish between the energy available for the stretch–fold operation and that available for the shear. Suppose that the energy made available for performing a stretch–fold part of the map at time n is $E_{SF}(n)$, and that the magnetic energy at this time is $E_M(n)$. We shall increase E_{SF} by an amount ΔE_{SF} per unit time. If there is sufficient energy at time n, a stretch–fold operation is performed and energy is converted into magnetic energy and also lost to heat by viscous dissipation. We take these viscous losses to be ΔE_{SF}, which ensures that in the absence of field the stretch–fold operations continue without unbounded build-up of E_{SF}. Since $4E_M(n)$ is the magnetic energy immediately after a stretch–fold map, we shall perform this map at time n whenever $E_{SF}(n) \geq 3E_M(n) + \Delta E_{SF}$. Otherwise no stretch–fold is performed.

Next, we carry out the shear part of the map. Let the energy available for this, $E_S(n)$ at time n, be increased by an amount ΔE_S per unit time. Recall that in the SFS model there is no y-field associated with the fold, so that shearing formally requires no energy. This is unrealistic, since in a physical model there would be field in the fold that would be stretched by shearing. To introduce an energetic cost of shearing, suppose that the structure of y-field in the fold is much like the x-field in the unit square, and introduce a certain fraction λ of $E_M(n)$ as the energy associated with the y-field. If this is sheared by an amount γ, its energy increases to $\lambda E_M(1 + \gamma^2)$ by line stretching.

Let $\alpha_n = \gamma_{n+1} - \gamma_n$ be the incremental shear at time n, γ being the total shear since the last stretch–fold. We also introduce a viscous dissipation quadratic in α. Given γ_n the incremental shear is determined by solving

$$\Delta E_S = \lambda E_M(n)(\gamma_{n+1}^2 - \gamma_n^2) + \nu(\gamma_{n+1} - \gamma_n)^2 \tag{12.3.1}$$

for γ_{n+1}. Following the shear, the field diffuses for time one and the cycle repeats. Whenever the stretch–fold is performed γ resets to zero.

To summarize: at the conclusion of each cycle E_{SF} and E_S are increased by fixed increments ΔE_{SF} and ΔE_S. The cycle consists of a shear determined by (12.3.1), and a diffusion for unit time, until $E_{SF} \geq 3E_M + \Delta E_{SF}$, at which time the full stretch–fold–shear is performed, γ and E_S reset to zero, E_{SF} is reduced by $3E_M + \Delta E_{SF}$, and E_M increases by a factor of 4. Once the field

---

[85]This section is based upon work done in collaboration with P.J. Morrison.

energy is stabilized the supply of energy for the SFS process is equal to the Joule dissipation plus viscous dissipation.

We now discuss some results. We show in Fig. 12.7 calculations at several diffusivities for a particular set of parameters. From Fig. 12.7(a) we observe that the equilibrium magnetic energy increases as ϵ decreases but the equilibrated range takes longer to set up. However Fig. 12.7(b) reveals that for all diffusivities the ratio $B_{\mathrm{LS}}/B_{\mathrm{SS}}$ rapidly assumes an $O(1)$ variation over the cycle. This ratio is zero immediately after the fold operation, but increases and varies in an irregular way over the cycle of shearing up to the next fold, the shearing being indicated in Fig. 12.7(c). The small-scale field does not develop to the dominance it has in the kinematic problem. The large-scale field is developed by the continual shearing between fold operations. The results of Fig. 12.7 are thus consistent with equilibration by suppression of stretch.

We can, however, also exhibit the other scenarios. For example, set $\lambda = 10$, which makes shear energetically expensive, and take $\nu = 1$ and $\epsilon = 10^{-6}$. An average of $B_{\mathrm{LS}}/B_{\mathrm{SS}}$ of only about 0.05 is developed. In this case very little shear acts, and we have an example of suppression of twist (here shear), with a significantly reduced mean field.

12.3.2 Low-dimensional Flows

While it seems clear that the main issues regarding dynamical equilibration of fast dynamos cannot be settled without extensive numerical simulation of MHD turbulence, it is of interest in the large Pr_{M} limit to exploit models which simplify the flow field but include the full range of magnetic structure. This approach can be thought of as 'laminar' in the flow but with typical fast dynamo structure in the magnetic field. We consider for simplicity a Navier–Stokes form of the momentum equation, driven only by a body force,

$$\rho\frac{d\mathbf{u}}{dt} + \nabla p + \mathbf{B} \times \mathbf{J} - \nu\nabla^2\mathbf{u} = \mathbf{F}, \qquad (\nabla \cdot \mathbf{u} = 0). \tag{12.3.2}$$

We assume that only the large-scale Lorentz force has any significant dynamical effect. This condition can be justified by making Re sufficiently large.

An example of this kind is considered by Gilbert et $al.$ (1993). Instead of appealing to a large-Re limit, a low-order truncation of the velocity field is imposed, of the form

$$\mathbf{u}(x, y, z, t) = \mathbf{v}(t)e^{ix} + \mathbf{v}^*(t)e^{-ix} + \mathbf{w}(t)e^{iy} + \mathbf{w}^*(t)e^{-iy}, \tag{12.3.3}$$

determined by the complex vectors $\mathbf{v} = (0, v_2, v_3)$, and $\mathbf{w} = (w_1, 0, w_3)$. The magnetic field is

$$\mathbf{B}(x, y, z, t) = \mathbf{b}(x, y, t)e^{ikz} + \mathbf{b}^*(x, y, t)e^{-ikz}. \tag{12.3.4}$$

The relation to the kinematic models of Chap. 2 is obvious. The restriction of $\mathbf{B}$ to one z-wavenumber k is unphysical, but is imposed to allow fast computations and easy exploration of these models.

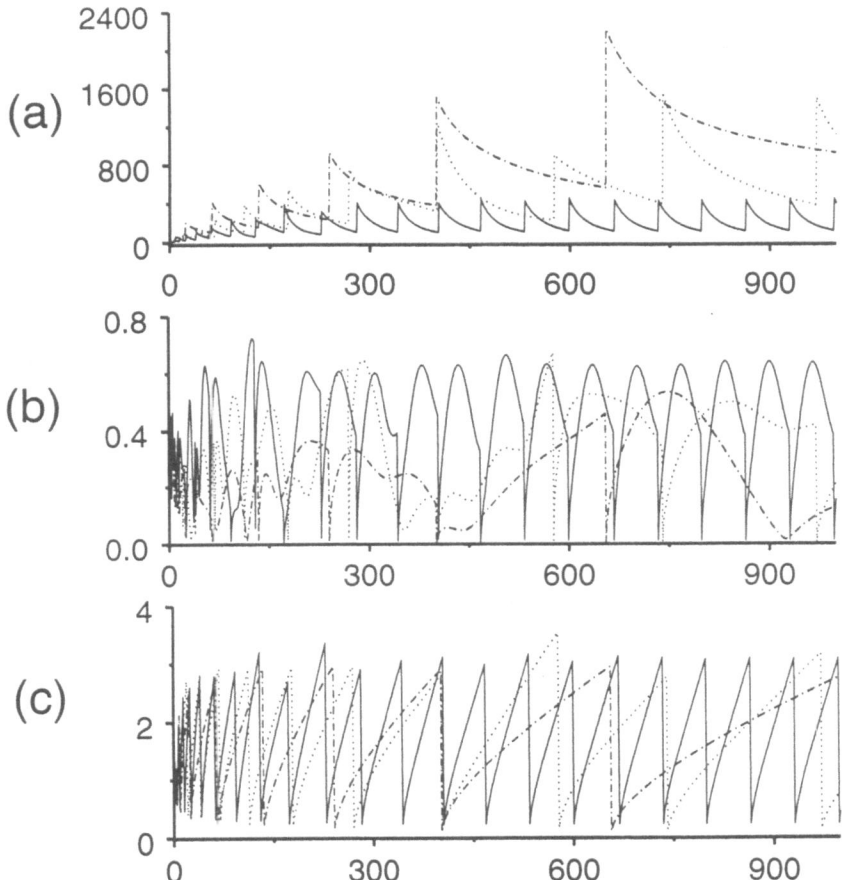

Fig. 12.7 The dynamical SFS model with $\Delta E_{\mathrm{SF}} = 20$, $\Delta E_{\mathrm{S}} = 4$, $\lambda = 0.15$ and $\nu = 10$. In each figure the diffusivity $\epsilon = 5 \times 10^{-5}$ (solid), 10^{-5} (dotted), and 10^{-6} (dot–dash). Plotted as a function of step n are (a) magnetic energy, (b) flux divided by the square root of total magnetic energy, and (c) shear variable γ.

The momentum equation (12.3.2) is now replaced by its projection onto the two modes of (12.3.3), and takes the form

$$(\partial_t + \nu)v_l = p_l + f_l, \qquad (\partial_t + \nu)w_m = q_m + g_m, \tag{12.3.5}$$

where $l = 2, 3$ and $m = 1, 3$. **f** and **g** are the modal projections of the driving force onto the modes e^{ix} and e^{iy} (cf. 12.3.3), and **p** and **q** the modal projections of the Lorentz force, which can be written as

$$p_l = <ib_1 b_l e^{-ix}>, \qquad q_m = <ib_2 b_m e^{-iy}>, \qquad\qquad (12.3.6)$$

where the brackets denote integration over the periodic box $\mathcal{D}$. The magnetic field in (12.3.6) has been scaled by $\sqrt{\mu\rho}$, so as to have the same dimensions as $\mathbf{u}$, and $\nu = 1$ in simulations.

In Gilbert *et al.* (1993) forcings were chosen to drive the flows MW± of Otani (1993), and the CP and LP flows of Galloway & Proctor (1992). Starting from a seed field the magnetic energy increases until saturation and equilibration; at this point a variety of behavior is observed for the different models, in terms of the time evolution and the distribution of magnetic energy. Also at equilibration the models show both transient and chaotic behavior over very long diffusive time-scales, as they explore the MHD attractor appropriate to the forcing; for this reason information about the equilibrated field, its structure and statistics, is limited. Of the equilibration options (i–iii) in §12.2.4, option (i) does not appear to be selected. In all cases the effective magnetic Reynolds number R^{eq} after equilibration remains large and of the same order as the kinematic value of R. Saturation is presumably by suppression of stretch or twist, or a combination. Within this class of models let us explore two that show markedly different behavior.

(1) **Randomly forced flow.** In this case the forcing is random and helical:

$$\begin{aligned}
\mathbf{f} = {}& 2\cos^2 t \, (0, \sin(x+\phi), \cos(x+\phi)) \\
& + 2\sin^2 t \, (\sin(y+\psi), 0, -\cos(y+\psi)).
\end{aligned} \qquad (12.3.7)$$

The random phase $\phi(t)$ is constant on each time interval $n - 1/2 < t/\pi < n + 1/2$ and the phase $\psi(t)$ is constant on $n < t/\pi < n + 1$, for integers n. The values of ϕ and ψ are chosen at random, uniformly in $[0, 2\pi]$, and independently on different intervals. The forcing is motivated by Otani's MW+ flow, and the random waves of §11.5. This random buffeting is designed to avoid the creation of organized structure in the magnetic field.

Equations (12.3.3–6) and the induction equation are simulated numerically for $k = 0.5$, $\nu = 1$ and $R \equiv 1/\epsilon$ from 100 to 2000, for t from 0 to 500π. The flow field comprises the four complex modes of (12.3.3), while the magnetic field is resolved to 65^2 or 97^2 modes. Starting with weak magnetic field, kinematic fast dynamo action occurs with growth insensitive to ϵ, until $t \simeq 300$ when magnetic and kinetic energy are comparable and saturation commences.

For each value of R the equilibrated kinetic energy is reduced by about 3, and so $R^{\mathrm{eq}} \approx R/\sqrt{3}$ remains large, ruling out saturation mechanism (i). At saturation the magnetic and kinetic energies do appear to be in equipartition with $\overline{E}_{\mathrm{M}}/\overline{E}_{\mathrm{V}} = 18\text{–}22$; although the magnetic energy is notably larger than the kinetic energy, there is no significant dependence on R for $R = 100$ to 2000. These average energies were calculated from $t = 500$ to 1570; the instantaneous ratio $E_{\mathrm{M}}/E_{\mathrm{V}}$ varies wildly because of zero-crossings of E_{V}.

Although equipartition appears to hold in the final state, the large-scale field is not subdominant. In Fig. 12.8 we plot the ratio $E_{\mathrm{M}}/E_{\mathrm{LS}}$, where

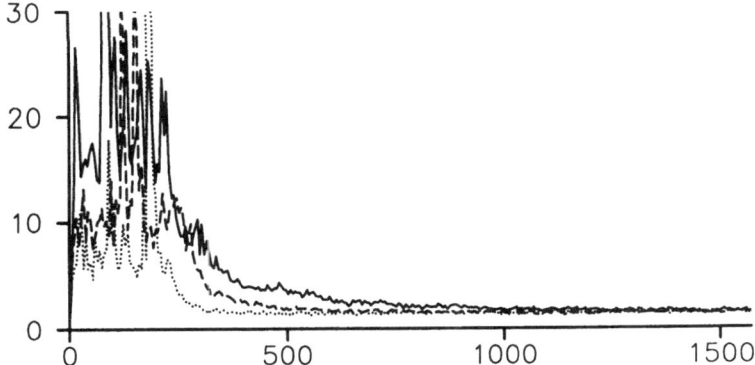

Fig. 12.8 Plot of E_M/E_{LS} against time for the randomly forced model (1) with $R = 100$ (dot), 500 (dash) and 2000 (solid).

$E_{LS} \equiv |<b>|^2$ is the contribution to the magnetic energy from the gravest $\exp(\pm ikz)$ modes of the system (see 12.3.4). In a kinematic regime, this would scale as $\epsilon^{-\varrho_E}$ from the post-kinematic estimate (12.2.1). This regime is given by $0 \leq t \lesssim 300$ and here the ratio E_M/E_{LS} fluctuates strongly as energy moves in and out of large scales. There is an increase in the typical ratio as R is increased. However as saturation commences the ratio E_M/E_{LS} drops to a value around 1–2 independent of R; post-kinematic estimates do not apply in the dynamical regime. The dynamo appears to saturate through mechanism (ii): the suppression of stretching leads to the extinction of small scales and the growth of a dominant large-scale field with $B_{LS} \sim B_{eq}$, as discussed in §12.2.4. The process can also be viewed as an inverse cascade of magnetic energy to large scales; see §12.4. Note that the time-scale of approach to the final equilibrium in Fig. 12.8 increases with R.

(2) **Dynamical MW+ flow.** In this case the forcing is chosen to maintain the MW+ flow in the absence of field. This case (under the name PW+) was studied in Gilbert *et al.* (1993) and is quite different in its behavior from the random model above. Fig. 12.9 shows the energy E_M as a function of time for $k = 1.0$ and $R = 1000$; after kinematic growth the energy peaks and then subsides, only to show occasional bursts during subsequent dynamical evolution. Eventually for $t \gtrsim 1500$, the energy shows chaotic time-evolution. The values of E_M/E_{LS} are large, around 70 most of the time, but rising during the bursts, and varying in a complex manner when $t \gtrsim 1500$.

In the bursting regime the large-scale field is very weak, and the field is confined to diffusive boundary layers reminiscent of the Roberts cell. Fig. 12.10 shows the evolution of field during one burst. The burst comprises a single folding operation, creating fine structure which then diffuses away very slowly.

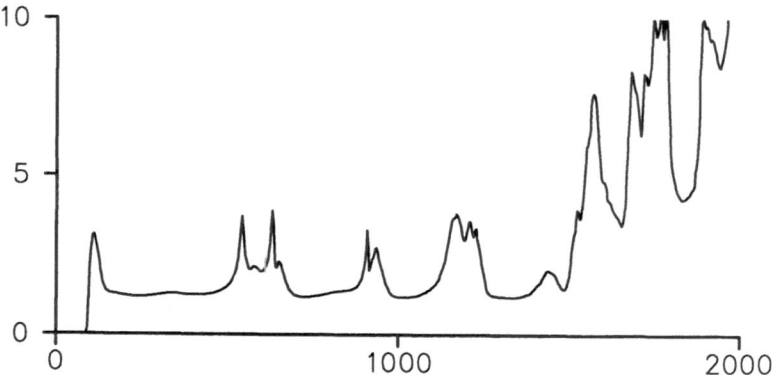

Fig. 12.9 Plot of E_M against time for the dynamical MW+ flow (2) with $k = 1.0$ and $R = 1000$.

This dynamical boundary layer dynamo does not fit easily into the saturation scenarios discussed in §12.2.4 above, but appears to be a combination of (ii) and (iii). The flow appears to have lost many of its stretching properties as in (ii), the chaotic folding only happening rarely. On the other hand the field is dominated by small-scale magnetic layers, as in (iii). These are maintained by stretching at the X-points, but this stretching is not chaotic. It appears that the Liapunov exponent Λ_{Liap} is virtually zero but the maximum Liapunov exponent Λ_{Liap}^{max} remains positive. The ratio E_M/E_{LS} increases from the kinematic value of 21 to around 70 or larger in the dynamical regime, again as would suggest (iii). Here the large-scale field is suppressed dynamically to a level lower than given by the post-kinematic estimates! This is a highly organized field responding to a deterministic forcing; in a more turbulent system the previous random model is more appropriate.

12.3.3 The MHD Planar Non-dynamo

Although dynamo action is not possible for planar flows, the problem of non-linear evolution from given initial conditions is relatively accessible analytically and numerically for a range of Pr_M. We consider an incompressible, non-rotating fluid without gravity. For planar flows with $\mathbf{u} = (\psi_y, -\psi_x)$ and $\mathbf{B} = (A_y, -A_x)$ the problem reduces to the two dimensionless equations

$$\partial_t \nabla^2 \psi + J(\psi, \nabla^2 \psi) - \delta \nabla^4 \psi - \beta^{-1} J(A, \nabla^2 A) = F, \tag{12.3.8}$$

$$\partial_t A + J(\psi, A) - \epsilon \nabla^2 A = 0, \tag{12.3.9}$$

(a)   *(b)*

(c)   *(d)*

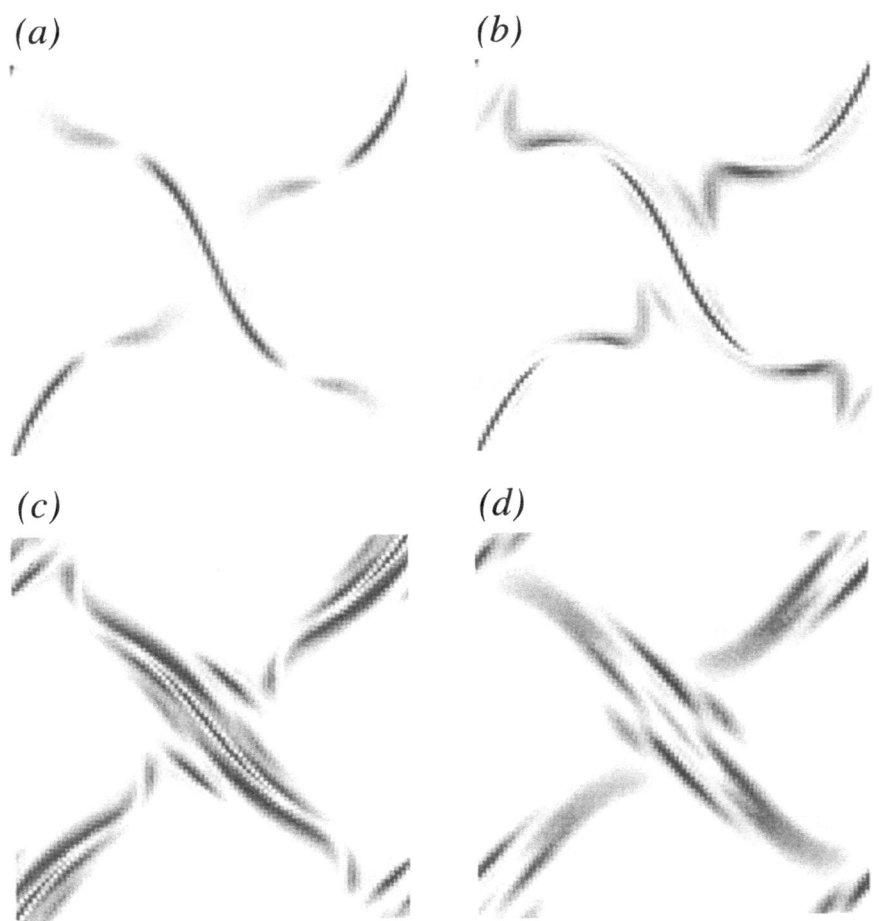

Fig. 12.10 Boundary layer bursting for the dynamical MW+ flow with $R = 1000$ and $k = 1.0$. The magnetic energy density at $z = 0$ is plotted on a grey scale for (a) $t = 288\pi$, (b) 289π, (c) 290π and (d) 292π.

where $J(\phi, \chi) = \phi_x \chi_y - \phi_y \chi_x$. Here δ^{-1} is the Reynolds number UL/ν (based on the flow with zero field) and $\beta = \mu\rho U^2/B_0^2$, where B_0 is a characteristic value of the initial magnetic field. The forcing function F drives the system.

The case of freely decaying two-dimensional MHD turbulence, $F \equiv 0$, $\epsilon \ll 1$ and $\delta \ll 1$, has received considerable attention, particularly when $Pr_M = \epsilon/\delta = O(1)$; see, e.g., Biskamp (1993). In simulations from smooth initial fields with comparable scales, strong current sheets are developed within a few eddy turnover times, and there is a rapid rise in the dissipation rate. As ϵ, $\delta \to 0$, the current sheets become thinner, the indication being that the thinning is exponential in an ideal fluid (Frisch *et al.* 1983). These results are consistent with the kinematics of flux expulsion and magnetic boundary layers (Chap. 5). However diffusive calculations indicate a subsequent break-up of sheets into smaller-scale structures, further enhancing dissipation.

If the system is forced at a scale $l \ll L$ and the initial magnetic field B_{LS} (of magnitude B_0) has scale L, the dynamical effect of large-scale field threading through a small-scale flow can be studied. We define $R = \epsilon^{-1}$ now as Ul/η. Cattaneo & Vainshtein (1991) find, in calculations with $\epsilon = 0.002$, $\delta = 0.001$ and random forcing with $L = 5l$, that a large-scale field can substantially extend the lifetime of the large-scale flow and field, by reducing the turbulent transport and dissipation of field by the small-scale eddies; see Fig. 12.8.

Following a few eddy turnover times, the field is wound up into highly sheared structures with scales of order $R^{-1/2}$ and field strengths of order $R^{1/2}B_{\mathrm{LS}}$. This is of course a purely kinematic estimate. At this stage the mean fields satisfy $<B^2> \approx <B_{\mathrm{SS}}^2> \approx RB_{\mathrm{LS}}^2$. In the kinematic theory, the field then decays abruptly, within a time of order $L^2/\eta R$. If it is assumed that turbulent transport, with an approximate diffusivity Ul, will be reduced when there is equipartition of energy between small-scale velocity and magnetic structures, we see that this can happen when $\beta \sim R$. This makes turbulent transport very sensitive to the addition of large-scale field when R is large.

With dynamical feedback the time of decay may be estimated as $L^2/\beta\eta$ since at equipartition the dissipation of $<A^2>$ should proceed at the rate $2\eta<\mathbf{B}^2> \sim \eta<B_{\mathrm{SS}}^2> \sim \beta\eta B_{\mathrm{LS}}^2$ once B_{SS} equilibrates. (B_{LS}, being the large-scale component of the field with magnetic potential A, satisfies $B_{\mathrm{LS}}^2 \sim <A^2>/L^2$, and $B_{\mathrm{LS}} \sim B_0$.) If the expression is modified to yield the molecular value for small β, the survival time of the field becomes $L^2/(\eta + \beta\eta)$. To make this go over to the kinematic case when $\beta = \infty$, we modify this again to

$$T_{\mathrm{decay}} = \frac{L^2}{\eta}\left(R^{-1} + \frac{1}{1+\beta}\right). \tag{12.3.10}$$

(Cattaneo & Vainshtein 1991). Note from Fig. 12.11 that the addition of mean field lowers the equilibrium magnetic energy, even as it extends the time of its survival.

Cattaneo, Hughes & Weiss (1991) have noted that the persistence of field in a non-dynamo according to (12.3.10) raises the question of how to detect dynamo action reliably from a numerical simulation over a finite time. It is reasonable to run for times long compared to the natural decay time of a large-scale field in a non-dynamo, as estimated in (12.3.10). To verify that this is done, Cattaneo et al. (1991) suggest tracking a time $T(t) = <B^2>/\eta<J^2>$ as a running estimate of the decay time of scales of size l. Running a simulation for times exceeding $(L/l)^2T$ by a factor of 2 or 3, they suggest, should allow dynamo action to be tested reliably. We emphasize however that this discussion relies on simulations with modest dissipation and Pr_{M} of order unity.

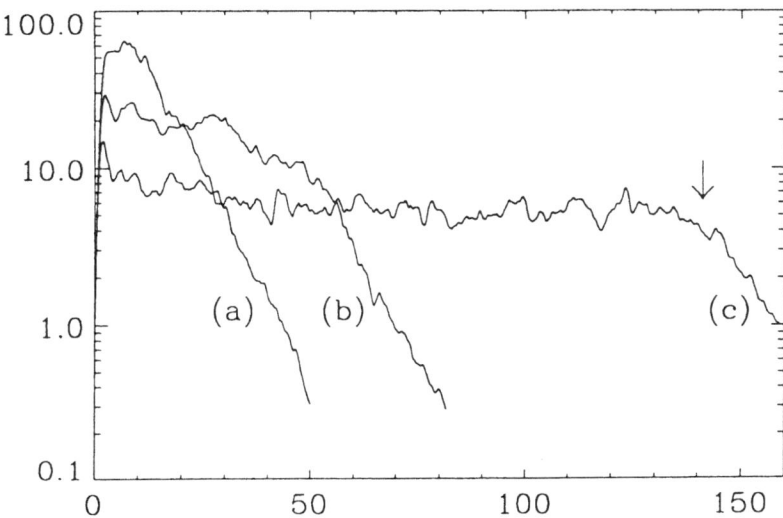

Fig. 12.11 Magnetic energy density in decaying two-dimensional forced flow plotted against time with $\beta = (a)$ ∞ (kinematic case), (b) 100, and (c) 30. (Fig. 3 of Cattaneo & Vainshtein (1991), reproduced with permission of the University of Chicago Press.)

There could be significant departures at small Pr_M and higher Reynolds numbers if instabilities occur which alter the scales and enhance dissipation.

The preceding results suggest that similar considerations might apply to dynamo action modelled by an alpha effect. From the nature of the attractor of Fig. 12.1, it is clear that, whenever the power to the dynamo is bounded and the excitation is expressed in terms of an alpha effect, the alpha tensor of an MHD fast dynamo must tend to zero at large magnetic energy. Otherwise magnetic energy would grow without limit and the attractor could not be bounded. When R is large, the needed suppression of alpha will depend on the dissipation which has to be overcome, and this is small for the large-scale field B_{LS}. Thus the extinction of dynamo action is a very sensitive issue at large R.

Vainshtein & Cattaneo (1992) estimate an effective α of the form $\alpha_{\text{eff}} \sim \alpha_0\beta/(\beta + R)$, where α_0 is the kinematic value and again $\beta = \mu\rho U^2/B_0^2 = B_{\text{eq}}^2/sbls^2$. The important point is the appearance of the factor $1/R$, which in effect confirms post-kinematic theory with $\varrho_E = 1$ as correct. Tao et $al.$ (1993) give some support for this conclusion with three-dimensional simulations at an R of about 130 and with a imposed mean magnetic field. The suppression of alpha has previously been studied by several authors, including Moffatt (see Moffatt 1978), in the context of helical waves in a rotating fluid. Rüdiger & Kichatinov (1993) give explicit results for non-isotropic suppression of alpha in a general turbulence model. This and other references are collected by Brandenburg et al. (1992) in a critique of weak mean-field equilibration.

The principal question is clearly what exponent ϱ_E actually pertains to the equilibrated fields. The analysis of this in the limit of large R is extraordinarily difficult, comparable in complexity to the calculation of turbulent transport at large Reynolds number. In the latter problem 'large' means values of approximately 10^6, and a similar estimate could well apply to the magnetic Reynolds number in the dynamo problem. The steady flux rope geometry studied by Galloway et al. (1978) and Jones & Galloway (1993a, b) (see §12.2.3) is the only equilibration model addressing the large-R limit. Time-dependent flow would presumably differ considerably from steady flux rope equilibrium in, say, the value of ϱ_E allowed by the dynamics.

We conclude that, although suppression of alpha is a necessary consequence of dynamo action, none of the studies carried out to date examines a turbulent flow at the values of R needed to draw firm conclusions applicable to solar or galactic dynamos. The calculation of the large-R behavior of effective alpha and diffusion coefficients in three dimensions remains a challenging problem with important implications for the dynamical theory.

12.3.4 A Case Study in Three Dimensions

We summarize now the results of the recent simulations by Nordlund et al. (1992) in three dimensions, extending calculations of Brandenburg et al. (1990) and dealing with dynamo action in a model of the solar convection zone. Additional recent results are given by Brandenburg et al. (1993). Other dynamical computations may be found in, for example, Gilman (1983), Glatzmaier (1985), Deluca & Gilman (1986, 1988), Meneguzzi & Pouquet (1989), Galanti, Sulem & Pouquet (1992), and Brandenburg et al. (1995b). A review of computational modelling is given by Brandenburg (1994).

The basic geometry falls within the model of §12.1.1, with periodicity in z replaced by boundary conditions on the planes $z = 0$ and $z = 2L$. The fluid is rotating and compressible (neither the Boussinesq nor the anelastic approximations are imposed), and the depth of the box is approximately four scale heights. The temperature of the top boundary is held fixed, with the gradient of internal energy supplied on the lower boundary. The domain is actually divided into two layers, the lower layer rendered convectively stable by increasing the conductivity K by a factor of 3. This models the stable 'overshoot' layer of the Sun, located at the base of the convection zone; see DeLuca & Gilman (1986, 1988) and Spiegel (1994). The Prandtl number $Pr = \nu/\kappa$ is equal to 0.2, and Pr_M is varied from 0.25 to 20. The conditions lead to convection at a Rayleigh number $Ra = gL^4 s'/\kappa\nu$, where s' is the initial entropy gradient, of about 10^6, or 50 times critical. Rotation is significant, as measured by a Taylor number $Ta = (2\Omega L^2/\nu)^2$ of about 10^5. The value of the fluid Reynolds number Re is about 300.

When a magnetic field with zero mean is introduced into the convective flow, dynamo action is tested using the criterion of Cattaneo *et al.* (1991) in §12.3.3. For $Pr_M = 1$ no growth of the magnetic energy occurs. Dynamo action is observed at $Pr_M = 2$ and 4, as η is lowered, with saturation occurring at a magnetic energy of less than $1/10$ the kinetic energy, well below equipartition. At a fixed Pr_M the same saturation magnetic energy is obtained from initial conditions smaller by a factor 10^{-2}. An interesting property of the equilibrium state is tested by removing various contributions to the Lorentz force. It is found that magnetic pressure is unimportant, and that the curvature stresses, contributing the component of $\mathbf{B} \cdot \nabla \mathbf{B}$ normal to $\mathbf{B}$, dominate the early phase of equilibration. It therefore appears that at these values of R and Pr_M, magnetic flux is carried about as material lines, until such time that the tension and curvature forces can resist the drag of relative motion between tube and fluid. At this stage the growth of magnetic energy subsides, while kinetic energy decreases only slightly. The saturation magnetic energy increases with Pr_M, since a given tube is carried more effectively by the flow.

The mechanism for dynamo action is deduced from three-dimensional vector plots and video animation. It appears that this involves narrow helical downdrafts as cold fluid falls and diverges at the bottom boundary, the spin of the column resulting from Coriolis forces. The boundary conditions do not allow flux to escape through the bottom boundary; field is returned by wide updrafts. The upward and downward motion is not symmetrical; the stretching and twisting of the field occurs mainly in narrow downdrafts, as tubes are caught within the descending vortical flow; field is thus pumped downwards into the overshoot layer, and magnetic buoyancy resisted.

The description is reminiscent of a Roberts cell or a Ponomarenko dynamo, but with the interesting difference that the descending helical flow is one of a random series of downdraft events. As the position of downdrafts shifts in the horizontal (either randomly or periodically) the resulting flow would become chaotic and exponential line stretching would result. Equilibration, in this picture, would come from the effect of field amplitude on the dynamo action realized in any one event, rather than on the pattern of events.

12.4 MHD Turbulence and Dynamo Saturation

In earlier discussion we have used single scales U and L for $\mathbf{u}$. We now consider dynamo action in turbulence. Then, in the absence of a magnetic field, it is well known that fully developed turbulence at $Re \gg 1$, created by forcing at wavenumber k_0, will contain an inertial range of wavenumbers k satisfying $k_0 \ll k \ll k_\nu$ where $k_\nu = (\varepsilon_V / \nu^3)^{1/4}$ is the Kolmogorov dissipation wavenumber and ε_V is the rate of dissipation of kinetic energy. In this inertial range the kinetic energy spectrum function, defined analogously to the magnetic energy spectrum,

$$E_V = \frac{\rho}{2} <\mathbf{u}^2> = \int_0^\infty E_V(k)\, dk, \tag{12.4.1}$$

has the Kolmogorov form

$$E_V(k, \varepsilon_V) = C\varepsilon_V^{2/3}\, F_V(k/k_\nu)\, k^{-5/3}, \quad (C \approx 1.5), \quad (k \gg k_0), \tag{12.4.2}$$

with $F_V(0) = 1$. This is the main information we have concerning $\mathbf{u}$ as a seed magnetic field develops. We must recognize two effects of the vorticity eddies in this early kinematic phase. First, each eddy can advect field of larger scale, and by twisting and folding create field of smaller scale. Secondly, dynamo action can occur, associated with eddies of sufficiently small size, which can amplify the magnetic field on larger scales. The kinetic energy is approximately conserved in the inertial range of the vorticity eddies, leading to the cascade of energy which creates the Kolmogorov spectrum. The magnetic field can 'cascade' to both large and small scales by the two processes just mentioned, although the magnetic energy is not an invariant of the ideal equations, this fact being the basis of dynamo action. Consequently the magnetic field and vorticity are quite different in their physical properties, and this is magnified at extreme values of Pr_M.

Batchelor (1950) noted the formal identity between the induction equation for $\mathbf{B}$ and the curl of the Navier–Stokes equation for vorticity in incompressible fluid of constant density with $Pr_M = 1$, and negligible Lorentz forces. If the fluid motion is sustained by forcing on scale L and the seed field is introduced with this scale, it is reasonable, at least for large R and Re, to postulate that the spectrum of $\mathbf{B}$ will develop an inertial range form identical to that for vorticity, equal to $k^2 E_V(k, \varepsilon_V) \sim k^{1/3}$ (corresponding to $\varrho_E = 2/3$ in §12.2), a point made by Moffatt (1961). This spectrum should, moreover, be insensitive to Prandtl number for k well below the smaller of the two dissipation wavenumbers. The analogy is not perfect, however. It omits the fact that the magnetic field is not forced, and that the vorticity is related to the velocity. The magnetic field can thus take up configurations denied to the vorticity field. Nevertheless the simulation of Brandenburg et al. (1993) of convective dynamo action at $Pr_M = 4$ (see §12.3.4) exhibits a $k^{1/3}$ magnetic spectrum during growth and an approximate k^{-1} spectrum at saturation.

Golitsyn (1960) and Moffatt (1961) have studied the smaller magnetic scales during this early phase for the case $Pr_M \ll 1$. Then an appropriate cutoff for the magnetic spectrum may be derived by analogy with the Kolmogorov scale, and we define $k_\eta = (\varepsilon_V/\eta^3)^{1/4}$. When $k_\eta \ll k \ll k_\nu$, $E_M(k)$ has a $k^{-11/3}$ behavior. The exponent can be understood in terms of the response of the seed field $\mathbf{B}_0$ to vorticity in the inertial range, once the local magnetic Reynolds number of the eddies drops below one, i.e., once k exceeds the dissipation wavenumber k_η. Vortical eddies then interact with $\mathbf{B}_0$ through an induction equation dominated by diffusion, yielding a field $\mathbf{B}(k) \approx -\eta^{-1}\nabla^{-2}\nabla \times (\mathbf{u}(k) \times \mathbf{B}_0)$, where k here selects a shell of Fourier modes. Thus we estimate $B(k) \sim (\eta k)^{-1} B_0 u(k)$, and so $E_M(k) \sim k^{-2} E_V(k) \sim k^{-11/3}$.

A small-scale field can lead to dynamo action at low wavenumbers through an α-effect, provided the helicity of the flow is non-zero; see Moffatt (1978), Krause & Rädler (1980), and Pouquet et al. (1976) (who also analyze the resulting inverse cascade using a spectral closure model). At the same time, small eddies with large local magnetic Reynolds numbers can act as fast dynamos with growth rates of order $ku(k)$. Thus the field amplitude can grow with time. When $E_M(k)$ and $E_V(k)$ become comparable at a given k, we may assume that growth ceases at that scale. This is likely to happen first at the small diffusive scales (Moffatt 1961). The crucial question is what happens subsequently.

Given the existence of dynamo action we may assume that *ultimately* $E_M(k) \approx E_V(k)$ within the velocity inertial range out to k_η where it is replaced by the $-11/3$ spectrum. In the equilibrated range the spectrum is no longer necessarily the hydrodynamic $-5/3$ inertial range. Independently Iroshnikov (1964), using field-theoretic methods, and Kraichnan & Nagarajan (1967), using an approximate closure (Kraichnan 1965), obtain a slightly shallower $-3/2$ spectrum for flow and magnetic field in non-helical turbulence, having $<\mathbf{u} \cdot \nabla \times \mathbf{u}> = 0$. Note that for this spectrum the magnetic energy is still dominated by large scales. The change can be viewed as a result of eddies interacting as propagating Alfvén waves (Biskamp 1993). Pouquet et al. (1976) verify this in a turbulence closure model and also study helical turbulence maintained by helical forcing. This has an inverse cascade of magnetic helicity set up by the dynamo action of the helical flow. As the field cascades to ever larger scales, E_M exceeds equipartition and E_M/E_V increases. Saturation is observed to involve equipartition of energy in an inertial range and they find evidence of a spectrum $E_M(k) \sim k^{-1}$ at large scales; see also Ruzmaikin & Shukurov (1982).

Although the details of the spectra of the saturated state are by no means clear, we must conclude that in a turbulent fast dynamo the dynamic saturation at small Pr_M changes substantially the magnetic energy spectrum, and therefore the partition of energy between large-scale and small-scale components will depart from the post-kinematic estimates. It seems clear that the dominance of small-scale field suggested by the vorticity analogy or by the post-kinematic estimates is at odds with the spectral decay at equipartition. Similar arguments should apply to Pr_M of order unity within the inertial range. The key observation is the existence of a non-linear inverse cascade of magnetic field. The time required to modify the post-kinematic spectrum and achieve the self-consistent equilibrium is of course extremely important.

We now turn to the case of large Pr_M. Here we cannot define k_η as $(\varepsilon_V/\eta^3)^{1/4}$, since the vorticity is dissipated away well before the smallest scale of the field is reached. During the kinematic growth phase, we assume a Kolmogorov spectrum $E_V(k)$ with smallest eddy of size $l_\nu = k_\nu^{-1}$. If an eddy of size l and velocity $(\varepsilon_\mathrm{V} l)^{1/3}$ operates as a fast dynamo, we expect to see magnetic field created down to scale $lR_l^{-1/2}$ where R_l is the local magnetic Reynolds number $\varepsilon_\mathrm{V}^{1/3} l^{4/3}/\eta$. But $R_{l_\nu} = Pr_\mathrm{M} \gg 1$ ensures that the smallest scale so developed is $l_\eta \equiv l_\nu Pr_\mathrm{M}^{-1/2}$, so that with $\varepsilon_\mathrm{V} = U^3/L$ we have

$$l_\eta = L Pr_\mathrm{M}^{1/4} R^{-3/4}, \qquad (R = UL/\eta). \tag{12.4.3}$$

This estimate of the size of the smallest magnetic structures in MHD turbulence at large Pr_M has been proposed by Brandenburg, Procaccia & Segel (1995a). It should replace the estimate $LR^{-1/2}$ for fully-developed turbulence, at least in the early stage of growth. Some modification during saturation is expected, if the Kolmogorov $k^{-5/3}$ spectrum is altered.

Note that large Pr_M allows very viscous laminar flows, with single scales U, L determined by the forcing, in which case the estimate $LR^{-1/2}$ may be retained and applies to the saturation phase. The dynamics of the flow are then determined by averaging the Lorentz stresses, as in §§12.2.3, 3.2.

Since $E_\mathrm{M}(k)$ now extends well beyond k_ν the saturated state cannot be characterized by local equipartition. The evidence from kinematic fast dynamos (§12.2.2) suggests a small-scale spectrum with a power-law behavior $E_\mathrm{M}(k) \sim \mu^{-1} B_\mathrm{LS}^2 (k/k_\eta)^{-q}$ for $k_\nu \ll k \ll k_\eta$, with q in the range 0–1. Biskamp (1993) suggests $q \sim 1$. The case of large Pr_M provides a direct link with kinematic theory at least in the ordering of the length scales of vorticity and magnetic field. This spectrum is reasonable for any fast dynamo, because of the creation of small-scale magnetic structure by even quite simple flows.

12.5 Astrophysical Implications

To understand the astrophysical significance of small-scale magnetic structure it is necessary to review some of the current ideas regarding cosmical magnetism. We consider galactic fields in §12.5.1, and solar fields in §12.5.2.

12.5.1 Galactic Magnetic Fields

There are at least three components of the galactic environment which should be modelled: the gas, having a density of about 2×10^{-24} g/cm^3, the magnetic field, of magnitude roughly 3×10^{-6} G, and the cosmic rays, estimated to exert a pressure of about 0.5×10^{-12} g/cm sec^2 (Parker 1979, Chap. 22, Zeldovich, Ruzmaikin & Sokoloff 1983, tables 2.2 and 3.1). Thus we have the estimate $B^2/2\mu \approx 0.4 \times 10^{-12}$ g/cm sec^2.[86] We deduce that cosmic rays are possibly important in the dynamics of the field. Equilibration with kinetic

---

[86]We take μ equal to its vacuum value, in which case to obtain magnetic pressure in g/cm sec^2 the square of B in Gauss should be divided by 8π.

energy occurs when U is equal to the Alfvén speed of about 6×10^5 cm/sec. A velocity scale of about 10^6 cm/sec is usually regarded as characteristic of the motions of the galactic gas. Zeldovich *et al.* (1983) obtain this value from the Kolmogorov spectrum by estimating length scales and energy dissipation. A typical scale of inhomogeneity is about $l \sim 100$ parsecs or 3×10^{20} cm, giving an eddy turnover time l/U of about 10^7 years. If a dimensionless dynamo growth rate of 0.1 is assumed, for a primordial field of 10^{-20} Gauss to grow to the size of the present field would require about 300 eddy turnovers.

Given that the galactic field will possess a broad range of scales, a key question is whether a given measurement is sensitive to B_{SS}, or instead performs in effect a flux average and determines B_{LS}. Any dynamical measure, such as a balance between field and cosmic ray pressure, invokes the Lorentz force and should evaluate B_{SS}. A common technique uses Faraday rotation of the radio emission from extra-galactic sources. Here the measurement is of the line integral from source to observer of the product of the local electron density and the the line-of-sight component of the magnetic field. This is a measure of B_{LS}. Both methods seem to indicate field of roughly the same size — a few microgauss; Zeldovich *et al.* (1983) estimate $B_{SS}/B_{LS} \approx 1.7$.

Are these observations compatible with the origin of the galactic field by fast dynamo action? There are three basic problems. Firstly, there is the issue of the time required for equilibration following the post-kinematic phase. To study the onset of saturation, Anderson & Kulsrud (1993) and Kulsrud & Anderson (1993) follow the kinematic growth of a weak seed field in large Pr_M turbulence, using a form of the closure model of Kraichnan & Nagarajan (1967) to formulate an initial-value problem. It contains no alpha effect to produce an inverse cascade in magnetic energy. Their conclusion is that the growth rate $U/l \approx 10^4$ years (based upon an estimate at the Kolmogorov scale l_ν with $\nu = 10^{21}$ cm^2/sec) will lead to rapid saturation by the small-scale field. Indeed a primordial field of 10^{-20} G would saturate in as little as 10^6 years. The model does not however consider the subsequent evolution of large-scale fields in the presence of equilibrated small-scale field. The issue is one of the time of equilibration because the only mechanism for creating sizable large-scale fields is the nonlinear inverse cascade and saturation over all scales; if this takes too long (exceeding the age of the universe, say), then dynamics cannot be invoked and a dynamo origin for the galactic field is in doubt.

A different problem concerns the observation of distant galaxies. If current fields at the microgauss level are accepted as the result of sustained growth, one would expect to see weaker fields in the younger galaxies. Kronberg (1994), using recently developed indirect methods for inferring field strengths at large redshifts, finds strengths at the microgauss level in galaxies as young as a few times 10^8 years. If this is accepted as a property of mean fields, given a typical eddy turnover time of 10^6 years, a dynamo theory of amplification of seed fields any smaller than 10^{-10} G becomes untenable.

One way in which large-scale field could be acquired rapidly is through turbulence associated with magnetic instabilities, a point recently emphasized by Brandenburg *et al.* (1994). These authors study dynamo action in a Keplerian shear flow, modelling an accretion disc, and a mean field is developed self-consistently to an amplitude of about one-half the mean magnetic energy.

Secondly, there is the issue of the size of B_{SS}/B_{LS} following equilibration. As we have noted above the observations suggest that a redistribution of energy, bringing large- and small-scale fields to comparable levels, has actually occurred. At issue here is the efficacy of fast dynamo action as well as the effect of Lorentz stresses on the structure of turbulence. We have seen that both effects can substantially reduce B_{SS}/B_{LS} in model problems. Plainly the sustained effects of fast dynamo action in dynamically equilibrated flow will have to be studied before the present observations can be understood.

Thirdly, based upon acceptance of strong small-scale fields at the onset of equipartition and saturation, there is the objection that there exist no effective mechanisms for diffusing away these fields. At large Pr_M the consequence of equipartition would be, in this view, such as to shut off the mixing processes responsible for both the dynamo and the cascade to small-scale magnetic structure. Without this cascade the dynamo shuts down and the growth of large-scale field to acceptable values is prevented. This argument, however, suffers from the internal inconsistency of including intense small-scale magnetic structure to invoke premature saturation, while needing to suppress small-scale processes to effect the quenching of the alpha effect. In fact the inertial range of MHD turbulence involves the interaction of weakly damped Alfvén waves, with consequences for dynamo action and energy transfer which are not fully understood. Equilibration could well proceed by turning off the flow of magnetic energy into small scales without fully arresting the growth of large-scale field, as we found to occur in models discussed in §12.3.

Also, the problem would be overcome immediately by as yet unrecognized processes for removing field effectively. The recent proposals of Parker (1992) provide examples of such approaches. Finally, we should not forget that the break-up of sunspots in a time of order months, rather than a diffusion time of order 10^9 years, is observed in the solar convection zone!

Of these three problems, the first may be the most worrisome to a dynamo theory of galactic magnetic fields. Although it is still premature to argue convincingly for or against a dynamo explanation of galactic fields, it seems certain that the full nonlinear dynamical equilibration, and the time it takes to happen, must first be understood. While the principles of fast dynamo action are fairly simple and robust, there are probably many realizations with different dynamical behavior. In our opinion developing understanding of equilibration takes precedence over a search for mechanisms of enhanced diffusion, or for non-dynamo origins of galactic fields.

12.5.2 The Solar Magnetic Field

We turn next to the modelling of the solar dynamo. Because of the measurements of this nearby magnetic activity and the recent helioseismic probing of the Sun's interior structure, the modelling of the solar magnetic cycle has received considerable attention in recent years. We cannot hope to do justice to these many developments in this brief summary, and instead make some observations and speculations with regard to the earlier discussion in this chapter. Useful references are Moffatt (1978), Parker (1979), Krause & Rädler (1980), Priest (1984) and Weiss (1994). For a review of the numerical simulations see Brandenburg (1994). At first sight the Sun's magnetic field is disorganized, as one might expect for a dynamo functioning at large magnetic Reynolds number. The field observed on the surface is highly intermittent and is dominated by fluctuations, of strength 1000–2000 G, while the mean (poloidal) field is relatively weak, of the order of 1–2 G. The term *fibrillation* is sometimes used to describe the arrangement of the magnetic field into intense flux tubes.

However, closer observation reveals order within this chaotic sea of fibrils. Firstly, there is the 11-year cycle in sunspot numbers (ignoring gaps such as the Maunder minimum). During this cycle sunspot pairs emerge at latitudes of about $\pm 30°$, travel towards the equator and subsequently disappear; this leads to the well-known butterfly diagram. Sunspot pairs appear to originate from a belt of toroidal field deeper in the convection zone. Buoyant tubes of flux, whose ends are connected to this deeper belt, rise and break the surface of the Sun. At the same time they are twisted because of the Coriolis force and take up a configuration of two spots of opposite sign of field, angled at about 10° to lines of constant latitude.

Secondly, the observations reveal a striking pattern to the *sign* of the average magnetic field in sunspots. In each hemisphere, to a good approximation (97%; see Priest 1984), the leading spots have the same sign of magnetic field; the sign is the opposite in the other hemisphere, and reverses with the magnetic half-cycle. The indication is that sunspots are flux tubes derived from a coherent belt of toroidal field, antisymmetric about the equator, which oscillates with the solar cycle. This is the second feature of order within the solar dynamo. A key problem in nonlinear dynamo theory is to explain how this organization can exist at $R \sim 10^8$, and how it coexists with the fluctuating surface fields.

In a kinematic fast dynamo the mean field is dominated by the fluctuations, and in this respect the theory is consistent with the fibrillated solar field. However this is clearly an example of applying kinematic theory beyond its intended scope. Also, numerical estimates are not especially convincing. If $R \sim 10^8$ and we assume an efficient kinematic fast dynamo flow (on the scale of the Sun) with say $\varrho_E = 0.24$ (as for MW+) we obtain $B_{LS} \sim B_{SS}/10$ from (12.2.2). This is easily satisfied by the poloidal field; but for the belt of toroidal field the situation is marginal, given that (12.2.2) is only a scaling law, and '$\sim$'

hides an unknown order-one constant. However including random convection on a smaller scale, or range of scales, is likely to decrease B_{LS}/B_{SS} further (Cattaneo *et al.* 1995), and so it appears difficult to model the solar field in terms of *kinematic* eigenfunctions for a fast dynamo (see Vainshtein *et al.* 1993*a*).

We thus have to take into account the nonlinear evolution of the solar field from a kinematic eigenfunction into the current dynamically equilibrated state. We have seen in simple models (§§12.3.1, 2) and theoretically (§12.2.4, §12.4) that the field can evolve into a state which defies the post-kinematic estimates and has a dominant mean field. The question then is, how can the belt of coherent large-scale toroidal field be compatible with the 11-year solar cycle (in which time this belt must be destroyed and created with the opposite sign) given that $R \sim 10^8$?

The difficulties of investigating this question are formidable, whether it is approached in terms of simulations of solar convective flows at large R, or in understanding the dynamo as an $\alpha\omega$-dynamo with nonlinear quenching and other transport effects. In theory knowledge of how α and ω are quenched by a large-scale field relates time scales of evolution to the ratio B_{LS}/B_{eq}; however the forms of these quenchings are not known with any certainty. Furthermore it may be that such transport effects are inappropriate for detailed modelling of the solar dynamo, despite their undoubted success in giving the correct overall picture. Their appropriateness depends on whether the solar dynamo is operating by means of a large-scale flow, on the scale of the Sun, stretching and folding field, or by the net transport effect of smaller-scale flows. If the former is true, it may be that quite particular flows are needed to organize the toroidal belt of field; for example, a narrow zone of strong shear at the base of the convection zone, the so-called *overshoot layer*, may be significant in shaping a coherent field at large R (Spiegel & Weiss 1980). In the overshoot layer fluid is stably stratified but the effects of the overlying convective field would presumably excite some small-scale motion which could produce a dynamo cycle.

The importance of the overshoot layer as a possible seat for the solar dynamo has been brought into prominence by recent discoveries of helioseismology (see, e.g., Weiss 1994) as well as from difficulties with conventional mean-field models of the convection zone. The older models of $\alpha\omega$ type were reasonably successful in simulating the butterfly diagram and the basic polarity cycle of the Sun, but are in conflict with properties of the azimuthal flow needed to achieve the observed equatorward migration of poloidal field components associated with sunspots (Krause & Rädler 1980). In recent years the prevailing view of the internal circulations of the Sun has been changed profoundly by the helioseismological measurements. A critical function of the theory is Ω_r, the radial derivative of the azimuthal velocity component of the Sun. The recent measurements indicate that in the upper 50 Mm layer Ω depends only upon the latitude, so the radial derivative vanishes and there is no sensible ω-effect. However at the lower boundary of this layer there is a thin sublayer, the *tachocline*, where Ω_r is large and a function of latitude. The maintenance and structure of the tachocline has been discussed by Spiegel & Zahn (1992). Below this sublayer the solar interior appears to be in solid-body rotation. Thus the sublayer becomes a likely candidate for the seat of the dynamo, where strong toroidal fields are held until magnetic buoyancy leads to instability and flux tubes are released into the convection zone (Galloway & Weiss 1981, Hughes & Proctor 1988, Hughes 1992). The exact nature of the complete nonlinear dynamo cycle in this picture remains obscure. On the other hand Prautzsch (1993) has incorporated the helioseismological results into what can be described as an $\alpha^2\omega$ kinematic model, which reproduces such essential features as the migration of poloidal waves toward the equator.

12.6 The Prospects for Modelling

In attempting to assess the possible application of fast dynamo theory as described in this monograph to the many instances of magnetism in the cosmos, we must first recognize that the basic $\alpha\omega$ model, as formulated by Parker (1955) and developed by many workers (see Krause & Rädler 1980), has been especially significant and influential. This is not only because of the sound physical basis of the model, and its relatively simple formulation, but also because it is widely applicable. Its constituent mechanisms can be realized in many ways and in a variety of physical systems. The omega effect epitomizes various ways of rapidly increasing field energy for a finite time, usually introducing large shear in the process, without achieving sustained dynamo action. In a planet, star, or galaxy, conditions are generally present for stretching field in a preferred, usually azimuthal direction. As an example, consider a thin flat galaxy near the (r, θ)-plane, in differential rotation about the z-axis. Then radial components of field will tend to be stretched out into the θ or azimuthal direction. Within the Earth's fluid core or the deep interior of the Sun, the axisymmetric toroidal component of the flow provides the differential rotation needed to draw poloidal field out into an intense toroidal field. Other mechanisms having a similar effect could arise from discrete jets, fluid plumes, and shock waves. Within young galaxies, outflows associated with intense periods of star formation, and the inflation of magnetic loops by cosmic rays are interesting processes in the same category (Parker 1992, Kronberg 1994). Regardless of the physical cause, the magnetic effects of importance can be understood from simple line stretching. The omega effect can be compared usefully to the stretch–fold element of the basic fast dynamo cycle.

It is in the precise description of the alpha effect that the main difficulties of a large-R dynamo theory reside. The reader will have noticed that the alpha effect has not played a prominent role in these pages, primarily because the theory we have summarized has not aimed at the derivation of equations governing the mean field. A different emphasis might take the calculation of the large-R form of α as the central problem of fast dynamo theory. It is certainly the main obstacle to analyzing kinematic fast dynamos within the conventional framework of mean-field theory (Childress & Soward 1985). If we compare α to the 'twist', or 'shear' of a fast dynamo map, we have an accurate reflection of the difficulty of following phase shifts (as in Chap. 8) or constructive folding in chaotic flows; this process extracts mean field from the fluctuations which arise from the stretch–fold or ω effects.

We thus come to a view of the principal problem now faced by the theory as one of better understanding constructive assembly of stretched field in terms of a conventional mean-field α. For the solar dynamo, progress is likely to involve maps based upon a combination of differential rotation and wave-like components. In this context the kinematic fast dynamos derived from a pair of pulsed waves (Chap. 2) would correspond to an α^2 mean-field model. For galactic modelling, on the other hand, given the possibility of an 'ω-effect' during the early stage of galaxy formation, the requisite 'α' would seem to amount to a gradual consolidation of stretched field toward the large scales, i.e., an inverse cascade associated with equilibrated MHD turbulence. How and why such a process arises dynamically, if indeed this is the implication of the observations, remains to be understood.

References

Abramowitz, M., Stegun, I.A. (1972): Handbook of Mathematical Functions. Dover.

Adler, R.L., Konheim, A.G., McAndrew, M.H. (1965): Topological entropy. Trans. Amer. Math. Soc. **114**, 309–319.

Alfvén, H. (1950): Discussion of the origin of the terrestrial and solar magnetic fields. Tellus **2**, 74–82.

Anderson, S.W., Kulsrud, R.M. (1993): Magnetic noise and the galactic dynamo. In: Solar and Planetary Dynamos (ed. M.R.E. Proctor, P.C. Matthews, A.M. Rucklidge), pp. 1–7. Cambridge University Press.

Anosov, D.V. (1962): Roughness of geodesic flows on compact Riemannian manifolds of negative curvature. Sov. Math. Doklady **3**, 1068–1070.

Anosov, D.V. (1963): Ergodic properties of geodesic flows on closed Riemannian manifolds of negative curvature. Sov. Math. Doklady **4**, 1153–1156.

Anosov, D.V. (1967): Geodesic flows on closed Riemannian manifolds with negative curvature. Proc. Steklov Inst. Math. **90**, 1.

Antonsen, T.M., Ott, E. (1991): Multifractal power spectra of passive scalars convected by chaotic fluid flows. Phys. Rev. A **44**, 851–857.

Anufriyev, A.P., Fishman, V.M. (1982): Magnetic field structure in the two-dimensional flow of a conducting fluid. Geomagn. Aeron. **22**, 245–248.

Apostol, T.M. (1957): Mathematical Analysis. Addison–Wesley.

Arnold, V.I. (1965): Sur la topologie des écoulements stationnaires des fluides parfaits. C. R. Acad. Sci. Paris **261**, 17–20.

Arnold, V.I. (1972): Notes on the three-dimensional flow pattern of a perfect fluid in the presence of a small perturbation of the initial velocity field. Prikl. Matem. Mekh. **36**(2), 255–262. [English transl.: Appl. Math. Mech. **36**, 236–242.]

Arnold, V.I. (1974): The asymptotic Hopf invariant and its applications. In Proc. Summer School in Differential Equations, Erevan 1974. Armenian SSR Acad. Sci. [English transl.: Sel. Math. Sov. **5** (1986) 327–345.]

Arnold, V.I. (1978): Mathematical Methods of Classical Mechanics. Springer–Verlag.

Arnold, V.I. (1984): On the evolution of magnetic field under the action of advection and diffusion. In: Some Questions of Present-day Analysis (ed. V.M. Tikhomirov), pp. 8–21. Moscow University Press.

Arnold, V.I. (1988): Geometrical Methods in the Theory of Ordinary Differential Equations. Springer–Verlag.

Arnold, V.I. (1994): Mathematical problems in classical physics. In: Trends and Perspectives in Applied Mathematics (ed. L. Sirovich), pp. 1–20. Springer–Verlag.

Arnold, V.I., Avez, A. (1967): Problèmes Ergodiques de la Mécanique Classique. Gauthier–Villars, Paris. [English transl.: Ergodic Problems of Classical Mechanics. Benjamin (1968).]

382   References

Arnold, V.I., Korkina, E.I. (1983): The growth of a magnetic field in a three-dimensional steady incompressible flow. Vest. Mosk. Un. Ta. Ser. 1, Matem. Mekh., no. 3, 43–46.

Arnold, V.I., Sinai, Ya.G. (1962): Small perturbations of the automorphisms of the torus. Dokl. Akad. Nauk SSSR **144**, 695–698. [English transl.: Sov. Math. **3** (1962) 783–787.]

Arnold, V.I., Zeldovich, Ya.B., Ruzmaikin, A.A., Sokoloff, D.D. (1981): A magnetic field in a stationary flow with stretching in Riemannian space. Zh. Eksp. Teor. Fiz. **81**, 2052–2058. [English transl.: Sov. Phys. JETP **54** (1981) 1083–1086.]

Arnold, V.I., Zeldovich, Ya.B., Ruzmaikin, A.A., Sokoloff, D.D. (1982): Steady-state magnetic field in a periodic flow. Dokl. Akad. Nauk SSSR **266**, 1357–1361. [English transl.: Sov. Phys. Dokl. **27** (1982) 814–816.]

Artuso, R. (1991): Diffusive dynamics and periodic orbits of dynamical systems. Phys. Lett. A **160**, 528–530.

Artuso, R., Aurell, E., Cvitanović, P. (1990a): Recycling of strange sets: I. Cycle expansions. Nonlinearity **3**, 325–359.

Artuso, R., Aurell, E., Cvitanović, P. (1990b): Recycling of strange sets: II. Applications. Nonlinearity **3**, 361–386.

Aubry, S., Abramovici, G. (1990): Chaotic trajectories in the standard map. The concept of anti-integrability. Physica D **43**, 199–219.

Aurell, E. (1992): On proving fast dynamo action using (singular) integral operators. Unpublished.

Aurell, E., Gilbert, A.D. (1993): Fast dynamos and determinants of singular integral operators. Geophys. Astrophys. Fluid Dyn. **73**, 5–32.

Avellaneda, M., Majda, A.J. (1991): An integral representation and bounds on the effective diffusivity in passive advection by laminar and turbulent flows. Comm. Math. Phys. **138**, 339–391.

Babcock, H.D. (1959): The Sun's polar magnetic field. Astrophys. J. **130**, 364–380.

Babiano, A., Boffetta, G., Provenzale, A., Vulpiani, A. (1994): Chaotic advection in point vortex models and two-dimensional turbulence. Phys. Fluids **6**, 2465–2474.

Backus, G. (1958): A class of self-sustaining dissipative spherical dynamos. Ann. Physics **4**, 372–447.

Backus, G. (1975): Gross thermodynamics of heat engines in deep interior of Earth. Proc. Nat. Acad. Sci. USA **72**, 1555–1558.

Balmforth, N.J., Cvitanović, P., Ierley, G.R., Spiegel, E.A., Vattay, G. (1993): Advection of vector fields by chaotic flows. In: Stochastic Processes in Astrophysics, pp. 148–160. Annals New York Acad. Sci., vol. 706.

Batchelor, G.K. (1950): On the spontaneous magnetic field in a conducting liquid in turbulent motion. Proc. R. Soc. Lond. A **201**, 405–416.

Batchelor, G.K. (1952): The effect of homogeneous turbulence on material lines and surfaces. Proc. R. Soc. Lond. A **213**, 349–366.

Batchelor, G.K. (1953): The Theory of Homogeneous Turbulence. Cambridge University Press.

Batchelor, G.K. (1956): A proposal concerning laminar wakes behind bluff bodies at large Reynolds number. J. Fluid Mech. **1**, 388–398.

Batchelor, G.K. (1967): An Introduction to Fluid Mechanics. Cambridge University Press.

Baxendale, P.H., Rozovskii, B.L. (1993): Kinematic dynamo and intermittence in a turbulent flow. Geophys. Astrophys. Fluid Dyn. **73**, 33–60.

Bayly, B.J. (1986): Fast magnetic dynamos in chaotic flow. Phys. Rev. Lett. **57**, 2800–2803.

Bayly, B.J. (1992a): Infinitely conducting dynamos and other horrible eigenprob-
lems. In: Nonlinear Phenomena in Atmospheric and Oceanic Sciences (ed. G.F.
Carnevale, R.T. Pierrehumbert), pp. 139–176, IMA Series in Mathematics and its
Applications, vol. 40. Springer–Verlag.

Bayly, B.J. (1992b): The solenoidality condition for weak solutions of the perfectly
conducting magnetic induction equation. Preprint.

Bayly, B.J. (1993): Scalar dynamo models. Geophys. Astrophys. Fluid Dyn. **73**,
61–74.

Bayly, B.J. (1994): Maps and dynamos. In: Lectures on Solar and Planetary Dynamos (ed. M.R.E. Proctor, A.D. Gilbert), pp. 305–329. Cambridge University
Press.

Bayly, B.J., Childress, S. (1987): Fast dynamo action in unsteady flows and maps
in three dimensions. Phys. Rev. Lett. **59**, 1573–1576.

Bayly, B.J., Childress, S. (1988): Construction of fast dynamos using unsteady flows
and maps in three dimensions. Geophys. Astrophys. Fluid Dyn. **44**, 211–240.

Bayly, B.J., Childress, S. (1989): Unsteady dynamo effects at large magnetic
Reynolds numbers. Geophys. Astrophys. Fluid Dyn. **49**, 23–43.

Bayly, B.J., Rado, A. (1993): Cancellation exponents of the stretch–fold–shear dynamo. J. Fluid Mech. **257**, 286–287. (Appendix B of Du & Ott 1993b.)

Beltrami, E. (1889): Opera Matematiche, vol. 4, p. 304.

Bensoussan, A., Lions, J.L., Papanicolaou, G.C. (1978): Asymptotic Analysis for
Periodic Structures. North–Holland.

Benton, E.R. (1979): Kinematic dynamo action with helical symmetry in an unbounded fluid conductor. Geophys. Astrophys. Fluid Dyn. **12**, 313–344.

Berger, M.A. (1990): Third-order link invariants. J. Phys. A Math. Gen. **23**, 2787–
2793.

Berger, M.A., Field, G.B. (1984): The topological properties of magnetic helicity. J.
Fluid Mech. **147**, 133–148.

Bertozzi, A.L. (1988): Heteroclinic orbits and chaotic dynamics in planar fluid flows.
SIAM J. Math. Analy. **19**(6), 1271–1294.

Bertozzi, A.L., Chhabra, A.B. (1994): Cancellation exponents and fractal scaling.
Phys. Rev. E **49**, 4716–4719.

Biskamp, D. (1993): Nonlinear Magnetohydrodynamics. Cambridge University Press.

Blennerhassett, P.J. (1979): A three-dimensional analogue of the Prandtl–Batchelor
closed streamline theory. J. Fluid Mech. **93**, 319–324.

Bollobás, B. (1990): Linear Analysis. Cambridge University Press.

Bondi, H., Gold, T. (1950): On the generation of magnetism by fluid motion. Mon.
Not. Roy. Astr. Soc. **110**, 607–611.

Boothby, W.M. (1986): An Introduction to Differentiable Manifolds and Riemannian
Geometry. Academic Press.

Boozer, A.H. (1983): Evaluation of the structure of ergodic fields. Phys. Fluids **26**,
1288–1291.

Boozer, A.H. (1992): Dissipation of magnetic energy in the solar corona. Astrophys.
J. **394**, 357–362.

Bowen, R. (1971): Entropy for group endomorphisms and homogeneous spaces.
Trans. Amer. Math. Soc. **153**, 401–414.

Bowen, R. (1978): On Axiom A Diffeomorphisms. CBMS Regional Conference Series in Mathematics, vol. 35, AMS Publications, Providence.

Boyland, P.L., Franks, J. (1989): Notes on Dynamics of Surface Homeomorphisms.
Lecture Notes, University of Warwick Nonlinear Systems Laboratory.

Braginsky, S.I. (1963): Structure of the F layer and reasons for convection in the Earth's core. Dokl. Akad. Nauk SSSR **149**, 1311–1314. [English transl.: Sov. Phys. Dokl. **149** (1963) 8–10.]

Braginsky, S.I. (1964): Self excitation of a magnetic field during the motion of a highly conducting fluid. Sov. Phys. JETP **20**, 148–150.

Braginsky, S.I., Roberts, P.H. (1995): Equations governing convection in the Earth's core and the geodynamo. Preprint.

Brandenburg, A. (1988): Hydrodynamic Green's functions for atmospheric oscillations. Astron. Astrophys. **203**, 154–161.

Brandenburg, A. (1994): Solar dynamos: computational background. In: Lectures on Solar and Planetary Dynamos (ed. M.R.E. Proctor, A.D. Gilbert), pp. 117–159. Cambridge University Press.

Brandenburg, A. (1995): Flux tubes and scaling in MHD dynamo simulations. Chaos, Solitons & Fractals **5**, to appear.

Brandenburg, A., Nordlund, Å., Pulkkinen, P., Stein, R.F., Tuominen, I. (1990): 3D simulation of turbulent cyclonic magnetoconvection. Astron. Astrophys. **232**, 277–291.

Brandenburg, A., Krause, F., Nordlund, Å., Ruzmaikin, A.A., Stein, R.F., Tuominen, I. (1992): On the magnetic fluctuations produced by a large scale magnetic field. Preprint.

Brandenburg, A., Jennings, R.L., Nordlund, Å., Rieutord, M., Stein, R.F., Tuominen, I. (1993): Magnetic structures in a dynamo simulation. Preprint.

Brandenburg, A., Procaccia, I., Segel, D. (1995a): The size and dynamics of magnetic flux structures in magnetohydrodynamic turbulence. Phys. Plasmas **2**, 1148–1156.

Brandenburg, A., Nordlund, Å., Stein, R.F., Torkelsson, U. (1995b): Dynamo generated turbulence and large scale magnetic fields in a Keplerian shear flow. Astrophys. J. **446**, 741–754.

Brillouin, L. (1946): Wave Propagation in Periodic Structures. Dover Publications.

Brown, S.N., Stewartson, K. (1978): The evolution of the critical layer of a Rossby wave. Part II. Geophys. Astrophys. Fluid Dyn. **10**, 1–24.

Busse, F. (1978): Magnetohydrodynamics of the Earth's dynamo. Ann. Rev. Fluid Mech. **10**, 435–462.

Cattaneo, F., Vainshtein, S.I. (1991): Suppression of turbulent transport by a weak magnetic field. Astrophys. J. **376**, L21–L24.

Cattaneo, F., Hughes, D.W., Weiss, N.O. (1991): What is a stellar dynamo? Mon. Not. Roy. Astr. Soc. **253**, 479–484.

Cattaneo, F., Kim, E., Proctor, M.R.E., Tao, L. (1995): Fluctuations in quasi-two-dimensional fast dynamos. Phys. Rev. Lett., submitted.

Chapman, C.J., Proctor, M.R.E. (1980): Nonlinear Rayleigh–Bénard convection with poorly-conducting boundaries. J. Fluid Mech. **101**, 759–782.

Childress, S. (1967): Construction of steady-state hydromagnetic dynamos. I. Spatially periodic fields. Report MF-53, Courant Institute of Mathematical Sciences.

Childress, S. (1970): New solutions of the kinematic dynamo problem. J. Math. Phys. **11**, 3063–3076.

Childress, S. (1979): Alpha-effect in flux ropes and sheets. Phys. Earth Planet. Int. **20**, 172–180.

Childress, S. (1984): An introduction to dynamo theory. In: Turbulence and Predictability in Geophysical Fluid Dynamics (ed. M. Ghil, R. Benzi, G. Parisi), pp. 200–225. North-Holland.

Childress, S. (1992): Fast dynamo theory. In: Topological Aspects of the Dynamics of Fluids and Plasmas (ed. H.K. Moffatt, G.M. Zaslavsky, P. Comte, M. Tabor), pp. 111–147. Kluwer Academic Publishers.

Childress, S. (1993 a): On the geometry of fast dynamo action in unsteady flows near the onset of chaos. Geophys. Astrophys. Fluid Dyn. **73**, 75–90.

Childress, S. (1993 b): Note on perfect fast dynamo action in a large-amplitude SFS map. In: Solar and Planetary Dynamos (ed. M.R.E. Proctor, P.C. Matthews, A.M. Rucklidge), pp. 43–50. Cambridge University Press.

Childress, S., Klapper, I. (1991): On some transport properties of baker's maps. J. Stat. Phys. **63**, 897–914.

Childress, S., Soward, A.M. (1985): On the rapid generation of magnetic fields. In: Chaos in Astrophysics (ed. J.R. Buchler, J.M. Perdang, E.A. Spiegel), pp. 223–244. D. Reidel.

Childress, S., Soward, A.M. (1989): Scalar transport and alpha-effect for a family of cat's-eye flows. J. Fluid Mech. **205**, 99–133.

Childress, S., Landman, M., Strauss, H.R. (1990 a): Steady motion with helical symmetry at large Reynolds number. In: Topological Fluid Dynamics (ed. H.K. Moffatt, A. Tsinober), pp. 216–224. Cambridge University Press.

Childress, S., Collet, P., Frisch, U., Gilbert, A.D., Moffatt, H.K., Zaslavsky, G.M. (1990 b): Report on workshop on: Small-diffusivity dynamos and dynamical systems, Observatoire de Nice, 25–30 June 1989. Geophys. Astrophys. Fluid Dyn. **52**, 263–270.

Cocke, W.J. (1969): Turbulent hydrodynamic line stretching: consequences of isotropy. Phys. Fluids **12**, 2488–2492.

Cook, A.E., Roberts, P.H. (1970): The Rikitake two-disc dynamo system. Proc. Camb. Phil. Soc. **68**, 547–569.

Courant, R., Hilbert, D. (1953): Methods of Mathematical Physics, Volume 1. Interscience.

Cowling, T.G. (1934): The magnetic field of sunspots. Mon. Not. Roy. Astr. Soc. **140**, 39–48.

Cowling, T.G. (1957 a): Magnetohydrodynamics. Interscience.

Cowling, T.G. (1957 b): The dynamo maintenance of steady magnetic fields. Quart. J. Mech. Appl. Math. **10**, 129–136.

Crisanti, A., Falcioni, M., Paladin, G., Vulpiani, A. (1990): Anisotropic diffusion in fluids with steady periodic velocity fields. J. Phys. A Math. Gen. **23**, 3307–3315.

Davis, L. (1958): A fluid self-excited dynamo. In: Electromagnetic Phenomena in Cosmical Physics (ed. B. Lehnert), pp. 27–31. Cambridge University Press.

Deluca, E.E., Gilman, P.A. (1986): Dynamo theory for the interface between the convection zone and the radiative interior of a star. Part I: Model equations and exact solutions. Geophys. Astrophys. Fluid Dyn. **37**, 85–127.

Deluca, E.E., Gilman, P.A. (1988): Dynamo theory for the interface between the convection zone and the radiative interior of a star. Part II: Numerical solutions of the nonlinear equations. Geophys. Astrophys. Fluid Dyn. **43**, 119–148.

Dinaburg, E.I. (1970): The relation between topological entropy and metric entropy. Dokl. Akad. Nauk SSSR **190**, 19–22. [English transl.: Sov. Math. Doklady **11** (1970) 13–16.]

Dittrich, P., Molchanov, S.A., Sokoloff, D.D., Ruzmaikin, A.A. (1984): Mean magnetic field in renovating random flow. Astron. Nachr. **305**, 119–125.

Dobrokhotov, S.Yu., Shafarevich, A.I., Martinez–Olivé, V., Ruzmaikin, A.A. (1993): Asymptotics of the kinematic dynamo. Internal report 142, Centro de Investigacion y de Estudios Avanzados del IPN, Departamento de Matematicas, Mexico.

Dombre, T. Frisch, U., Greene, J.M., Hénon, M., Mehr, A., Soward, A.M. (1986): Chaotic streamlines in the ABC flows. J. Fluid Mech. **167**, 353–391.

Drazin, P.G., Reid, W.H. (1981): Hydrodynamic Stability. Cambridge University Press.

Drummond, I.T. (1992): Stretching and bending of line elements in random flows. J. Fluid Mech. **252**, 479–498.

Drummond, I.T., Horgan, R.R. (1986): Numerical simulation of the α-effect and turbulent magnetic diffusion with molecular diffusivity. J. Fluid Mech. **163**, 425–438.

Drummond, I.T., Münch, W.H.P. (1990): Turbulent stretching of line and surface elements. J. Fluid Mech. **215**, 45–59.

Drummond, I.T., Duane, S., Horgan, R.R. (1983): The stochastic method for numerical simulations: higher order corrections. Nuclear Phys. B **220**, 119–136.

Drummond, I.T., Duane, S., Horgan, R.R. (1984): Scalar diffusion in simulated helical turbulence with molecular diffusivity. J. Fluid Mech. **138**, 75–91.

Du, Y., Ott, E. (1993a): Fractal dimensions of fast dynamo magnetic fields. Physica D **67**, 387–417.

Du, Y., Ott, E. (1993b): Growth rates for fast kinematic dynamo instabilities of chaotic fluid flows. J. Fluid Mech. **257**, 265–288.

Du, Y., Ott, E., Finn, J.M. (1994a): Linear evolution of high Reynolds number flows toward fractal vorticity distributions. Preprint.

Du, Y., Tél, T., Ott, E. (1994b): Characterization of sign singular measures. Physica D **76**, 168–180.

Eddy, S.A. (1976): The Maunder minimum. Science **192**, 1189–1202.

Ellis, R.S. (1985): Entropy, Large Deviations, and Statistical Mechanics. Springer–Verlag.

Falcioni, M., Paladin, G., Vulpiani, A. (1988): Regular and chaotic motion of fluid particles in two-dimensional fluid. J. Phys. A Math. Gen. **21**, 3451–3462.

Falcioni, M., Paladin, G., Vulpiani, A. (1989): Intermittency and multifractality in magnetic dynamos. Europhys. Lett. **10**, 201–206.

Falconer, K.J. (1990): Fractal Geometry. John Wiley & Sons.

Fannjiang, A., Papanicolaou, G. (1994): Convection enhanced diffusion for periodic flows. SIAM J. Appl. Math. **54**, 333–408.

Feingold, M., Kadanoff, L.P., Piro, O. (1988): Passive scalars, three-dimensional volume-preserving maps, and chaos. J. Stat. Phys. **50**, 529–565.

Finn. J.M., Ott, E. (1988a): Chaotic flows and magnetic dynamos. Phys. Rev. Lett. **60**, 760–763.

Finn, J.M., Ott, E. (1988b): Chaotic flows and fast magnetic dynamos. Phys. Fluids **31**, 2992–3011.

Finn, J.M., Ott, E. (1990): The fast kinematic magnetic dynamo and the dissipationless limit. Phys. Fluids B **2**, 916–926.

Finn, J.M., Hanson, J.D., Kan, I., Ott, E. (1989): Do steady fast dynamos exist? Phys. Rev. Lett. **62**, 2965–2968.

Finn, J.M., Hanson, J.D., Kan, I., Ott, E. (1991): Steady fast dynamo flows. Phys. Fluids B **3**, 1250–1259.

Flanders, H. (1963): Differential Forms. Academic Press.

Freedman, M.H. (1988): A note on topology and magnetic energy in incompressible perfectly conducting fluids. J. Fluid Mech. **194**, 549–551.

Friedlander, S., Vishik. M.M. (1992): Instability criteria in fluid dynamics. In: Topological Aspects of the Dynamics of Fluids and Plasmas (ed. H.K. Moffatt, G.M. Zaslavsky, P. Comte, M. Tabor), pp. 535–549. Kluwer Academic Publishers.

Friedlander, S., Gilbert, A.D., Vishik, M.M. (1993): Hydrodynamic instability for certain ABC flows. Geophys. Astrophys. Fluid Dyn. **73**, 97–107.

Friedman, B. (1956): Principles and Techniques of Applied Mathematics. John Wiley & Sons.

Friedrichs, K.O. (1965): Perturbation of Spectra in Hilbert Space. Lectures in Applied Mathematics, vol. III. AMS Publications, Providence.

Frisch, U., Pouquet, A., Sulem, P.L., Meneguzzi, M. (1983): Dynamics of two-dimensional ideal MHD. J. Méch. Théor. Appl. (special issue on two-dimensional turbulence), 191–216.

Furstenberg, H. (1963): Noncommuting random products. Trans. Am. Math. Soc. **108**, 377–428.

Galanti, B., Sulem, P.L., Pouquet, A. (1992): Linear and nonlinear dynamos associated with ABC flows. Geophys. Astrophys. Fluid Dyn. **66**, 183–208.

Galanti, B., Pouquet, A., Sulem, P.-L. (1993): Influence of the period of an ABC flow on its dynamo action. In: Solar and Planetary Dynamos (ed. M.R.E. Proctor, P.C. Matthews, A.M. Rucklidge), pp. 99–104. Cambridge University Press.

Galloway, D.J., Frisch, U. (1984): A numerical investigation of magnetic field generation in a flow with chaotic streamlines. Geophys. Astrophys. Fluid Dyn. **29**, 13–18.

Galloway, D.J., Frisch, U. (1986): Dynamo action in a family of flows with chaotic streamlines. Geophys. Astrophys. Fluid Dyn. **36**, 53–83.

Galloway, D.J., Frisch, U. (1987): A note on the stability of a family of space-periodic Beltrami flows. J. Fluid Mech. **180**, 557–564.

Galloway, D.J., Moore, D.R. (1979): Axisymmetric convection in the presence of a magnetic field. Geophys. Astrophys. Fluid Dyn. **12**, 73–105.

Galloway, D.J., O'Brian, N.R. (1993): Numerical calculations of dynamos for ABC and related flows. In: Solar and Planetary Dynamos (ed. M.R.E. Proctor, P.C. Matthews, A.M. Rucklidge), pp. 105–113. Cambridge University Press.

Galloway, D.J., Proctor, M.R.E. (1992): Numerical calculations of fast dynamos for smooth velocity fields with realistic diffusion. Nature **356**, 691–693.

Galloway, D.J., Weiss, N.O. (1981): Convection and magnetic fields in stars. Astrophys. J. **243**, 945–953.

Galloway, D.J., Proctor, M.R.E., Weiss, N.O. (1978): Magnetic flux ropes and convection. J. Fluid Mech. **87**, 243–261.

Ghil, M., Childress, S. (1987): Topics in Geophysical Fluid Dynamics, Atmospheric Dynamics, Dynamo Theory, and Climate Dynamics. Springer–Verlag.

Gilbert, A.D. (1988): Fast dynamo action in the Ponomarenko dynamo. Geophys. Astrophys. Fluid Dyn. **44**, 214–258.

Gilbert, A.D. (1991a): Fast dynamo action in a steady chaotic flow. Nature **350**, 483–485.

Gilbert, A.D. (1991b): Note on the symmetries of the 3-d flow of Galloway and Proctor. Unpublished.

Gilbert, A.D. (1992): Magnetic field evolution in steady chaotic flows. Phil. Trans. R. Soc. Lond. A **339**, 627–656.

Gilbert, A.D. (1993): Towards a realistic fast dynamo: models based on cat maps and pseudo-Anosov maps. Proc. R. Soc. Lond. A **443**, 585–606.

Gilbert, A.D., Bayly, B.J. (1992): Magnetic field intermittency and fast dynamo action in random helical flows. J. Fluid Mech. **241**, 199–214.

Gilbert, A.D., Childress, S. (1990): Evidence for fast dynamo action in a chaotic web. Phys. Rev. Lett. **65**, 2133–2136.

Gilbert, A.D., Sulem, P.-L. (1990): On inverse cascades in alpha effect dynamos. Geophys. Astrophys. Fluid Dyn. **51**, 243–261.

Gilbert, A.D., Otani, N.F., Childress, S. (1993): Simple dynamical fast dynamos. In: Solar and Planetary Dynamos (ed. M.R.E. Proctor, P.C. Matthews, A.M. Rucklidge), pp. 129–136. Cambridge University Press.

Gilman, P.A. (1983): Dynamically consistent nonlinear dynamos driven by convection in a rotating spherical shell. II. Dynamos with cycles and strong feedbacks. Astrophys. J. Suppl. **53**, 243–268.

Glatzmaier, G.A. (1985): Numerical simulations of stellar convective dynamos. II. Field propagation in the convection zone. Astrophys. J. **192**, 300–307.

Gold, T. (1968): Rotating neutron stars as the origin of the pulsating radio sources. Nature **218**, 731–732.

Golitsyn, G.S. (1960): Fluctuations of the magnetic field and current density in a turbulent flow of a weakly conducting fluid. Sov. Phys. Dokl. **5**, 536–539.

Gough, D.O. (1969): The anelastic approximation for thermal convection. J. Atmos. Sci. **26**, 448–456.

Grebogi, C., Ott, E., Yorke, J.A. (1988): Unstable periodic orbits and the dimensions of multifractal chaotic attractors. Phys. Rev. A **37**, 1711–1724.

Grebogi, C., Hammel, S.M., Yorke, J.A., Sauer, T. (1990): Shadowing of physical trajectories in chaotic dynamics: containment and refinement. Phys. Rev. Lett. **65**, 1527–1530.

Guckenheimer, J., Holmes, P. (1983): Nonlinear Oscillations, Dynamical Systems, and Bifurcations of Vector Fields. Springer–Verlag.

Gutzwiller, M.C. (1990): Chaos in Classical and Quantum Mechanics. Springer–Verlag.

Hamermesh, M. (1962): Group Theory. Addison–Wesley.

Hardy, G.H., Littlewood, J.E., Pólya, G. (1967): Inequalities. Cambridge University Press.

Henningson, D.S., Reddy, S.C. (1994): On the role of linear mechanisms in transition to turbulence. Phys. Fluids **6**, 1396–1398.

Hénon, M. (1966): Sur la topologie des lignes de courant dans un cas particulier. C.R. Acad. Sci. Paris A **262**, 312–314.

Herzenberg, A. (1958): Geomagnetic dynamos. Phil. Trans. R. Soc. Lond. A **250**, 543–583.

Hewitt, J.M., McKenzie, D.D., Weiss, N.O. (1975): Dissipative heating in convective flows. J. Fluid Mech. **68**, 721–738.

Hide, R. (1988): Towards the interpretation of Uranus's eccentric magnetic field. Geophys. Astrophys. Fluid Dyn. **44**, 207–209.

Hide, R., Roberts, P.H. (1961): The origin of the main geomagnetic field. In: Physics and Chemistry of the Earth (ed. L.H. Ahrens, K. Rankama, S.K. Runcorn), pp. 25–98. Pergamon Press.

Hollerbach, R., Galloway, D.J., Proctor, M.R.E. (1995): Numerical evidence of fast dynamo action in a spherical shell. Phys. Rev. Lett. **74**, 3145–3148.

Hoyng, P. (1987a): Turbulent transport of magnetic fields: I. A simple mechanical model. Astron. Astrophys. **171**, 348–356.

Hoyng, P. (1987b): Turbulent transport of magnetic fields: II. The role of fluctuations in kinematic theory. Astron. Astrophys. **171**, 357–367.

Hoyng, P. (1988): Turbulent transport of magnetic fields. III. Stochastic excitation of global magnetic modes. Astrophys. J. **332**, 857–871.

Hughes, D.W. (1992): The formation of flux tubes at the base of the convection zone. In: Sunspots: Theory and Observations (ed. J.H. Thomas, N.O. Weiss), pp. 371–384. Kluwer.

Hughes, D.W., Proctor, M.R.E. (1988): Magnetic fields in the solar convection zone. Ann. Rev. Fluid Mech. **20**, 187–223.

Ince, E.L. (1926): Ordinary Differential Equations. Reprinted by Dover Publications (1956).

Iroshnikov, P.S. (1964): Turbulence of a conducting fluid in a strong magnetic field. Sov. Astron. **7**, 566–571.

Isaacson, E., Keller, H.B. (1966): Analysis of Numerical Methods. John Wiley & Sons.

Isichenko, M.B. (1992): Percolation, statistical topography, and transport in random media. Rev. Mod. Phys. **64**, 961–1043.

Ivers, D.J. (1984): Antidynamo theorems. PhD Thesis, University of Sydney.

Ivers, D.J., James, R.W. (1984): Axisymmetric antidynamo theorems in compressible non-uniform conducting fluids. Phil. Trans. R. Soc. Lond. A **312**, 179–218.

James, R.W., Roberts, P.H., Winch, D.E. (1980): The Cowling anti-dynamo theorem. Geophys. Astrophys. Fluid Dyn. **15**, 149–160.

Jones, C.A., Galloway, D.J. (1993*a*): Axisymmetric magnetoconvection in a twisted field. J. Fluid Mech. **253**, 297–326.

Jones, C.A., Galloway, D.J. (1993*b*): Alpha-quenching in cylindrical magnetoconvection. In: Solar and Planetary Dynamos (ed. M.R.E. Proctor, P.C. Matthews, A.M. Rucklidge), pp. 161–170. Cambridge University Press.

Kadanoff, L., Tang, C. (1984): Escape from strange repellers. Proc. Nat. Acad. Sci. USA **81**, 1276–1279.

Kaiser, R., Schmitt, B.J., Busse, F.H. (1994): On the invisible dynamo. Geophys. Astrophys. Fluid Dyn. **77**, 93–109.

Kaplun, S. (1967): Fluid Mechanics and Singular Perturbations. Academic Press.

Kato, T. (1966): Perturbation Theory of Linear Operators. Springer–Verlag.

Kazantsev, A.P. (1967): Enhancement of a magnetic field by a conducting fluid. Zh. Eksp. Teor. Fiz. **53**, 1806–1813. [English transl.: Sov. Phys. JETP **26** (1968) 1031–1034.]

Keldysh, M.V. (1951): On the eigenvalues and eigenfunctions of certain non-self-adjoint equations. Dokl. Akad. Nauk SSSR **77**, 11–14.

Kerswell, R., Childress, S. (1992): Equilibrium of a magnetic flux tube in a compressible flow. Astrophys. J. **385**, 746–757.

Klapper, I. (1991): Chaotic fast dynamos. PhD Thesis, Courant Institute of Mathematical Sciences, New York University.

Klapper, I. (1992*a*): Shadowing and the role of small diffusivity in the chaotic advection of scalars. Phys. Fluids A **4**, 861–864.

Klapper, I. (1992*b*): A study of fast dynamo action in chaotic helical cells. J. Fluid Mech. **239**, 359–381.

Klapper, I. (1992*c*): Kinematic fast dynamo action in a time-periodic chaotic flow. In: Topological Aspects of the Dynamics of Fluids and Plasmas (ed. H.K. Moffatt, G.M. Zaslavsky, P. Comte, M. Tabor), pp. 563–571. Kluwer Academic Publishers.

Klapper, I. (1993): Shadowing and the diffusionless limit in fast dynamo theory. Nonlinearity **6**, 869–884.

Klapper, I., Young, L.S. (1995): Bounds on the fast dynamo growth rate involving topological entropy. Comm. Math. Phys., to appear.

Knobloch, E. (1977): The diffusion of scalar and vector fields by homogeneous stationary turbulence. J. Fluid Mech. **83**, 129–140.

Koch, D.L., Cox, R.G., Brenner, H., Brady, J.F. (1989): The effect of order on dispersion in porous media. J. Fluid Mech. **200**, 173–188.

Kraichnan, R.H. (1965): Inertial-range spectrum of hydromagnetic turbulence. Phys. Fluids **8**, 1385–1387.

Kraichnan, R.H. (1970): Diffusion by a random velocity field. Phys. Fluids **13**, 22–31.

Kraichnan, R.H. (1974): Convection of a passive scalar by a quasi-uniform random straining field. J. Fluid Mech. **64**, 737–762.

Kraichnan, R.H. (1976*a*): Diffusion of weak magnetic fields by isotropic turbulence. J. Fluid Mech. **75**, 657–676.

Kraichnan, R.H. (1976*b*): Diffusion of passive-scalar and magnetic fields by helical turbulence. J. Fluid Mech. **77**, 753–768.

Kraichnan, R.H. (1979): Consistency of the α-effect turbulent dynamo. Phys. Rev. Lett. **42**, 1677–1680.

Kraichnan, R.H., Nagarajan, S. (1967): Growth of turbulent magnetic fields. Phys. Fluids **10**, 859–870.

Krause, F., Rädler, K.-H. (1980): Mean-field magnetohydrodynamics and dynamo theory. Pergamon Press.

Kronberg, P.P. (1994): Extragalactic magnetic fields. Rep. Prog. Phys. **57**, 325–382.

Kulsrud, R.M., Anderson, S.W. (1992): The spectrum of random magnetic fields in the mean field dynamo theory of the galactic magnetic field. Astrophys. J. **396**, 606–630.

Kulsrud, R.M., Anderson, S.W. (1993): Magnetic fluctuations in fast dynamos. In: Solar and Planetary Dynamos (ed. M.R.E. Proctor, P.C. Matthews, A.M. Rucklidge), pp. 195–202. Cambridge University Press.

Lagerstrom, P.A. (1975): Solutions of the Navier–Stokes equations at large Reynolds number. SIAM J. Appl. Math. **28**, 202–214.

Lamb, H. (1932): Hydrodynamics. Cambridge University Press.

Larmor, J. (1919): How could a rotating body such as the Sun become a rotating magnet? Rep. Brit. Assoc. Adv. Sci. (1919) 159–160.

Latour, J., Speigel, E.A , Toomre, J., Zahn, J.-P. (1976): Stellar convection theory I. The anelastic modal equations. Astrophys. J. **207**, 233–243.

Latushkin, Yu.D., Stepin, A.M. (1991): Weighted translation operators and linear extensions of dynamical systems. Uspekhi Mat. Nauk **46**(2), 85–143. [English transl.: Russian Math. Surveys **46**(2) (1991) 95–165.]

Lau, Y.-T., Finn, J.M. (1993a): Fast dynamos with finite resistivity in steady flows with stagnation points. Phys. Fluids B **5**, 365–375.

Lau, Y.-T., Finn, J.M. (1993b): The magnetic field structures of a class of fast dynamos. In: The Cosmic Dynamo (ed. F. Krause, K.-H. Rädler, G. Rüdiger), pp. 231–235. Kluwer Academic Publishers.

Lefshetz, S. (1930): Topology. Am. Math. Soc. Coll. Ser., vol. 12. AMS Publications, Providence.

Lichtenberg, A.J., Lieberman, M.A. (1983): Regular and Stochastic Motion. Springer-Verlag.

Lifschitz, A.E. (1989): Magnetohydrodynamics and Spectral Theory. Kluwer Academic Publishers.

Lighthill, M.J. (1962): Introduction to Fourier Analysis and Generalized Functions. Cambridge University Press.

de la Llave, R. (1993): Hyperbolic dynamical systems and generation of magnetic fields by perfectly conducting fluids. Geophys. Astrophys. Fluid Dyn. **73**, 123–131.

Loper, D.E., Roberts, P.H. (1983): Compositional convection. In: Stellar and Planetary Magnetism (ed. A.M. Soward), pp. 297–327. Gordon & Breach Science Publishers.

Lortz, D. (1968): Exact solutions of the hydrodynamic dynamo problem. Plasma Physics **10**, 967–972.

Mackay, R.S., Meiss, J.D., Percival, I.C. (1984): Transport in Hamiltonian systems. Physica D **13**, 55–81.

Majda, A.J., McLaughlin, R.M. (1993): The effect of mean flows on enhanced diffusivity in transport by incompressible periodic velocity fields. Stud. Appl. Math. **89**, 245–279.

Margulis, G.A. (1967): Y-flow on three-dimensional manifolds. Appendix to: Some smooth ergodic systems (by D.V. Anosov, Ya.G. Sinai). Usp. Math. Nauk **22**(5) 107–172. [English transl.: Russ. Math. Surveys **22**(5) (1967) 103–167.]

Markus, A.S., Matsaev, V.I. (1984): Comparison theorems for spectra of linear operators and spectral asymptotics. Trans. Moscow Math. Soc. **45**, 139–187.

Mather, J.N. (1968): Characterization of Anosov diffeomorphisms. Indagat. Math. **30**, 479–483.

McKean, H.P. (1969): Stochastic Integrals. Academic Press.

McMillan, J. (1988): A quasi-dynamical dynamo model. Woods Hole Oceanographic Inst. Tech. Rep. WHOI-88-16, pp. F7.1–21.

Melnikov, V.K. (1963): On the stability of the center for time-periodic perturbations. Trans. Moscow Math. Soc. **12**, 1–57.

Meneguzzi, M., Pouquet, A. (1989): Turbulent dynamos driven by convection. J. Fluid Mech. **205**, 297–318.

Milne-Thomson, L.M. (1955): Theoretical Hydrodynamics. The Macmillan Company, New York

Moffatt, H.K. (1961): The amplification of a weak applied magnetic field by turbulence in fluids of moderate conductivity. J. Fluid Mech. **11**, 625–635.

Moffatt, H.K. (1969): The degree of knottedness of tangled vortex lines. J. Fluid Mech. **35**, 117–129.

Moffatt, H.K. (1974): The mean electromotive force generated by turbulence in the limit of perfect conductivity. J. Fluid Mech. **65**, 1–10.

Moffatt, H.K. (1978): Magnetic Field Generation in Electrically Conducting Fluids. Cambridge University Press.

Moffatt, H.K. (1985): Magnetostatic equilibria and analogous Euler flows of arbitrarily complex topology. Part I. Fundamentals. J. Fluid Mech. **159**, 359–378.

Moffatt, H.K. (1986): Magnetostatic equilibria and analogous Euler flows of arbitrarily complex topology. Part II. Stability considerations. J. Fluid Mech. **166**, 359–378.

Moffatt, H.K. (1989): Stretch, twist and fold. Nature **341**, 285–286.

Moffatt, H.K., Kamkar, H. (1983): On the time-scale associated with flux expulsion. In: Stellar and Planetary Magnetism (ed. A.M. Soward), pp. 91–97. Gordon & Breach Science Publishers.

Moffatt, H.K., Proctor, M.R.E. (1985): Topological constraints associated with fast dynamo action. J. Fluid Mech. **154**, 493–507.

Molchanov, S.A., Ruzmaikin, A.A., Sokoloff, D.D. (1983): Equations of dynamo in random velocity field with short correlation time. Magn. Gidrodin., no. 4, 67–72. [English transl.: Magnetohydrodynamics **19** (1983) 402–407.]

Molchanov, S.A., Ruzmaikin, A.A., Sokoloff, D.D. (1984): A dynamo theorem. Geophys. Astrophys. Fluid Dyn. **30**, 242–259.

Molchanov, S.A., Ruzmaikin, A.A., Sokoloff, D.D. (1985): Kinematic dynamo in random flow. Usp. Fiz. Nauk **145**, 593–628. [English transl.: Sov. Phys. Usp. **28** (1985) 307–327.]

Morgan, F. (1988): Geometric Measure Theory. Academic Press.

Newhouse, S.E. (1980): Lectures on Dynamical Systems. In: Dynamical Systems. Progress in Mathematics, no. 8, pp. 1–114. Birkhäuser, Boston.

Newhouse, S.E. (1987): Entropy and volume as measures of orbit complexity. In: The Physics of Phase Space (ed. Y.S. Kim, W.W. Zachary), pp. 2–8. Springer-Verlag.

Nordlund, Å., Brandenburg, A., Jennings, R.L., Rieutord, M., Ruokolainen, J., Stein, R.F., Tuominen, I. (1992): Dynamo action in stratified convection with overshoot. Astrophys. J. **392**, 647–652.

Novikov, V.G., Ruzmaikin, A.A., Sokoloff, D.D. (1983): Kinematic dynamo in a reflection-invariant random field. Zh. Eksp. Teor. Fiz. **85**, 909–918. [English transl.: Sov. Phys. JETP **58** (1983) 527–532.]

Núñez, M. (1993): Localized magnetic fields in a perfectly conducting fluid. In: Solar and Planetary Dynamos (ed. M.R.E. Proctor, P.C. Matthews, A.M. Rucklidge), pp. 225–229. Cambridge University Press.

Núñez, M. (1994): Localized eigenmodes of the induction equation. SIAM J. Appl. Math. 5, 1254–12267.

Orszag, S.A. (1970): Comments on 'Turbulent hydrodynamic line stretching: consequences of isotropy.' Phys. Fluids 13, 2203–2204.

Oseledets, V.I. (1968): A multiplicative ergodic theorem: Liapunov characteristic numbers for dynamical systems. Trudy. Moskov. Mat. Obusuc. 19, 179. [English transl.: Trans. Moscow Math. Soc. 19 (1968) 197–231.]

Oseledets, V.I. (1984): Liapunov entropy and the spectral radius of the dynamo operator. Sixth International Symposium on Information Theory, Tashkent, 1984, Abstracts, Part 3, pp 162–163. IPI.

Oseledets, V.I. (1993): Fast dynamo problem for a smooth map on a two-torus. Geophys. Astrophys. Fluid Dyn. 73, 133–145.

Otani, N.F. (1988): Computer simulation of fast kinematic dynamos. EOS, Trans. Am. Geophys. Union 69(44), abstract no. SH51–15, p. 1366.

Otani, N.F. (1993): A fast kinematic dynamo in two-dimensional time-dependent flows. J. Fluid Mech. 253, 327–340.

Ott, E., Antonsen, T.M. (1988): Chaotic fluid convection and the fractal nature of passive scalar gradients. Phys. Rev. Lett. 61, 2839–2842.

Ott, E., Antonsen, T.M. (1989): Fractal measures of passively convected vector fields and scalar gradients in chaotic fluid flows. Phys. Rev. A 39, 3660–3671.

Ott, E., Vainshtein, S.I. (1995): On dynamo generation of magnetic flux and fractal properties of the field. In: Research Trends in Physics: Plasma Astrophysics (ed. R.M. Kulsrud, G. Burbidge, V. Stefan), to appear. Am. Inst. Phys., New York.

Ott, E., Du, Y., Sreenivasan, K.R., Juneja, A., Suri, A.K. (1992): Sign-singular measures: fast magnetic dynamos and high Reynolds number fluid turbulence. Phys. Rev. Lett. 69, 2654–2657.

Ottino, J.M. (1989): The Kinematics of Mixing: Stretching, Chaos, and Transport. Cambridge University Press.

Ozorio de Almeida, A.M. (1988): Hamiltonian Systems: Chaos and Quantization. Cambridge University Press.

Parker, E.N. (1955): Hydromagnetic dynamo models. Astrophys. J. 122, 293–314.

Parker, R.L. (1966): Reconnexion of lines of force in rotating spheres and cylinders. Proc. R. Soc. Lond. A 291, 60–72.

Parker, E.N. (1979): Cosmical Magnetic Fields. Clarendon Press.

Parker, E.N. (1992): Fast dynamos, cosmic rays, and the galactic magnetic field. Astrophys. J. 401, 137–145.

Peckover, R.S., Weiss, N.O. (1978): On the dynamic interaction between magnetic fields and convection. Mon. Not. Roy. Astr. Soc. 182, 189–208.

Perkins, F.W., Zweibel, E.G. (1987): A high magnetic Reynolds number dynamo. Phys. Fluids 30, 1079–1084.

Pitskel, B. (1991): Poisson limit law for Markov chains. Ergod. Th. & Dynam. Sys. 11, 501–513.

Plante, J.F., Thurston, W.P. (1972): Anosov flows and the fundamental group. Topology 11, 147–150.

Poénaru, V. (1979): Présentation d'ensemble des théorèmes de Thurston sur les surfaces. Astérisque 66–67, 5–20.

Poirier, J.P. (1988): Transport properties of liquid metals and viscosity of the Earth's core. Geophys. J. 92, 99–105.

Ponomarenko, Y.B. (1973): On the theory of hydromagnetic dynamos. Zh. Prikl. Mekh. & Tekh. Fiz. (USSR) 6, 47–51.

Ponty, Y., Pouquet, A., Rom–Kedar, V., Sulem, P.-L. (1993): Dynamo action in a nearly-integrable chaotic flow. In: Solar and Planetary Dynamos (ed. M.R.E. Proctor, P.C. Matthews, A.M. Rucklidge), pp. 241–248. Cambridge University Press.

Ponty, Y., Pouquet, A., Sulem, P.-L. (1995): Dynamos in weakly chaotic two-dimensional flows. Geophys. Astrophys. Fluid Dyn., in press.

Pouquet, A., Frisch, U., Léorat, J. (1976): Strong MHD helical turbulence and the nonlinear dynamo effect. J. Fluid Mech. **77**, 321–354.

Prandtl, L. (1904): Überflüssigkeitsbewegung bei sehr kleiner reibung. International Mathematical Congress, Heidelberg, 484–491. See Gesammelte Abhandlungen (1961), 575–584.

Prandtl, L. (1952): Essentials of Fluid Mechanics. Hafner Publishing Company.

Prautzsch, T. (1993): The dynamo mechanism in the deep convection zone of the Sun. In: Solar and Planetary Dynamos (ed. M.R.E. Proctor, P.C. Matthews, A.M. Rucklidge), pp. 249–256. Cambridge University Press.

Priest, E.R. (1984): Solar Magnetohydrodynamics. Reidel Publishing Company.

Proctor, M.R.E. (1981): Planform selection by finite-amplitude thermal convection between poorly conducting slabs. J. Fluid Mech. **113**, 469–485.

Rado, A. (1993): Onset and intermittency in the SFS map. MS Thesis, Courant Institute of Mathematical Sciences, New York University.

Rechester, A.B., White, R.B. (1980): Calculation of turbulent diffusion for the Chirikov–Taylor model. Phys. Rev. Lett. **44**, 1586–1589.

Reddy, S.C., Henningson, D.S. (1993): Energy growth in viscous channel flows. J. Fluid Mech. **252**, 209–238.

Rhines, P.B., Young, W.R. (1983): How rapidly is a passive scalar mixed within closed streamlines? J. Fluid Mech. **133**, 133–145.

Robbins, K.A. (1977): A new approach to subcritical instability and turbulent transition in a simple dynamo. Proc. Camb. Philos. Soc. **82**, 309–325.

Roberts, P.H. (1960): Characteristic value problems posed by differential equations arising in hydrodynamics and hydromagnetics. J. Math. Anal. Appl. **1**, 195–214.

Roberts, P.H. (1967): An Introduction to Magnetohydrodynamics. American Elsevier, New York.

Roberts, G.O. (1970): Spatially periodic dynamos. Phil. Trans. R. Soc. Lond. A **266**, 535–558.

Roberts, P.H. (1971): Dynamo theory. In: Mathematical Problems in the Geophysical Sciences, vol. 2 (ed. W.H. Reid), pp. 129–206. Lectures in Applied Mathematics, vol. 14. AMS Publications, Providence.

Roberts, G.O. (1972): Dynamo action of fluid motions with two-dimensional periodicity. Phil. Trans. R. Soc. Lond. A **271**, 411–454.

Roberts, P.H. (1987): Dynamo theory. In: Irreversible Phenomena and Dynamical Systems Analysis in Geosciences (ed. C. Nicolis, G. Nicolis), pp. 73–133. D. Reidel.

Roberts, P.H., Soward, A.M. (1992): Dynamo theory. Ann. Rev. Fluid Mech. **24**, 459–512.

Rosenbluth, M.N., Berk, H.L., Doxas, I., Horton, W. (1987): Effective diffusion in laminar convective flows. Phys. Fluids **30**, 2636–2647.

Rüdiger, G., Kichatinov, L.L. (1993): Alpha-effect and alpha-quenching. Astron. Astrophys. **269**, 581–588.

Ruelle, D. (1978): Thermodynamic Formalism. Addison–Wesley.

Ruelle, D. (1989a): Chaotic Evolution and Strange Attractors. Cambridge University Press.

Ruelle, D. (1989b): The thermodynamic formalism for expanding maps. Comm. Math. Phys. **125**, 239–262.

Ruelle, D. (1989*c*): Une extension de la théorie de Fredholm. C. R. Acad. Sci. Paris, Série I **309**, 309–310.

Rugh, H.H. (1994): On the asymptotic form and the reality of spectra of Perron–Frobenius operators. Nonlinearity **7**, 1055–1066.

Ruzmaikin, A.A., Shukurov, A.M. (1982): Spectrum of the galactic magnetic field. Astrophys. Spa. Sci. **82**, 397–407.

Ruzmaikin, A.A., Sokoloff, D.D. (1980): Helicity, linkage, and dynamo action. Geophys. Astrophys. Fluid Dyn. **16**, 73–82.

Ruzmaikin, A.A., Sokoloff, D.D., Shukurov, A.M. (1988): A hydromagnetic screw dynamo. J. Fluid Mech. **197**, 39–56.

Ruzmaikin, A.A., Liewer, P.C., Feynman, J. (1993): Random cell dynamo. Geophys. Astrophys. Fluid Dyn. **73**, 163–177.

Sattinger, D.H. (1970): The mathematical problem of hydrodynamic stability. J. Math. Mech. **19**, 797–817.

Schmidt, G., Chernikov, A.A., Rogalsky, A.V. (1993): Magnetic field generation in convective cells. Ann. New York Acad. Sci. **706**, 161–169.

Schüssler, M. (1990): Theoretical aspects of small-scale photospheric magnetic fields. In: Solar Photosphere: Structure, Convection and Magnetic Fields, IAU Symp. 138 (ed. J.O. Stenflo), pp. 161–179. Kluwer Academic Publishers.

Schutz, B.F. (1980): Geometrical Methods of Mathematical Physics. Cambridge University Press.

Shannon, C.E., Weaver, W. (1964): The Mathematical Theory of Communication. University of Illinois Press.

Shraiman, B.I. (1987): Diffusive transport in a Rayleigh–Bénard convection cell. Phys. Rev. A **36**, 261–267.

Smale, S. (1967): Differentiable dynamical systems. Bull. Amer. Math. Soc. **73**, 199–206.

Soward, A.M. (1971): Nearly symmetric advection. J. Math. Phys. **12**, 2052–2062.

Soward, A.M. (1972): A kinematic theory of large magnetic Reynolds number dynamos. Phil. Trans. R. Soc. Lond. A **272**, 431–462.

Soward, A.M. (1987): Fast dynamo action in a steady flow. J. Fluid Mech. **180**, 267–295.

Soward, A.M. (1988): Fast dynamos with flux expulsion. In: Secular, Solar and Geomagnetic Variation in the Last 10,000 Years (ed. F.R. Stephenson, A.W. Wolfendale), pp. 79–96. NATO ASI Series C: Mathematical and Physical Sciences, vol. 236. Kluwer Academic Publishers.

Soward, A.M. (1989): On dynamo action in a steady flow at large magnetic Reynolds number. Geophys. Astrophys. Fluid Dyn. **49**, 3–22.

Soward, A.M. (1990): A unified approach to a class of slow dynamos. Geophys. Astrophys. Fluid Dyn. **53**, 81–107.

Soward, A.M. (1993*a*): An asymptotic solution of a fast dynamo in a two-dimensional pulsed flow. Geophys. Astrophys. Fluid Dyn. **73**, 179–215.

Soward, A.M. (1993*b*): Analytic fast dynamo solution for a two-dimensional pulsed flow. In: Solar and Planetary Dynamos (ed. M.R.E. Proctor, P.C. Matthews, A.M. Rucklidge), pp. 275–286. Cambridge University Press.

Soward, A.M. (1994*a*): On the role of stagnation points and periodic particle paths in a two-dimensional pulsed flow fast dynamo model. Physica D **76**, 181–201.

Soward, A.M. (1994*b*): Fast dynamos. In: Lectures on Solar and Planetary Dynamos (ed. M.R.E. Proctor, A.D. Gilbert), pp. 181–217. Cambridge University Press.

Soward, A.M., Childress, S. (1990): Large magnetic Reynolds number dynamo action in spatially periodic flow with mean motion. Phil. Trans. R. Soc. Lond. A **331**, 649–733.

Spiegel, E.A. (1994): The chaotic solar cycle. In: Lectures on Solar and Planetary Dynamos (ed. M.R.E. Proctor, A.D. Gilbert), pp. 245–265. Cambridge University Press.

Spiegel, E.A., Weiss, N.O. (1980): Magnetic activity and variations in solar luminosity. Nature **287**, 5783–5784.

Spiegel, E.A., Zahn, J.-P. (1992): The solar tachocline. Nature **287**, 616–617.

Steenbeck, M., Krause, F. (1969): On the dynamo theory of stellar and planetary magnetic fields. I. A.C. dynamos of solar type. Astron. Nachr. **291**, 49–84.

Steenbeck, M., Krause, F., Rädler, K.-H. (1966): A calculation of the mean electromotive force in an electrically conducting fluid in turbulent motion, under the influence of Coriolis forces. Z. Naturforsch. **21a**, 369–376.

Strang, G. (1989): Wavelets and dilation equations: a brief introduction. SIAM Review **31**, 614–627.

Strauss, H.R. (1986): Resonant fast dynamo. Phys. Rev. Lett. **57**, 2231–2233.

Tao, L., Cattaneo, F., Vainshtein, S.I. (1993): Evidence for the suppression of the alpha-effect by weak magnetic fields. In: Solar and Planetary Dynamos (ed. M.R.E. Proctor, P.C. Matthews, A.M. Rucklidge), pp. 303–310. Cambridge University Press.

Taylor, A.E. (1958): Introduction to Functional Analysis. John Wiley & Sons.

Tennekes, H., Lumley, J.L. (1989): A First Course in Turbulence. MIT Press.

Thompson, M. (1990): Kinematic dynamo in random flows. Mat. Aplic. Comp. **9**, 213–245.

Trefethen, L.N. (1992): Pseudospectra of matrices. In: Numerical Analysis 1991 (ed. D.F. Griffiths, G.A. Watson), pp. 1–33. Longman.

Trefethen, L.N., Trefethen, A.E., Reddy, S.C., Driscoll, T.A. (1993): Hydrodynamic stability without eigenvalues. Science **261**, 578–584.

Vainshtein, S.I. (1970): The generation of a large-scale magnetic field by a turbulent fluid. Zh. Eksp. Teor. Fiz. **58**, 153–159. [English transl.: Sov. Phys. JETP **31** (1970) 87–89.]

Vainshtein, S.I. (1982): Theory of small-scale magnetic fields. Zh. Eksp. Teor. Fiz. **83**, 161–175. [English transl.: Sov. Phys. JETP **56** (1982) 86–94.]

Vainshtein, S.I., Cattaneo, F. (1992): Nonlinear restrictions on dynamo action. Astrophys. J. **393**, 199–203.

Vainshtein, S.I., Kichatinov, L.L. (1986): The dynamics of magnetic fields in a highly conducting turbulent medium and the generalized Kolmogorov–Fokker–Planck equations. J. Fluid Mech. **168**, 73–87.

Vainshtein, S.I., Rosner, R. (1991): On turbulent diffusion of magnetic fields and the loss of magnetic flux from stars. Astrophys. J. **376**, 199–203.

Vainshtein, S.I., Zeldovich, Ya.B. (1972): Origin of magnetic fields in astrophysics. Usp. Fiz. Nauk **106**, 431–457. [English transl.: Sov. Phys. Usp. **15** (1972) 159–172.]

Vainshtein, S.I., Parker, E.N., Rosner, R. (1993a): On the generation of 'strong' magnetic fields. Astrophys. J. **404**, 773–780.

Vainshtein, S.I., Tao, L., Cattaneo, F., Rosner, R. (1993b): Turbulent magnetic transport effects and their relation to magnetic field intermittency. In: Solar and Planetary Dynamos (ed. M.R.E. Proctor, P.C. Matthews, A.M. Rucklidge), pp. 311–320. Cambridge University Press.

Van Dyke, M. (1975): Perturbation Methods in Fluid Mechanics. Parabolic Press, Palo Alto, California.

Varadhan, S.R.S. (1984): Large Deviations and Applications. SIAM Press.

Városi, F., Antonsen, T.M., Ott, E. (1991): The spectrum of fractal dimensions of passively convected scalar gradients in chaotic fluid flows. Phys. Fluids A **3**, 1017–1028.

Vishik, M.M. (1988): Magnetic field generation by a three-dimensional stationary flow of a conducting fluid at large magnetic Reynolds numbers. Izv., Akad. Nauk SSSR, Fiz. Zemli, no. 3, 3–12. [English transl.: Izv., Acad. Sci. USSR, Phys. Solid Earth **24**(3) (1988) 173–180.]

Vishik, M.M. (1989): Magnetic field generation by the motion of a highly conducting fluid. Geophys. Astrophys. Fluid Dyn. **48**, 151–167.

Vishik, M.M. (1992): Lecture given at Isaac Newton Institute, University of Cambridge.

Walters, P. (1982): An Introduction to Ergodic Theory. Springer–Verlag.

Weiss, N.O. (1966): The expulsion of magnetic flux by eddies. Proc. R. Soc. Lond. A **293**, 310–328.

Weiss, N.O. (1971): The dynamo problem. Q. J. Roy. Astr. Soc. **12**, 432–446.

Weiss, N.O. (1994): Solar and stellar dynamos. In: Lectures on Solar and Planetary Dynamos (ed. M.R.E. Proctor, A.D. Gilbert), pp. 59–95. Cambridge University Press.

Wiggins, S. (1992): Chaotic Transport in Dynamical Systems. Springer–Verlag.

Woltjer, L. (1958): A theorem on force-free magnetic fields. Proc. Nat. Acad. Sci. USA **44**, 489–491.

Yomdin, Y. (1987): Volume growth and entropy. Israel J. Maths. **57**, 285–300.

Yoshida, Z. (1994): A remark on the Hamiltonian form of the magnetic field line equations. Phys. Plasmas **1**, 208–209.

Young, L.S. (1977): Entropy of continuous flows on compact 2-manifolds. Topology **16**, 469–471.

Yudovich, V. (1989): The Linearization Method in Hydrodynamical Stability Theory. AMS Translations of Mathematical Monographs, vol. 74.

Zaslavski, G.M., Sagdeev, R.Z., Usikov, D.A., Chernikov, A.A. (1991): Weak Chaos and Quasi-Regular Patterns. Cambridge University Press.

Zeldovich, Ya.B. (1957): The magnetic field in the two-dimensional motion of a conducting turbulent liquid. Sov. Phys. JETP **4**, 460–462.

Zeldovich, Ya.B., Ruzmaikin, A.A. (1980): The magnetic field in a conducting fluid in two-dimensional motion. Zh. Eksp. Teor. Fiz. **78**, 980–986. [English transl.: Sov. Phys. JETP **51** (1980) 493–497.]

Zeldovich, Ya.B., Ruzmaikin, A.A., Sokoloff, D.D. (1983): Magnetic Fields in Astrophysics. The Fluid Mechanics of Astrophysics and Geophysics, vol. 3. Gordon & Breach Science Publishers.

Zeldovich, Ya.B., Ruzmaikin, A.A., Molchanov, S.A., Sokoloff, D.D. (1984): Kinematic dynamo problem in a linear velocity field. J. Fluid Mech. **144**, 1–11.

Zeldovich, Ya.B., Molchanov, S.A., Ruzmaikin, A.A., Sokoloff, D.D. (1987): Intermittency in random media. Usp. Fiz. Nauk **152**, 3–32. [English transl.: Sov. Phys. Usp. **30** (1987) 353–369.]

Zeldovich, Ya.B., Molchanov, S.A., Ruzmaikin, A.A., Sokoloff, D.D. (1988): Intermittency, diffusion, and generation in a nonstationary medium. Sov. Sci. Rev. C Math. Phys. **7**, 1–110.

Zheligovsky, V.A. (1993): A kinematic magnetic dynamo sustained by a Beltrami flow in a sphere. Geophys. Astrophys. Fluid Dyn. **73**, 217–254.

Index

Lecture Notes in Physics

For information about Vols. 1–425
please contact your bookseller or Springer-Verlag

New Series m: Monographs